THE ENGINEERING OF CHEMICAL REACTIONS

TOPICS IN CHEMICAL ENGINEERING
A Series of Textbooks and Monographs

THE ENGINEERING OF CHEMICAL REACTIONS

SECOND EDITION

LANNY D. SCHMIDT

University of Minnesota

New York   Oxford
OXFORD UNIVERSITY PRESS
2005

OXFORD UNIVERSITY PRESS

Oxford   New York
Auckland Bangkok Buenos Aires Cape Town
Chennai Dar es Salaam Delhi Hong Kong Istanbul
Karachi Kolkata Kuala Lumpur Madrid Melbourne
Mexico City Mumbai Nairobi São Paulo
Shanghai Taipei Tokyo Toronto

Published by Oxford University Press, Inc.
198 Madison Avenue, New York, New York 10016
www.oup.com

Oxford is a registered trademark of Oxford University Press

Library of Congress Cataloging-in-Publication Data

Schmidt, Lanny D., 1938–
    The engineering of chemical reactions / Lanny D. Schmidt.—
2nd ed.
        p.   cm. — (Topics in chemical engineering)
    ISBN 0-19-516925-5 (cloth)
    1. Chemical reactors.   I. Title.   II. Topics in chemical
engineering (Oxford University Press)
TP157.S32   2004
660'.2832—dc22   2004043285

Printing number: 9 8 7 6 5 4

Printed in the United States of America
on acid-free paper

CONTENTS

3   SINGLE REACTIONS IN CONTINUOUS ISOTHERMAL REACTORS   *87*

4   MULTIPLE REACTIONS IN CONTINUOUS REACTORS   *148*

PART II: APPLICATIONS   *327*

8   NONIDEAL CHEMICAL REACTORS   *332*

9   REACTIONS OF SOLIDS   *353*

PREFACE TO THE SECOND EDITION

The main change that has been made for the second edition of *The Engineering of Chemical Reactions* is the addition of two new chapters: Chapter 12 on biological reaction engineering and Chapter 13 on environmental reaction engineering. The increased prominence of biological and environmental reactions in chemical engineering makes it important that these subjects receive adequate coverage in any course in chemical reaction engineering. I personally find it exciting to add consideration of biological processes and environmental issues to a text that introduces students to the field.

The other changes made for the second edition include the addition of a few new sections, figures, and tables, the numbering of equations and headings, the inclusion of references at the end of many chapters, and the correction of errors that appeared in early printings of the first edition.

It has been my intent to keep the text as short as possible while covering all fundamentals and providing enough information on important applications to allow instructors to cover the material as time and interests dictate. I suggest that this book, while designed for use in an introductory undergraduate course, may be suitable for use in an advanced undergraduate or even a graduate course with the addition of selected material that the instructor wishes to emphasize.

I retain the firm belief that a book on chemical reaction engineering must contain enough information about the chemistry of different types of processes to help students see the richness of applications and their relevance to the analytical tools of reaction engineering. Otherwise chemical reaction engineering will be in danger of becoming an exercise whose relevance and excitement is lost on chemical engineering students.

PREFACE TO THE FIRST EDITION

I learned about chemical reactors at the knees of Rutherford Aris and Neal Amundson, when, as a surface chemist, I taught recitation sections and then lectures in the Reaction Engineering undergraduate course at Minnesota. The text was Aris' *Elementary Chemical Reactor Analysis*, a book that was obviously elegant but at first did not seem at all elementary. It described porous pellet diffusion effects in chemical reactors and the intricacies of nonisothermal reactors in a very logical way, but to many students it seemed to be an exercise in applied mathematics with dimensionless variables rather than a description of chemical reactors.

We later used Octave Levenspiel's book *Chemical Reaction Engineering*, which was written with a delightful style and had many interesting figures and problems that made teaching from it easy. Levenspiel had chapters on reactions of solids and on complex reactors such as fluidized beds, topics to which all chemical engineering students should be introduced. However, the book had a notation in which all problems were worked in terms of the molar feed rate of one reactant F_{Ao} and the fractional conversion of this reactant X. The "fundamental equations" for the PFTR and CSTR given by Levenspiel were $V = F_{Ao} \int dX/r_A(X)$ and $V = F_{Ao}X/r_A(X)$, respectively. Since the energy balance is conventionally written in terms of spatial variations of properties (as is the general species balance), there was no logical way to solve mass and energy balance equations simultaneously, as we must do to consider nonisothermal and nonideal reactors. This notation also prohibits the correct handling of multiple reaction systems because there is no obvious X or r_A with multiple reactions, and Levenspiel could only describe selectivity and yield qualitatively. In that notation, reactors other than the perfect plug flow and the perfectly stirred reactor could not be handled because it did not allow consideration of properties versus position in the reactor. However, Levenspiel's books describe complex multiphase reactors much more thoroughly and readably than any of its successors, certainly more than will be attempted here.

We next used the texts of Hill and then Fogler in our chemical reactors course. These books use similar notation and organization to that of Levenspiel, although they reduced or omitted reactions of solids and complex reactors, and their notation required fairly qualitative consideration of nonisothermal reactors. It was our opinion that these texts actually made diffusion in porous pellets and heat effects seem more complicated than they need be because they were not sufficiently logically or mathematically based. These texts also had an unnecessary affinity for the variable density reactor such as $A \rightarrow 3B$ with ideal gases where the solutions require dealing with high-order polynomials and partial fractions. In contrast, the assumption of constant density (any liquid-phase reactor or gases with diluent) generates easily solved problems.

At the same time, as a chemist I was disappointed at the lack of serious chemistry and kinetics in reaction engineering texts. All beat $A \rightarrow B$ to death without much mention that irreversible isomerization reactions are very uncommon and never very interesting. Levenspiel and its progeny do not handle the series reactions $A \rightarrow B \rightarrow C$ or parallel reactions $A \rightarrow B, A \rightarrow C$ sufficiently to show students that these are really the prototypes of all multiple reaction systems. It is typical to introduce rates and kinetics in a reaction engineering course with a section on analysis of data in which log–log and Arrhenius plots are emphasized with the only purpose being the determination of rate expressions for single reactions from batch reactor data. It is typically assumed that any chemistry and most kinetics come from previous physical chemistry courses.

Up until the 1950s there were many courses and texts in chemical engineering on "Industrial Chemistry" that were basically descriptions of the industrial processes of those times. These texts were nearly devoid of mathematics, but they summarized the reactions, process conditions, separation methods, and operating characteristics of chemical synthesis processes. These courses in the chemical engineering curriculum were all replaced in the 1950s by more analytical courses that organized chemical engineering through "principles" rather than descriptions because it was felt that students needed to be able to understand the principles of operation of chemical equipment rather than just memorize pictures of them. Only in the Process Design course does there remain much discussion of the processes by which chemicals are made.

While the introduction of principles of chemical engineering into the curriculum undoubtedly prepared students to understand the underlying equations behind processes, succeeding generations of students rapidly became illiterate regarding these processes and even the names and uses of the chemicals that were being produced. We became so involved in understanding the *principles of* chemical engineering that we lost interest in and the capability of dealing with *processes*.

In order to develop the processes of tomorrow, there seems to be a need to combine principles and mathematical analysis along with applications and synthesis of these principles to describe processes. This is especially true in today's changing market for chemical engineers, where employers no longer are searching for specialists to analyze larger and larger equipment but rather are searching for engineers to devise new processes to refurbish and replace or retrofit old, dirty, and unsafe ones. We suggest that an understanding of how and why things were done in the past present is essential in devising new processes.

Students need to be aware of the following facts about chemical reactors.

1. The definition of a chemical engineer is one who handles the engineering of chemical reactions. Separations, fluid flow, and transport are details (admittedly sometimes very important) in that task. Process design is basically reactor design, because the chemical reactors control the sizes and functions of other units.

2. The most important reactor by far in twentieth century technology is the fluidized catalytic cracker. It processes more chemicals than any other reactor (except the automotive catalytic converter), the products it creates are the raw materials for most of chemical technology, and this reactor is undoubtedly the largest and most complex piece of equipment in our business. Yet it is very possible that a student can receive a B.S. degree in chemical engineering without ever hearing of it.

3. Most industrial processes use catalysts. Homogeneous single reaction systems are fairly rare and unimportant. The most important homogeneous reaction systems in fact involve free radical chains, which are very complex and highly nonlinear.

4. Energy management in chemical reactors is essential in reactor design.

5. Most industrial reactors involve multiple phases, and mass transfer steps between phases are essential and usually control the overall rates of process.

6. Polymers and their monomers are the major commodity and fine chemicals we deal with; yet they are considered mostly in elective polymer chemistry and polymer properties courses for undergraduates.

7. Chemical engineering is rapidly changing such that petroleum processing and commodity chemical industries are no longer the dominant employers of chemical engineers. Polymers, bioprocesses, microelectronics, foods, films, and environmental concerns are now the growth industries needing chemical engineers to handle essential chemical processing steps.

8. The greatest safety hazard in chemical engineering operations is without question caused by uncontrolled chemical reactions, either within the chemical reactor or when flammable chemicals escape from storage vessels or pipes. Many undergraduate students are never exposed to the extremely nonlinear and potentially hazardous characteristics of exothermic free radical processes.

It is our belief that a course in chemical reaction engineering should introduce all undergraduate students to all these topics. This is an ambitious task for a one-semester course, and it is therefore essential to focus carefully on the essential aspects. Certainly, each of these subjects needs a full course to lay out the fundamentals and to describe the reaction systems peculiar to them. At the same time, we believe that a course that considers chemical reactors in a unified fashion is essential to show the common features of the diverse chemical reactors that our students will be called on to consider.

Perhaps the central idea to come from Minnesota is the notion of modeling in chemical engineering. This is the belief that the way to understand a complex process is to construct the simplest description that will allow one to solve the problem at hand. Sometimes a single equation gives this insight in a back-of-the-envelope calculation, and sometimes a complete simulation on a supercomputer is necessary. The chemical engineer must be prepared to deal with problems at whatever level of sophistication is required. We want to show students how to do simple calculations by capturing the essential principles without getting lost in details. At the same time, it is necessary to understand the complex problem with sufficient clarity that the further steps in sophistication can be undertaken with confidence. A modeling approach also reveals the underlying beauty and unity of dealing with the engineering of chemical reactions.

Chemical reaction engineering has acquired a reputation as a subject that has become too theoretical and impractical. In fact, we believe that reaction engineering holds the key in improving chemical processes and in developing new ones, and it requires the greatest skills in both analysis and intuition. Students need to see these challenges and be equipped to solve the next generation of challenges.

OVERALL ORGANIZATION

The book starts with a review of kinetics and the batch reactor in Chapter 2, and the material becomes progressively more complex until Chapter 12, which describes all the types of multiphase reactors we can think of. This is the standard, linear, boring progression followed in essentially all textbooks.

In parallel with this development, we discuss the chemical and petroleum industries and the major processes by which most of the classical products and feedstocks are made. We begin in Chapter 2 with a section on "The Real World," in which we describe the reactors and reactions in a petroleum refinery and then the reactions and reactors in making polyester. These are all catalytic multiphase reactors of almost unbelievable size and complexity. By Chapter [14] the principles of operation of these reactors will have been developed.

Then throughout the book the reactions and reactors of the petroleum and commodity chemical industries are reintroduced as the relevant principles for their description are developed.

Along with these topics, we attempt a brief historical survey of chemical technology from the start of the Industrial Revolution through speculations on what will be important in the twenty-first century. The rise of the major petroleum and chemical companies has created the chemical engineering profession, and their current downsizing creates significant issues for our students' future careers.

Projection into the future is of course the goal of all professional education, and we at least mention the microelectronic, food, pharmaceutical, ceramic, and environmental businesses which may be major employers of chemical engineering students. The notion of evolution of technology from the past to the future seems to be a way to get students to begin thinking about their future without faculty simply projecting our prejudices of how the markets will change.

Finally, our goal is to offer a compact but comprehensive coverage of all topics by which chemical reactors are described and to do this in a single consistent notation. We want to get through the fundamental ideas as quickly and simply as possible so that the larger issues of new applications can be appreciated. It is our intent that an instructor should then have time to emphasize those topics in which he or she is especially knowledgeable or regards as important and interesting, such as polymerization, safety, environment, pharmaceuticals, microelectronics, ceramics, foods, etc.

At Minnesota we cover these topics in approximately [45] lectures and 20 recitations. This requires two to four lectures per chapter to complete all chapters. Obviously some of the material must be omitted or skimmed to meet this schedule. We assume that most instructors will not cover all the industrial or historical examples but leave them for students to read.

We regard the "essential" aspects of chemical reaction engineering to include multiple reactions, energy management, and catalytic processes; so we regard the first seven chapters as the core material in a course. Then the final five chapters consider topics such as environmental, polymer, solids, biological, and combustion reactions and reactors, subjects that may be considered "optional" in an introductory course. We recommend that an instructor attempt to complete the first seven chapters within perhaps 3/4 of a term to allow time to select from these topics and chapters. The final chapter on multiphase reactors is of course very important, but our intent is only to introduce some of the ideas that are important in [their] design.

We have tried to disperse problems on many subjects and with varying degrees of difficulty throughout the book, and we encourage assignment of problems from later chapters even if they were not covered in lectures.

The nonlinearities encountered in chemical reactors are a major theme here because they are essential factors, both in process design and in safety. These generate polynomial equations for isothermal systems and transcendental equations for nonisothermal systems. We consider these with graphical solutions and with numerical computer problems. We try to keep these simple so students can see the qualitative features and be asked significant questions on exams. We insert a few computer problems in most chapters, starting with $A \rightarrow B \rightarrow C \rightarrow D \rightarrow \ldots$, and continuing through the wall-cooled reactor with diffusion and mass and heat transfer effects. We keep these problems very simple, however, so that students can write their own programs or use a sample Basic or Fortran program in the appendix. Graphics [are] essential for these problems, because the evolution of a solution versus time can be used as a "lab" to visualize what is happening.

The use of computers in undergraduate courses is continuously evolving, and different schools and instructors have very different capabilities and opinions about the level and methods that should be used. The choices are between (1) Fortran, Basic, and spread-sheet programming by students, (2) equation-solving programs such as Mathematica and MathCad, (3) specially written computer packages for reactor problems, and (4) chemical engineering flowsheet packages such as Aspen. We assume that each instructor will decide and implement specific computer methods or allow students to choose their own methods to solve numerical problems. At Minnesota we allow students to choose, but we introduce Aspen flowsheets of processes in this course because this introduces the idea of reactor-separation and staged processes in chemical processes before they see them in Process Design. Students and instructors always seem most uncomfortable with computer problems, and we have no simple solutions to this dilemma.

One characteristic of this book is that we repeat much material several times in different chapters to reinforce and illustrate what we believe to be important points. For example, petroleum refining processes, NO_x reactions, and safety are mentioned in most chapters as we introduce particular topics. We do this to tie the subject together and show how complex processes must be considered from many angles. The downside is that repetition may be regarded as simply tedious.

This text is focused primarily on chemical reactors, not on chemical kinetics. It is common that undergraduate students have been exposed to kinetics first in a course in physical chemistry, and then they take a chemical engineering kinetics course, followed by a reaction engineering course, with the latter two sometimes combined. At Minnesota we now have three separate courses. However, we find that the physical chemistry course contains less kinetics every year, and we also have difficulty finding a chemical kinetics text that covers the material we need (catalysis, enzymes, polymerization, multiple reactions, combustion) in the chemical engineering kinetics course.

Consequently, while I jump into continuous reactors in Chapter 3, I have tried to cover essentially all of conventional chemical kinetics in this book. I have tried to include all the kinetics material in any of the chemical kinetics texts designed for undergraduates, but these are placed within and at the end of chapters throughout the book. The descriptions of reactions and kinetics in Chapter 2 do not assume any previous exposure to chemical kinetics. The simplification of complex reactions (pseudosteady-state and equilibrium step approximations) are covered in Chapter 4, as are theories of unimolecular and bimolecular reactions. I mention the need for statistical mechanics and quantum mechanics in interpreting reaction rates but do not go into state-to-state dynamics of reactions. The kinetics with catalysts (Chapter 7), solids (Chapter 9), combustion (Chapter 10), polymerization (Chapter 11), [biological reactions (Chapter 12), environmental reactions (Chapter 13), and reactions between phases (Chapter 14)] are all given sufficient treatment that their rate expressions can be justified and used in the appropriate reactor mass balances.

I suggest that we may need to be able to teach all of chemical kinetics within chemical engineering and that the integration of chemical kinetics within chemical reaction engineering may have pedagogical value. I hope that these subjects can be covered using this text in any combination of courses and that, if students have had previous kinetics courses, this material can be skipped in this book. However, chemistry courses and texts give so little and such uneven treatment of topics such as catalytic and polymerization kinetics that reactors involving them cannot be covered without considering their kinetics.

[Many] texts strive to be encyclopedias of a subject from which the instructor takes a small fraction in a course and that are to serve as a future reference when a student later needs to learn in detail about a specific topic. This is emphatically not the intent of this text. First, it seems impossible to encompass all of chemical reaction engineering with less than a Kirk–Othmer encyclopedia. Second, the student needs to see the logical flow of the subject in an introductory course and not become bogged down in details. Therefore, we attempt to write a text that is short enough that a student can read all of it and an instructor can cover most of it in one course. This demands that the text and the problems focus carefully. The obvious pitfall is that short can become superficial, and the readers and users will decide that difference.

Many people assisted in the writing of this book. Marylin Huff taught from several versions of the manuscript at Minnesota and at Delaware and gave considerable help. John Falconer and Mark Barteau added many suggestions. All of my graduate students have been forced to work problems, find data and references, and confirm or correct derivations. Most important, my wife Sherry has been extremely patient about my many evenings spent at the Powerbook.

THE ENGINEERING OF
CHEMICAL REACTIONS

PART I

FUNDAMENTALS

Chapter 1

INTRODUCTION

1.1 CHEMICAL REACTORS

The chemical reactor is the heart of any chemical process. Chemical processes turn inexpensive chemicals into valuable ones, and chemical engineers are the only people technically trained to understand and handle them. While separation units are usually the largest components of a chemical process, their purpose is to purify raw materials before they enter the chemical reactor and to purify products after they leave the reactor.

Here is a very generic flow diagram of a chemical process.

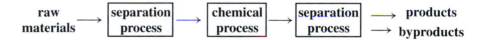

Raw materials from another chemical process or purchased externally must usually be purified to a suitable composition for the reactor to handle. After leaving the reactor, the unconverted reactants, any solvents, and all byproducts must be separated from the desired product before it is sold or used as a reactant in another chemical process.

The key component in any process is the chemical reactor; if it can handle impure raw materials or not produce impurities in the product, the savings in a process can be far greater than if we simply build better separation units. In typical chemical processes the capital and operating costs of the reactor may be only 10 to 25% of the total, with separation units dominating the size and cost of the process. Yet the performance of the chemical reactor totally controls the costs and modes of operation of these expensive separation units, and thus the chemical reactor largely controls the overall economics of most processes. Improvements in the reactor usually have enormous impact on upstream and downstream separation processes.

Design of chemical reactors is also at the forefront of new chemical technologies. The major challenges in chemical engineering involve

1. searching for alternate processes to replace old ones,

2. finding ways to make a product from different feedstocks, or

3. reducing or eliminating a troublesome byproduct.

The search for alternate technologies will certainly proceed unabated into the next century as feedstock economics and product demands change. Environmental regulations create continuous demands to alter chemical processes. As an example, we face an urgent need to reduce the use of chlorine in chemical processes. Such processes (propylene to propylene oxide, for example) typically produce several pounds of salt (containing considerable water and organic impurities) per pound of organic product that must be disposed of in some fashion. Air and water emission limits exhibit a continual tightening that shows no signs of slowing down despite recent conservative political trends.

1.2   CHEMICAL REACTION ENGINEERING

Since before recorded history, we have been using chemical processes to prepare food, ferment grain and grapes for beverages, and refine ores into utensils and weapons. Our ancestors used mostly batch processes because scaleup was not an issue when one just wanted to make products for personal consumption.

The throughput for a given equipment size is far superior in continuous reactors, but problems with transients and maintaining quality in continuous equipment mandate serious analysis of reactors to prevent expensive malfunctions. Large equipment also creates hazards that backyard processes do not have to contend with.

Not until the industrial era did people want to make large quantities of products to sell, and only then did the economies of scale create the need for mass production. Not until the twentieth century was continuous processing practiced on a large scale. The first practical considerations of reactor scaleup originated in England and Germany, where the first large-scale chemical plants were constructed and operated, but these were done in a trial-and-error fashion that today would be unacceptable.

The systematic consideration of chemical reactors in the United States originated in the early twentieth century with DuPont in industry and with Walker and his colleagues at MIT, where the idea of reactor "units" arose. The systematic consideration of chemical reactors was begun in the 1930s and 1940s by Damkohler in Germany (reaction and mass transfer), Van Heerden in Holland (temperature variations in reactors), and by Danckwerts and Denbigh in England (mixing, flow patterns, and multiple steady states). However, until the late 1950s the only texts that described chemical reactors considered them through specific industrial examples. Most influential was the series of texts by Hougen and Watson at Wisconsin, which also examined in detail the analysis of kinetic data and its application in reactor design. The notion of mathematical modeling of chemical reactors and the idea that they can be considered in a systematic fashion were developed in the 1950s and 1960s in a series of papers by Amundson and Aris and their students at the University of Minnesota.

In the United States two major textbooks helped define the subject in the early 1960s. The first was a book by Levenspiel that explained the subject pictorially and included a large range of applications, and the second was two short texts by Aris that concisely described the mathematics of chemical reactors. While Levenspiel had fascinating updates in the *Omnibook* and the *Minibook*, the most-used chemical reaction engineering texts in the 1980s were those written by Hill and then Fogler, who modified the initial book of Levenspiel, while keeping most of its material and notation.

The major petroleum and chemical companies have been changing rapidly in the 1980s and 1990s to meet the demands of international competition and changing feedstock supplies and prices. These changes have drastically altered the demand for chemical engineers and the skills required of them. Large chemical companies are now looking for people with greater entrepreneurial skills, and the best job opportunities probably lie in smaller, nontraditional companies in which versatility is essential for evaluating and comparing existing processes and designing new processes. The existing and proposed new chemical processes are too complex to be described by existing chemical reaction engineering texts.

The first intent of this text is to update the fundamental principles of the operation of chemical reactors in a brief and logical way. We also intend to keep the text short and cover the fundamentals of reaction engineering as briefly as possible.

Second, we will attempt to describe the chemical reactors and processes in the chemical industry, not by simply adding homework problems with industrially relevant molecules, but by discussing a number of important industrial reaction processes and the reactors being used to carry them out.

Third, we will add brief historical perspectives to the subject so that students can see the context from which ideas arose in the development of modern technology. Further, since the job markets in chemical engineering are changing rapidly, the student may perhaps also be able to see from its history where chemical reaction engineering might be heading and the causes and steps by which it has evolved and will continue to evolve.

Every student who has just read that this course will involve *descriptions* of industrial process and the *history* of the chemical process industry is probably already worried about what will be on the tests. Students usually think that problems with numerical answers (5.2 liters and 95% conversion) are somehow easier than anything where memorization is involved. We assure you that most problems will be of the numerical answer type. However, by the time students become seniors, they usually start to worry (properly) that their jobs will not just involve simple, well-posed problems but rather examination of messy situations where the boss does not know the answer (and sometimes doesn't understand the problem). You are employed to think about the big picture, and numerical calculations are only occasionally the best way to find solutions. Our major intent in discussing descriptions of processes and history is to help you see the contexts in which we need to consider chemical reactors. Your instructor may ask you to memorize some facts or use facts discussed here to synthesize a process similar to those here. However, even if your instructor is a total wimp, we hope that reading about what makes the world of chemical reaction engineering operate will be both instructive and interesting.

1.3   WHAT DO WE NEED TO KNOW?

There are several aspects of chemical reaction engineering that are encountered by the chemical engineer that in our opinion are not considered adequately in current texts, and we will emphasize these aspects here.

The chemical engineer almost never encounters a single reaction in an ideal single-phase isothermal reactor. Real reactors are extremely complex with multiple reactions, multiple phases, and intricate flow patterns within the reactor and in inlet and outlet streams. An engineer needs enough information from this course to understand the basic concepts of reactions, flow, and heat management and how these interact so that she or he can begin to assemble simple analytical or intuitive models of the process.

The chemical engineer almost never has kinetics for the process she or he is working on. The problem of solving the batch or continuous reactor mass-balance equations with known kinetics is much simpler than the problems encountered in practice. We seldom know reaction rates in useful situations, and even if these data were available, they frequently would not be particularly useful.

Many industrial processes are mass-transfer limited so that reaction kinetics are irrelevant or at least thoroughly disguised by the effects of mass and heat transfer. Questions of catalyst poisons and promoters, activation and deactivation, and heat management dominate most industrial processes.

Logically, the subject of designing a chemical reactor for a given process might proceed as shown in the following sequence of steps.

bench-scale batch reactor $\rightarrow$ bench-scale continuous $\rightarrow$ pilot plant $\rightarrow$ operating plant

The conversions, selectivities, and kinetics are ideally obtained in a small batch reactor, the operating conditions and catalyst formulation are determined from a bench-scale continuous reactor, the process is tested and optimized in a pilot plant, and finally the plant is constructed and operated. While this is the ideal sequence, it seldom proceeds in this way, and the chemical engineer must be prepared to consider all aspects simultaneously.

The chemical engineer usually encounters an existing reactor that may have been built decades ago, has been modified repeatedly, and operates far from the conditions of initial design. Very seldom does an engineer have the opportunity to design a reactor from scratch. Basically, the typical tasks of the chemical engineer are to

1. maintain and operate a process,
2. fix some perceived problem, or
3. increase capacity or selectivity at minimum cost.

While no single course could hope to cover all the information necessary for any of these tasks, we want to get to the stage where we can meaningfully consider some of the key ideas.

Real processes almost invariably involve multiple reactors. These may be simply reactors in series with different conversions, operating temperatures, or catalysts in each reactor. However, most industrial processes involve several intermediates prepared and purified between initial reactants and final product. Thus we must consider the flow diagram of the overall process along with the details of each reactor.

One example is the production of aspirin from natural gas. Current industrial technology involves the steps

natural gas $\rightarrow$ methane $\rightarrow$ syngas $\rightarrow$ methanol $\rightarrow$ acetic acid $\rightarrow$ acetylsalicylic acid

Although a gas company would usually purify the natural gas, a chemical company would buy methane and convert it to acetic acid, and a pharmaceutical company would make and sell aspirin.

An engineer is typically asked to solve some problem as quickly as possible and move on to other problems. Learning about the process for its own sake is frequently regarded as *unnecessary* or even *harmful* because it distracts the engineer from solving other more important problems. However, we regard it as an essential task to show the student how to construct models of the process. We need simple analytical tools to estimate with numbers how and why the reactor is performing as it is so that we can estimate how it might be modified quickly and cheaply. Thus modeling and simulation will be constant themes throughout this text.

The student must be able to do back-of-the-envelope computations very quickly and confidently, as well as know how to make complete simulations of the process when that need arises. Sufficient computational capabilities are now available that an engineer should be able to program the relevant equations and solve them numerically to solve problems that happen not to have analytical solutions.

Analysis of chemical reactors incorporates essentially all the material in the chemical engineering curriculum. A "flow sheet" of these relationships is indicated in the diagram.

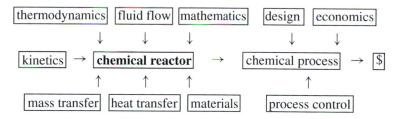

In this course we will need to use material from thermodynamics, heat transfer, mass transfer, fluid mechanics, and especially chemical kinetics. We assume that the student has had some exposure to these topics, but we will attempt to define concepts when needed so that those unfamiliar with particular topics can still use them here.

We regard the subject of chemical reactors as the final topic in the fundamental chemical engineering curriculum. This course is also an introduction to process design where we consider the principles of the design of a chemical reactor. Chemical reaction engineering precedes process control, where the operation and control of existing reactors is a major topic, and the process design course, where economic considerations and integration of components in a chemical plant are considered.

1.4   INDUSTRIAL PROCESSES

In parallel with an analytical and mathematical description of chemical reactors, we will attempt to survey the petroleum and chemical industries and related industries in which chemical processing is important. We can divide the major processes into petroleum refining, commodity chemicals, fine chemicals, food processing, materials, and pharmaceuticals. Their plant capacities and retail prices are summarized in Table 1–1.

The quantities in Table 1–1 have only qualitative significance. Capacity means the approximate production of that product in a single, large, modern, competitive plant that would be operated by a major oil, chemical, food, or pharmaceutical company. However, the table indicates the wide spread between prices and costs of different chemicals, from gasoline to insulin, that chemical engineers are responsible for making. There is a tradeoff

TABLE 1–1
Some Chemicals, Plant Sizes, Prices, and Waste Produced

Category	Typical plant capacity	Price	Waste/product
petroleum refining	10^6–10^8 tons/year	$0.1/lb	0.1
commodity chemicals	10^4–10^6	0.1–2	1–3
fine chemicals	10^2–10^4	2–10	2–10
foods		1–50	
materials		0–∞	
pharmaceuticals	10–10^3	10–∞	10–100

between capacity and price so that most chemical plants generate a cash flow between 10^7 and 10^9 per year.

The last column indicates the approximate amounts of waste products produced per amount of desired products (perhaps with each measured in pounds). The petroleum industry wastes very little of its raw material, but produces the largest amount of product, while the pharmaceutical industry produces large amounts of waste while making small amounts of very valuable products.

The engineer's task is quite different in each of these categories. In petroleum and commodity chemicals, the costs must be very carefully controlled to compete internationally, because every producer must strive to be the "low-cost producer" of that product or be threatened with elimination by its competitors. In fine chemicals the constraints are frequently different because of patent protection or niche markets in which competitors can be kept out. In foods and pharmaceuticals, the combination of patents, trademarks, marketing, and advertising usually dominate economics, but the chemical engineer still has a role in designing and operating efficient processes to produce high-quality products.

In addition to processes in which chemical engineers make a particular product, there are processes in which the chemical engineer must manage a chemical process such as pollution abatement. While waste management and sewage treatment originated with the prehistoric assembly of our ancestors, government regulations make the reduction of air and water pollution an increasing concern, perhaps the major growth industry in chemical engineering.

Throughout this text we will attempt to describe some examples of industrial processes that are either major processes in the chemical and petroleum industries or are interesting examples of fine chemicals, foods, or pharmaceuticals. The processes we will consider in this book are listed in Table 1–2.

Our discussion of these processes will necessarily be qualitative and primarily descriptive. We will describe raw materials, products, process conditions, reactor configurations, catalysts, etc., for what are now the conventional processes for producing these products. We will expect the student to show basic familiarity with these processes by answering simple and qualitative questions about them on exams. This will necessarily require some memorization of facts, but these processes are sufficiently important to all of chemical technology that we believe all chemical engineers should be literate in their principles.

Listed in Table 1–2 are most of the processes we will be concerned with in this book, both in the text and in homework problems.

TABLE 1–2

Industrial Chemicals Described in This Book

Petroleum	Chapters 2, 7, and 14
catalytic cracking (FCC)	
catalytic reforming	
hydrotreating	
alkylation	
Commodity chemicals	Chapters 3 and 4
olefins	
nitric acid	
chlorine	Chapter 9
alkalis	Chapter 9
ethylene oxide (EO)	Chapter 5
propylene oxide (PO)	
acetone	
phenol	Chapter 10
acetic acid	
syngas	
hydrogen	
CO	
Cl_2	
Polymers	Chapters 2, 11, and 14
polyolefins	
ethylene	Chapter 3 and 4
propylene	
butadiene	
styrene	Chapter 10
vinyl chloride	
vinyl acetate	
methyl methacrylate (MMA)	
acrylonitrile	
polyesters	
terephthalic acid	
ethylene glycol	
propylene glycol	
1,3–propanediol	
1,4–butanediol	
polyactic acid	Chapter 13
Nylon	
adipic acid (AA)	
hexamethylene diamine (HMD)	
caprolactam	
polyurethanes	Chapter 11
isocyanates	
maleic anhydride	
polycarbonate	Chapter 11
Fuels	
syngas	Chapter 4
methanol	Chapter 3
synthetic diesel fuel	Chapter 11
coal	Chapter 9
Environment	Chapter 13
NO_x	
SO_x	
chlorine	
ozone	

Continued

TABLE 1–2
Continued

heavy metals	
PCBs	
dioxin	
Specialty chemicals	Chapter 12
ethanol	
ibuprofen	
aspirin	
insulin	
Food	Chapters 9 and 12
HFCS	
Wheaties	
Cheerios	
Materials	Chapters 9 and 14
microcircuits	
iron	
nickel	
TiO_2	
photographic film	
catalysts	Chapter 7
fibers	
crystals	

1.5 MODELING

Students will also discover another difference between this course and the "Principles" courses taken previously. In this course we are interested mainly in *simple approximations to complex processes*. In more "fundamental" courses rigor is essential so that we can deal with any situation accurately. However, chemical reactors are so complex that we cannot begin to solve the relevant mass, energy, and momentum balance equations exactly, even on the largest supercomputers. Instead, we frequently need "back-of-the-envelope" estimates of reactor performance. Chemical engineers usually earn their living on these quick feasibility calculations. We need to know the details of thermodynamics and heat and mass transfer, for example, but we will usually assume that all properties (heat capacity, thermal conductivity, viscosity, diffusivity, etc.) are constants for a given calculation. All gases will be assumed to be ideal mixtures of ideal gases, and all liquids will be ideal solutions at constant density.

We will attempt to keep the mathematical details as brief as possible so that we will not lose sight of the principles of the design and operation of chemical reactors. The student will certainly see more *applied mathematics* here than in any other undergraduate course except Process Control. However, we will try to indicate clearly where we are going so students can see that the mathematical models developed here are essential for describing the application at hand.

Further, we want to be able to work problems with numerical solutions. This will require simplifying assumptions wherever possible so that the equations we need to solve are not too messy. This will require that fluids are at constant density so that we can use concentrations in moles/volume. This is a good approximation for liquid solutions but not for gases, where a reaction produces a changing number of moles, and temperatures and pressures can change. We will mention these complications but seldom solve problems without assuming constant density.

On the other hand, we will be downright sloppy about dimensions of quantities, frequently switching between English engineering and metric units. This is because one important task of the chemical engineer is in *language translation* between technical and nontechnical people, be they managers or customers. In U.S. industry you will hear amounts in pounds or tons, temperatures in degrees Fahrenheit, and pressures in psi gauge almost exclusively, with many practicing engineers not even knowing the meaning of kilograms, kelvins, and pascals. We will refer to energies in calories, power in watts, and amounts of material in gram moles, pounds, and tons without apology. Volumes in liters, cubic feet, or gallons and lengths in centimeters or miles must be handled without effort to effectively communicate with one's colleagues.

However, two types of systems are sufficiently important that we can use them almost exclusively: (1) liquid aqueous solutions and (2) ideal gas mixtures at atmospheric pressure. In aqueous solutions we assume that the density is 1 g/cm^3, the specific heat is 1 cal/g K, and at any solute concentration, pressure, or temperature there are ~55 moles/liter of water. In gases at one atmosphere and near room temperature we assume that the heat capacity per mole is $\frac{7}{2}R$, the density is 1/22.4 moles/liter, and all components obey the ideal gas equation of state. Organic liquid solutions have constant properties within ~20%, and nonideal gas solutions seldom have deviations larger than these.

We try to make calculations come out in round numbers; so in many problems the feed concentrations are 2 moles/liter, conversions are 90%, reactor volumes 100 liters, and feed temperatures 300 or 400 K. We further assume that all heats of reaction and heat capacities are independent of temperature, pressure, and composition. We sometimes even assume the ideal gas constant $R = 2$ cal/mole K, just because it makes it easier to remember than 1.98. . . .

We can then work out many numerical answers without even using a calculator, although several problems distributed throughout the course will be assigned where computer solutions and graphics are required. Some problems (the most interesting ones) cannot be worked so simply, and we must resort to numerical solutions. There are computer problems interspersed throughout the text, and your instructor will tell you exactly what programs and methods you should use to solve them.

1.6 SOURCES

The "game" is thus to make chemicals that can be sold for high prices from inexpensive raw materials. This involves finding a chemical reactor system that will do this better than the competition, finding cheap and abundant raw materials, finding a good and reliable market for the product, and disposing of byproducts.

A working chemical engineer needs continuous information on prices, markets, and processes on which to base calculations and estimations. A readily accessible source is *Chemical and Engineering News*, a weekly magazine published by the American Chemical Society that contains the latest gossip in chemistry and chemical engineering and some information about trends. Much more reliable are *Chemical Marketing Reporter* and *Chemical Weekly*. These magazines provide considerable and reliable information on prices and markets for industrial chemicals. Table 1–3 lists the top 50 chemicals in the country

TABLE 1–3
Top 50 Chemicals*

Rank	Chemical	Billions of lb	Rank	Chemical	Billions of lb
1	sulfuric acid	89.20	26	ethylene oxide	6.78
2	nitrogen	67.54	27	toluene	6.75
3	oxygen	67.54	28	hydrochloric acid	6.71
4	ethylene	49.67	29	p-xylene	6.23
5	lime	48.52	30	ethylene glycol	5.55
6	ammonia	38.35	31	cumene	5.16
7	propylene	37.93	32	ammonium sulfate	5.08
8	sodium hydroxide	25.83	33	phenol	4.05
9	phosphoic acid	25.26	34	acetic acid	3.82
10	chlorine	24.20	35	propylene oxide	3.70
11	sodium carbonate	20.56	36	butadiene	3.40
12	ethylene dichloride	18.70	37	carbon black	3.31
13	nitric acid	17.65	38	potash	3.14
14	ammonium nitrate	17.61	39	acrylonitrile	3.08
15	urea	16.13	40	vinyl acetate	3.02
16	vinyl chloride	14.81	41	acetone	2.77
17	benzene	14.66	42	titanium dioxide	2.74
18	methyl-*tert*-butyl ether	13.67	43	aluminum sulfate	2.30
19	ethylbenzene	11.87	44	sodium silicate	2.13
20	styrene	11.27	45	cyclohexane	2.11
21	carbon dioxide	10.99	46	adipic acid	1.80
22	methanol	10.81	47	caprolactam	1.68
23	xylene	9.06	48	bisphenol A	1.48
24	terephthalic acid	8.64	49	n-butyl alcohol	1.45
25	formaldehyde	7.94	50	isopropyl alcohol	1.39

*Data for 1994, from *Chemical and Engineering News*. Productions and rankings change continuously, but these indicate major chemicals.

from *Chemical and Engineering News* and a list of wholesale prices of chemicals in August 1995 taken from a many-page list in *Chemical Marketing Reporter*. Some of these are listed in Table 1–4 with their 1997 prices.

Finally, the practicing engineer needs to remain continuously informed about technologies and processes, both existing and new. Here one needs to find relevant books, journals, and monographs in technical libraries. One of the best of these is the *Kirk-Othmer Encyclopedia of Chemical Technology*, a multivolume set that describes existing processes for manufacturing many chemicals.

TABLE 1–4
Commodity Chemical Prices

Chemical	Description	Price	Price/lb
crude oil		$15–26/barrel	
	sweet light	22/barrel	$0.076
	heavy sour	18/barrel	0.056
diesel fuel	0.05% S	0.59/gal	0.083
gasoline	unleaded regular	0.62/gal	0.087
propane	fuel	0.38/gal	0.078
methane		2.0/MMBtu	0.046
coal		1.6/MMBtu	0.020
oxygen			0.015

Continued

TABLE 1–4
Continued

Chemical	Description	Price	Price/lb
hydrogen			
CO			
chlorine		200/ton	0.10
NaOH	50% liquid	315/ton	$0.16
	beads		0.31
ammonia		230/ton	0.12
nitric acid		125/ton	0.11
sulfuric acid		75/ton	0.038
phosgene			0.76
ethane			0.068
propane			0.076
butanes			0.076
isobutane			0.090
naphtha		0.75/gal	0.13
cyclohexane		1.33/gal	0.20
benzene		1.1/gal	0.15
toluene		0.92/gal	0.13
cumene			0.22
ethylbenzene			0.25
xylenes			0.13
p-xylene			0.19
methanol			0.083
ethanol	fuel grade		0.22
acetic acid			0.38
ethylene oxide			0.56
propylene oxide			0.64
ethylene glycol	polyester		0.34
maleic anhydride			0.47
tetrahydrofuran			1.35
acetone			0.4
phenol			0.36
formaldehyde			0.18
ethylene			0.26
propylene			0.19
isobutylene			0.31
butadiene			0.21
styrene			0.31
acetylene			0.50
polyethylene	low density		0.56
	linear low density		0.47
	high density		0.47
polystyrene			0.45
polypropylene			0.43
PET	beads		0.70
adipic acid			0.72
cyclohexanone			0.73
caprolactam			0.96
toluene diisocyanate			0.95
terephthalic acid			0.32
aspirin			3.6
ibuprofin			11
insulin			100,000
methamphetamine			
tetrahydrocanabinol			

These are average wholesale prices on Gulf Coast in Spring 1997. Do not use any of them for serious calculations. Recent prices, producers, and uses of these chemicals can be found on several websites.

1.7 REFERENCES

Listed below are some references to books that relate to the material in this text. Relevant references are repeated at the end of many chapters.

Allen, G., and Bevington, J. C., eds., *Comprehensive Polymer Science*, Pergamon, 1989.

Aris, Rutherford, *Introduction to the Analysis of Chemical Reactors*, Prentice-Hall, 1965.

Aris, Rutherford, *Elementary Chemical Reactor Analysis*, Prentice-Hall, 1969.

Bailey, T. J., and Ollis, D., *Biochemical Engineering*, 2nd ed., McGraw-Hill, 1987.

Biesenherger, J. A., and Sebastian, D. H., *Principles of Polymerization Engineering*, Wiley, 1983.

Boudart, Michel, *The Kinetics of Chemical Processes*, Prentice-Hall, 1968.

Butt, John B., *Reaction Kinetics and Reactor Design*, Prentice-Hall, 1980.

Carberry, James J., *Chemical and Catalytic Reaction Engineering*, McGraw-Hill, 1976.

Clark, Alfred, *The Theory of Adsorption and Catalysis*, Academic Press, 1970.

Cooper, C. D., and Alley, F. C., *Air Pollution Control*, 2nd ed., Waveland Press, 1994.

Crowl, Daniel A., and Louvar, Joseph, *Chemical Process Safety: Fundamentals with Applications*, Prentice-Hall, 1990.

Denbigh, Kenneth G., *Chemical Reactor Theory*, Cambridge University Press, 1965.

Doraiswamy, L. K., and Sharma, M. M., *Heterogeneous Reactions, Volume I: Gas–Solid and Solid–Solid Reactions*, Wiley, 1984.

Doraiswamy, L. K., and Sharma, M. M., *Heterogeneous Reactions, Volume II: Fluid–Fluid–Solid Reactions*, Wiley, 1984.

Dotson, N. A., Galvan, R., Lawrence, R. L., and Tirrell, M. V., *Polymerization Process Modeling*, VCH, 1996.

Farrauto, R. J., and Bartholomew, C. H., *Fundamentals of Catalytic Processes*, Chapman and Hall, 1997.

Fogler, H. Scott, *Elements of Chemical Reaction Engineering*, 2nd ed., Prentice-Hall, 1992.

Frank-Kamenetskii, D. A., *Diffusion and Heat Exchange in Chemical Kinetics*, Princeton University Press, 1955.

Froment, Gilbert F., and Bischoff, Kenneth B., *Chemical Reactor Analysis and Design*, 2nd ed., Wiley, 1990.

Gardiner, William C., Jr., *Combustion Chemistry*, Springer-Verlag, 1984.

Gates, Bruce C., *Catalytic Chemistry*, Wiley, 1992.

Gates, Bruce C., Katzer, James R., and Schuit, G. C. A., *Chemistry of Catalytic Processes*, McGraw-Hill, 1979.

Gupta, S. K., and Kumar, A., *Reaction Engineering of Step-Growth Polymerization*, Plenum, 1987.

Heaton, A., *The Chemical Industry*, 2nd ed., Blackie Academic and Professional, 1986.

Hill, Charles G., Jr., *An Introduction to Chemical Engineering Kinetics and Reactor Design*, Wiley, 1977.

Hougen, O. A., and Watson, K. M., *Chemical Process Principles, Volume III: Kinetics and Catalysis*, Wiley, 1947.

Kirk-Othmer Encyclopedia of Chemical Technology, 27 vols., Wiley, 1991–1998.

Kletz, Trevor A., *What Went Wrong? Case Studies of Process Plant Disasters*, Gulf Publishing, 1985.

Laidler, Keith J., *Chemical Kinetics*, 3rd ed., Harper and Row, 1987.

Lee, Hong H., *Heterogeneous Reactor Design*, Butterworths, 1985.

Levenspiel, Octave, *Chemical Reaction Engineering*, Wiley, 1962.

Levenspiel, Octave, *The Chemical Reactor Minibook*, OSU Book Stores, 1979.

Levenspiel, Octave, *The Chemical Reactor Omnibook*, OSU Book Stores, 1984.

Lewis, B., and von Elbe, G., *Combustion, Flames, and Explosions of Gases*, Academic Press, 1987.

Lowrance, William W., *Of Acceptable Risk: Science and the Determination of Safety*, William Kaufmann Inc., 1976.

McKetta, John J., *Encyclopedia of Chemical Processing and Design*, Marcel Dekker, 1987.

Meyer, Robert A., *Handbook of Chemical Production Processes*, McGraw-Hill, 1986.

Missen, Ronald R., Mims, Charles A., and Saville, Bradley A., *Introduction to Chemical Reaction Engineering and Kinetics*, Wiley, 1999.

Odian, G., *Principles of Polymerization*, 3rd ed., Wiley, 1991.

Peterson, E. E., *Chemical Reaction Analysis,* Prentice-Hall, 1965.

Pilling, Michael J., and Seakins, Paul W., *Reaction Kinetics*, Oxford University Press, 1995.

Ramachandran, P. A., and Chaudari, R. V., *Three-Phase Catalytic Reactors*, Gordon and Breach, 1983.

Remmp, P., and Merrill, E. W., *Polymer Synthesis*, Huthig and Wepf, 1991.

Sandler, H. J., and Lukiewicz, E. T., *Practical Process Engineering*, Ximix, 1993.

Satterfield, Charles N., *Heterogeneous Catalysis in Practice*, McGraw-Hill, 1980.

Smith, J. M., *Chemical Engineering Kinetics*, 3rd ed., McGraw-Hill, 1981.

Spitz, Peter H., *Petrochemicals: The Rise of an Industry*, Wiley, 1988.

Thomas, J. M., and Thomas, W. J., *Introduction to the Principles of Heterogeneous Catalysis*, Academic Press, 1967.

Twigg, Martyn V., *Catalyst Handbook*, Wolfe Publishing, 1989.

Ullmann's Encyclopedia of Industrial Chemistry, 5th ed., 37 vols., VCH, 1985–1996.

Unger, Stephen H., *Controlling Technology: Ethics and the Responsible Engineer*, Wiley, 1994.

Warnatz, J., Maas, U., and Dibble, R. W., *Combustion*, Springer-Verlag, 1996.

Wei, J., Russel, T. W. F., and Swartzlander, M. W., *The Structure of the Chemical Process Industries*, McGraw-Hill, 1979.

Westerterp, K. R., van Swaaij, W. P. M., and Beenackers, A. A. C. M., *Chemical Reactor Design and Operation*, Wiley, 1993.

The following books will not help you work any homework problems, but they may provide insight into technologies related to chemical engineering.

Brandt, E. N., *Growth Company: Dow Chemical's First Century*, Michigan State University Press, 1997.

Diamond, Jarred, *The Third Chimpanzee*, HarperCollins, 1992.

Diamond, Jarred, *Guns, Germs, and Steel*, Norton and Company, 1999.

Florman, Samuel C., *The Existential Pleasures of Engineering*, St. Martin's Press, 1976.

Florman, Samuel C., *Blaming Technology: The Irrational Search for Scapegoats*, St. Martin's Press, 1981.

Gaines, Ann, *Wallace Carothers and the Story of DuPont Nylon*, Mitchell Lane, 2001.

Goran, Morris, *The Story of Fritz Haber*, University of Oklahoma Press, 1967.

Petroski, Henry, *The Engineer Is Human: The Role of Failure in Successful Design*, St. Martin's Press, 1985.

Petroski, Henry, *The Pencil: A History of Design and Circumstances*, Knopf, 1990.

Petroski, Henry, *Design Paradigms: Case Histories of Error and Judgment in Engineering*, Cambridge University Press, 1994.

Petroski, Henry, *Invention by Design: How Engineers Get from Thought to Thing*, Harvard University Press, 1996.

Watson, James D., *The Double Helix,* Simon & Schuster, 2001.

Yergen, Daniel, *The Prize*, Simon & Schuster, 1991.

1.1 From what you now know about chemical reactions, guess the (1) major uses and (2) reaction processes that will produce the following chemicals starting only from natural gas (CH_4 and C_2H_6), air, water, and salt.

(a) CO

(b) H_2

(c) CH_3OH

(d) ethylene

(e) Cl_2 and NaOH

(f) phosgene

(g) urea

(h) toluene

(i) HCHO

(j) C_2H_5OH

(k) vinyl chloride

(l) acetic acid

Throughout this course we will be examining reactions such as these. The objective of the chemical industry is to produce high-purity molecules in large quantities as cheaply as possible. Many of your guesses will probably not be as effective as those currently practiced, and almost all processes use catalysts to increase the reaction rates and especially to increase the rate of the desired reaction rate and thus provide a high selectivity to the desired product.

1.2 Write out the chemical formulas for the organic chemicals shown in Table 1–2. What are the IUPAC names of these chemicals?

1.3 There are several books that describe chemical processes, and these are excellent and painless places to begin learning about how a particular chemical is made. One of the best of these is the *Kirk-Othmer Encyclopedia of Chemical Technology*, an excellent multivolume set. Choose one of the following chemicals and describe one or more processes and reactors by which it is currently made.

Some bioprocesses:
   citric acid
   lysine
   fructose
   ethanol

Some microelectronic and ceramic precursors and materials:
   silicon
   SiH_4
   $SiCl_4$
   GaAs
   AsH_3
   TiC coatings
   TiO_2 pigment
   cement

Some polymers and adhesives
   silicones

epoxy glue
urethane varnish
latex paint
polycarbonate plastics

Your instructor will specify the length of the writeup (<1 page each is suggested) and the number of items required. He or she may suggest additional topics or allow you to choose some that interest you.

1.4 From the list of wholesale prices in Table 1–4, calculate the value of the ingredients in a bottle of aspirin or ibuprofen (whichever bottle you have in your medicine cabinet). How much per bottle is costs in processing, packaging, distribution, retail markup, and advertising? Which of these is largest?

1.5 From the list of wholesale prices in Table 1–4, calculate the wholesale price differences per mole of the following processes.

(a) propane to propylene

(b) ethylene to polyethylene

(c) cyclohexane to benzene

(d) ethylbenzene to styrene

(e) styrene to polystyrene

(f) propane to acetone

All costs in a process obviously have to be less than these differences.

1.6 Our ancestors made vinegar by aerobic bacterial fermentation of alcohol, which is derived from sugar, while it is now made by carbonylation of methanol, which is derived by reaction of synthesis gas, which is obtained by steam reforming of methane.

(a) Write out these reactions.

(b) Compare the industrial acetic acid price per pound with its price (in dilute water solution) in the grocery store.

1.7 Our ancestors made alcohol by anaerobic fermentation of sugar, while industrial ethanol is made by hydration of ethylene, which is obtained by dehydrogenation of ethane.

(a) Write out these reactions.

(b) Compare the industrial price per pound of ethanol with its price (in dilute water solution) in the grocery store after subtracting taxes.

1.8 Ethane costs $0.05/lb, and ethylene sells for $0.18/lb. A typical ethylene plant produces 1 billion pounds/year.

(a) What are the annual sales?

(b) If we had a perfect process, what must be the cost of producing ethylene if we want a profit of 10% of sales?

(c) The actual process produces about 0.8 moles of ethylene per mole of ethane fed (the yield of the process is 80%). What is the cash flow of the process? What must be the cost of producing ethylene if we want a profit of 10% of sales?

1.9 What are price differences in manufacturing the following chemicals? Use prices in Table 1–4 and assume that any O_2, H_2, or CO reactants are free and any H_2 produced has no value.

(a) methanol from methane

(b) acetic acid from methane

(c) formaldehyde from methane

(d) cumene from benzene and propane

(e) acetone and phenol from cumene

(f) ethylene from ethane

(g) ethylene glycol from ethane

(h) benzene from cyclohexane

(i) PET from ethylene glycol and terephthalic acid

(j) PET from ethane and xylenes

(k) polystyrene from ethylene and benzene

What must be the cost per pound of manufacturing each of these chemicals if we need to make a profit ot 10% of sales, and all processes have 80% yield?

1.10 Most chemical processes involve multiple stages. What must be the relative costs of each stage for the following processes assuming the prices in the table?

(a) ethane to ethylene glycol

(b) acetone and phenol from cyclohexane and propane

(c) caprolactam from benzene and methane (to produce ammonia)

(d) acetic acid from methane

Note that the prices of chemical intermediates are in fact determined by the costs of the individual prices.

1.11 Shown in Figure 1–1 are formulas of some organic chemicals that are produced by the chemical and pharmaceutical industries. Some of these are molecules that you eat or use every day and some of these you really do not want to be near. From your previous courses in organic chemistry and biochemistry and by discussing with fellow students, name the compounds.

1. sucrose
2. glucose
3. fructose
4. fat
5. soap
6. detergent
7. vitamin C (ascorbic acid)
8. 2,4,5–T (Agent Orange)
9. dioxin
10. DDT
11. ibuprofen
12. aspirin
13. Tylenol (acetaminophen)
14. Contac (phenylpropanolamine)
15. Valium (diazepam)
16. amphetamine
17. methamphetamine

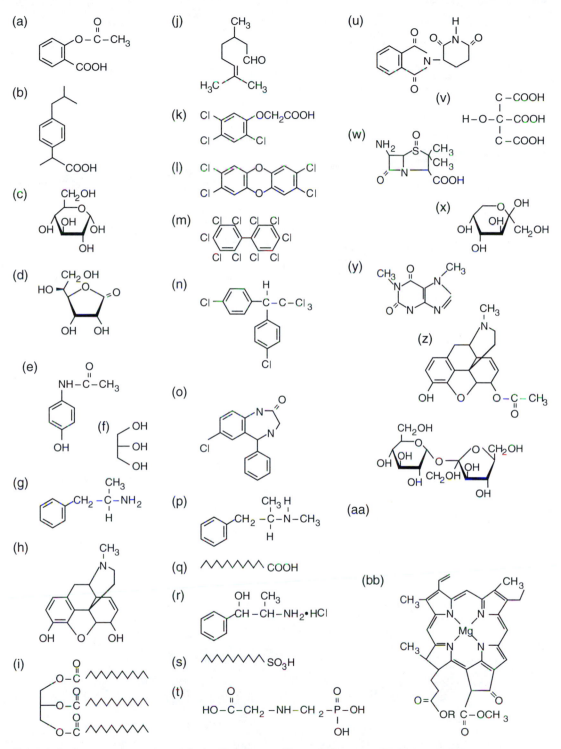

Figure 1–1 Some very important organic chemicals that are either made in nature or by us.

18. Roundup (glyphosphate)
19. PCB
20. chlorophyll
21. citronella
22. morphine
23. heroin
24. thalidomide
25. penicillin
26. glycerin
27. caffeine
28. citric acid

1.12 The following problems relate to the molecules in Figure 1–1.

(a) How does one convert salicylic acid to aspirin?

(b) How does one convert morphine to heroin?

(c) Phenyl acetone is a harmless chemical, yet it is a controlled substance. What can it be easily converted into? How would you run this reaction? [Do not try part b or c at home.]

(d) What is the reaction by which sucrose is converted into fructose and glucose? This reaction occurs in your stomach.

(e) What is the reaction by which glucose is converted into fructose? It is probably easier to visualize this reaction if the molecules are opened up into linear chains. The production of high-fructose corn syrup in soft drink sweetener is a major chemical process using enzyme catalysts.

(f) Thalidomide has one chiral center. One isomer is a tranquilizer, while the other causes serious birth defects. What are these isomers?

(g) What is the reaction that converts fat into soap?

(h) Detergents are made by reacting α-olefins with sulfuric acid. What are the reactions?

(i) Agent Orange is a fairly harmless herbicide that was used as a defoliant in the Vietnam War, and dioxin is a minor but very troublesome byproduct of manufacturing Agent Orange. What is the reaction that converts Agent Orange into dioxin?

(j) Why was PCB a popular heat transfer fluid and transformer oil?

Chapter 2

REACTION RATES, THE BATCH REACTOR, AND THE REAL WORLD

In the previous chapter we discussed some topics that chemical engineers need to know about chemical reactions and chemical reactors. In this chapter we will begin to define quantities, formulate and solve mass-balance equations, and consider some examples. We will then completely change topics and summarize some of the major reactors in the petroleum and chemical industries.

We will consider only the batch reactor in this chapter. This is a type of reactor that does not scale up well at all, and continuous reactors dominate the chemical industry. However, students are usually introduced to reactions and kinetics in physical chemistry courses through the batch reactor (one might conclude from chemistry courses that the batch reactor is the only one possible); so we will quickly summarize it here. As we will see in the next chapter, the equations and their solutions for the batch reactor are in fact identical to the plug flow tubular reactor, which is one of our favorite continuous reactors; so we will not need to repeat all these definitions and derivations in the section on the plug flow tubular reactor.

In parallel with these definitions and equations and their solutions, we will describe in this chapter some examples of important processes in the chemical engineering industry. This material will initially be completely disconnected from the equations, but eventually (by Chapter 14) we hope students will be able to relate the complexities of industrial practice to the simplicity of these basic equations.

Much of this chapter will be a review for those who have had courses in chemical kinetics. In this chapter we will also review some aspects of thermodynamics that are important in considering chemical reactors. For students who have not had courses in kinetics and in the thermodynamics of chemical reactions, this chapter will serve as an introduction to those topics. This chapter will also introduce the notation we will use throughout the book.

2.1 CHEMICAL REACTIONS

We first describe our representation of chemical reactions. Consider the isomerization reaction of cyclopropane to propylene,

$$\text{cyclo-C}_3\text{H}_6 \rightarrow \text{C}_3\text{H}_6 \tag{2.1}$$

or in symbols

$$\Delta \rightarrow \hspace{0.3em} ^=\backslash \tag{2.2}$$

Propylene will not significantly transform back to cyclopropane, and we call this reaction *irreversible*. The irreversible first-order reaction

$$A \rightarrow B \tag{2.3}$$

is the most used example in chemical kinetics, and we will use it throughout this book as the prototype of a "simple" reaction. However, the ring opening of cyclopropane to form propylene (which has absolutely no industrial significance) is one of only a handful of irreversible isomerization reactions of the type $A \rightarrow B$.

Next consider the motion of the double bond in butylenes

$$\textit{trans-}2\text{-C}_4^= \rightarrow \textit{cis-}2\text{-C}_4^= \tag{2.4}$$

$$\textit{cis-}2\text{-C}_4^= \rightarrow \textit{trans-}2\text{-C}_4^= \tag{2.5}$$

$$1\text{-C}_4^= \rightarrow \textit{trans-}2\text{-C}_4^= \tag{2.6}$$

$$\textit{iso-}\text{C}_4^= \rightarrow \textit{trans-}2\text{-C}_4^= \tag{2.7}$$

[Throughout this book we will refer to chemicals and processes by their common names and by chemical symbols. We believe that it is essential for chemical engineers to be literate in all the designations used by organic chemists and by customers. One of the important skills of a successful engineer is the ability to deal with colleagues in their own language, rather than being confined to one set of units and notation. Examples of names and symbols are butylenes versus butenes, and styrene versus phenylethene. (What is the common name of polymerized chloroethene? Answer: PVC.) Students unfamiliar with particular notations may want to review them from organic chemistry texts. The material in those courses really is important, in spite of what you probably thought when you were taking them.]

The above reactions all have one reactant and one product, and we write them as

$$A \rightarrow B \tag{2.8}$$

However, the reactions of butylenes will all proceed among each other and are described as *reversible*; so there are 12 reactions among the four butylenes.

We write each of these as

$$A \rightleftarrows B \tag{2.9}$$

signifying that the process involves both $A \rightarrow B$ and $B \rightarrow A$. We can write the complete set of butylene isomerization reactions as shown in Figure 2–1. Since each of these molecules can isomerize to all the others, this is a set of 12 chemical reactions.

Next consider the formation of nitric oxide from air

$$\tfrac{1}{2}\text{N}_2 + \tfrac{1}{2}\text{O}_2 \rightarrow \text{NO} \tag{2.10}$$

Figure 2–1 The 12 reactions among 4 isomers of butylenes.

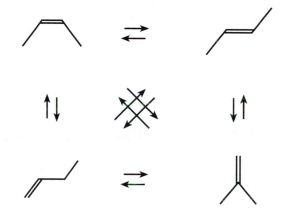

This is an extremely important reaction to which we will refer throughout this book. It is responsible for all NO_x formation in the atmosphere (the brown color of the air over large cities) as well as nitric acid and acid rain. This reaction only occurs in high-temperature combustion processes and in lightning bolts, and it occurs in automobile engines by free-radical chain reaction steps, which will be the subject of Chapter 10. It is removed from the automobile exhaust in the automotive catalytic converter, which will be considered in Chapter 7.

The above reaction can also be written as

$$N_2 + O_2 \rightarrow 2NO \tag{2.11}$$

We could generalize these reactions as

$$\tfrac{1}{2}A + \tfrac{1}{2}B \rightarrow C \tag{2.12}$$

or

$$A + B \rightarrow 2C \tag{2.13}$$

with $A = N_2$, $B = O_2$, and $C = NO$. The coefficients of the chemical symbols in a reaction are termed *stoichiometric coefficients*, and it is evident that one can multiply all of them by a constant and still preserve mass conservation.

We need to distinguish between stoichiometric coefficients of reactants and products. For this we move all terms to the left with appropriate signs, so that the preceding reaction becomes

$$-A - B + 2C = 0 \tag{2.14}$$

Further, since the alphabet is limited, we will increase our capacity for naming chemicals by naming a species j as A_j, so that this reaction is

$$-A_1 - A_2 + 2A_3 = 0 \tag{2.15}$$

with $A_1 = N_2$, $A_2 = O_2$, and $A_3 = NO$. In this notation the *generalized single reaction* becomes

$$\boxed{\sum_{j=1}^{s} \nu_j A_j = 0}$$ (2.16)

The quantity ν_j is the stoichiometric coefficient of species j, and by convention $\nu_j < 0$ for a reactant and $\nu_j > 0$ for a product.

All reactions must satisfy mass conservation. The reaction $A \rightarrow B$ must be an isomerization reaction because the molecular weights of A and B must be identical. Also, one should add to these relations the requirements that the number of atoms of each element must be conserved, but this is usually intuitively obvious for most reaction systems. We do this whenever we balance a chemical equation.

2.1.1   High-fructose corn syrup

There is at least one important reversible isomerization reaction in chemical engineering, the conversion of glucose into fructose. All soft drinks are essentially sugar water with a small amount of flavoring added (usually synthetic), and you like to drink them basically because they taste sweet. The natural sugar in sugar cane and sugar beets is basically sucrose, a disaccharide consisting of one molecule of glucose and one molecule of fructose. Plants form these molecules as food reserves for themselves, and we harvest the plants and extract sucrose in water solution from them. We have bred strains of cane and beets that produce much more sucrose than they would by natural selection.

Sucrose is rapidly dissociated into glucose and fructose by the enzymes in your mouth and in your stomach, and your taste receptors sense sweetness. A problem (for Coca-Cola) is that fructose tastes five times as sweet as glucose; so 40% of the sucrose they purchase is wasted compared to pure fructose. A problem for you is that both sugars have the same calories, and the soft drink companies want to advertise lower calories for an acceptable sweetness.

Within the past 25 years chemical engineers figured out how to run the reaction to convert glucose into fructose

$$\text{glucose} \rightleftarrows \text{fructose}$$ (2.17)

which when drawn out looks as shown in Figure 2–2.

The equilibrium in this isomerization reaction is 58% fructose and 42% glucose. An enzyme was discovered called *glucose isomerase*, which isomerizes the molecule by

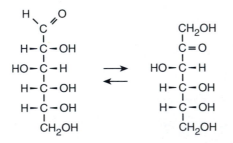

Figure 2–2 Isomerization reaction that converts glucose into fructose. An enzyme causes this reaction to run selectively near room temperature. This process, the largest bioprocess in the chemical industry, makes high-fructose corn syrup for the beverage industry.

exchanging the end aldehyde group with the neighboring OH group to convert glucose into fructose. This enzyme cannot isomerize any other bonds in these molecules. This is fortunate because only this isomer tastes sweet and is digestible, so no manmade catalysts could hope to be successful in this process.

The high-fructose corn syrup (HFCS) industry produces $\sim$10 million tons of fructose from glucose annually (how many pounds do you purchase per year?), and this is the major industrial bioengineering processes today, at least in volume.

This process could not work without finding an effective and cheap way to run this biological reaction and to separate fructose from glucose (sucrose contains 50% glucose). These were accomplished by finding improved strains of the enzyme, by finding ways to immobilize and stabilize the enzyme on solid beads to keep them in the reactor, and by finding adsorbents to separate fructose from glucose.

These are accomplished in large fermenters in chemical plants, mostly in the Midwest. The cheapest feedstock is starch rather than sugar (other enzymes convert starch to glucose), and corn from the Midwest is the cheapest source of starch.

Artificial sweeteners have also been developed to give the taste of sweetness without the calories. These chemicals have sweetness many times that of sugar, so they sell for high prices as low-calorie sweeteners. Many artificial flavors have also been developed to replace natural biological flavors. In all cases we search for processes that convert inexpensive raw materials into chemicals that taste or smell like natural chemicals, either by producing the same chemical synthetically or by producing a different chemical that can replace the natural chemical.

In this book we will consider mostly the simpler chemical reaction processes in the petroleum and commodity chemicals industries because they are more "central" to chemical engineering. However, the same principles and strategies apply in the pharmaceutical and food industries, and students may need these principles for these or other applications later in their careers.

2.2 MULTIPLE REACTIONS

Every chemical process of practical interest, such as the isomerization of butylenes written previously, forms several products (some undesired) and involves *multiple reactions*. Consider next the reaction system

$$N_2 + O_2 \rightarrow 2NO \tag{2.18}$$

$$2NO + O_2 \rightarrow 2NO_2 \tag{2.19}$$

In NO_x smog formation (NO_x is a mixture of NO, N_2O, NO_2, N_2O_4, and N_2O_5) the NO is produced by reaction of N_2 and O_2 at the high temperatures of combustion in automobiles and fossil fuel power plants, and NO_2 and the other NO_x species are produced by subsequent low-temperature oxidation of NO in air. NO is colorless, but NO_2 absorbs visible radiation and produces brown haze. We write these reactions as a set of two reactions among four species,

$$-A_1 - A_2 + 2A_3 = 0 \tag{2.20}$$

$$-2A_3 - A_2 + 2A_4 = 0 \tag{2.21}$$

where A_1 is N_2, A_2 is O_2, etc. This is a simplification of all the possible reactions involving N and O atoms, since we have not included stable molecules such as N_2O, N_2O_2, NO_3, and N_2O_5.

We can generalize this notation to any set of reactions as

$$\sum_{j=1}^{S} v_{ij} A_j = 0, \qquad i = 1, 2, \ldots, R \qquad (2.22)$$

as a set of R reactions among S species, with $j (= 1$ to $S)$ the index of chemical species and $i (= 1$ to $R)$ the index of the reaction. Thus we call v_{ij} the stoichiometric coefficient of species j in reaction i. We will use this standard notation throughout this book, and we suggest that you always give the is and js these meanings and never reverse them.

2.3 REACTION RATES

We next need to describe the rates of transformations of chemical species among each other. For this we use the symbol r for a single reaction and r_i for the ith reaction in a multiple reaction system. Reaction rates are basically empirical expressions that describe the dependence of the rate of transformation on the parameters in the system.

2.3.1   Rate of a single irreversible reaction

It is found by experiment that rates almost always have power-law dependences on the densities (such as concentration, density on a surface, or partial pressure) of chemical species. For example, our first example of the homogeneous reaction of cyclopropane to propylene exhibits a rate of decomposition that can be written as

$$r = k \text{ [cyclopropane]} \qquad (2.23)$$

while the homogeneous NO_2 formation from NO and oxygen has a rate

$$r = k[NO]^2[O_2] \qquad (2.24)$$

In these expressions we write the rate as a positive quantity, designated by lowercase r, with dimensions of amount converted per unit volume per unit time.

We usually use the number of gram moles N_j of species j in a reactor of volume V to describe the amount of that species. For density we will usually use concentration (moles per volume),

$$C_j = \frac{N_j}{V} \qquad (2.25)$$

which is usually expressed in moles per liter (one liter equals one cubic decimeter or dm^3 in SI units). This unit is especially useful for densities of species in liquid solution, but we can also use it for gaseous and solid solutions. For symbols of concentration we use the following notations for species A

$$C_A = [A] = \text{[cyclopropane]} \qquad (2.26)$$

or for species j

$$C_j = [A_j] \tag{2.27}$$

In gases the most used quantity for the density of species j is the partial pressure P_j. This can be related to concentration and mole fraction y_j through the relations

$$y_j = \frac{N_j}{\sum N_j} = \frac{N_j}{N} \tag{2.28}$$

and

$$P_j = y_j P = C_j RT \tag{2.29}$$

where P is the total pressure and N is the total number of moles in the system. In this equation we assume ideal gases ($PV = NRT$) to relate partial pressure to concentration, while for nonideal gases (not considered here) we would need an equation of state to describe the density of each chemical species.

For liquid solutions we could use

$$x_j = \frac{N_j}{\sum N_j} \tag{2.30}$$

where x_j is the mole fraction of species j in liquid solution. However, we will use only the concentration, $C_j = N_j/V$, throughout this text.

For an irreversible reaction we can frequently describe the rate to a good approximation as

$$\boxed{r = k \prod_{j=1}^{S} C_j^{m_j}} \tag{2.31}$$

where m_j is the order of the reaction with respect to the jth species and the product extends over all species in the reaction with $m_j = 0$ for species that do not affect the rate of reaction. If the rate is proportional to the concentration of a species raised to a power (m_j), we say that this form of the rate expression is described by "power-law kinetics." This empirical function is frequently used to describe reaction rates, but it frequently is not accurate, especially with surface or enzyme-catalyzed reactions, which we will consider later.

Several alternate definitions of the reaction rate are used in different texts. In our notation we will always write a chemical reaction as an equation and then define the rate of that reaction as the positive rate of change for that particular stoichiometry. Consider the reaction

$$2A \rightarrow B + 3C, \quad r = k[A]^2 \tag{2.32}$$

In an alternate definition of reaction rates, one writes a rate as *the rate of formation of each species*. In that notation one would define r_A, r_B, and r_C, with the definitions

$$r_A = -2k[A]^2 = -2r \tag{2.33}$$

$$r_B = +k[A]^2 = r \tag{2.34}$$

$$r_C = +3k[A]^2 = +3r \tag{2.35}$$

These rates are the rates of *production* of species A, B, and C ($r_j = v_j r$); so these rates are written as negative quantities for reactants and positive quantities for products. This notation quickly becomes cumbersome for complex reaction stoichiometry, and the notation is not directly usable for multiple reaction systems.

We will consider the rate r as a single positive quantity describing the rate of a particular reaction. Note that the rate can now only be defined *after* we write the chemical reaction. In our two ways of writing the NO formation reaction previously, the rate would be smaller by a factor of two when the stoichiometric coefficients are multiplied by a factor of two.

2.3.2   Rates of reversible reactions

If the reaction is reversible, we frequently find that we can write the rate as a difference between the rate of the forward reaction r_f and the reverse (or back) reaction r_b,

$$r = r_f - r_b = k_f \prod_{j=1}^{S} C_j^{m_{fj}} - k_b \prod_{j=1}^{S} C_j^{m_{bj}}$$ (2.36)

where m_{fj} and m_{bj} are the orders of the forward and reverse (or back) reactions with respect to the jth species, and k_f and k_b are the rate coefficient of the forward and reverse reactions.

2.3.3   Rates of multiple reactions

We also need to describe the rates of multiple-reaction systems. We do this in the same way as for single reactions with each of the i reactions in the set of R reactions being described by a rate r_i, rate coefficient k_i, order of the forward reaction m_{fij} with respect to species j, etc.

We repeat that the procedure we follow is first to write the reaction steps with a consistent stoichiometry and then to express the rate of each reaction to be consistent with that stoichiometry. Thus, if we wrote a reaction step by multiplying each stoichiometric coefficient by two, the rate of that reaction would be smaller by a factor of two, and if we wrote the reaction as its reverse, the forward and reverse rates would be switched.

For a multiple-reaction system with reversible reactions, we can describe each of the R reactions through a reaction rate r_i,

$$r_i = r_{fi} - r_{bi} = k_{fi} \prod_{j=1}^{S} C_j^{m_{fij}} - k_{bi} \prod_{j=1}^{S} C_j^{m_{bij}}$$ (2.37)

with symbols having corresponding definitions to those used for single reactions.

This looks like a maze of notation, but for most examples this notation merely formalizes what are usually simple and intuitively obvious expressions. However, we need this formal notation for situations where intuition fails us, as is the case for most industrial reaction processes. Whenever a reaction is irreversible, we use k and m_j as rate coefficient and order, respectively, omitting the subscript f.

2.4 APPROXIMATE REACTIONS

Consider the hydrolysis or saponification of an ester (ethyl acetate) into an alcohol (ethanol) and an acid (acetic acid),

$$CH_3COOC_2H_5 + H_2O \rightleftarrows CH_3COOH + C_2H_5OH \tag{2.38}$$

This reaction and its reverse take place readily in basic aqueous solution. We write this reaction as

$$A + B \rightleftarrows C + D \tag{2.39}$$

Next consider the addition of water to an olefin to form an alcohol,

$$RCH = CH_2 + H_2O \rightleftarrows RCH_2CH_2OH \tag{2.40}$$

which can be written as

$$A + B \rightleftarrows C \tag{2.41}$$

In many situations we carry out these reactions in dilute aqueous solutions, where there is a large excess of water. The concentration of pure liquid water is 55 moles/liter, and the concentration of water in liquid aqueous solutions is nearly constant even when the above solutes are added up to fairly high concentrations. The rates of these forward reactions are

$$r = k[CH_3COOC_2H_5][H_2O] \tag{2.42}$$

and

$$r = k[RCH = CH_2][H_2O] \tag{2.43}$$

respectively. However, since the change of the concentration of water $[H_2O]$ is usually immeasurably small whenever water is a solvent, we may simplify these reactions as

$$CH_3COOC_2H_5 \rightarrow CH_3COOH + C_2H_5OH, \qquad r = k[CH_3COOC_2H_5] \tag{2.44}$$

and

$$RCH = CH_2 \rightarrow RCH_2CH_2OH, \qquad r = k[RCH = CH_2] \tag{2.45}$$

These reactions do not satisfy total mass conservation because the mole of water is omitted as a reactant. We have also redefined a new rate coefficient as $k = k[H_2O]$ by grouping the nearly constant $[H_2O]$ with k. After grouping the concentration of the solvent $[H_2O]$ into the rate coefficient, we say that we have a pseudo-first-order rate expression.

It is fairly common to write reactions in this fashion omitting H_2O from the chemical equation and the rate; so these reactions become of the type

$$A \rightarrow C + D, \qquad r = kC_A \tag{2.46}$$

and

$$A \rightarrow C, \qquad r = kC_A \tag{2.47}$$

respectively. Thus, besides isomerization, there are in fact a number of reactions that we write approximately as $A \rightarrow B$ or $A \rightarrow$ products; so our use of these simple rate expressions is in fact appropriate for a large number of reaction systems.

Reaction rate expressions are always empirical, which means that we use whatever expression gives an accurate enough description of the problem at hand. No reactions are as simple as these expressions predict if we need them to be correct to many decimal places. Further, all reaction systems in fact involve multiple reactions, and there is no such thing as a truly irreversible reaction if we could measure all species to sufficient accuracy. If we need a product with impurities at the parts per billion (ppb) level, then all reactions are in fact reversible and involve many reactions.

Chemical engineers get paid to make whatever approximations are reasonable to find answers at the level of sophistication required for the problem at hand. If this were easy, our salaries would be lower.

2.5 RATE COEFFICIENTS

We next consider the ks in the above expressions. We will generally call these the *rate coefficients* (the coefficient of the concentration dependences in r). They are sometimes called *rate constants*; they are independent of concentrations, but rate coefficients are almost always strong functions of temperature.

It is found empirically that these coefficients frequently depend on temperature as

$$k(T) = k_o e^{-E/RT} \qquad (2.48)$$

where E is called the *activation energy* for the reaction and k_o is called (unimaginatively) the *pre-exponential factor*.

This relation is credited to Svante Arrhenius and is called the *Arrhenius temperature dependence*. Arrhenius was mainly concerned with thermodynamics and chemical equilibrium. Some time later Michael Polanyi and Eugene Wigner showed that simple molecular arguments lead to this temperature dependence, and this form of the rate is frequently called the *Polanyi–Wigner relation*. They described chemical reactions as the process of crossing a potential energy surface between reactants and products (see Figure 2–3), where E_f and E_b are the energy barriers for forward and reverse reactions, ΔH_R is the heat of the reaction to be discussed later, and the horizontal scale is called the *reaction coordinate*, an ill-defined

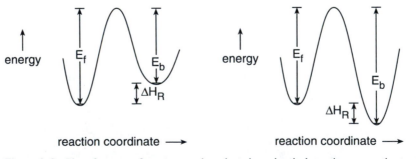

Figure 2–3 Plot of energy of reactants and products in a chemical reaction versus the reaction coordinate. The activation energy for the forward reaction is E_f, for the back reaction E_b, and the heat of the reaction is $\Delta H_R = E_f - E_b$. The curve at the left is for an endothermic reaction ($E_f > E_b$), while the curve at the right is for an exothermic reaction ($E_f < E_b$).

distance that molecules must travel in converting between reactants and products. Polanyi and Wigner first showed from statistical mechanics that the rates should be described by expressions of the form as given in the boxed equation by a Boltzmann factor, $\exp(-E/RT)$, which is the probability of crossing a potential energy barrier between reactant and product molecules. In fact, it is very rare ever to find reaction-rate coefficients that are not described with fair accuracy by expressions of this form.

This functional form of $k(T)$ predicts a very strong dependence of reaction rates on temperature, and this fact is central in describing the complexities of chemical reactions, as we will see throughout this book.

Example 2–1   How much does a reaction rate with an activation energy of 15,000 cal/mole vary when the temperature is increased from 300 to 310 K? From 300 to 400 K?

The ratio of the rate of this reaction at 310 K to that at 300 K,

$$\frac{k_{310}}{k_{300}} = \frac{e^{-E/RT_1}}{e^{-E/RT_2}} = \frac{\exp[-15,000/(2 \times 310)]}{\exp[-15,000/(2 \times 300)]} = 2.24$$

(We use the approximation of $R = 2$ cal/mole K). Between 300 and 400 K this ratio is very large,

$$\frac{k_{400}}{k_{300}} = \frac{e^{-E/RT_1}}{e^{-E/RT_2}} = \frac{\exp[-15,000/(2 \times 400)]}{\exp[-15,000/(2 \times 300)]} = 517$$

This shows that for this activation energy an increase of temperature by 10 K approximately doubles the rate and an increase of 100 K increases it by more than a factor of 500.

This example shows why the temperature is so important in chemical reactions. For many nonreacting situations a 10 K increase in T is insignificant, but for our example it would decrease by a factor of two the size of the reactor required for a given conversion. A decrease in the temperature by 100 K would change the rate so much that it would appear to be zero, and an increase by this amount would make the rate so fast that the process would be difficult or impossible to handle.

Let us consider finally the units of k. Basically, we choose units to make the rate (in moles liter^{-1} time^{-1}) dimensionally correct. For $r = kC_A^n$, k has units of liter^{n-1} mole^{1-n} time^{-1}, which gives k(time^{-1}) for $n = 1$ and k(liters/mole time) for $n = 2$. The units of k for some common orders of reactions are shown in Table 2–1.

TABLE 2–1
Units of Reaction-Rate Coefficients

Order	Units of k
1	time^{-1}
0	moles liter^{-1} time^{-1}
2	liter mole^{-1} time^{-1}
3	liter^2 mole^{-2} time^{-1}
0.5	mole$^{0.5}$ liter$^{-0.5}$ time^{-1}
n	(liter/mole)$^{n-1}$ time^{-1}

2.6 ELEMENTARY REACTIONS

We emphasize again that rate expressions are basically *empirical* representations of the dependence of rates of reaction on concentrations and temperature, $r(C_j, T)$. From the preceding examples and from intuition one can *guess* the order of a reaction from its stoichiometry. The forward rates appear to be proportional to the concentrations of reactant raised to powers m_{fj} equal to their stoichiometric coefficients, while the backward rates appear to be proportional to the concentrations of products raised to the powers m_{bj} of their stoichiometric coefficients, or, more formally,

$$m_{fj} = -v_j, \qquad v_j < 0 \tag{2.49}$$

$$m_{fj} = 0, \qquad v_j > 0 \tag{2.50}$$

$$m_{bj} = v_j, \qquad v_j > 0 \tag{2.51}$$

$$m_{bj} = 0, \qquad v_j < 0 \tag{2.52}$$

This set of relations between reaction orders and stoichiometric coefficients defines what we call an *elementary reaction*, one whose kinetics are consistent with stoichiometry. We later will consider another restriction on an elementary reaction that is frequently used by chemists, namely, that the reaction as written also describes the *mechanism* by which the process occurs. We will describe complex reactions as a sequence of elementary steps by which we will mean that the molecular collisions among reactant molecules cause chemical transformations to occur in a single step at the molecular level.

We note again that there is an arbitrariness in writing a reaction because all these v_j coefficients can be multiplied by a constant without changing the nature of the reaction or violating mass conservation. Thus, for NO decomposition, written previously as two different reaction equations, we would be tempted to write either $r = k[\text{NO}]^{1/2}$ or $r = k[\text{NO}]^1$. However, one of these must be incorrect, and in some situations neither equation correctly describes the experimentally observed rates. Only in some simple situations are reactions described by elementary kinetics.

On a molecular level, reactions occur by collisions between molecules, and the rate is usually proportional to the density of each reacting molecule. We will return to the subject of reaction mechanisms and elementary reactions in Chapter 4. Here we define elementary reactions more simply and loosely as reactions whose kinetics "agree with" their stoichiometry. This relationship between stoichiometry and kinetics is sometimes called the *Law of Mass Action*, although it is by no means a fundamental law of nature, and it is frequently invalid.

2.7 STOICHIOMETRY

Molecules are lost and formed by reaction, and mass conservation requires that amounts of species are related. In a closed (batch) system the change in the numbers of moles of all molecular species N_j are related by reaction stoichiometry.

For our NO decomposition example $\text{NO} \rightarrow \frac{1}{2}\text{N}_2 + \frac{1}{2}\text{O}_2$, from an O atom balance we see that

$$\tfrac{1}{2}N_{\text{NO}} + N_{\text{O}_2} = \text{constant} \tag{2.53}$$

while an N atom balance gives

$$\tfrac{1}{2}N_{\text{NO}} + N_{\text{N}_2} = \text{constant} \tag{2.54}$$

and subtracting these, we obtain

$$N_{\text{N}_2} - N_{\text{O}_2} = \text{constant} \tag{2.55}$$

in a closed system. The stoichiometry of the molecules requires that the moles of these species are related by these relations.

Subtracting the initial values (subscript o), we obtain relations such as

$$N_{\text{NO}} - N_{\text{NO}_0} = 2N_{\text{N}_{2o}} - 2N_{\text{N}_2} \tag{2.56}$$

or

$$\Delta N_{\text{NO}} = -2\Delta N_{\text{N}_2} \tag{2.57}$$

In general, for any single reaction we can write

$$\boxed{\frac{N_j - N_{jo}}{\nu_j} = \chi, \qquad \text{all } j} \tag{2.58}$$

and we can write $S - 1$ independent combinations of these relations among S chemical species in a reaction to relate the change in number of moles of all species to each other.

There is therefore always a *single composition variable* that describes the relationship among all species in a single reaction. In the preceding equation we defined χ as the relation between the N_js,

$$N_j = N_{jo} + \nu_j \chi \tag{2.59}$$

and we will call the quantity χ the *number of moles extent.*

For simple problems we most commonly use one of the reactants as the concentration variable to work with and label that species A to use C_A as the variable representing composition changes during reaction. We also make the stoichiometric coefficient of that species ν_A equal to -1.

Another way of representing a single reaction is *fractional conversion X*, a dimensionless quantity going from 0 with no reaction to 1 when reaction is complete. We define X through the relation

$$N_A = N_{Ao}(1 - X) \tag{2.60}$$

To make $0 < X < 1$ we also have to choose species A as the *limiting reactant* so that this reactant disappears and X approaches unity when the reaction is complete. We can then define all species through the relation

$$\frac{N_j - N_{jo}}{\nu_j} = N_{Ao}X \tag{2.61}$$

Recall our NO decomposition example, $NO \rightarrow \frac{1}{2}N_2 + \frac{1}{2}O_2$. Starting with $N_{NO} = N_{NO_0}$ initially, and all other species absent, we can relate moles of all species through the relation

$$N_{NO} = N_{NO_0}(1 - X) \tag{2.62}$$

$$N_{N_2} = \tfrac{1}{2}N_{NO_0}X + N_{N_2 0} \tag{2.63}$$

$$N_{O_2} = \tfrac{1}{2}N_{NO_0}X + N_{O_2 0} \tag{2.64}$$

For multiple reactions we need a variable to describe each reaction. Further, we cannot in general find a single key reactant to call species A in the definition of X. However, it is straightforward to use the number of moles extent χ_i for each of the R reactions. Thus we can define the change in the number of moles of species j through the relation

$$N_j = N_{jo} + \sum_{i=1}^{R} \nu_{ij} \chi_i \tag{2.65}$$

To summarize, we can always find a *single concentration variable* that describes the change in all species for a single reaction, and for R simultaneous reactions there must be R independent variables to describe concentration changes. For a single reaction, this problem is simple (use either C_A or X), but for a multiple-reaction system one must set up the notation quite carefully in terms of a suitably chosen set of R concentrations or χ_is.

2.8 REACTION RATES NEAR EQUILIBRIUM

In thermodynamics we learned how to describe the composition of molecules in chemical equilibrium. For the generalized single reaction

$$\sum_{j=1}^{S} \nu_j A_j = 0 \tag{2.66}$$

it can be shown that the free energy change in a system of chemically interacting species is related to the chemical potentials μ_j of each species through the relationship

$$\Delta G = \sum_{j=1}^{S} \nu_j G_j = \sum_{j=1}^{S} \nu_j \mu_j \tag{2.67}$$

where G_j is the Gibbs free energy per mole of species j and ΔG is the Gibbs free energy change per mole in the reaction. We call $\mu_j = G_j$ (actually $\partial G / \partial N_j$, the partial molar free energy) the *chemical potential* of species j. The chemical potential of species j is related to its chemical potential in the standard state (the state in which the activity a_j of species j is unity) by the relation

$$\mu_j = \mu_j^{\circ} + RT \ln a_j \tag{2.68}$$

At chemical equilibrium at constant temperature and pressure the Gibbs free energy of the system is a minimum and $\Delta G = 0$. Therefore, we have

$$\sum_{j=1}^{S} v_j \mu_j = 0 = \sum_{j=1}^{S} v_j \mu_j^o + \sum_{j=1}^{S} v_j RT \ln a_j \tag{2.69}$$

at chemical equilibrium. In this expression a_j is the activity of species j, which is a measure of the amount of a species defined such that $a_j = 1$ in the standard state where $\mu_j = \mu_j^o$. For gases the standard state is usually defined as the ideal-gas state at 1 bar (1 bar=1.023 atm), while for liquids it may be either the pure material or the material in a solution at a concentration of unity. The definitions of standard state and activity are somewhat arbitrary, but they are uniquely related by the definition of unit activity in the standard state. Once the standard state is defined, the situation is well-defined.

Next, dividing the preceding equation by RT and taking exponentials on both sides, we obtain

$$\exp\left(-\sum \frac{v_j \mu_j^o}{RT}\right) = \prod_{j=1}^{S} a_j^{v_j} \tag{2.70}$$

Since we define $\sum v_j \mu_j^o = \Delta G_R^o$, the Gibbs free energy change of the reaction in the standard state, we obtain

$$\prod_{j=1}^{S} a_j^{v_j} = \exp\left(-\frac{\Delta G_R^o}{RT}\right) = K_{eq} \tag{2.71}$$

where K_{eq} is the equilibrium constant as defined by this equation. [We note in passing that this notation is misleading in that the "equilibrium constant" is constant only for fixed temperature, and it usually varies strongly with temperature. To be consistent with our definition of the "rate coefficient," we should use "equilibrium coefficient" for the equilibrium constant, but the former designation has become the accepted one.]

We define the standard state of a liquid as $a_j = 1$ and for gases as an ideal gas pressure of 1 bar, $P_j = 1$. For ideal liquid solutions (activity coefficients of unity), we write $a_j = C_j$; so at chemical equilibrium

$$\boxed{K_{eq} = \exp\left(-\frac{\Delta G_R^o}{RT}\right) = \prod_{j=1}^{S} C_j^{v_j}} \tag{2.72}$$

and for gases

$$\boxed{K_{eq} = \exp\left(-\frac{\Delta G_R^o}{RT}\right) = \prod_{j=1}^{S} P_j^{v_j}} \tag{2.73}$$

where the difference between these K_{eq}s is that they are defined from μ_js at $C_j = 1$ and $P_j = 1$, respectively. In these expressions K_{eq} is dimensionless, while P_j has dimensions; this equation is still correct because we implicitly write each partial pressure as $P_j/1$ and $C_j/1$, which are dimensionless.

We can now return to our reaction system and examine the situation near chemical equilibrium. For a reversible reaction we have

$$r = r_f - r_b = k_f \prod_{\text{reactants}} C_j^{m_{fj}} - k_b \prod_{\text{products}} C_j^{m_{bj}} = 0 \qquad (2.74)$$

because the rate is zero at chemical equilibrium. Rearranging this equation, we obtain

$$\frac{k_f}{k_b} = \frac{\prod_j C_j^{m_{bj}}}{\prod_j C_j^{m_{fj}}} = \prod_{j=1}^{S} C_j^{m_{bj} - m_{fj}} \qquad (2.75)$$

Since we just noted that

$$K_{eq} = \exp\left(-\frac{\Delta G_R^o}{RT}\right) = \prod_{j=1}^{S} C_j^{\nu_j} \qquad (2.76)$$

we can immediately identify terms in these equations,

$$K_{eq} = \frac{k_f}{k_b} \qquad (2.77)$$

and

$$\nu_j = m_{bj} - m_{fj} \qquad (2.78)$$

at equilibrium. There is an apparent problem in the preceding equations in that K_{eq} is dimensionless, while k_f and k_b can have different dimensions if the orders of forward and back reactions are not identical. However, as noted, we implicitly divide all concentrations by the standard state values of 1 mole/liter, so that all these expressions become dimensionless.

From the preceding equations it can be seen that the rate coefficients and the equilibrium constant are related. Recall from thermodynamics that

$$\Delta G_R^o = \Delta H_R^o - T \, \Delta S_R^o \qquad (2.79)$$

where ΔH_R^o is the standard state enthalpy change and ΔS_R^o is the standard state entropy change in the reaction. Both ΔH_R^o and ΔS_R^o are only weakly dependent on temperature. We can therefore write

$$K_{eq} = \exp(-\Delta G_R^o / RT) = \exp(\Delta S_R^o / R) \exp(-\Delta H_R^o / RT) \qquad (2.80)$$

$$= \frac{k_f}{k_b} = \frac{k_{fo}}{k_{bo}} \exp[-(E_f - E_b / RT] \qquad (2.81)$$

Therefore, we can identify

$$E_f - E_b = \Delta H_R^o \qquad (2.82)$$

and

$$\frac{k_{fo}}{k_{bo}} = \exp(\Delta S_R^o / R) \qquad (2.83)$$

[While ΔG_R and ΔG_R^o can have very different values, depending on T and P, ΔH_R and ΔH_R^o are frequently nearly independent of T and P, and we will use ΔH_R from now on to designate the heat of a reaction in any state. We will therefore frequently omit the superscript o on ΔH_R.]

These relationships require that reactions be elementary, and it is always true that *near equilibrium* all reactions obey elementary kinetics. However, we caution once again that in general kinetics are empirically determined. These arguments show that near equilibrium the kinetics of reactions must be consistent with thermodynamic equilibrium requirements.

Note also that the description of a reaction as irreversible simply means that the equilibrium constant is so large that $r_f \gg r_b$. The notion of an irreversible reaction is an operational one, assuming that the reverse reaction is sufficiently small compared to the forward reaction so that it can be neglected. It is frequently a good approximation to assume a reaction to be irreversible when $\Delta G_R^o \ll 0$.

Returning to the energy diagrams of the previous figure, we see that the difference between E_f and E_b is the energy difference between reactants and products, which is the heat of the reaction ΔH_R. The heat of reaction is given by the relation

$$\Delta H_R = \sum_j \nu_j H_j \qquad (2.84)$$

where H_j is the heat of formation of species j. We necessarily described the energy scale rather loosely, but it can be identified with the enthalpy difference ΔH_R in a reaction system at constant pressure, an expression similar to that derived from classical thermodynamics.

These relations can be used to estimate rate parameters for a back reaction in a reversible reaction if we know the rate parameters of the forward reaction and the equilibrium properties ΔG_R^o and ΔH_R^o.

We emphasize several cautions about the relationships between kinetics and thermodynamic equilibrium. First, the relations given apply only for a reaction that is close to equilibrium, and what is "close" is not always easy to specify. A second caution is that kinetics describes the rate with which a reaction approaches thermodynamic equilibrium, and this rate cannot be predicted from its deviation from the equilibrium composition.

A fundamental principle of reaction engineering is that we may be able to find a suitable *catalyst* that will accelerate a desired reaction while leaving others unchanged or an *inhibitor* that will slow reaction rates. We note the following important points about the relations between thermodynamics and kinetics:

1. Thermodynamic equilibrium requires that we cannot go from one side of the equilibrium composition to the other in a single process.

2. Kinetics predicts the rates of reactions and which reactions will go rapidly or slowly towards equilibrium.

One never should try to make a process violate the Second Law of Thermodynamics, but one should never assume that ΔG_R^o alone predicts what will happen in a chemical reactor.

2.9   REACTOR MASS BALANCES

We need reaction-rate expressions to insert into species mass-balance equations for a particular reactor. These are the equations from which we can obtain compositions and other quantities that we need to describe a chemical process. In introductory chemistry courses students are introduced to first-order irreversible reactions in the batch reactor, and the impression is sometimes left that this is the only mass balance that is important in chemical reactions. In practical situations the mass balance becomes more complicated.

We will write all reactor mass and heat balances as

$$\boxed{\text{accumulation} = \text{flow in} - \text{flow out} + \text{generation by reaction}}$$

an expression we will see many times in mass and energy balances throughout this book. We remark before proceeding that equations such as

$$\frac{dC_A}{dt} = -kC_A \qquad (2.85)$$

and its integrated form

$$C_A = C_{Ao}e^{-kt} \qquad (2.86)$$

arise from a *very special situation* requiring both a *single first-order irreversible reaction* and a *constant-volume isothermal batch reactor*. This example is almost *trivial*, although we will use it frequently as a comparison with more interesting and accurate examples. The assumption of first-order kinetics is a simple first guess for kinetics and a good starting point before more elaborate calculations.

We note before proceeding that we must formulate and solve many mass-balance equations. We strongly encourage the student not to memorize anything except the basic defining relations. We stress that you should be able to derive every equation from these definitions as needed. This is because (1) only by being able to do this will you understand the principles of the subject; and (2) we need to make many different approximations, and remembering the wrong equation is disastrous.

2.10 THE BATCH REACTOR

A batch reactor is defined as a closed spatially uniform system which has concentration parameters that are specified at time zero. It might look as illustrated in Figure 2–4. This requires that the system either be stirred rapidly (the propeller in Fig. 2–4) or started out spatially uniform so that stirring is not necessary. Composition and temperature are therefore independent of position in the reactor, so that the number of moles of species j in the system N_j is a function of time alone. Since the system is closed (no flow in or out), we can write simply that the change in the total number of moles of species j in the reactor is equal to the stoichiometric coefficient v_j multiplied by the rate multiplied by the volume of the reactor,

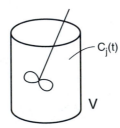

Figure 2–4.   Sketch of a uniform closed container for running chemical reactions, which we call a batch reactor.

$$[\text{accumulation}] = [\text{generation by reaction}]$$

or

$$\frac{dN_j}{dt} = V v_j r \qquad (2.87)$$

for a single reaction or

$$\frac{dN_j}{dt} = V \sum_i v_{ij} r_i \qquad (2.88)$$

for multiple reactions. These are simple *integral mass balances* on species j integrated over the entire reactor of volume V in a closed batch reactor for a single reaction and for multiple reactions, respectively.

The number of moles of species j in a batch reactor is simply the reactor volume V times the concentration

$$N_j = V C_j \qquad (2.89)$$

so the above equation becomes

$$\frac{d(V C_j)}{dt} = V \frac{dC_j}{dt} + C_j \frac{dV}{dt} = V v_j r(C_j) \qquad (2.90)$$

If the reactor is at constant volume, then we can divide each term by V to yield

$$\frac{dC_j}{dt} = v_j r(C_j) \qquad (2.91)$$

which is usually thought of as the "mass balance for a single reaction in a batch reactor," although it is only valid if the volume of the reactor does not change, as we will discuss later.

2.10.1   The first-order irreversible reaction

Let us immediately apply this equation to the first-order irreversible reaction

$$A \rightarrow B, \qquad r = kC_A \qquad (2.92)$$

The mass-balance equation on species A in a constant-density batch reactor is

$$\frac{dC_A}{dt} = v_A r = -kC_A \qquad (2.93)$$

since $v_A = -1$. We need an initial condition to solve a first-order differential equation, and for this system we assume that the reactor is charged initially with reactant A to give $C_A = C_{Ao}$ at $t = 0$. The variables can be separated

$$\frac{dC_A}{C_A} = -k \, dt \qquad (2.94)$$

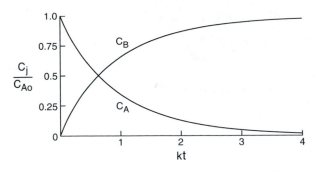

Figure 2–5 Plots of $C_A(t)$ and $C_B(t)$ versus kt for the first-order irreversible reaction $A \rightarrow B, r = kC_A$ in a batch reactor.

and integrated from $C_A = C_{Ao}$ at $t = 0$ to C_A at $t = t$

$$\int_{C_{Ao}}^{C_A} \frac{dC_A}{C_A} = -k \int_{t=0}^{t} dt \tag{2.95}$$

to give

$$\ln \frac{C_A}{C_{Ao}} = -kt \tag{2.96}$$

or

$$C_A = C_{Ao}e^{-kt} \tag{2.97}$$

The above two simple equations are those you will use most often in this course.

The solution of $C_A(t)$ for the first-order irreversible reaction is plotted in Figure 2–5. For a first-order irreversible reaction C_A decreases from C_{Ao} at $t = 0$ to C_{Ao}/e at $t = 1/k$ and to C_{Ao}/e^2 at $t = 2/k$. For these kinetics doubling the reaction time increases the conversion by a factor of 10.

Example 2–2 The reaction $A \rightarrow B$ has $k = 0.01 \text{ sec}^{-1}$. For $C_{Ao} = 2.0$ moles/liter, what time is required for 90% conversion in a constant-volume batch reactor? For 99%? For 99.9%?

Simple application of the preceding equation for 90% conversion ($C_A = 0.2$) yields

$$t = -\frac{1}{k} \ln \frac{C_A}{C_{Ao}} = +\frac{1}{k} \ln \frac{C_{Ao}}{C_A} = \frac{1}{0.01} \ln \frac{2}{0.2} = 100 \ln 10$$

$$= 100 \times 2.303 = 230 \text{ sec}$$

For 99% conversion ($C_A = 0.02$), we obtain

$$t = \frac{1}{0.01} \ln 100 = 100 \times 2 \times 2.303 = 460 \text{ sec}$$

For 99.9% conversion ($C_A = 0.002$), we obtain

$$t = \frac{1}{0.01} \ln 1000 = 100 \times 3 \times 2.303 = 690 \text{ sec}$$

These problems are easy! Note that the reactor residence time (proportional to reactor size) increases markedly as the required conversion increases. Note also that for this example (first-order kinetics) we did not need even to specify C_{Ao} because the equation involves only the ratio C_{Ao}/C_A.

We can use this solution for any first-order irreversible reaction

$$A \rightarrow \text{products}, \qquad r = kC_A \tag{2.98}$$

where "products" could be B, $2B$, $B/2$, $B+C$, or any other combination of product species. This is valid as long as the density of the fluid does not change significantly with composition, and this is a good approximation whenever the fluid is a liquid or a gas in which the volume is kept constant as the pressure changes so that there is a negligible change in number of moles as the reaction proceeds.

2.10.2 Second-order irreversible reaction

For second-order kinetics

$$A \rightarrow \text{products}, \qquad r = kC_A^2 \tag{2.99}$$

we obtain

$$\frac{dC_A}{dt} = -kC_A^2 \tag{2.100}$$

and separate variables to yield

$$\frac{dC_A}{C_A^2} = -kdt \tag{2.101}$$

With $C_A = C_{Ao}$ at $t = 0$, this equation can be integrated to give

$$\int_0^t dt = -k \int_{C_{Ao}}^{C_A} \frac{dC_A}{C_A^2} \tag{2.102}$$

or

$$t = \frac{1}{k}\left(\frac{1}{C_A} - \frac{1}{C_{Ao}}\right) \tag{2.103}$$

After rearranging and solving for $C_A(t)$, the expression becomes

$$C_A = \frac{C_{Ao}}{1 + C_{Ao}kt} \tag{2.104}$$

We could also write the above reaction as

$$2A \to 2B, \qquad r = kC_A^2 \qquad (2.105)$$

which looks like the standard form we suggested with $m_A = -\nu_A$. However, these are exactly the same reactions except by the latter definition the rate coefficient k in this expression is $\frac{1}{2}$ of that above because $\nu_A = -2$. This emphasizes that all stoichiometric coefficients can be multiplied or divided by an arbitrary constant, but the rate expression must be consistent with the stoichiometry chosen. The reaction and the corresponding rate must be consistent, and we recommend that one always write them out together as

$$\cdots \to \cdots, \qquad r = \cdots$$

so that the reaction and the rate are defined together.

Example 2–3   The reaction $A \to B$ obeys second-order kinetics with $k = 0.01$ liter $\mathrm{mole}^{-1} \mathrm{sec}^{-1}$. The initial concentration is $C_{Ao} = 2$ moles/liter. What time is required for 90% conversion in a batch reactor? For 99%? For 99.9%?

Application of the equation yields

$$t = \frac{1}{k}\left(\frac{1}{C_A} - \frac{1}{C_{Ao}}\right) = \frac{1}{0.01}\left(\frac{1}{0.2} - \frac{1}{2}\right) = 100 \times (5 - 0.5)$$

$$= 450 \text{ sec} = 7.5 \text{ min}$$

for 90% conversion,

$$t = \frac{1}{0.01}\left(\frac{1}{0.02} - \frac{1}{2}\right) = 100(50 - 0.5) = 4950 \text{ sec} = 1.38 \text{ h}$$

for 99% conversion, and

$$t = \frac{1}{0.01}\left(\frac{1}{0.002} - \frac{1}{2}\right) = 100(500 - 0.5) = 49,950 \text{ sec} = 13.9 \text{ h}$$

for 99.9% conversion. Once we have an equation for the solution, we can frequently solve the problem even without a hand calculator.

2.10.3   The nth-order irreversible reaction

For the nth-order irreversible reaction

$$A \to \text{products}, \qquad r = kC_A^n \qquad (2.106)$$

we obtain

$$\frac{dC_A}{dt} = -kC_A^n \qquad (2.107)$$

We now separate the equation to yield

$$C_A^{-n}\, dC_A = -k\, dt \qquad (2.108)$$

Figure 2–6 Plot of reactant concentration C_A versus dimensionless time $(n-1)kC_{Ao}^{n-1}t$ for the nth-order irreversible reactions $A \rightarrow$ products, $r = kC_A^n$. The expression is indeterminate for $n = 1$.

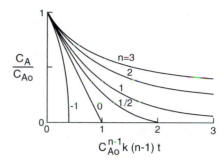

This can be integrated

$$\int_{C_{Ao}}^{C_A} C_A^{-n}\, dC_A = -\int_0^t k\, dt \tag{2.109}$$

to obtain

$$\frac{1}{-n+1}\left(C_A^{-n+1} - C_{Ao}^{-n+1}\right) = -kt \tag{2.110}$$

or

$$\boxed{C_A(t) = C_{Ao}[1 + (n-1)kC_{Ao}^{n-1}t]^{1/(1-n)}} \tag{2.111}$$

This equation is valid for any irreversible reaction of order n, but it cannot be used for $n = 1$ because the expression becomes indeterminate.

The concentrations of species A can be plotted for an irreversible nth-order reaction versus a dimensionless time scale $C_{Ao}^{n-1}kt$ (is this dimensionless?) as shown in Figure 2–6.

2.10.4 Low-order and negative-order reactions

The situation for $n \leq 1$ can be handled by this expression, but it requires modification at high conversions. For $n = 0$, the mass-balance equation is

$$\frac{dC_A}{dt} = -k \tag{2.112}$$

which when integrated from C_{Ao} at $t = 0$ yields

$$C_A = C_{Ao} - kt \tag{2.113}$$

However, $C_A = 0$ at $t = C_{Ao}/k$ and the concentration becomes *negative* for longer times. Therefore, this expression must be modified to become

$$C_A = C_{Ao} - kt, \qquad t \leq C_{Ao}/k \tag{2.114}$$

$$C_A = 0, \qquad t \geq C_{Ao}/k \tag{2.115}$$

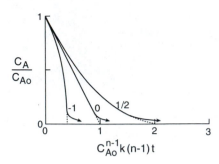

Figure 2–7   Plots of C_A versus t for an irreversible reaction for $n = 0, \frac{1}{2}$, and -1. The kinetics for all reactions must approach first order as the reactant concentration approaches zero to be consistent with equilibrium requirements.

Similar expressions must be used for any orders zero or fewer because reactant concentrations go to zero at finite time and can never be negative.

In fact, any kinetics of this type must be an approximation of a more complicated rate expression. We will show later that catalytic reactions frequently obey expressions such as

$$A \rightarrow \text{ products}, \qquad r = \frac{kKC_A}{1 + KC_A} \qquad (2.116)$$

where k and K are temperature-dependent coefficients. In fact, K is an adsorption–desorption equilibrium constant, as we will consider in Chapter 7. Note that whenever $KC_A \gg 1$, this expression becomes $r \approx k$, to give zeroth-order kinetics. However, as $C_A \rightarrow 0$, the rate becomes approximately

$$r \approx kKC_A \qquad (2.117)$$

and the reaction approaches first-order kinetics so that the solution for $C_A(t)$ in a batch reactor varies smoothly for all times.

Similarly the reaction

$$A \rightarrow \text{ products}, \qquad r = \frac{kKC_A}{(1 + KC_A)^2} \qquad (2.118)$$

obeys negative-order kinetics,

$$r = \frac{k}{KC_A} \qquad (2.119)$$

if $KC_A \gg 1$, but again approaches first-order kinetics,

$$r = kKC_A \qquad (2.120)$$

if $KC_A \ll 1$.

We will encounter similar rate expressions of this type when we consider surface or enzyme-catalyzed reactions in Chapter 7. These rate expressions are called Langmuir–Hinshelwood and Michaelis–Menten kinetics, respectively.

These rates versus time would be plotted as shown in Figure 2–7.

2.10.5 Bimolecular reactions

Consider next a bimolecular reaction

$$A + B \rightarrow 3C, \qquad r = kC_A C_B \qquad (2.121)$$

The mass balance on A is

$$\frac{dC_A}{dt} = -kC_A C_B \tag{2.122}$$

which cannot be solved without eliminating C_B. We showed previously that the number of moles of all species in a batch reactor are related by the relation

$$\frac{N_{jo} - N_j}{\nu_j} = \text{const} \tag{2.123}$$

or

$$\frac{C_{jo} - C_j}{\nu_j} = \text{const} \tag{2.124}$$

for a constant-density system. Therefore, for this reaction the loss in A is equal to the loss in B, which is equal to $\frac{1}{3}$ the gain in C, or in symbols

$$C_A - C_{Ao} = C_B - C_{Bo} = \tfrac{1}{3}(C_{Co} - C_C) \tag{2.125}$$

Therefore, we can immediately solve for the concentration of B,

$$C_B = C_{Bo} - C_{Ao} + C_A \tag{2.126}$$

Substitution of C_B in terms of C_A yields

$$\frac{dC_A}{C_A(C_{Bo} - C_{Ao} + C_A)} = -k \, dt \tag{2.127}$$

If $C_{Ao} = C_{Bo}$ at $t = 0$, then $C_A = C_B$ at all times, and this expression is identical to the expression for a second-order reaction

$$\frac{dC_A}{dt} = -kC_A^2 \tag{2.128}$$

This equation can be integrated to yield

$$C_A(t) = \frac{C_{Ao}}{1 + C_{Ao}kt} \tag{2.129}$$

and we can find $C_B(t) = C_A(t)$ and $C_C(t)$ by substitution of $C_A(t)$ into the preceding relations.

We could also solve this equation in terms of the fractional conversion X by expressing both C_A and C_B in terms of X. We can write $C_A = C_{Ao}(1 - X)$, which gives $dC_A = -C_{Ao} \, dX$. If $C_{Ao} = C_{Bo}$, then the mass balance becomes

$$-C_{Ao}\frac{dX}{dt} = -kC_A^2 = kC_{Ao}^2(1 - X)^2 \tag{2.130}$$

so that

$$\frac{dX}{dt} = kC_{Ao}(1 - X)^2 \tag{2.131}$$

This equation can be integrated to give $X(t) = kC_{Ao}t/(1 + kC_{Ao}t)$ and substituted back into $C_A = C_B = C_{Ao}(1 - X)$ and $C_C = C_{Ao}X$ to find $C_A(t)$, $C_B(t)$, and $C_B(t)$. However, for most simple reaction systems it is just as convenient (and there is less chance of a mathematical error) to work directly in concentration units.

2.10.6 Reversible reactions

The next more complex rate expression is the reversible reaction

$$A \rightleftarrows B, \qquad r = k_f C_A - k_b C_B \qquad (2.132)$$

If $C_A = C_{Ao}$ and $C_B = 0$ at $t = 0$, we have

$$\int_{C_{Ao}}^{C_A} \frac{dC_A}{k_f C_A - k_b(C_{Ao} - C_A)} = -\int_0^t dt = -t$$

$$= \frac{1}{k_f + k_b}\{\ln[k_f C_A - k_b(C_{Ao} - C_A)] - \ln[k_f C_{Ao}]\} \qquad (2.133)$$

Rearranging and solving for $C_A(t)$, we obtain

$$C_A(t) = C_{Ao}\left(1 - \frac{k_f}{k_f + k_b}(1 - e^{-(k_f + k_b)t})\right) \qquad (2.134)$$

We can proceed to examples of increasing complexity. For example,

$$A \rightleftarrows 2B, \qquad r = k_f C_A - k_b C_B^2 \qquad (2.135)$$

can be written as a function of C_A alone from the relation $C_B = 2(C_{Ao} - C_A)$. Substitution gives the integral

$$t = -\int_{C_{Ao}}^{C_A} \frac{dC_A}{k_f C_A - 4k_b(C_{Ao} - C_A)^2} \qquad (2.136)$$

which must be solved by factoring the denominator of the integrand and solving by partial fractions.

Example 2–4 Solve the preceding example by partial fractions to find $C_A(t)$.

You have seen partial fraction solutions to integral equations in math courses. We will find many situations where rate equations can only be integrated by partial fractions; so it is worth reviewing this procedure. The solutions are simple but they require some algebraic manipulation. The integrand of the previous integral can be written as

$$t = -\int_{C_{Ao}}^{C_A} \frac{dC_A}{k_f C_A - 4k_b(C_{Ao} - C_A)^2} = \frac{1}{4k_b}\int_{C_{Ao}}^{C_A} \frac{dC_A}{C_A^2 - (\frac{k_f}{4K_b} + 2C_{Ao})C_A + C_{Ao}^2}$$

$$= \frac{-1}{4K_b(x_2 - x_1)}\int_{C_{Ao}}^{C_A}\left[\frac{1}{C_A - x_1} - \frac{1}{C_A - x_2}\right]dC_A$$

$$= \frac{1}{4k_b(x_2 - x_1)}\left[\ln\frac{C_{Ao} - x_1}{C_A - x_1} - \ln\frac{C_{Ao} - x_2}{C_A - x_2}\right]$$

where

$$x_1, x_2 = \frac{1}{2}\left\{2C_{Ao} + \frac{k_f}{4k_b} \pm \left[(2C_{Ao} + \frac{k_f}{4K_h})^2 + 4C_{Ao}^2\right]^{1/2}\right\}$$

are roots of the polynomial

$$C_A^2 - (2C_{Ao} + \frac{k_f}{4k_b}C_A + C_{Ao}^2 = 0$$

It is obvious from this example why we quickly lose interest in solving mass-balance equations when the kinetics become high order and reversible. However, for any single reaction the mass-balance equation is *always separable* and soluble as

$$t = -\int_{C_{Ao}}^{C_A} \frac{dC_A}{r(C_A)} \tag{2.137}$$

Since $C_A = C_{Ao}(1 - X)$, we can also write this in terms of the fractional conversion X,

$$t = C_{Ao}\int_0^X \frac{dX}{r(X)} \tag{2.138}$$

Since $r(C_A)$ or $r(X)$ is usually a polynomial in C_A or X, one can solve this analytically for $t(C_A)$ or $t(X)$ by factoring the denominator of the integrand and solving the resulting partial fractions for $C_A(t)$ or $X(t)$. This can always be done, but the solution is frequently neither simple nor instructive.

One can always solve these problems *numerically* for particular values of C_{Ao} and k, and we will do this for many situations. However, it is still frequently desirable to have analytical expressions for $C_A(t)$ because we can then solve this expression by just substituting particular parameters into the equation. In any numerical solution one has to solve the differential equation again for each set of parameters.

Another reason for searching for analytical solutions is that we can only solve numerically a problem that is well posed mathematically. We must program a valid mathematical expression of the problem on the computer or the answers may be nonsense. The need for proper descriptions of the equations, initial and boundary conditions, and stoichiometric relations among the variables is the same whether one is interested in an analytical or a numerical solution.

2.11 VARIABLE DENSITY

The previous examples are all simple problems to integrate for $C_A(t)$ or at least for $t(C_A)$. We assumed that the density was constant (the volume in the equations). This would be true in any batch reactor where the volume is held constant, but not in a constant-pressure reaction with gases if the number of moles changes with the reaction.

Batch reactors are usually operated at constant volume because it is easy to construct a constant-volume closed container (as long as the pressure does not increase enough to burst the vessel). However, in flow reactors the density frequently changes as the reaction proceeds, even though the reactor volume is constant, and we need to be able to handle this situation.

For reactions such as $A \rightarrow B$ and $A + B \rightarrow 2C$ with ideal gases the density clearly does not change as the reaction proceeds if P and T remain constant. It is frequently also a good approximation with reactions among gases with changing numbers of moles if the reactants are diluted by an inert solvent. Constant density is also a good approximation for most liquid solutions because the density of a liquid solution usually does not change much as the reaction proceeds. As noted previously, the concentration of liquid water is $\sim$55 moles/liter, and in almost any aqueous reaction the reactant will be diluted by many moles of water per mole of reactants or products.

An important situation in which we must be concerned with variable density is with nonideal gases or in which one of the reactants or products condenses or evaporates. For example, the hydration of ethylene

$$C_2H_4 + H_2O \rightarrow C_2H_5OH \tag{2.139}$$

involves gases and liquids at typical temperatures and pressures. These systems can be very complex to describe because any gases are usually very far from ideal and because they involve both phase and reaction equilibrium considerations in addition to chemical reaction rates. We will not consider these complicated situations until Chapter 14.

If the volume V of a batch reactor depends on conversion or time, then the derivations of all of the previous equations are incorrect. We could find $V(C_A)$ and integrate the mass-balance equation as before, but it is usually more convenient to use a different variable such as the fractional conversion X. We finally write $C_A = N_A/V$ and then substitute for $N_A(X)$ and $V(X)$ to find $C_A(X)$.

Consider the reaction

$$A \rightarrow 3B, \qquad r = kC_A \tag{2.140}$$

in a constant-pressure batch reactor. This is a situation with first-order kinetics but with 3 moles of product formed for every mole of reactant decomposed. We assume that we start with N_{Ao} moles of pure A. If all species are ideal gases at constant pressure at initial volume V_o, then at completion the volume of the reactor will be $3V_o$. When the reaction has proceeded to a conversion X, the number of moles of A and B are given by the relations

$$N_A = N_{Ao}(1 - X) \tag{2.141}$$

$$N_B = 3N_{Ao}X \tag{2.142}$$

$$\sum N = N_{Ao}(1 + 2X) \tag{2.143}$$

and the volume V occupied by this number of moles will be

$$V = V_o(1 + 2X) \tag{2.144}$$

To solve the batch-reactor mass-balance equation, we write

$$C_A = \frac{N_A}{V} = \frac{N_{Ao}(1 - X)}{V_o(1 + 2X)} = C_{Ao}\frac{1 - X}{1 + 2X} \tag{2.145}$$

so that we have written C_A as a function of X and constants.

For this problem the equation $dC_A/dt = -kC_A$ is not appropriate, and we must solve the equation

$$\frac{dN_A}{dt} = -VkC_A = V_o(1 + 2X)kC_{Ao}\frac{1 - X}{1 + 2X} \tag{2.146}$$

Since $dN_A = -N_{Ao}\,dX$, this mass-balance equation on species A can be converted to

$$\frac{dX}{dt} = +k\frac{(1 - X)(1 + 2X)}{1 + 2X} = k(1 - X) \tag{2.147}$$

This equation can be separated to yield

$$\frac{dX}{1 - X} = k\,dt \tag{2.148}$$

and integrated from $X = 0$ at $t = 0$ to give

$$t = \int_{X=0}^{X}\frac{dX}{r(X)} = \frac{1}{k}\int_{X=0}^{X}\frac{dX}{1 - X} \tag{2.149}$$

$$t = -\frac{1}{k}\,\ln(1 - X) \tag{2.150}$$

This equation can be solved for $X(t)$,

$$X(t) = 1 - e^{-kt} \tag{2.151}$$

Finally, substituting back into $C_A(X)$, we obtain

$$C_A(t) = C_{Ao}\frac{1 - X}{1 + 2X} = C_{Ao}\frac{e^{-kt}}{1 + 2(1 - e^{-kt})} = C_{Ao}\frac{e^{-kt}}{3 - 2e^{-kt}} \tag{2.152}$$

Next consider the preceding as an nth-order irreversible reaction

$$A \to mB, \qquad r = kC_A^n \tag{2.153}$$

The mass balance is

$$\frac{dN_A}{dt} = -VkC_A^n \tag{2.154}$$

or

$$\frac{dX}{dt} = kC_{Ao}^{n-1}(1 - X)^n[1 + (m - 1)X]^{-n+1} \tag{2.155}$$

and this equation can be separated to yield

$$t = \frac{1}{kC_{Ao}^{n-1}}\int_{0}^{X}\frac{[1 + (m - 1)X]^{n-1}}{(1 - X)^n}\,dX \tag{2.156}$$

This equation can be solved for $t(X)$ or $t(C_A)$ by partial fractions, but the solution is not particularly simple to solve explicitly for $X(t)$ or $C_A(t)$.

For the reversible reaction

$$A + B \rightleftarrows C, \qquad r = k_fC_AC_B - k_bC_C \tag{2.157}$$

among ideal gases, if $C_{Ao} = C_{Bo}$ and $C_{Co} = 0$, the substitution of $C_j s$ in terms of X yields

$$r = k_f C_{Ao}^2 \frac{(1-X)^2}{(1-\frac{1}{2}X)^2} - k_b C_{Ao} \frac{X}{1-\frac{1}{2}X} \tag{2.158}$$

which again can be integrated for $X(t)$ and substituted back to find C_A, C_B, and C_C.

Example 2–5   Find the conversion versus time for the reaction

$$A \rightarrow 3B, \qquad r = kC_A^2$$

in a constant-pressure batch reactor assuming A and B are ideal gases starting with pure A at C_{Ao}.

The number of moles of A and B are found by setting up the following mole table:

Species	Moles initially	Moles
A	N_{Ao}	$N_{Ao}(1-X)$
B	0	$3N_{Ao}X$
total moles	N_{Ao}	$N_{Ao}(1+2X)$
volume	V_o	$V_o(1+2X)$

Therefore, the concentrations are

$$C_A = \frac{N_A}{V} = \frac{N_{Ao}(1-X)}{V_o(1+2X)} = C_{Ao}\frac{1-X}{1+2X}$$

and

$$C_B = \frac{N_B}{V} = C_{Ao}\frac{3X}{1+2X}$$

In a constant-pressure batch reactor the mass balance on A becomes

$$\frac{dN_A}{dt} = -N_{Ao}\frac{dX}{dt} = -rV = -kC_A^2 V$$

$$= -kC_{Ao}^2 \left(\frac{1-X}{1+2X}\right)^2 V_o(1+2X)$$

so that

$$\frac{dX}{dt} = kC_{Ao}\frac{(1-X)^2}{1+2X}$$

This equation can be separated to yield

$$t = \frac{1}{kC_{Ao}} \int_0^X \frac{1+2X}{(1-X)^2} dX$$

which must be integrated by partial fractions to give

$$t = \frac{1}{kC_{Ao}}\left[\frac{3X}{1-X} + 2\ln(1-X)\right]$$

This equation cannot be inverted to find $X(t)$ explicitly, and so we can only find $C_A(t)$ numerically.

2.12 CHEMICAL REACTORS

The chemical reactor is the "unit" in which chemical reactions occur. Reactors can be operated in batch (no mass flow into or out of the reactor) or flow modes. Flow reactors operate between limits of completely unmixed contents (the plug-flow tubular reactor or PFTR) and completely mixed contents (the continuous stirred tank reactor or CSTR). A flow reactor may be operated in steady state (no variables vary with time) or transient modes.

The properties of continuous flow reactors will be the main subject of this course, and an alternate title of this book could be "Continuous Chemical Reactors." The next two chapters will deal with the characteristics of these reactors operated isothermally. We can categorize chemical reactors as shown in Figure 2–8.

We will define these descriptions of reactors later, with the steady-state PFTR and CSTR being the most considered reactors in this course.

Figure 2–8   A "flow sheet" of possible reactor configurations and modes of operation.

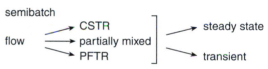

Example 2–6   Consider the situation where the reactants at constant density are fed continuously into a pipe of length L instead of a tank of volume V as in the batch reactor. The reactants react as they flow down the tube with a speed u, and we assume that they flow as a plug without mixing or developing the laminar flow profile. Show that the conversion of the reactants is *exactly the same* in these very different reactor configurations.

A molecule flowing with a speed u has traveled a distance z after it has flowed in the reactor for time

$$t = \frac{z}{u}$$

and the time τ it requires to flow down the reactor is

$$\tau = \frac{L}{u}$$

as sketched in Figure 2–9. Therefore, the increment of time dt to travel a distance dz is given by

$$dt = \frac{dz}{u}$$

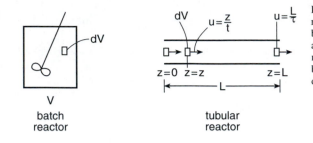

Figure 2–9 Sketch showing correspondence between time in a batch reactor and position a in a plug-flow tubular reactor. The mass-balance equations describe both reactors for the constant-density situations.

batch
reactor

tubular
reactor

and the PFTR equation we will solve most often is

$$\frac{dC_j}{d\tau} = u\frac{dC_j}{dz} = v_j r(C_j)$$

We can therefore replace dt by dz/u in all of the preceding differential equations for the mass balance in the batch reactor and use these equations to describe reactions during flow through a pipe. This reactor is called the plug-flow tubular reactor, which is the most important continuous reactor encountered in the chemical industry.

The preceding example shows that all the previous equations for the batch reactor can be immediately transformed into the plug-flow tubular reactor simply by replacing $dt \to dz/u$ in the differential equation or by replacing $t \to L/u$ in the integrated equation.

We do not have to solve these equations again for this very important flow reactor! It is important to note, however, that this transformation $t \to z/u$ is only valid if the velocity in the tube is constant. This requires that the tube diameter be constant and that there be no change in the fluid density as it moves down the tube because of pressure drops, temperature changes, or changes in the number of moles due to reaction.

Example 2–7 The reaction $A \to B$ with $k = 0.01$ sec^{-1} takes place in a continuous-plug-flow tubular reactor What residence time in the tube is required for 90% conversion? For 99%? For 99.9%?

Simple application of the equation given previously yields

$$\tau = \frac{1}{k} \ln \frac{C_{Ao}}{C_A} = \frac{1}{0.01} \ln 10 = 100 \times 2.303 = 230 \text{ sec}$$

for 90% conversion

$$\tau = \frac{1}{0.01} \ln 100 = 100 \times 2 \times 2.303 = 460 \text{ sec}$$

for 99% conversion, and

$$\tau = \frac{1}{0.01} \ln 1000 = 100 \times 3 \times 2.303 = 690 \text{ sec}$$

for 99.9% conversion.

Flow-reactor problems are just as simple as batch-reactor problems. In fact, they are the *same mathematical problem* even though the reactor configuration and operation are totally different.

2.13 THERMODYNAMICS AND REACTORS

A chemical reaction cannot be made to produce a conversion beyond that of chemical equilibrium, $\Delta G = 0$, where $r = 0$, as we discussed previously. This is an application of the Second Law of Thermodynamics.

Chemical reactors can liberate or absorb very large amounts of energy, and the handling of this energy is a major concern in reaction engineering. This topic is an application of the First Law of Thermodynamics, which says that mechanical and thermal energy is conserved in any process. When we describe a chemical reaction, we designate its rate, but

TABLE 2–2
Standard State Enthalpy and Free Energy Changes for Some Important Reactions Discussed in This Book

Reaction	ΔH^o_R, 298	ΔG^o_R, 298
$NO \rightarrow \frac{1}{2}N_2 + \frac{1}{2}O_2$	-90.37	-86.69
$2NO \rightarrow N_2 + O_2$	-180.75	-173.4
$N_2 + O_2 \rightarrow 2NO$	$+180.75$	$+173.4$
$2NO + O_2 \rightarrow 2NO_2$	-114.14	-70.785
trans-2-$C_4 =\rightarrow$ *cis*-2-$C_4 =,$		
$1 - C_4 =\rightarrow$ *trans*-2-$C_4 =$		
$CH_4 + 2O_2 \rightarrow CO_2 + 2H_2O$	-802.32	-800.78
$CH_4 + H_2O \rightarrow CO + 3H_2$	$+206.15$	$+142.12$
$CH_4 + \frac{1}{2}O_2 \rightarrow CO + 2H_2$	-35.677	-86.475
$C_2H_6 \rightarrow C_2H_4 + H_2$	$+136.95$	$+101.01$
$C_2H_6 + \frac{1}{2}O_2 \rightarrow C_2H_4 + H_2O$	-209.75	-255.17
$C_2H_4 + 3O_2 \rightarrow 2CO_2 + 2H_2O$	-929.45	-919.69
$C_2H_4 + \frac{1}{2}O_2 \rightarrow C_2H_4O$	-218.47	-199.75
$C_2H_4 + H_2O \rightarrow C_2H_5OH$	-28.987	-7.8529
$CO + 2H_2 \rightarrow CH_3OH$	-90.642	-24.317
$CO + \frac{1}{2}O_2 \rightarrow CO_2$	-282.99	-257.12
$CO + H_2O \rightarrow CO_2 + H_2$	-41.163	-28.522
$H_2 + \frac{1}{2}O_2 \rightarrow H_2O$	-241.83	-228.59
$\frac{1}{2}N_2 + \frac{3}{2}H_2 \rightarrow NH_3$	-45.857	-16.33
$NH_3 + \frac{5}{4}O_2 \rightarrow NO + \frac{3}{2}H_2O$	-226.51	-239.87
$NH_3 + \frac{3}{4}O_2 \rightarrow \frac{1}{2}N_2 + \frac{3}{2}H_2O$	-316.88	-326.56

Note: All values are in kJ/mole of the first species listed.

we should also be very concerned about the heat of the reaction; so we need to specify ΔH_R^o and ΔG_R^o for every reaction.

$$\sum_{j=1}^{S} \nu_j A_j = 0, \qquad r = k_f \prod_{j=1}^{S} C_j^{m_{fj}} - k_b \prod_{j=1}^{S} C_j^{m_{bj}}, \qquad \Delta H_R^o, \qquad \Delta G_R^o \qquad (2.159)$$

Table 2–2 lists some important chemical reactions along with their standard state enthalpies and free energies of reaction, all in kJ/mole.

These are all industrially important reactions that we will discuss throughout this book in the text and in homework problems. The student might want to try to identify the type of reaction represented by each equation and why it is important.

These values of ΔH_R are standard state enthalpies of reaction (all gases in ideal-gas states) evaluated at 1 atm and 298 K. All values of ΔH_R are in kilojoules per mole of the first species in the equation. When ΔH_R is negative, the reaction liberates heat, and we say it is *exothermic*, while, when ΔH_R is positive, the reaction absorbs heat, and we say it is *endothermic*. As Table 2–2 indicates, some reactions such as isomerizations do not absorb or liberate much heat, while dehydrogenation reactions are fairly endothermic and oxidation reactions are fairly exothermic. Note, for example, that combustion or total oxidation of ethane is highly exothermic, while partial oxidation of methane to synthesis gas ($CO + H_2$) or ethylene (C_2H_4) are only slightly exothermic.

Simple examination of ΔH_R of a reaction immediately tells us how much heat will be absorbed or liberated in the reaction. This is the amount of heat Q that must be added or extracted to maintain the reactor isothermal. (This heat is exactly the enthalpy change in any flow reactor or in a batch reactor at constant pressure, and it is close to this for other conditions.)

2.14 ADIABATIC REACTOR TEMPERATURE

It is also important to estimate the temperature increase or decrease in a reactor in which no heat is added or removed, which is called an *adiabatic reactor*. From the First Law of Thermodynamics, we can construct a thermodynamic cycle to estimate the ΔT in going from reactants at temperature T_1 to products at temperature T_2, as shown in Figure 2–10. Assume that reaction occurs at T_o with heat of reaction ΔH_R per mole of a key reactant that is irreversible and goes to completion. The heat evolved Q is zero in an adiabatic process, and this requires that N moles of product are produced per mole of this reactant and then heated to temperature T_1.

$$Q = \int_{T_1}^{T_o} N_{reactants} C_{p,reactants} \, dT + \Delta H_R^o + \int_{T_o}^{T_2} N_{products} C_{p,products} \, dT = 0 \qquad (2.160)$$

where $C_{p,products}$ is the average heat capacity *per mole* of product. We note that "products" means all species in the final mixture including unreacted reactants and inerts. If we assume that ΔH_R and C_p are both independent of temperature (something seldom allowed in thermodynamics courses but almost always assumed in this course), then ΔH_R is independent of T, and we can write

$$Q = \Delta H_R^o + N_{products} C_{p,products} (T_2 - T_1) = 0 \qquad (2.161)$$

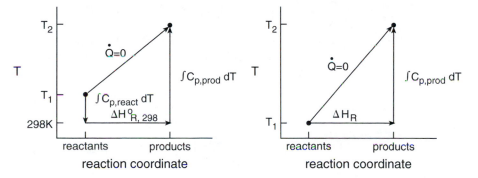

Figure 2–10   Energy diagram illustrating the temperature change in a reaction in which no heat is exchanged with the surroundings, a situation called an adiabatic reactor.

and ΔT becomes

$$\Delta T_{\text{adiabatic}} = T_2 - T_1 = \frac{-\Delta H_R}{N C_{p,\text{products}}} \qquad (2.162)$$

Example 2–8   Suppose the room you are in now actually contains 9.5% CH_4. [You could not detect the presence of CH_4 in air unless the natural gas contained a mercaptan odorizer.] If someone turned on the light switch and created a spark, what would be the temperature and pressure in the room before the windows and walls burst?

The methane combustion reaction is

$$CH_4 + 2O_2 \rightarrow CO_2 + 2H_2O$$

and this reaction has a heat of reaction of -192 kcal/mole of methane. We assume that there are 4 moles of N_2 per mole of O_2 (air actually contains 21% O_2, but 20% makes the calculation a bit simpler). This composition is the stoichiometric mixture of CH_4 in air to CO_2 and H_2O. For a basis of 1 mole of CH_4, there are 2 moles of O_2 and 8 moles of N_2 before reaction. After reaction there are 1 mole of CO_2, 2 moles of H_2O, and 8 moles of N_2 for total of 11 moles of product per mole of methane reacted.

If we assume the heat capacity C_p is equal to $\frac{7}{2}R$ (a reasonable approximation for small molecules such as N_2 at moderate temperatures), this predicts an adiabatic reaction temperature of

$$\Delta T_{\text{adiabatic}} = \frac{-\Delta H_R}{N C_p} = \frac{192 \text{ kcal/mole}}{(11)(\frac{7}{2})R} = 2500 \text{ K}$$

From the ideal-gas law $PV = NRT$ we have

$$\frac{P_2}{P_1} = \frac{T_2}{T_1} = \frac{2500 + 300}{300} = 9 \text{ atm}$$

This problem actually should be calculated for $\Delta U = 0$ rather than $\Delta H = 0$ for a constant-volume process, and at these temperatures there are many additional products because CO_2 and H_2O will dissociate significantly at this temperature. However, the final result would not make much difference, especially if you were in the room.

Recompute the final temperature and pressure if the methane were 5% in air.

This mixture has 1/2 the stoichiometric amount of CH_4, so the products will be 1 mole of CO_2, 2 moles of H_2O, 2 moles of O_2 remaining, and 16 moles of N_2, for a total of 21 moles. The final temperature rise will therefore be approximately half of that above, but you will still leave the room very quickly.

Recompute the final temperature and pressure if the methane were 3% in air.

A mixture of methane in air is flammable only between 5 and 15%; so no reaction occurs and the temperature and pressure are unchanged. You might ask the person who turned on the lights to open the window because the room seems a bit stuffy. We will discuss combustion processes and explosions more in Chapter 10.

We usually estimate the adiabatic temperature changes for exothermic reactions by assuming ΔH_R evaluated at 298 K and C_p for air at 298 K. These are calculated with stoichiometric reactant mixtures assuming complete reaction. Even mildly exothermic reactions have adiabatic temperature rises much above 100 K, and total combustion processes for stoichiometric mixtures in air have adiabatic temperatures above 2000 K. These are not accurate calculations for high final temperatures because properties vary with temperature and because of other reactions that will occur at high temperatures. Note that large N produces a smaller ΔT, and thus ΔT will be larger for pure O_2 instead of air and smaller if the reactant mixture is diluted with an inert. While these calculations give only approximate final temperatures because we assumed constant heat capacities, they indicate the size of the thermal hazard of exothermic chemical reactions from a very simple calculation.

For reactions in liquid solution we write an enthalpy balance on a 1 liter volume of the liquid

$$Q = \Delta H_R(C_{Ao} - C_A) + \int_{T_o}^{T} \rho C_p \, dT = 0 \qquad (2.163)$$

where now each term has units of calories/liter. If this reaction goes to completion, $C_A = 0$ and the equation becomes

$$T = T_o + \frac{-\Delta H_R}{\rho C_p} C_{Ao} \qquad (2.164)$$

Example 2–9 A reaction with a heat of reaction of −25 kcal/mole takes place in aqueous solution with a reactant concentration of 2 moles/liter and T_o =50°C. What is the temperature if this reaction goes to completion in an adiabatic reactor? What is the final temperature?

The specific heat of water is 1 cal/cm³ K or 1000 cal/liter K. Therefore, the final adiabatic temperature is

$$T = 50 + \frac{25000}{1000} 2 = 50 + 50 = 100°C$$

Anyone designing a reactor (and indeed anyone working near one) should be very aware of the heat absorbed or liberated in the chemical reactor and should be especially aware of the adiabatic temperature predicted. This heat must be released whenever a chemical reaction occurs, and this temperature will be attained if this heat is not removed.

The chemical reactor is the most hazardous unit in any chemical plant because most accidents occur by uncontrolled reaction, either within the reactor or after reactants have escaped the reactor and perhaps reacted with oxygen in air. Obviously no reactor or piping can withstand the temperatures and pressures of total combustion unless designed specifically for these conditions. We will consider the energy balance and temperature variations in continuous reactors in more detail in Chapters 5 and 6, while flames and explosions will be considered in Chapter 10.

Thus we see why it is essential to consider the energy balance very carefully in designing chemical reactors. The isothermal reactor assumption, while a good starting point for estimating reactor performance (the next two chapters), is seldom adequate for real reactors, and neglect of heat release and possible temperature increases can have very dangerous consequences.

2.15 CHEMICAL EQUILIBRIUM

No reactor can produce yields of products beyond those predicted by chemical equilibrium, and the second calculation anyone should perform on a process (after calculating the adiabatic temperature for safety considerations) is the equilibrium composition.

We discussed thermodynamic equilibrium previously in relating kinetics to reversible reactions. One should always estimate these quantities and keep them in mind before performing more detailed design of reactors and separation units. We also note that one must consider all chemical reactions that may occur for a given feed, not just the one desired.

To obtain the equilibrium conversion for a single reaction, we need to solve the equation

$$\prod_{j=1}^{S} a_j^{v_j} = \exp(-\Delta G_R^o/RT) = K_{eq} \tag{2.165}$$

for the relevant activities a_j. For gases we usually define ΔG^o with a standard state ($a_j = 1$) as the ideal gas at 1 atm; so the above expression becomes

$$\prod_{j=1}^{S} P_j^{v_j} = \exp(-\Delta G_R^o/RT) \tag{2.166}$$

We need to determine ΔG_R° at the temperature of the calculation. This is obtained from the van't Hoff equation

$$\frac{d(-\Delta G_R^\circ/RT)}{dT} = \frac{\Delta H_R^\circ}{RT^2} = \frac{d \ln K_{eq}}{dT} \qquad (2.167)$$

which relates equilibrium composition to temperature. This equation can be integrated from 298 K to any temperature T to yield

$$\boxed{\ln K_T = \ln K_{298} + \int_{298}^{T} \frac{\Delta H_R}{RT^2} \, dT} \qquad (2.168)$$

The enthalpy change of reaction varies with temperature as

$$\Delta H_R = \sum v_j H_j = \sum v_j \left(H_{fj,298} + \int_{298}^{T} C_{pj}(T) \, dT \right) \qquad (2.169)$$

where $H_{fj,298}$ is the heat of formation of species j at 298 K. The variation of H_{fj} with T is determined by the variation of C_{pj} with T, as calculated by this integral.

It is a good approximation for estimations to ignore the C_{pj} term so that K_T is given approximately by the expression

$$\boxed{\ln K_T = \ln K_{298} - \frac{\Delta H_{R298}}{R} \left(\frac{1}{T} - \frac{1}{298} \right)} \qquad (2.170)$$

While ΔH_R does not vary strongly with T, ΔG_R° and K_{eq} are strong functions of T. This is illustrated in Figure 2–11, which shows that ΔG_R° varies by orders of magnitude from 298 to 1000 K. Note also that K_{eq} either increases or decreases, depending on the sign of ΔH_R. Endothermic reactions have favorable equilibrium compositions at high temperatures, while exothermic reactions have favorable equilibrium compositions at low temperatures. You probably learned of these effects through *Le Chatelier's Principle*, which states that, if a reaction liberates heat, its equilibrium conversion is more favorable at low temperature, and if a reaction causes an increase in the number of moles, its equilibrium is more favorable at low pressure (and the converse situations). While these equations permit you to calculate these effects, this simple principle permits you to check the sign of the change expected.

For multiple reactions we have simultaneous equilibrium equations

$$\prod_{j=1}^{S} a_j^{v_{ij}} = \exp(-\Delta G_{Ri}^\circ/RT) = K_i, \qquad i = 1, 2, \ldots, R \qquad (2.171)$$

which is a set of R polynomials that must be solved simultaneously, as is considered in thermodynamics courses.

Three commonly used sets of units for describing densities of fluids are partial pressure (P_j), mole fraction (x_j), and concentration (C_j). For these units the standard state is defined as unit activity a_j, which is typically $P_j = 1$ atm and 298 K, or $x_j = 1$ for pure liquid at 1 atm and 298 K, or $C_j = 1$ mole/liter at 298 K, respectively. Students have seen the first two

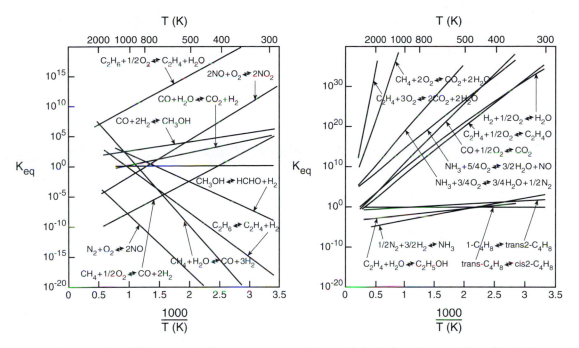

Figure 2–11   Plots of equilibrium constant K_T versus temperature for several chemical reactions we will consider in this text. All these reactions give nearly straight lines when plotted as log K_T versus $1/T$. Exothermic reactions have large K_T at low T, while endothermic reactions have large K_T at high T.

of these for gases and liquids in thermodynamics. We will use concentration units wherever possible in this course, and the natural standard state would be a 1 molar solution. However, data are usually not available in this standard state, and therefore to calculate equilibrium composition at any temperature and pressure, one usually does the calculation with P_j or x_j and then converts to C_j.

As a final remark on the importance and use of thermodynamics in chemical reactors, we note that the preceding equations no longer need be solved in practice because thermo-dynamics computer programs and databases allow one to compute all these quantities by simply listing the molecules and conditions into the computer programs. The data in Table 2–2 were calculated using CHEMEQ from Sandler's thermodynamics text, and sophisticated chemical process programs such as ASPEN contain databases for most of the complex molecules and conditions that one might encounter in industrial practice.

The variations in K_T with temperature are shown in Figure 2–11 for the reactions in Table 2–2. These should be straight lines when plotted as log K_T versus $1/T$ if ΔH_R is independent of temperature, and the curves show that this approximation is accurate for all these reaction systems.

2.16   PETROLEUM REFINING

Next, before we proceed to consider flow reactors, we will completely change topics and discuss two examples: the refining of petroleum and the production of polyester. Our first purpose is to describe these reactions and the reactors that are used in order to acquaint students with these very important processes.

Our second purpose in discussing these processes is to indicate that the reactions and reactors used for these processes have almost *mind-boggling complexity*. The chemical engineer is in charge of designing and operating these processes, and we must be aware of this complexity and see through it to discover the principles by which these processes operate (and to fix them if they are not operating satisfactorily).

The following sections will be *purely descriptive* with no equations except for the chemical reactions, and even the chemical reactions listed will be highly simplified. As we proceed through this course, the principles by which these reactors are operated will become clearer, such that by the final chapter we will have considered most of these reactors and the principles of their operation in more detail.

2.16.1   Energy sources

The processing of crude oil into gasoline, other fuels, and chemicals has always been the bread-and-butter employer of chemical engineers. The discipline of chemical engineering had its origins in producing fuels and chemicals, and these tasks remain the dominant topic of the chemical engineer.

The energy sources we use are listed in Table 2–3.

The widespread use of coal started the Industrial Revolution, and in the nineteenth century wood and then coal were the major energy and chemical sources, while whale oil was our energy source for lighting as lamps replaced candles. Then in the late nineteenth and early twentieth century petroleum replaced all of these. Now natural gas is becoming very important, and coal may someday see a revival.

Renewable energy sources such as trees and plants have always had a niche market in energy and chemicals. In fact, all energy except nuclear energy is ultimately derived from solar energy in some form or other.

Energy for lighting has an interesting history as compared to energy for power and for chemicals. Campfires using solid fuel were replaced by candles and then by lamps that burned liquid oil from animal fats and then whale oil to produce portable light sources. As whale oil was depleted (as were the whales), lamps supplied by pipeline distribution systems

TABLE 2–3
Energy Sources and the Times When They Were Important

Energy source	Years dominant
solar, wind, and water	always
wood	always
animal fat	<1900
whale oil	1700–1900
coal	>1700
crude oil	>1880
natural gas	>1950
nuclear energy	>1950
shale oil	>1970
tar sands	>1970

were developed for large cities. These were fueled first by "town gas" (a mixture of CO and H_2), and then these distribution networks were replaced by today's natural gas lines. Now these same pipeline distribution systems are used mostly for heating. Finally, the generation of electricity from coal, hydroelectricity, and nuclear fuel has made the electric light the only significant source of lighting. Power and electricity generation have become topics in mechanical engineering rather than in chemical engineering, although all but nuclear power require combustion and control of resultant pollution, which mechanical engineers are not equipped to handle (in the opinion of chemical engineers).

It was discovered in the late nineteenth century that coal can be incompletely burned to yield a gas consisting primarily of CO and H_2, and many people were undoubtedly asphyxiated and killed by explosions before these processes were harnessed successfully. We will see later that the use of a $CO + H_2$ mixture (now called synthesis gas) for the production of chemicals has had an important role in chemical synthesis (it was very important for explosives and synthetic fuels in both World Wars), and it is now one of the most promising routes to convert natural gas and coal into liquid diesel fuel and methanol. We will describe these processes in more detail in later chapters.

2.16.2   Cracking

Initially the petroleum (tar) that oozed from the ground in eastern Pennsylvania could be burned as easily as whale oil and animal fats, although much of it was too heavy to burn without processing. Processing was done initially by pyrolyzing the oil in retorts and extracting the volatile components when the oil molecules cracked into smaller ones, leaving tar at the bottom. (Many people were killed as this process evolved before they learned about the volatility and flammability of the various hydrocarbon fractions produced.) Retorts (batch reactors) were soon replaced by continuous cracking units (stills), in which the crude oil was passed through heated tubes. One problem in these processes is that some of the hydrocarbon is cracked down to methane, which cannot be liquefied easily. A much more serious problem was that some also turns to tar and coke, a black solid mass (mess) that coats and eventually plugs the tube furnace.

Reactions in hydrocarbon cracking may be represented as

$$\text{hydrocarbons} \rightarrow \text{smaller hydrocarbons} + \text{coke}$$

The hydrocarbons in crude oil are alkanes, olefins, aromatics, polyaromatics, and organic compounds containing S, N, O, and heavy metals. Since there are many isomers of all of these types of molecules, the reactions implied by the preceding equations rapidly approach infinity. A representative reaction of these might be the cracking of hexadecane (number 3 heating oil) into octane and octene (components in gasoline),

$$n\text{--}C_{16}H_{34} \rightarrow n\text{--}C_8H_{18} + n\text{--}C_8H_{16}$$

and

$$n\text{--}C_{16}H_{34} \rightarrow 16C_s + 17H_2$$

These reactions are all very endothermic, and heat must be supplied to the retorts or tube furnaces, not only to heat the reactants to a temperature of 600 to 1000°C where they will react at an acceptable rate, but also to provide the heat of the reactions, which is several times the sensible heat.

Sometime in the early twentieth century it was found that if the steel tubes in the furnace had certain kinds of dirt in them, the cracking reactions were faster and they produced less methane and coke. These clays were acting as *catalysts*, and they were soon made synthetically by precipitating silica and alumina solutions into aluminosilicate cracking catalysts. The tube furnace also evolved into a more efficient reactor, which performs *fluidized catalytic cracking* (FCC), which is now the workhorse reactor in petroleum refining.

2.16.3   Petroleum refinery

A modern petroleum refinery in the United States processes between 100,000 and 500,000 barrels/day of crude oil. The incoming crude is first desalted and then passed through an atmospheric pressure distillation column that separates it into fractions, as shown in Figure 2–12.

The streams from distillation are classified roughly by boiling point, with names and boiling ranges shown in Figure 2–12. The lightest are the overheads from the distillation column, which are the lowest-molecular-weight components. Then come the low-boiling liquids, which are called naphtha; these compounds range from C_5 to C_{10} and have the appropriate vapor pressure for gasoline, although their octane rating would be low. Next comes gas oil with 8 to 16 carbons, which is appropriate for diesel fuel and heating fuel. Finally come the bottoms from the distillation column, and this fraction is usually separated again by vacuum distillation into a component that will boil and one that will decompose (crack) before boiling.

Crude oils vary greatly in boiling range and types of molecular structures. The best now comes only from the Middle East and is mostly naphtha that needs little refining. The worst now comes from the United States, is very high molecular weight, and consists largely of polyaromatic compounds containing S, N, and heavy metals, particularly V and Ni. The

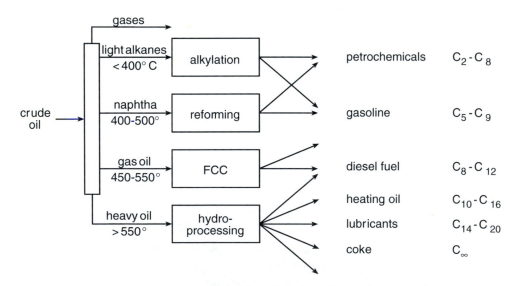

Figure 2–12   Qualitative flow sheet of reactants, reactors, and products in petroleum refining.

TABLE 2–4
The Four Major Reactors in Petroleum Refining

Reactor	Feed	Products	T(°C)	P (atm)
catalytic cracking	gas oil	lower-boiling alkanes	550	2
hydroprocessing	heavy oil	lower-boiling alkanes and aromatics	500	20–10
catalytic reforming	naphtha	high octane, aromatics	450	10–30
alkylation	olefins, alkanes	branched alkanes, alkyl aromatics	0	1

C:H ratio of these ranges also varies from perhaps 1:3 for light fractions to 1:1 for vacuum residual oil.

The task of the refinery is to turn crude oil into valuable products. The demands for products vary greatly with region and with season. The Gulf Coast needs more petrochemical feedstocks, while the Upper Midwest needs more heating fuel. Gasoline accounts for approximately one-half of the total petroleum used in the United States, but demand for gasoline is highest in the spring and summer vacation season.

These fractions from primary distillation are sent to reactors to crack them further and isomerize them into products with more value, such as gasoline and petrochemicals. The refinery is a massive blending system in which streams into and out of reactors are recycled and blended into products with the desired properties and amounts.

There are basically four types of reactors in refining. Fluidized catalytic cracking reactors process perhaps 30% of all the crude, the gas oil fraction. The heavier fraction must have hydrogen added, and these processes are called *hydroprocessing*. Gasoline must have certain isomers for octane rating, and this is done in catalytic reforming. Some products have too low a molecular weight and must be recombined into larger molecules, a process that is called *alkylation*. We list these reactor units in Table 2–4.

2.16.4   Catalytic cracking

This is the primary chemical process in the refinery. The heavy gas oil stream is cracked into smaller hydrocarbons suitable for gasoline. The empty tube furnace was first replaced with tubes filled with aluminosilicate catalyst pellets. Then it was found that the tubes could be replaced by a series of tanks with interstage heating to maintain the desired temperature. In all cases it was necessary to burn the coke out of the reactor by periodically shutting down and replacing the feed by air, a complicated and expensive process that lowers the capacity of the reactor.

The problems in fixed bed cracking reactors are (1) heat must be supplied to heat the reactants to the desired temperature and overcome the endothermicities of the cracking reactions; and (2) the reactor must be shut down periodically for coke removal. Both of these problems were overcome by the development in the 1940s and 1950s of a fluidized bed system that replaces the need to shut down to remove coke from the catalyst. This process is called *fluidized catalytic cracking*, or FCC.

The FCC reactor is really two reactors with solid catalyst pellets cycled between them. The vaporized gas oil is fed along with fresh catalyst to the first, called the reactor, and the

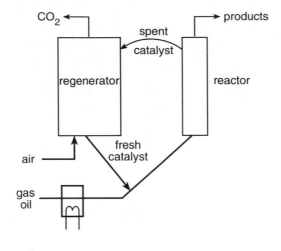

Figure 2–13 Sketch of a fluidized catalytic cracking reactor (FCC) for cracking heavy petroleum fractions into the boiling range needed for gasoline.

spent catalyst is separated from the products in a cyclone and sent to the regenerator, where air (now sometimes O_2) is added to oxidize the carbon. The flows of reactants, products, air, and catalyst are indicated in Figure 2–13. The reactor cracks the hydrocarbon and forms coke on the catalyst. Then in the regenerator the coke is burned off and the catalyst is sent back into the reactor.

The "magic" of FCC is that the reactor–regenerator combination solves both the heat management and coking problems simultaneously. Burning off the coke is strongly exothermic, and this reaction in the regenerator supplies the heat (carried with the hot regenerated catalyst particles) for the endothermic cracking reactions in the reactor.

Typical operating conditions of these components of the FCC reactor are indicated in Table 2–5. The residence time in the regenerator is longer than in the reactor, and it is therefore considerably larger.

Many reactor–regenerator configurations have been developed with quite different shapes as this technology has evolved and FCC units have become larger. A modern FCC reactor resembles the Space Shuttle (but one hopes it stays on the ground) in size, tank configurations, and engineering complexity. The FCC reactor is without question the most complex and important equipment in chemical engineering.

TABLE 2–5
Operating Conditions of the FCC Reactor

	Reactor	Regenerator
feed	gas oil	air
products	alkanes, olefins, H_2	CO_2
temperature	550°C	650°C
pressure	2 atm	2 atm
gas residence time	<1 sec	several sec
catalyst residence time	<1 sec	several sec
flow pattern	PFTR (riser)	CSTR
heat	absorbed	generated

2.16.5   Heavy oil processing

The "bottom of the barrel" contains heavy, smelly compounds that have polyaromatic rings and that contain up to several percent of S and N in aromatic rings and in side chains sulfides and amines. This fraction will not boil below temperatures where the molecules begin to crack, and it is called "residual" oil or "vacuum resid" if it boils at reduced pressure. This fraction also contains perhaps 0.1% of heavy metals tied up as porphyrin rings in the polyaromatics. All these species are severe poisons to either FCC or catalytic reforming catalysts, but fortunately the lighter naphtha and gas oil feeds to these reactors contain negligible S, N, and metals.

The other problem with this petroleum fraction is that it is deficient in hydrogen. We need C:H = 7:16 for 2,2,3-trimethylbutane (TMB), but this ratio is greater than 1:1 for many heavy feeds. Therefore, we need to add hydrogen in the refining process, and we could describe them generically as the reaction

$$(CH)_n + \frac{n}{2}H_2 \rightarrow (CH_2)_n$$

and the need to add H_2 to obtain usable products gives the prefix "hydro" to process names such as hydroprocessing, hydrotreating, hydrocracking, hydrorefining, hydrodesulfurization, etc.

These cracking and H-addition processes also require catalysts, and a major engineering achievement of the 1970s was the development of hydroprocessing catalysts, in particular "cobalt molybdate" on alumina catalysts. The active catalysts are metal sulfides, which are resistant to sulfur poisoning. One of the major tasks was the design of porous pellet catalysts with wide pore structures that are not rapidly poisoned by heavy metals.

Modern processes operate with fixed beds with a 20:1 to 50:1 H_2:HC ratio at pressures typically 50 atm at 550°C. Since the reactants must be operated below their boiling points, the catalyst is a solid, and the H_2 is a gas, all these reactors involve three phases in which the catalyst is stationary, the gas moves upward through the reactor, and the liquid fraction flows down the reactor. This reactor type is called a *trickle bed*, and it is a very important chemical engineering unit in refining heavy crude feedstocks and turning them into the molecular weight suitable for gasoline. The flow of gas and liquid through a trickle bed reactor is shown in Figure 2–14.

Figure 2–14   Sketch of a trickle bed reactor used for hydroprocessing of the residual oil fraction of crude oil into the boiling range used for gasoline and diesel fuel.

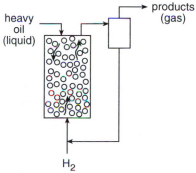

2.16.6 Catalytic reforming

The catalytic cracking and hydroprocessing reactors produce a large fraction of hydrocarbons with a molecular weight suitable for gasoline, C_5 to C_9. However, these products have a low octane number. In the spark ignition engine some isomers ignite appropriately, while some tend to ignite before the piston has reached the top of the cylinder, and this causes the engine to knock. We will discuss hydrocarbon combustion reactions in Chapter 10 because they are free-radical chain reactions. Highly branched alkanes and aromatics are superior in this to linear alkanes, and hydrocarbons are compared by their *octane number*. The molecule 2,2,3-trimethyl butane is arbitrarily given an octane number of 100, and *n*-heptane is given an octane number of 0. [For unknown reasons, the molecule 2,2,3-trimethyl butane is called "isooctane" even though it is actually a heptane.]

An arbitrary mixture of hydrocarbons is compared to a mixture of these two compounds, with its octane number that equal to the appropriate mixture of these standard compounds. Some molecules and their octane ratings are indicated in Table 2–6. Aromatics have a high octane number (toluene is 120), and some compounds such as tetraethyl lead have a strong octane enhancement when added to other mixtures (blending octane number). Oxygenates such as ethanol and ethers (MTBE) have fairly high octane numbers and supposedly produce less pollution, either alone or blended with hydrocarbons.

The products from the catalytic cracker and from the hydroprocessor contain too many linear isomers and cyclic aliphatics, and the isomerization of these linear alkanes to branched alkanes and dehydrogenation of cyclics to aromatics would enhance the octane enormously. In the 1950s it was found that a catalyst consisting of Pt on γ-Al_2O_3 would almost miraculously achieve very large octane enhancements without cracking products undesirably. This process is called *catalytic reforming*, because it "reforms" the skeleton of molecules without cracking C–C bonds into smaller molecules. Reforming also produces high yields of aromatics that have high octane and are needed as petrochemical feedstocks.

This process is similar to catalytic cracking in some ways. Because of the higher pressure required, it uses fixed beds rather than fluidized beds that are now used for catalytic cracking. However, now we want primarily to *isomerize* rather than *crack* the hydrocarbons because the naphtha feed has the desired molecular weight. Therefore, in catalytic reforming the temperature is lower and H_2 is added to suppress coke formation, and thus regeneration needs to be done much less frequently.

TABLE 2–6
Octane Ratings of Hydrocarbons

Molecule	Research octane number (RON)
n-heptane	0 (defined)
2,2,3-trimethylbutane	100 (defined)
toluene	120
benzene	110
2-methylhexane	42
ethanol	106
methanol	118
methyl-*t*-butyl ether (MTBE)	135

Figure 2–15 Reactions in catalytic reforming of n-heptane to 2,2,3-trimethylbutane and toluene.

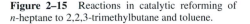

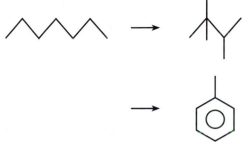

Modern catalytic reformers operate as either fixed beds in series, in which reactants flow through several tanks filled with catalyst, or as moving beds, in which catalyst flows slowly down a tube as the reactants move up. A major concern is the minimization of the formation of cracking products such as coke that poison the catalyst. Catalysts inevitably deactivate by coke formation, and this must be burned off by reacting with air. In fixed-bed reactors this is done by varying the flow through the tanks such that one reactor is periodically in "swing" where the coke is burned off. In the moving-bed reactors some of the catalyst is continuously withdrawn from the bottom and the carbon is oxidized before it is reinserted in the top of the reactor.

Reactions in catalytic reforming can be represented as

$$n-C_7H_{16} \rightarrow 2,2,3\text{-trimethylbutane}$$

$$n-C_7H_{16} \rightarrow \text{toluene} + 3H_2$$

$$n-C_7H_{16} \rightarrow 7C + 8H_2$$

These can also be written as shown in Figure 2–15.

The first reaction is the *isomerization* from a zero-octane molecule to an alkane with 100 octane; the second is the *dehydrocyclization* of heptane to toluene with 120 octane, while the third is the undesired formation of coke. To reduce the rate of cracking and coke formation, the reactor is run with a high partial pressure of H_2 that promotes the reverse reactions, especially the coke removal reaction. Modern catalytic reforming reactors operate at 500 to 550°C in typically a 20:1 mole excess of H_2 at pressures of 20–50 atm. These reactions are fairly endothermic, and interstage heating between fixed-bed reactors or periodic withdrawal and heating of feed are used to maintain the desired temperatures as reaction proceeds. These reactors are sketched in Figure 2–16.

2.16.7 Alkylation

The final refinery reactor on our list joins back together molecules that were cracked too much in the previous processes. The process can be thought of as adding an alkyl group to an alkane or aromatic, and the process is therefore called *alkylation*.

An important alkylation reaction is the combination of propylene to isobutane,

$$C_3H_6 + i\text{-}C_4H_{10} \rightarrow 2,2,3\text{-TMB}$$

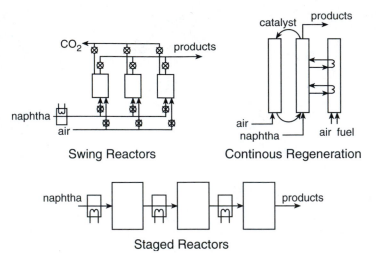

Swing Reactors

Continous Regeneration

Figure 2–16 Several reactor configurations for catalytic reforming to produce high-octane gasoline.

Staged Reactors

which makes a 100-octane molecule. Another important alkylation reaction is the addition of propylene to benzene to make isopropyl benzene,

$$C_3H_6 + C_6H_6 \rightarrow C_6H_5C_3H_7$$

This molecule is called cumene, and it is easily oxidized to acetone and phenol, both important commodity chemicals,

$$C_6H_5C_3H_7 \rightarrow (CH_3)_2CO + C_6H_5OH$$

Alkylation of benzene and ethylene produces ethyl benzene, which is dehydrogenated to styrene for polystyrene.

Alkylation is an association reaction that is exothermic. Therefore, it has a favorable equilibrium only at low temperatures. The process is catalyzed by liquid acids of solid $AlCl_3$, and modern alkylation reactors use sulfuric acid or liquid HF as catalysts operating at 0°C in a refrigerated reactor that is stirred rapidly to dissolve and create bubbles of the hydrocarbons in the acid.

Picture a reactor that is a very large kitchen blender with pipes coming in and out. The many tons of liquid HF in alkylation processes in refineries represent significant health hazards if the reactor ever leaks, and considerable research and engineering are being devoted to replace HF by safer catalysts.

2.17 POLYESTER FROM REFINERY PRODUCTS AND NATURAL GAS

The largest-volume organic chemical is ethylene, an intermediate building block in many chemicals and polymers. Polyethylene is the largest of these uses, and polyester is a very important polymer for fabrics that uses ethylene in its synthesis. In this section we will describe the processes by which components of natural gas and crude oil are turned into these products. Natural gas sells for about $0.10/lb, while clothing sells for about $20/lb, and the job of the chemical engineer is to turn these reactants into products as efficiently and cheaply as possible. [What is the price per pound of gasoline if one gallon weighs 8 pounds?]

Figure 2–17 Reaction steps to prepare polyethylene terephthalate from ethane and naphtha.

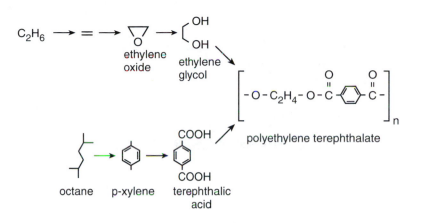

octane   p-xylene   terephthalic acid

However, in contrast to fuels, petrochemicals intermediates must be produced at extremely high purities. For example, CO at ppm levels will poison polyethylene catalysts, and acetylene in ethylene at this level will produce a crosslinked polymer that will have unsatisfactory properties. Therefore, the chemical engineer must produce these intermediates with extremely high purities, and this requires both careful attention to minor reactor products and to efficient separation of them from the desired product. These factors are also important in the economics of petrochemicals.

A flow sheet of the steps in forming polyester is as follows:

ethane → ethylene → ethylene oxide → ethylene glycol →

polyethylene terephthalate

crude → naphtha → xylenes → *p*-xylene → terephthalic acid →

In stick notation we write these reactions as shown in Figure 2–17. We described the steps by which crude oil is converted into aromatics such as *p*-xylene previously. Here we discuss the production of ethylene glycol from ethane.

Natural gas is primarily methane, but it contains 5–10% of natural gas liquids, primarily ethane, propane, and butane. These are available at essentially the cost of natural gas heating except for the cost of condensing these less volatile components from methane. Natural gas liquids are a major source of chemicals, along with petroleum, and we will describe a number of similar examples in this book.

2.17.1   Ethylene from ethane

Ethylene is made primarily by steam cracking of ethane and higher alkanes. These alkanes undergo dehydrogenation and cracking reactions. A higher-molecular-weight fraction of alkanes, C_4 to C_7, is called naphtha, which in steam cracking reacts as

$$naphtha \rightarrow olefins + H_2 + smaller\ alkanes$$

When ethane is used as a feedstock, the major reaction is simple dehydrogenation,

$$C_2H_6 \rightarrow C_2H_4 + H_2$$

In modern steam cracking processes using ethane as a feedstock approximately 80% ethylene selectivity is attained at approximately 40% ethane conversion. This reaction is highly endothermic, and more than 25 kcal/mole of ethane converted must be supplied in the reactor.

Equilibrium in the reactions shown is unfavorable except at high temperatures, and modern plants operate at temperatures of 850–900°C. At these temperatures reaction takes place homogeneously, but it is far from elementary and proceeds by a chain reaction involving many free-radical intermediates, which we will consider in Chapter 10. Reaction is roughly first order in ethane,

$$r = k[C_2H_6]$$

In-free radical reactions many products are formed in small quantities such as propylene, methane, butadiene, acetylene, and benzene. A major cost of an olefin plant is the separation of these byproducts from ethylene and also separation of unreacted ethane, which is recycled and fed back to the reactor.

Another major cost in producing ethylene results from the formation of carbon on the walls of the reactor. This occurs through reactions such as

$$C_2H_4 \rightarrow 2C_s + 2H_2$$

although ethane and all hydrocarbons can also decompose to carbon. Carbon formation from pure alkane feed would plug the reactor very quickly, and in olefin plants a large excess of water is added to suppress the cracking of large hydrocarbons into smaller ones. The process is called *steam cracking* because water is added to suppress coke formation. Water does not react directly with the hydrocarbon, but it reacts with carbon in the reaction

$$C_s + H_2O \rightarrow CO + H_2$$

a process called *steam gasification*. The addition of steam reduces the rate of coke formation such that the tubes of the reactor only slowly develop significant pressure drops because of carbon buildup. In an olefin plant the process must be shut down approximately once a month for a day or so while air is blown through the tubes to oxidize the carbon,

$$C_s + O_2 \rightarrow CO_2$$

a very exothermic gas–solid reaction.

Steam-cracking reactors typically consist of several steel tubes, perhaps 100 m long and 4 in. in diameter in a tube furnace with reactants and steam fed through the several tubes in parallel. The ceramic lined furnace is heated by burning natural gas at the walls to heat the tubes to 900°C by radiation. The reactor is fed by ethane and steam in a ratio of 1:1 to 1:3 at just above atmospheric pressure. The residence time in a typical reactor is approximately 1 sec, and each tube produces approximately 100 tons/day of ethylene. We will return to olefins and steam cracking in Chapter 4.

2.17.2 Polyethylene

Ethylene is used to produce many products, the largest volume of which is polyethylene (PE)

$$nC_2H_4 \rightarrow (CH_2)_{2n}$$

which is produced by free-radical homogeneous reactions (low-density PE) and also by a titanium catalyst (high-density PE). We will consider these processes further in Chapter 11.

2.17.3   Ethanol

Ethylene also readily reacts with H_2O to make ethanol,

$$C_2H_4 + H_2O \rightarrow C_2H_5OH$$

in liquid acids such as sulfuric acid at temperatures up to 200°C. The water molecule can be thought of as adding across the double bond.

This is the most economical process to produce ethanol, but laws prohibit drinking "synthetic" ethanol; so beverages are made much more expensively by fermentation of sugar or carbohydrates. Recent laws also mandate the addition of 10% ethanol in gasoline in cities during the winter, supposedly to reduce pollution. However, grain processors lobbied to require "renewable resources"; so fermentation is required to produce this fuel alcohol.

2.17.4   Ethylene glycol

Ethylene glycol (EG) has two OH groups; so it will polymerize as a linear polymer in polyesters, polyurethanes, or polyethers. Ethylene glycol is also water soluble and has a low melting point; so it is used in antifreeze.

Ethylene glycol was formerly made by adding water to acetylene,

$$C_2H_2 + 2H_2O \rightarrow HOC_2H_4OH$$

which can be thought of as adding two water molecules across the triple bond in acetylene. However, acetylene is expensive to produce, and EG is now made from ethylene in a two-stage process using an ethylene oxide intermediate.

2.17.5   Ethylene oxide

Ethylene oxide was formerly made in a two-stage process by first adding HOCl to ethylene and then removing HCl. However, in the 1960s Scientific Design, Union Carbide, and Shell Oil developed a one-step direct oxidation process that has largely replaced the old chlorohydrin process.

To make ethylene glycol, ethylene is first oxidized to ethylene oxide (EO)

$$C_2H_4 + \tfrac{1}{2}O_2 \rightarrow C_2H_4O$$

which is then reacted with water in acid to form EG,

$$C_2H_4O + H_2O \rightarrow HOC_2H_4OH$$

Ethylene can be oxidized to EO over a silver-on-alumina catalyst in 1-in.-diameter tubes approximately 20 ft long. A modern EO plant produces 200 tons/day, with a typical reactor consisting of 1000 tubes with an EO selectivity of 80% with a 4:1 C_2H_4:O_2 ratio at approximately 50% conversion of O_2. EO formation is mildly exothermic, while the competing complete combustion reaction

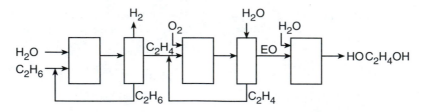

Figure 2–18   Flow sheet of process to produce ethylene glycol from ethane.

$$C_2H_4 + 3O_2 \rightarrow 2CO_2 + 2H_2O$$

is extremely exothermic. The selectivity decreases rapidly if the temperature increases above 250°C, and heat release will then increase strongly (and catastrophically). Therefore, it is essential to thermostat the tubes in a configuration that looks like a large heat exchanger in which boiling organic hydrocarbon on the shell side circulates to a heat exchanger, where it condenses to carry the heat generated on the shell side to a steam generator. An EO reactor configuration is shown in Figure 2–18.

We will consider ethylene oxide reactors in more detail in later chapters, because the cooling of these reactors is a major design consideration.

2.17.6   Polyethylene terephthalate

The most important polyester is polyethylene terephthalate (PET), which has many uses, from fabrics to milk bottles. The polymerization reaction between ethylene glycol (EG) and terephthalic acid (TPA),

$$EG + TPA \rightarrow PET + H_2O$$

is carried out in a multistage reactor.

The bifunctional acid is terephthalic acid (TPA), which is made by oxidizing *para*-xylene in the reaction

$$CH_3C_6H_4CH_3 + 3O_2 \rightarrow HOOCC_6H_4COOH + 2H_2O$$

This is another partial oxidation process that is usually carried out in several reactors in series with both air and HNO_3 as oxidants. The student might guess that this reaction is far from the only one, and much CO, CO_2, alcohols, aldehydes, and other acids are also produced that must be separated from the TPA before it is reacted to form PET.

Para-xylene is a product of petroleum refining, particularly in the catalytic reformer that produces *ortho*, *meta*, and *para* isomers in nearly equal amounts. Only the *para* isomer is useful for other than fuel, and the mixture of xylenes is isomerized into primarily *para*-xylene in a reactor filled with a zeolite, called ZSM5.

Example 2–10   The monomers for nylon are either the amino acid $H_2N-(CH_2)_5COOH$, which forms the polymer called Nylon 6 (six carbon atoms in the backbone) or two monomers adipic acid $HOOC-(CH_2)_4COOH$ and hexamethylene diamine $H_2N-(CH_2)_6NH_2$, which form the polymer called Nylon 66 (six carbon atoms in the backbone of each of the monomers). It is possible to make the 6-carbon atom monomers very easily with high purity, but the production of nylons with five, seven, or other carbons would be very difficult. Why?

Cyclohexane. It can be prepared with high purity by distillation of a mixture of alkanes from petroleum refining or by reduction of benzene.

Cyclohexane can be partially oxidized by just attacking one C–C bond to open the ring with functional groups on each end of the six-carbon chain to produce the six-carbon amino acid or adipic acid. Sketch the intermediates to these products starting with cyclohexanone.

2.18 "WHAT SHOULD I DO WHEN I DON'T HAVE REACTION RATES?"

Unfortunately, there are no tables of chemical reaction rates in this book, and you won't find them in other books either. It would be very useful if we could construct tables of rates such as

Reaction	Preexponential	Activation energy	Orders	Range of validity

and then list all the important reactions we may be interested in running. We would then just look up the reaction, find the rate that is applicable for the conditions at which we want to operate, insert these equations into the mass and energy balances, and solve them to predict performance. While you can find useful data tables in any text on thermodynamics, heat and mass transfer, or separations, reaction-rate data do not exist for most technologically interesting processes.

In thermodynamics the properties of pure materials are given to five significant figures, and correlations for mixtures are frequently accurate to better than 1%. Diffusivities of many binary systems are known to within a few percent, and, even for complex mixtures, they are probably available to within 5%. Mass and heat transfer texts give correlations that are usually good to a few percent, depending on the mixture and conditions. Separation process efficiencies can be computed to similar accuracies using empirical correlations and notions of equilibrium stages.

However, in reaction engineering we cannot even begin to do this. If someone claims to have a general correlation of reaction rates, the prudent engineer should be suspicious. The major problem is that most interesting reaction systems involve *multiple reactions*, and one would have to somehow list rates of forming many products from several reactants; so we would need a lot of data in tables to cover a process. Second, most interesting reactions are *catalytic*, either on a solid surface or by an enzyme. Different catalyst systems behave quite differently with different catalyst formulations, and they are notoriously sensitive to trace *impurities*, which can poison the catalyst or promote one rate over the others. Catalytic reaction rates also do not usually obey simple power-law rate expressions (such as $r = kC_A$), and one frequently finds that effective parameters such as orders and activation energies vary with conditions. Catalytic processes also exhibit *deactivation*, and all properties will typically vary with the time the processes have been in operation, and a major engineering question is how long will be required before the process must be shut down to reactivate or replace the catalyst.

This is the fun (and frustration) of chemical reaction engineering. While thermodynamics, mass and heat transfer, and separations can be said to be "finished" subjects for many engineering applications, we have to reexamine every new reaction system from first principles. You can find data and construct process flowsheets for separation units using sophisticated computer programs such as ASPEN, but for the chemical reactors in a process these programs are not much help unless you give the program the kinetics or assume equilibrium yields.

These complications show why we emphasize simple and qualitative problems in this course. In reactor engineering the third decimal place is almost always meaningless, and even the second decimal place is frequently suspect. Our answers may be in error by several orders of magnitude through no fault of our own, as in our example of the temperature dependence of reaction rates. We must be suspicious of our calculations and make estimates with several approximations to place bounds on what may happen. Whenever a chemical process goes badly wrong, we are blamed. This is why chemical reaction engineers must be clever people. The chemical reactor is the least understood and the most complex "unit" of any chemical process, and its operation usually dominates the overall operation and controls the economics of most chemical processes.

2.19 REACTION-RATE DATA

We consider next the acquisition of data on the kinetics of a chemical reaction. We want rate expressions $r(C_j, T)$ which we can insert into the relevant mass balance to predict reactor performance. The methods of acquiring these data, in order of increasing difficulty and expense, are

1. Literature values. If the process is simple and well known, there may be rate expressions in the literature that can be used.

2. Estimations. If one can find a process similar to the one of interest, then rates can be estimated from these data. For example, if one finds a reactor for which a specified conversion is obtained with a specified reactant composition and temperature, then one may guess the orders of the reaction with respect to each species (guess first order) and proceed to formulate a reasonable rate expression.

3. Theoretical rate calculations. Statistical mechanics permits one in principle to compute reaction-rate expressions from first principles if one knows the "potential energy surface" over which the reaction occurs, and quantum mechanics permits one to calculate this potential energy surface. In Chapter 4 we consider briefly the theory of reaction rates from which reaction rates would be calculated. In practice, these are seldom simple calculations to perform, and one needs to find a colleague who is an accomplished statistical mechanic or quantum mechanic to do these calculations, and even then considerable computer time and costs are usually involved.

4. Kinetics measurements. When detailed literature data are not available and when one badly needs accurate kinetics, then the only recourse is to obtain kinetic data in the laboratory.

2.19.1 Reaction chemistry

First one tries to find out what reactions should be expected with the given reactants and process conditions. Here one must know enough chemistry to decide what reactions can occur. If one is not very experienced in the chemistry of the process in question, one must examine the relevant literature or ask experts (organic chemists for organic reactions, polymer chemists for polymer reactions, material scientists for solids reactions, etc.).

Once one has formulated a list of possible reactions, one should then look up the relevant free energies ΔG_R^o in order to calculate the *equilibrium composition* for the desired feed and temperature. If ΔG_R^o is sufficiently negative, that reaction should be expected to be *irreversible*. If the equilibrium constant is sufficiently small so that equilibrium product concentrations are small for the conditions used, then that reaction can be neglected compared to others.

2.19.2 Batch-reactor data

Reaction kinetics are most easily and inexpensively obtained in a small batch reactor. With liquids this is frequently just a mixture of liquids in a beaker or flask placed on a hot plate or in a thermostatted water or sand bath. With gases the experiment would involve filling a container with gases and heating appropriately. One starts the process at $t = 0$ with C_{jo} and records $C_i(t)$.

The most difficult aspect of these experiments is finding a suitable method of analyzing the composition of the reactor versus time. Gas and liquid chromatography are by far the most used techniques for analyzing chemical composition. Spectroscopic methods (IR, visible, UV, NMR, ESR, etc.) can be used in some situations, and mass spectrometry is a versatile but difficult technique. For reactions in gases with a mole number change in a constant-volume batch reactor, the conversion can be determined by simply measuring the pressure change as the reaction occurs. All these techniques require calibration of the instruments under the conditions and the composition range of the experiment.

One measures $C_j(t, T)$ for given C_{jo} and then finds a suitable method of analyzing these data to find a suitable rate expression that will fit them. For liquid solutions the typical method is to obtain *isothermal* batch-reactor data with different C_{jo}s and continues to gather these data as a function of temperature to find a complete rate expression. For a simple irreversible reaction we expect that the rate should be describable as

$$r(C_j, T) = k(T) \prod_{j=1}^{S} C_j^{m_j} \tag{2.172}$$

Thus we expect the rate to be given by a *power* dependence on the concentrations and an exponential dependence on temperature $k_f(T) = k_{fo}e^{-E/RT}$. This form of the rate expression is not always accurate, especially for catalytic and enzyme reactions for which Langmuir–Hinshelwood and Michaelis–Menten expressions are required to fit experimental data.

Assuming that we have an irreversible reaction with a single reactant and power-law kinetics, $r = kC_A^n$, the concentration in a constant-volume isothermal batch reactor is given by integrating the expression

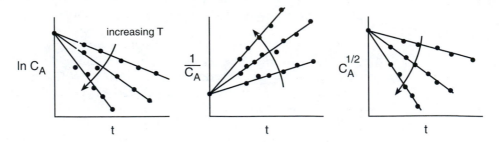

Figure 2–19 Plots of C_A versus time, which give straight lines for an nth-order irreversible reaction. The slopes of these lines give the rate coefficient k.

$$\frac{dC_A}{dt} = -kC_A^n \tag{2.173}$$

to obtain

$$\frac{1}{-n+1}(C_A^{-n+1} - C_{Ao}^{-n+1}) = -kt \tag{2.174}$$

for $n \neq 1$.

For second-order kinetics, the integrated rate expression is

$$\frac{1}{C_A} - \frac{1}{C_{Ao}} = kt \tag{2.175}$$

so that a plot of $1/C_A$ versus t should give a straight line whose slope is $-k$. For first-order kinetics, the appropriate plot is one of $\ln C_A$ versus t, and the slope of this plot is $-k$. [What plot will give a straight line for half-order kinetics?]

One thus obtains a family of these isothermal lines from batch-reactor data for a given C_{Ao} for different temperatures, as shown in the graphs of Figure 2–19 for $n = 1, 2$, and $\frac{1}{2}$.

2.19.3   Temperature dependence

We expect that the slopes k from this graph should depend on T as

$$k(T) = k_o \exp(-E/RT) \tag{2.176}$$

Taking the logarithm of both sides, we obtain

$$\ln k = \ln k_o - \frac{E}{RT} \tag{2.177}$$

so that a plot of $\ln k$ versus $1/T$ should give a straight line whose slope is $-E/R$, as shown in Figure 2–20. If we extrapolate this Arrhenius plot to $1/T = 0$ ($T = \infty$), the value of k is the preexponential k_o. We frequently plot these data on a basis of $\log_{10}$, for which the slope is $-E/2.303R$.

Thus from this procedure we have the simplest method to analyze batch-reaction data to obtain a rate expression $r(C_A, T)$ if the reaction is irreversible with a single reactant and obeys power-law kinetics with the Arrhenius temperature dependence.

If the reaction has multiple reactant species such as

$$A + B \rightarrow \text{ products}, \qquad r = kC_A C_B \tag{2.178}$$

Figure 2–20 Plot of log k (obtained from isothermal reaction data in the previous figure) versus $1/T$. The slope of this line is $-E/R$, and the intercept where $1/T = 0$ is the preexponential factor k_0.

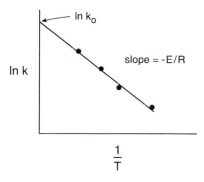

then the orders with respect to C_A and to C_B can be obtained by using a *large excess* of one component. In this example one might use $C_{Bo} \gg C_{Ao}$ so that C_B remains nearly constant in a batch-reactor experiment. Thus the rate would become $r = (kC_{Bo})C_A$, and a plot of $\ln C_A$ versus t would have a slope $-kC_{Bo}$. One would then repeat this experiment with $C_{Ao} \gg C_{Bo}$ where a plot of C_B versus t would have a slope of kC_{Ao}.

2.19.4 Differential-reactor data

Another method to obtain kinetic data is to use a differential reactor in which the concentration does not change much from the initial concentration C_{Ao}. In this case the differential rate expression

$$\frac{dC_A}{dt} = kC_A^{m_A} \tag{2.179}$$

can be written approximately as $\Delta C_A / \Delta t$. Taking logarithms, we obtain

$$\ln\left(\frac{\Delta C_A}{\Delta t}\right) = \ln k + m_A \ln C_A \tag{2.180}$$

so that a plot of $\ln(\Delta C_A/\Delta t)$ versus $\ln C_A$ has a slope of m_A and an intercept of $\ln k$, as shown in Figure 2–21. By plotting lines such as these versus temperature, the values of E and k_0 can be obtained from differential batch-reactor data.

Figure 2–21 Differential reactor data analysis. Plots of $\Delta C_A/\Delta T$ versus C_A on a log–log scale should give straight lines for fixed T. The slope of this line is the order m_A, while the value of the vertical scale where $C_A = 1$ is the rate coefficient k.

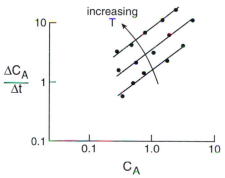

2.19.5   Statistical analysis of data

The accuracy of data obtained by these methods must be analyzed very carefully to determine the statistical confidence of rate parameters m_j, E, and k_o obtained. One must have data over a sufficient range of C_A, t, and T for accurate values, and data should be analyzed by methods such as least-squares analysis to assess its accuracy.

This analysis assumes *random error* in data acquisition, and an equally important problem in any experiment is *systematic error*, in which measurements are inaccurate because of measurement difficulties such as calibration errors, meter zero errors, or simple human error (writing down the wrong number). Statistical analysis can never be substituted for common sense and caution in acquiring and analyzing reaction-rate data to obtain reaction-rate parameters.

2.19.6   Data acquisition

Obtaining kinetic data is very tedious, and it requires great care to avoid both random and systematic errors. For this reason, it is very common to assemble computer-based data acquisition systems, frequently with simple personal computers equipped with data acquisition analog-to-digital capabilities and graphics. These computers can be programmed patiently to acquire the necessary data, make frequent calibrations, vary parameters such as temperature and concentration, analyze data statistics, and print out parameters.

However, since any process evaluation is no better than the rate data fed into it, the chemical engineer must always be suspicious of the validity of rate data, whether from the literature or obtained in house. A particular concern is the *extrapolation* of rates obtained under one set of conditions to different operating conditions. Perhaps one must process liquids at high temperatures and pressures, while lab data are easiest to acquire in a glass flask at atmospheric pressures and lower temperatures. The prudent engineer would want at least to spot check any rate expressions at actual operating conditions or risk trouble.

2.19.7   Complications

Reaction systems are seldom (never) as simple as those implied by an expression such as $r = kC_A^n$. While the ideas presented show how data are generally obtained, considerable ingenuity is required for the complex reactions encountered in practice. Among these complications are:

1. Real processes almost always involve multiple reactions. This is the subject of Chapter 4, where we will see that selectivity to form a desired product is a crucial issue and consequently there are many ks and orders that we need to know.

2. Real processes frequently involve catalysts or enzymes, which usually give rate expressions much more complex than these.

3. Catalysts and enzymes also can vary significantly between batches and exhibit activation and deactivation, so that reaction rates may be expected to vary with time. Thus it is not unusual to find that a reaction activation energy increases with the time that a process

has been onstream, which one might need to fit by assuming that E is a function of time. As you might expect, problems such as these require careful consideration and caution. We will consider catalytic reactions and their kinetics in Chapter 7.

4. It is frequently difficult to maintain reactors strictly isothermal because all reactions liberate or absorb considerable heat. These effects can be minimized by diluting reactants and using low temperatures, thus making reaction rates sufficiently slow that the system can be thermostatted accurately. However, kinetics under these conditions are not those desired in a reactor, and one must be careful of the necessary extrapolation to operating conditions. We will discuss heat effects in detail in Chapters 5 and 6.

5. Most reaction systems involve many simultaneous reactions, and the analysis of these systems can be very difficult or impossible. For a large set of reactions with unknown kinetics, it is common simply to assume reactions and rates (for example, with all reactions first order) and make *empirical fits* of experimental data to these rate expressions. Such rate expressions have little fundamental significance, but empirical reaction rates allow one to formulate models of the reactor behavior. Special caution must be used in extrapolating these empirical rate expressions beyond the range of data from which they were obtained. Many industrial reaction processes such as in the petroleum refinery are analyzed through such empirical models. These are developed over long periods of time to fit plant behavior, and each large producer of a given process has carefully guarded empirical models with which to optimize plant performance for given feedstocks and customer product demands.

6. Kinetic data are frequently acquired in continuous reactors rather than batch reactors. These data permit one to determine whether a process has come to steady state and to examine activation and deactivation processes. These data are analyzed in a similar fashion to that discussed previously for the batch reactor, but now the process variables such as reactant flow rate (mean reactor residence time) are varied, and the composition will not be a function of time after the reactor has come to steady state. Steady-state reactors can be used to obtain rates in a differential mode by maintaining conversions small. In this configuration it is particularly straightforward to vary parameters individually to find rates. One must of course wait until the reactor has come to steady state after any changes in feed or process conditions.

Small steady-state reactors are frequently the next stage of scaleup of a process from batch scale to full commercial scale. Consequently, it is common to follow batch experiments in the laboratory with a laboratory-scale continuous-reactor process. This permits one both to improve on batch kinetic data and simultaneously to examine more properties of the reaction system that are involved in scaling it up to commercial size. Continuous processes almost by definition use much more reactants because they run continuously. One quickly goes from small bottles of reactants to barrels in switching to continuous reactors. Because of higher reactant costs and the fact that these reactors must frequently be operated day and night, the costs of running continuous processes are much higher than batch processes.

Batch reactors are frequently operated by chemists who are responsible for obtaining the (boring) preliminary kinetic data on a process. Chemical engineers get involved when a continuous process is being considered because chemists do not understand anything beyond batch reactors. Steady-state continuous reactors are the subject of the next chapter.

2.20 SUMMARY

In this chapter we have defined some of the quantities we will need in considering chemical reactors. All these concepts have been developed in previous courses that most of you have taken, and none is particularly complicated. For students who may not have had these courses or who have forgotten this material, the development here should be adequate for our uses.

Then we switched topics completely to consider the chemical reactors that have always dominated the chemical engineering industries. These are extremely complicated and appear to have little relation to the simple batch reactors that you have seen previously.

In the rest of this book we will apply these ideas to increasingly complex situations, so that by the last chapter you should have seen all the ideas necessary to deal with these reactions and reactors. More important, these ideas should permit you be able to understand the even more complex reactions and reactors that you will have deal with to develop new processes for future technologies.

2.21 REFERENCES

Cooper, C. D., and Alley, F. C., *Air Pollution Control*, 2nd ed., Waveland Press, 1994.

Farrauto, R. J., and Bartholomew, C. H., *Fundamentals of Catalytic Processes*, Chapman and Hall, 1997.

Heaton, A., *The Chemical Industry*, 2nd ed., Blackie Academic and Professional, 1986.

Hougen, O. A., and Watson, K. M., *Chemical Process Principles, Volume III: Kinetics and Catalysis*, Wiley, 1947.

Kirk-Othmer Encyclopedia of Chemical Technology, 27 vols., Wiley, 1991–1998.

Kletz, Trevor A., *What Went Wrong? Case Studies of Process Plant Disasters*, Gulf Publishing, 1985.

Levenspiel, Octave, *The Chemical Reactor Omnibook*, OSU Book Stores, 1984.

Lowrance, William W., *Of Acceptable Risk: Science and the Determination of Safety*, William Kaufmann Inc., 1976.

McKetta, John J., *Encyclopedia of Chemical Processing and Design*, Marcel Dekker, 1987.

Meyer, Robert A., *Handbook of Chemical Production Processes*, McGraw-Hill, 1986.

Sampson, Anthony, *The Seven Sisters: The 100 Year Battle for the World's Oil Supply*, Viking, 1975.

Sandler, H. J., and Lukiewicz, E. T., *Practical Process Engineering*, Ximix, 1993.

Spitz, Peter H., *Petrochemicals: The Rise of an Industry,* Wiley, 1988.

Ullmann's Encyclopedia of Industrial Chemistry, 5th ed., 37 vols., VCH, 1985–1996.

Wei, J., Russel, T. W. F., and Swartzlander, M. W., *The Structure of the Chemical Process Industries*, McGraw-Hill, 1979.

PROBLEMS

2.1 [This is one of several "descriptive" problems (no numbers) that will be assigned throughout the course. You should be able to work these problems by just thinking about them, referring to your old texts, and discussing them with classmates.]

Surfactants are fairly large molecules with molecular weights of 100–200 amu that make oil particles soluble in water.

(a) What is the structure of these molecules and what is the structure of the oil+surfactant+water solution?

(b) Your great-grandmother made soap by cooking animal fat with wood ashes and water in a large pot open to the air for several days and then casting the product into bars. What reactions was she carrying out in her multiphase batch reactor?

(c) In the early twentieth century, companies such as Procter and Gamble began replacing the reactants by olefins, NaOH, and phosphates to scale up grandma's process, to reduce raw material costs, and to attain better quality control so they could sell many specialty products at high prices. Where does each of these raw materials come from?

(d) Carboxylic acids make too much foam, branched alkyl groups are not attacked by bacteria, and the phosphate builders in soaps are nutrients in lakes, so soaps were largely replaced by detergents in the mid-twentieth century. What are detergents?

(e) Biodegradable detergents are made by reacting α-olefins with alkaline sulfates. The α-olefins can be made by polymerizing olefins, or forming large olefins from smaller ones over a catalyst. Sketch how the successive reactions of ethylene with small α-olefins should produce exclusively α-olefins.

(f) Zeolites are crystalline synthetic clays that are porous and contain anionic groups in their lattice. The largest use of zeolites industrially is in the soap and detergent business, where a few percent of the Na salt of Zeolite A is added to most detergents. Why?

2.2 In ethanol production by fermentation of corn, hydrolyzed ground corn in a water suspension containing 50% corn by weight is mixed with sufficient enzyme to produce ethanol by the approximate reaction

$$C_6H_{12}O_6 \rightarrow 2C_2H_5OH + 2CO_2$$

When the solution reaches 12% ethanol, the yeast dies and the reaction stops.

(a) How many bushels of corn are required to produce 500 tons of pure ethanol per day? Assume 1 bushel of corn weighs 56 lb.

(b) The process requires 72 hours to go to completion. What size batch-reactor tank or tanks are required for this process?

2.3 The onset of World War II caused major expansion and technology innovation in the petroleum and petrochemical industries, with major increases in demands for (1) branched aliphatic and aromatic hydrocarbons, (2) high-purity toluene, (3) butadiene, and (4) nylon. What uses produced each of these demands? [Some hints: The jet engine was important only after 1950, rubber plantations were located in Malaysia, and cotton becomes weak and mildews when wet.]

2.4 You wish to design a plant to produce 100 tons/day of ethylene glycol from ethane, air, and water. The plant has three reactor stages, ethane dehydrogenation, ethylene oxidation, and ethylene oxide hydration.

(a) What are the reactions?

(b) Both dehydrogenation and hydration have nearly 100% selectivity (with recycle of unreacted reactants), but ethylene to ethylene oxide has only 70% selectivity with an old catalyst and 90% selectivity with a new and expensive catalyst. How many tons/day of ethane do we need to supply to this plant with each of these catalysts?

(c) Ethylene to EO has a heat of reaction of -25 kcal/mole, and the undesired byproducts are exclusively CO_2 and H_2O. [You can look up the heat of combustion of ethylene.] What is the rate of heat removal in watts with the two catalysts?

(d) If all this heat is used to produce low-pressure steam (from 25°C at 1 atm), approximately how many tons of steam per day can be produced?

(e) Could the heat from this reaction be used to provide heat in the other two reactions? Where else could it be used?

(f) Sketch a flow diagram of this plant including reactors and separation units.

2.5 We want to hydrolyze 500 lb per day of an ester at an initial concentration of 5 molar (the ester has a molecular weight of 120) in aqueous basic solution in a batch process, and we need product that is 99% hydrolyzed. In benchtop experiments in a flask we find that 50% of the ester hydrolyzes in 15 min for initial ester concentrations of either 1 or 5 molar. We also find that, when we react for 8 h, all of the ester has hydrolyzed. It takes 1 h to empty the reactor and refill and heat it to start another batch.

(a) What size reactor will we need?

(b) What size reactor will we need if we can tolerate 90% conversion?

2.6 (a) The previous process was found to have an activation energy of 12 kcal/mole, and we had been operating at 40°C. What reactor volumes would we need if we can operate at 80°C?

(b) This hydrolysis reaction is exothermic with $\Delta H_R = -8$ kcal/mole. What must be the average rate of cooling (in watts) during the reaction to maintain the reaction isothermal?

(c) If we started the batch reactor at 40°C but forgot to turn on the cooling, what would be the final temperature if the reactor were adiabatic (and the vessel would withstand the pressure)? Assume the heat capacity of the solution to be that of water, 1 cal/cm³ K.

(d) What cautions do you recommendations regarding operation at 80°C?

2.7 An aqueous ester hydrolysis reaction $A \rightarrow B + C$ has $k = 0.02$ min^{-1} and an equilibrium constant of 10 with all concentrations in moles/liter.

(a) Starting with $C_{Ao} = 1$ mole/liter, and $C_{Bo} = C_{Co} = 0$, what is the equilibrium composition?

(b) What is the reverse rate constant in the above reaction?

(c) Find $C_A(t)$, $C_B(t)$, and $C_C(t)$ in a batch reactor for these initial conditions.

2.8 (a) Sketch the steps, reactions, and flow sheet by which ethylene glycol is produced from ethane by the direct oxidation process.

(b) Ethylene glycol was made by the chlorohydrin route until the 1960s, when Union Carbide and Shell Oil developed the direct oxidation process. In the chlorohydrin process hypochlorous acid, HOCl, is reacted with ethylene to produce ethylene chlorohydrin. Then HCl is eliminated and the product is hydrated. Sketch these reactions and the flow sheet, starting from NaCl, ethane, water, air, and electricity.

(c) If the HCl must be disposed of by reacting it with NaOH, how many pounds of salt must be disposed of per pound of ethylene glycol produced, assuming all reaction steps have 100% efficiency?

(d) Write out the ethane to ethylene glycol reactions in our standard notation, $\sum v_{ij} A_j = 0, i = 1, 2, \ldots, R$.

2.9 Summarize the argument that all chemical energy sources are derived from solar energy. What is the single chemical reaction and what is the catalyst by which all of this chemical energy is produced? [Look it up in your biology text.] What is a reasonable definition of renewable and nonrenewable energy sources? What is the major reaction by which CO_2 is removed from and added to the biological cycle?

2.10 Before synthetic polymers were developed to produce fibers and sheets, your ancestors found many natural sources of polymers for clothing, housing, and tools. Make a list of these divided into soft materials and hard materials and into one-dimensional and two-dimensional polymers. List the source, uses, and chemical composition for each. Don't forget violin bows, guitar strings, writing surfaces, and wine jugs.

2.11 We have a process that reacts 67% CH_4 in O_2 at 10 atm to form syngas ($\Delta H_R = -8.5$ kcal/mole CH_4).

(a) Estimate the adiabatic reactor temperature at completion if we produce 100% syngas with a feed temperature of 400°C. Assume $C_p = \frac{7}{2}R = 7$ cal/mole K.

(b) Estimate the adiabatic reactor temperature if we suddenly begin producing 100% total combustion products ($\Delta H_R = -192$ kcal/mole).

(c) What do we have to be concerned with regarding reactor construction materials and pressure relief capabilities to design for this possibility?

2.12 We wish to produce B in the reaction system

$$A \rightarrow B, \qquad r_1 = k_1 C_A$$
$$A \rightarrow C, \qquad r_2 = k_2 C_A$$

A costs $1.20/lb, B sells for $2.50/lb, and C costs $0.50/lb to dispose.

(a) What relation between the ks will give a positive cash flow if the reactor and operating costs are ignored?

(b) What relation between the ks will give a 30% profit on sales if the reactor and operating costs are ignored?

2.13 Sterilization is a batch process by which bacteria and molds in food are killed by heat treatment. A can of vegetables that originally contained 10,000 viable spores has 100 spores remaining after heating for 10 min at 250°F. Assuming first-order kinetics, how long must this can be heated at this temperature to reduce the probability of having a single spore to an average of 1 in 1 million cans?

2.14 Methanol is synthesized commercially from CO and H_2 by the reaction

$$CO + 2H_2 \rightleftarrows CH_3OH$$

over a Cu/ZnO catalyst at ~350°C at 50–100 atm. The standard-state thermodynamic data on this reaction are $\Delta H_{R,298}^o = -90.64$ kJ/mole and $\Delta G_{R,298}^o = -24.2$ kJ/mole.

(a) What is the equilibrium conversion of a stoichiometric mixture of CO and H_2 to methanol at 298 K at 1 atm?

(b) Assuming ΔH^o independent of T, what are K_{eq} and the equilibrium conversion at 1 atm and 350°C?

(c) The actual value of K_{eq} at 350°C is 1.94×10^{-4}. How large is the error assuming constant ΔH?

(d) At what pressure will the equilibrium conversion be 50% at 350°C?

2.15 A fluidized-bed catalytic cracker operates nearly adiabatically. In the reactor, all cracking and dehydrogenation reactions are strongly endothermic, while in the regenerator, strongly exothermic oxidation reactions occur. Set up an energy balance for the reactor and regenerator assuming a highly simplified set of reactions

$$n\text{--}C_{16}H_{34} \rightarrow n\text{--}C_8H_{18} + n\text{--}C_8H_{16}$$

$$n\text{--}C_{16}H_{34} \rightarrow 16C + 17H_2$$

$$C + O_2 \rightarrow CO_2$$

(a) What fraction of the hydrocarbon must form coke for the overall reactor to be adiabatic? Heats of formation of $n\text{--}C_{16}H_{34}$, $n\text{--}C_8H_{18}$, and $n\text{--}C_8H_{16}$ are -89.23, -49.82, and -19.82 in kcal/mole, respectively, and ΔH_R of C oxidation is -94.05 kcal/mole.

(b) Indicate how you would determine the energy balance for a catalytic cracker knowing the feed and product composition of the reactor.

(c) It is important that the temperature of the reactor and regenerator remain constant, independent of reactions and conversions. How might the reactor be controlled to maintain constant temperatures? There are many options.

2.16 The enthalpies of some hydrocarbon reactions are shown here:

Reaction	ΔH_R (kcal/mol)
$C_2H_6 \rightarrow 2C_s + 3H_2$	$+20.2$
$C_2H_4 \rightarrow 2C_s + 2H_2$	-12.5
$C_2H_2 \rightarrow 2C_s + H_2$	-54.5
$C_2H_4 + H_2 \rightarrow C_2H_6$	-7.5
$2C_2H_4 \rightarrow C_4H_8$	-24.7

Estimate the final temperatures if a container of these gases at 25°C suddenly goes to complete reaction. Assume C_s has a negligible C_p. [One answer: $\sim 8000°C$!]

2.17 (a) What safety precaution is usually taken in storing acetylene?

(b) How much excess H_2 would have to be used in a gas-phase ethylene polymerization process $(n\ C_2H_4 \rightarrow (C_2H_4)_n, \Delta H_R/n = -25$ kcal/mole) to have the adiabatic temperature rise not exceed 300°C? List at least two reasons why isobutane is a preferred diluent instead of H_2 to prevent thermal runaway in this reaction.

2.18 Polyurethanes are used for most plastic parts in automobiles such as bumpers and nonmetal body panels. They are made from toluene diisocyanate and ethylene glycol. We are designing a chemical plant that makes TDIC and need to evaluate the safety of the units. Phosgene is made by reacting CO and Cl_2 in a 20 atm reactor with a residence time of 10 sec. Toluene is nitrated by reacting with concentrated nitric acid solution at 140°C with a 1 min residence time, and the aqueous and organic phases are separated. The organic phase is distilled to separate dinitrotoluene. The dinitrotoluene liquid is hydrogenated with H_2 at 5 atm at 120°C with a liquid residence time of 20 sec. The product is reacted with phosgene in a vapor-phase reaction at 150°C, $\tau = 2$ min, and the TDIC is distilled. The TDIC is finally reacted with ethylene glycol in a reaction injection molding process to form the bumper.

(a) This process consists of 5 reactors. Write down the reactions that occur in each.

(b) Sketch the flow sheet of the 5 reactors and 3 separation units with feeds of CO, Cl_2, toluene, H_2, and ethylene glycol.

(c) The process is to produce 10 tons/h of polymer. Assume that each reactor produces 100% conversion and that the nitration process produces equal amounts of the three nitrates. What are the sizes of each of these reactors? Approximately how much chemical do they contain?

(d) These 8 units are all connected by valves that will close quickly to isolate all units in the event of an upset of any kind. Describe the potential safety hazards of each unit and the possible things that might go wrong. Include runaway reaction and dangers from venting including chemical toxicity, fire, and explosion. Write ~1/4 page of text on each.

(e) Order the units in terms of decreasing hazard. Which units do you request not to be assigned to?

2.19 How many pentene isomers exist? How many isomerization reactions would describe their isomerizations?

2.20 Nitric acid is made from methane by first reacting CH_4 with H_2O to form syngas. Water is added to the $CO + H_2$ mixture to form H_2 and CO_2 which is easily separated. Hydrogen is reacted with N_2 from a liquid air plant. Ammonia is oxidized to NO over a Pt catalyst, and additional air is added to form NO_2, and the product is dissolved in water.

(a) Write out the reactions in this process.

(b) Sketch a flowsheet of a nitric acid plant.

2.21 Polyacrylonitrile is an important polymer for carpets and for vapor barrier films in food packaging.

(a) Until approximately 1975 most acrylonitrile was made by reacting acetylene with HCN. Acetylene is made by dehydrogenation of ethane. HCN is made by reacting CH_4, NH_3, and air over a Pt catalyst. Write out these reactions.

(b) In the 1970s Sohio (now British Petroleum) discovered a process by which propylene and NH_3 react over a bismuth molybdate catalyst in a process called ammoxidation to make acrylonitrile in one step in a fluidized bed reactor. Now nearly 100% of acrylonitrile is made by this process. Write out this reaction.

2.22 Alpha olefins with typically 12 C atoms have wide use as detergents, polymer additives, and many other chemicals. These α-olefins (linear molecules with the double bond only at the end carbon) can be made by the oligomerization (dimerization) of smaller olefins starting with ethylene or by successive carbonylation of smaller α-olefins. Write out these reactions, showing why they do not produce branched alkyls or double bonds at other than the end position.

2.23 An irreversible aqueous reaction gave 90% conversion in a batch reactor at 40°C in 10 min and required 3 min for this conversion at 50°C.

(a) What is the activation energy for this reaction?

(b) At what temperature can 90% conversion be obtained at 1 min?

(c) Find the rate coefficient assuming first-order kinetics.

(d) Assuming first-order kinetics, find the times for 99% conversion at 40 and at 50°C.

(e) Assuming first-order kinetics, find the temperature to obtain 99% conversion in a time of 1 minute.

(f) Assuming second-order kinetics with $C_{Ao} = 1$ mole/liter, find the times for 99% conversion at 40°C and at 50°C.

(g) Assuming second-order kinetics with $C_{Ao} = 1$ mole/liter, find the temperature to obtain 99% conversion in a time of 1 min.

2.24 Our winery has discovered that all of last year's crop now being stored for aging contained 200 ppm of a chemical which gives it a garlic flavor. However, we find that the concentration of this chemical is 50 ppm one year after it was put in storage. Taste tests in which this chemical was added to good wine shows that it can be detected only if its concentration is greater than 10 ppm. How long will we have to age this wine before we can sell it?

2.25 Our winery chemist has just found that the reaction by which the garlic flavored compound of the previous problem disappears is by a reaction that produces a tasteless dimer of the chemical.

   (a) With this mechanism, how long must we age our wine before we can sell it?

   (b) Since we don't entirely trust his judgment (why would a competent chemist work for a winery?), we decide to test his results by analyzing the wine after the second year of aging. What percentages should we find if the reaction is first or second order?

2.26 The carton of milk in your refrigerator contains 20 cells per glass of a strain of *E. coli* which makes people sick. This strain doubles every 2 days, and your immune system can destroy up to 1000 cells per glass without noticeable effects.

   (a) When should the carton of milk be poured down the drain?

   (b) At room temperature this strain doubles every 8 h. For how long is the milk drinkable if left on the counter?

2.27 Your neighbor just sneezed on your desk, and the drop contains 200 virus particles. Half of the virus particles are destroyed every 10 min by air oxidation. Can the person who sits at your desk next hour catch his cold?

Chapter **3**

SINGLE REACTIONS IN CONTINUOUS ISOTHERMAL REACTORS

3.1 CONTINUOUS REACTORS

In this chapter we consider the fundamentals of reaction in continuous isothermal reactors. Most industrial reactors are operated in a continuous mode instead of batch because continuous reactors produce more product with smaller equipment, require less labor and maintenance, and frequently produce better quality control. Continuous processes are more difficult to start and stop than batch reactors, but they make product without stopping to change batches and they require minimum labor.

Batch processes can be tailored to produce small amounts of product when needed. Batch processes are also ideal to measure rates and kinetics in order to design continuous processes: Here one only wants to obtain *information* rapidly without generating too much product that must be disposed of. In pharmaceuticals batch processes are sometimes desired to assure quality control: Each batch can be analyzed and certified (or discarded), while contamination in a continuous processes will invariably lead to a lot of worthless product before certifiable purity is restored. Food and beverages are still made in batch processes in many situations because biological reactions are never exactly reproducible, and a batch process is easier to "tune" slightly to optimize each batch. Besides, it is more romantic to produce beer by "beechwood aging," wine by stamping on grapes with bare feet, steaks by charcoal grilling, and similar batch processes.

We will develop mass balances in terms of mixing in the reactor. In one limit the reactor is stirred sufficiently to mix the fluid completely, and in the other limit the fluid is completely unmixed. In any other situation the fluid is partially mixed, and one cannot specify the composition without a detailed description of the fluid mechanics. We will consider these "nonideal" reactors in Chapter 8, but until then all reactors will be assumed to be either completely mixed or completely unmixed.

3.2 THE CONTINUOUS STIRRED TANK REACTOR

Here we consider the situation where mixing of fluids is sufficiently rapid that the composition does not vary with position in the reactor. This is a "stirred-tank" or "backmix

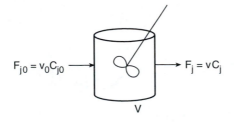

Figure 3–1 The continuous stirred tank reactor (CSTR) of volume V with inlet molar flow rate F_{jo} and outlet molar flow rate F_j.

reactor," which we will call the continuous stirred-tank reactor or CSTR (Figure 3–1). Our picture is that of a tank with a stirring propeller that is fed and drained by pipes containing reactants and products, respectively. In this situation the crucial feature is that the composition is identical everywhere in the reactor and in the exit pipe. Nothing is a function of position except between the inlet pipe and the reactor entrance, where mixing is assumed to occur instantly!

The idea that the composition is identical everywhere in the reactor and in the exit pipe requires some thought. It might seem that, since the concentration changes instantly at the entrance where mixing occurs, reaction occurs there and nothing else happens in the reactor because nothing is changing. However, reaction occurs throughout the reactor, but mixing is so rapid that nothing appears to change with time or position.

The "completely mixed" limit is in fact rather easy to achieve with ordinary mixing techniques. The approximation can be thought of in terms of "mixing time" τ_{mix} versus residence time τ of the fluid in the reactor. If

$$\tau_{mix} \ll \tau \tag{3.1}$$

then the reactor is totally mixed to a good approximation. Temperature variations within some large reactors also produce enough natural convection to help stir the contents. The approximation of a well-mixed reactor requires merely that the region of variable composition (and variable rate) near the entrance be small compared to the volume of the reactor.

Since the reactor is assumed to be uniform in composition everywhere, we can make an *integral mass balance* on the number of moles N_j of species j in a reactor of volume V. This gives

$$[\text{accumulation}] = [\text{flow in}] - [\text{flow out}] + [\text{generation}] \tag{3.2}$$

or

$$\boxed{\frac{dN_j}{dt} = F_{jo} - F_j + V\nu_j r} \tag{3.3}$$

where F_{jo} and F_j are *molar flow rates* of species j (in moles/time) in the inlet and outlet, respectively. Each term in this equation has dimensions of moles/time. This equation as written is *exact* as long as the reactor is completely mixed. We will develop several versions of this equation for different conditions, specifically depending on (1) steady-state versus transient conditions and (2) constant density versus variable density.

It is evident that this equation looks identical to the batch-reactor equation in the previous chapter except for the flow terms F_{jo} and F_j, which are of course zero in the batch reactor. In fact, the batch reactor and the CSTR share the characteristic that properties are identical everywhere in the reactor. However, the solutions in batch and CSTR are quite different except for transients, and, as we will see, the performance of the batch reactor is in fact much closer to the plug-flow tubular reactor than to the mixed reactor.

We can relate N_j and C_j by the relation

$$N_j = V C_j \tag{3.4}$$

We can also relate the molar flow rates F_{jo} and F_j of species j to the concentration by the relationships

$$F_{jo} = v_o C_{jo} \tag{3.5}$$

and

$$F_j = v C_j \tag{3.6}$$

respectively, where v_o and v are the volumetric flow rates into and out of the reactor.

For reactions among liquids and among gases where the total number of moles does not change, the *density* of the system does not change with composition, and therefore $v_o = v$. If we assume that V is constant and the density does not change with composition, differentiation of N_j yields

$$\frac{dN_j}{dt} = V \frac{dC_j}{dt} \tag{3.7}$$

If the density of the fluid is constant, then the volumetric flow rates in and out of the reactor are equal, $v = v_o$. The mass-balance equation then simplifies to become

$$\boxed{V \frac{dC_j}{dt} = v(C_{jo} - C_j) + V v_j r} \tag{3.8}$$

Next we assume that compositions are independent of time (steady state) and set the time derivative equal to zero to obtain

$$v(C_{jo} - C_j) + V v_j r = 0 \tag{3.9}$$

We call the volume divided by the volumetric flow rate the *reactor residence time*

$$\boxed{\tau = \frac{V}{v}} \tag{3.10}$$

[We caution that we have not yet proven that this is the true average residence time, and we will not do this until Chapter 8. Also, whenever v or V is a functions of conversion, we cannot treat τ as a constant that is independent of conversion.]

With these approximations we write the steady-state mass balance on species j in the CSTR as

$$\boxed{C_{jo} - C_j = -\tau v_j r} \tag{3.11}$$

This will be the most used form of the mass-balance equation in the CSTR in this book. Students should either memorize this equation or preferably be able to derive it from an integral mass balance on the reactor. This equation requires three major assumptions:

1. a steady state,
2. a single reaction, and
3. no density change with reaction.

3.3 CONVERSION IN A CONSTANT-DENSITY CSTR

For a reactant species $A(\nu_A = -1)$ the steady-state mass balance becomes

$$C_{Ao} - C_A = \tau r(C_A) \tag{3.12}$$

and we will now solve this equation for some simple rate expressions.

3.3.1 Irreversible reactions

Consider the nth-order irreversible reaction

$$A \rightarrow \text{products}, \qquad r = kC_A^n \tag{3.13}$$

For first-order kinetics, $n = 1$, the mass balance becomes

$$C_{Ao} - C_A = \tau k C_A \tag{3.14}$$

which can be rearranged to solve for C_A

$$C_A = \frac{C_{Ao}}{1 + k\tau} \tag{3.15}$$

and then solved explicitly for τ

$$\tau = \frac{C_{Ao} - C_A}{k C_A} \tag{3.16}$$

Note that we simply had to solve an algebraic equation to find $C_A(\tau)$ or $\tau(C_A)$.

In a CSTR with any single reaction rate $r(C_A)$, we can always solve explicitly for τ as

$$\tau = \frac{C_{Ao} - C_A}{r(C_A)} \tag{3.17}$$

If this reaction is $A \rightarrow B$, we can find C_B by simply substituting from stoichiometry,

$$C_B = C_{Ao} - C_A = \frac{C_{Ao} k\tau}{1 + k\tau} \tag{3.18}$$

if $C_{Bo} = 0$.

Example 3–1   The reaction $A \to B$, $r = kC_A$ occurs in CSTR with 90% conversion. If $k = 0.5$ min^{-1}, $C_{Ao} = 2$ moles/liter, and $v = 4$ liter/min, what residence time and reactor volume will be required?

From the preceding equation we have

$$\tau = \frac{C_{Ao} - C_A}{r(C_A)} = \frac{C_{Ao} - C_A}{kC_A} = \frac{2.0 - 0.2}{0.5 \times 0.2} = \frac{1.8}{0.1} = 18 \text{ min}$$

The reactor volume is

$$V = v\tau = 4 \times 18 = 72 \text{ liters}$$

Note that this problem is even easier than for a batch reactor because for the CSTR we just have to solve an algebraic equation rather than a differential equation.

For second-order kinetics, $r = kC_A^2$, the CSTR mass-balance equation becomes

$$C_{Ao} - C_A = \tau k C_A^2$$

so we must solve a quadratic in C_A to obtain

$$C_A = \frac{-1 + (1 + 4k\tau C_{Ao})^{1/2}}{2k\tau}$$

Throughout this book the reaction we will focus on for many examples will be variations of the preceding reaction: $A \to B$, $r = kC_A$, $C_{Ao} = 2$ moles/liter, $k = 0.5$ min^{-1}, $v_o = 4$ liter/min. We will compare it for several reactors in this chapter, and in Chapters 5 and 6 we will examine it for nonisothermal reactors. Watch for it.

Example 3–2   The reaction $A \to B$, $r = kC_A^2$, occurs in CSTR with 90% conversion. If $k = 0.5$ liter mole^{-1} min^{-1}, $C_{Ao} = 2$ moles/liter, and $v = 4$ liter/min, what residence time and reactor volume will be required?

From the above equation we have

$$\tau = \frac{C_{Ao} - C_A}{r(C_A)} = \frac{C_{Ao} - C_A}{kC_A^2} = \frac{2.0 - 0.2}{0.5 \times (0.2)^2} = \frac{1.8}{0.02} = 90 \text{ min}$$

The reactor volume is

$$V = v\tau = 4 \times 90 = 360 \text{ liters}$$

For nth-order kinetics in a CSTR we can easily solve for τ

$$\tau = \frac{C_{Ao} - C_A}{kC_A^n}$$

but this equation becomes an nth-order polynomial,

$$\tau k C_A^n + C_A - C_{Ao} = 0$$

and to solve for C_A we must find the proper root among the n roots of an nth-order polynomial.

We took the $+$ sign on the square root term for second-order kinetics because the other root would give a *negative concentration*, which is physically unreasonable. This is true for any reaction with nth-order kinetics in an isothermal reactor: There is only one real root of the isothermal CSTR mass-balance polynomial in the physically reasonable range of compositions. We will later find solutions of similar equations where multiple roots are found in physically possible compositions. These are true multiple steady states that have important consequences, especially for stirred reactors. However, for the nth-order reaction in an isothermal CSTR there is only one physically significant root ($0 < C_A < C_{Ao}$) to the CSTR equation for a given τ.

3.3.2   Fractional conversion

As shown in the previous chapter, we can use another variable, the *fractional conversion* X, which we defined as

$$C_A = C_{Ao}(1 - X) \qquad (3.19)$$

as long as the system is at constant density. In order for this variable to go from 0 to 1, it is necessary that we base it in terms of a reactant A whose concentration is C_A (the limiting reactant with $\nu_A = -1$). We can now write the mass-balance equation for the CSTR in terms of X as

$$C_{Ao} - C_A = C_{Ao}X = \tau k C_A(X) \qquad (3.20)$$

and solve for $X(\tau)$ rather than $C_A(\tau)$. For the first-order irreversible reaction this equation becomes

$$C_{Ao}X = \tau k C_{Ao}(1 - X) \qquad (3.21)$$

which can be solved for τ

$$\tau = \frac{1}{k}\frac{X}{1 - X} \qquad (3.22)$$

or for X,

$$X(\tau) = \frac{k\tau}{1 + k\tau} \qquad (3.23)$$

From the definitions of C_A and C_B in terms of X, we can use this $X(\tau)$ to find the same expressions as above for $C_A(\tau)$,

$$C_A(\tau) = C_{Ao}(1 - X) = C_{Ao}\left(1 - \frac{k\tau}{1 + k\tau}\right) = \frac{C_{Ao}}{1 + k\tau} \qquad (3.24)$$

and for $C_B(\tau)$, and

$$C_B(\tau) = C_{Ao}X = \frac{C_{Ao}k\tau}{1 + k\tau} \qquad (3.25)$$

which are the same answers we obtained by solving for C_A directly.

We could continue to write all these mass-balance equations as $X(\tau)$ instead of $C_A(\tau)$, but the solutions are not particularly instructive unless the rate expressions become so complicated that it is cumbersome to write an expression in terms of one concentration C_A. For simplicity, we will use $C_A(\tau)$ rather than $X(\tau)$ wherever possible.

3.3.3 Reversible reactions

Next consider the reaction

$$A \rightleftarrows B, \qquad r = k_f C_A - k_b C_B \tag{3.26}$$

Since from stoichiometry $C_B = C_{Bo} + C_{Ao} - C_A$, the rate can be written as

$$r = k_f C_A - k_b(C_{Ao} + C_{Bo} - C_A) \tag{3.27}$$

and the solution with $C_{Bo} = 0$ is

$$\tau = \frac{C_{Ao} - C_A}{r(C_A)} = \frac{C_{Ao} - C_A}{k_f C_A - k_b(C_{Ao} - C_A)} \tag{3.28}$$

or

$$C_A(\tau) = C_{Ao} \frac{1 + k_b \tau}{1 + (k_f + k_b)\tau} \tag{3.29}$$

Note that as $\tau \to \infty$, this gives the *equilibrium composition*

$$\frac{C_B}{C_A} = \frac{k_f}{k_b} \tag{3.30}$$

as required by thermodynamics. Also note that if $k_b = 0$, the expression becomes that for the *irreversible* first-order reaction as expected.

We can continue to consider more complex rate expressions for reversible reactions, but these simply yield more complicated *polynomials* $r(C_A)$ that have to be solved for $C_A(\tau)$. In many situations it is preferable to write $\tau(C_A)$ and to find C_A by trial and error or by using a computer program that finds roots of polynomials for known k_f, k_b, and feed composition.

3.4 THE PLUG-FLOW TUBULAR REACTOR

The CSTR is completely mixed. The other limit where the fluid flow is simple is the plug-flow tubular reactor, where the fluid is completely unmixed and flows down the tube as a plug. Here we picture a pipe through which fluid flows without dispersion and maintains a constant velocity profile, although the actual geometry for the plug-flow approximation may be much more complicated. Simple consideration shows that this situation can never exist exactly because at low flow rates the flow profile will be laminar (parabolic), while at high flow rates turbulence in the tube causes considerable axial mixing. Nevertheless, this is the limiting case of no mixing, and the simplicity of solutions in the limit of perfect plug-flow makes it a very useful model. In Chapter 8 we will consider the more complex situations and show that the error in the plug-flow approximation is only a few percent.

We must develop a differential mass balance of composition versus position and then solve the resulting differential equation for $C_A(z)$ and $C_A(L)$ (Figure 3–2). We consider a tube of length L with position z going from 0 to L. The molar flow rate of species j is F_{jo} at the inlet ($z = 0$), $F_j(z)$ at position z, and $F_j(L)$ at the exit L.

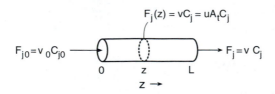

Figure 3-2   The plug-flow tubular reactor (PFTR). The length of the reactor is L, the inlet molar flow rate of species j is F_{jo}, and the outlet flow rate of species j is F_j.

We first assume that the tube has constant diameter D and also that the density does not vary with position (either liquids or gases with no mole number change, pressure drop, or temperature change with gases). In this case the linear velocity u with which the fluid flows through the tube is equal to the volumetric flow rate v divided by the cross-sectional tube area $A_t(A_t = \pi D^2/4$ for a cylindrical tube),

$$u = \frac{v}{A_t} = \frac{4v}{\pi D^2} \tag{3.31}$$

at any position.

A steady-state shell balance on species j in the element of length dz between z and $z + dz$ yields

$$F_j(z) - F_j(z + dz) + A_t \, dz \, v_j r = 0 \tag{3.32}$$

The molar flow rate of species j is related to these quantities by the relation

$$F_j = A_t u C_j \tag{3.33}$$

Therefore, the mass balance on species j becomes

$$A_t u[C_j(z) - C_j(z + dz)] + A_t \, dz \, v_j r = 0 \tag{3.34}$$

We next make a Taylor series expansion of the difference in C_j between z and $z + dz$ and let $dz \to 0$, keeping only the lead term,

$$\lim_{dz \to 0} [C_j(z) - C_j(z + dz)] = -\left(\frac{dC_j}{dz}\right)_z dz + \cdots \tag{3.35}$$

[We obtain the same result by just noting that this difference when divided by dz is simply the definition of a derivative.] Both A_t and dz can be canceled in each term; so the mass balance on species j becomes

$$\boxed{u \frac{dC_j}{dz} = v_j r} \tag{3.36}$$

which is the form of the PFTR equation we will most often use. Note again that this expression assumes

1. plug flow,
2. steady state,
3. constant density,

4. constant tube diameter, and

5. a single reaction.

This equation is not appropriate if all five of these conditions are not met. We can relax the third and fourth restrictions for the PFTR by considering the differential element of volume $dV = A_t\, dz$ rather than the differential element of length dz. The mass-balance equation at a position where the fluid has moved from volume V to volume $V + dV$ then becomes

$$F_j(V) - F_j(V + dV) + dV\ \nu_j r = 0 \tag{3.37}$$

and taking the limit $dV \to 0$ and dividing by dV, we obtain the expression

$$\frac{dF_j}{dV} = \nu_j r \tag{3.38}$$

This equation can also be used in situations where the density and tube cross section are not constants. The equation

$$\frac{dF_A}{dV} = -r_A \tag{3.39}$$

is described as the "fundamental equation" for the mass balance in a PFTR in the texts by Levenspiel and by Fogler (with $\nu_A r$ replaced by r_A). However, this equation cannot be simply modified to handle transients, nor can it be used to consider other than perfect plug flow, because for all of these situations we need equations in which the position z is the dependent variable. Since situations such as laminar flow and dispersion caused by turbulence are very important in all real tubular reactors, we prefer to use the constant-cross-section, constant-density version of this equation so that we can easily see how it must be modified to handle these situations.

3.5 CONVERSION IN A CONSTANT-DENSITY PFTR

We now consider solutions to the preceding equation for simple kinetics. For reactant species $A(\nu_A = -1)$ the equation becomes

$$u\,\frac{dC_A}{dz} = -r(C_A) \tag{3.40}$$

For $r = kC_A$ substitution yields

$$u\,\frac{dC_A}{dz} = -kC_A \tag{3.41}$$

and after separation we obtain the differential equation

$$\frac{dC_A}{C_A} = -k\,\frac{dz}{u} \tag{3.42}$$

This equation must be integrated between $z = 0$, where $C_A = C_{Ao}$, to position z, where $C_A = C_A(z)$, to position L, where $C_A = C_A(L)$,

$$\int_{C_{Ao}}^{C_A} \frac{dC_A}{C_A} = -\int_0^z k\,\frac{dz}{u} \tag{3.43}$$

This gives

$$\ln \frac{C_A(z)}{C_{Ao}} = -\frac{kz}{u} \tag{3.44}$$

Solving for C_A, we obtain

$$C_A(z) = C_{Ao}e^{-kz/u} \tag{3.45}$$

at z and

$$\boxed{C_A(L) = C_{Ao}e^{-kL/u} = C_{Ao}e^{-k\tau}} \tag{3.46}$$

at $z = L$.

Example 3–3   The reaction $A \rightarrow B$, $r = kC_A$ occurs in PFTR with 90% conversion. If $k = 0.5$ min^{-1}, $C_{Ao} = 2$ moles/liter, and $v = 4$ liters/min, what residence time and reactor volume will be required?

From the previous equation we have

$$\tau = \frac{V}{v} = -\frac{1}{k} \ln \frac{C_A(L)}{C_{Ao}} = +\frac{1}{0.5} \ln \frac{2.0}{0.2}$$

$$= 2 \ln 10 = 4.61 \text{ min}$$

so that

$$V = v\tau = 4 \times 4.61 = 18.4 \text{ liters}$$

Note that this answer is totally different than for this reaction and conversion in a CSTR and that the residence time and reactor volume required are considerably smaller in a PFTR than in a CSTR.

How long a 2-cm-diameter tube would be required for this conversion and what would be the fluid velocity?

The length is the volume divided by the cross-sectional area or

$$L = \frac{4V}{\pi D^2} = \frac{4 \times 18.4 \times 10^3}{\pi \times 2 \times 2} = 5860 \text{ cm} = 58.6 \text{ m}$$

and

$$u = \frac{L}{\tau} = \frac{5860}{4.61} = 1270 \text{ cm/min} = 21 \text{ cm/sec}$$

Thus, while the PFTR reactor volume is much smaller than the CSTR for this conversion, the PFTR tube length may become impractical, particularly when pumping costs are considered.

Example 3–4   The reaction $A \rightarrow B$, $r = kC_A^2$ occurs in PFTR with 90% conversion. If $k = 0.5$ liter mole^{-1} min^{-1}, $C_{Ao} = 2$ moles/liter, and $v = 4$ liters/min, what residence time and reactor volume will be required?

From the previous equation we have

$$\tau = \int_{C_A}^{C_{Ao}} \frac{dC_A}{kC_A^2} = \frac{1}{k}\left(\frac{1}{C_A} - \frac{1}{C_{Ao}}\right) = \frac{1}{0.5}\left(\frac{1}{0.2} - \frac{1}{2}\right) = 9 \text{ min}$$

so that

$$V = 4 \times 9 = 36 \text{ liters}$$

3.5.1 Comparison with batch reactor

We could proceed as before to write out the expressions for the irreversible reaction

$$A \rightarrow \text{ products,} \qquad r = kC_A^n \tag{3.47}$$

for different values of n, but in fact we have already worked these problems in the previous chapter.

In a PFTR the time t that a molecule has spent in the reactor is z/u, and the time for the molecule to leave the reactor is L/u, which is the total time that a molecule has spent in the reactor,

$$\tau_{PFTR} = t_{batch} \tag{3.48}$$

Therefore, to find the behavior of a PFTR for kinetics that we have solved in a batch reactor, all we have to do is make the transformation $t_{batch} \rightarrow \tau_{PFTR}$. The solution for the nth-order irreversible reaction from Chapter 2 is

$$\boxed{C_A = C_{Ao}[1 + (n-1)kC_{Ao}^{n-1}\tau]^{1/(1-n)}} \tag{3.49}$$

(except for $n = 1$), where all we did was replace t for the batch reactor by τ for the PFTR.

We can write for the residence time in a constant-density, constant-cross-section PFTR

$$\tau_{PFTR} = V/v = L/u \tag{3.50}$$

because, for constant reactor cross section A_t, we have $V = A_t L$ and $v = u A_t$.

The PFTR was in fact assumed to be in a *steady state* in which no parameters vary with time (but they obviously vary with position), whereas the batch reactor is assumed to be spatially uniform and vary only with time. In the argument we switched to a *moving coordinate system* in which we traveled down the reactor with the fluid velocity u, and in that case we follow the change in reactant molecules undergoing reaction as they move down the tube. This is identical to the situation in a batch reactor!

We can show this more formally by writing $dt \rightarrow dz/u$,

$$\frac{dC_A}{dt} \rightarrow u\frac{dC_A}{dz} = -r(C_A) \tag{3.51}$$

which shows that the performance of batch reactor and PFTR are identical, with the reaction time t in a batch reactor corresponding to the residence time τ in a PFTR. Again, we note that a PFTR "acts like" a batch reactor, while a CSTR "looks like" a batch reactor.

Example 3–5 Compare the reactor volumes necessary to attain the conversions in the previous examples for first and second order irreversible reactions in a CSTR with a CSTR.

For a first-order irreversible reaction

$$V_{CSTR} = 72 \text{ liters}$$

and

$$V_{PFTR} = 18.4 \text{ liters}$$

For a second-order irreversible reaction

$$V_{CSTR} = 360 \text{ liters}$$

and

$$V_{PFTR} = 36 \text{ liters}$$

The PFTR is the clear "winner" in this comparison if reactor volume is the only criterion. The choice is not that simple because of the costs of the reactors and pumping costs.

3.6 COMPARISON BETWEEN BATCH, CSTR, AND PFTR

We have now developed mass balance equations for the three simple reactors in which we can easily calculate conversion versus time t_{batch}, residence time τ, or position L for specified kinetics. For a first-order irreversible reaction with constant density we have solved the mass balance equations to yield

$$\tau_{PFTR} = t_{batch} = -\frac{1}{k} \int_{C_{Ao}}^{C_A} \frac{dC_A}{C_A} = -\frac{1}{k} \ln \frac{C_A}{C_{Ao}} = +\frac{1}{k} \ln \frac{C_{Ao}}{C_A} \tag{3.52}$$

and

$$\tau_{CSTR} = \frac{C_{Ao} - C_A}{kC_A} \tag{3.53}$$

Obviously, batch and PFTR will give the same conversion, but the CSTR gives a lower conversion for the same reaction time (batch) or residence time (continuous).

We can immediately see major reactor design considerations between batch, CSTR, and PFTR. Table 3–1 shows the first of many situations where we are interested in the design of a reactor. We may be interested in choosing minimum volume or many other process variables in designing the best reactor for a given process.

In spite of the simplicity and continuous operation of the CSTR, it usually requires a longer residence time for a given conversion. For first-order kinetics at identical conversions and volumetric flow rates, this ratio is

$$\frac{\tau_{CSTR}}{\tau_{PFTR}} = \frac{V_{CSTR}}{V_{PFTR}} = \frac{C_{Ao} - C_A}{C_A \ln (C_{Ao}/C_A)} = \frac{C_{Ao}/C_A - 1}{\ln (C_{Ao}/C_A)} = \frac{X}{(1 - X) \ln [1/(1 - X)]} \tag{3.54}$$

We indicate this ratio for different values of fractional conversion in Table 3–2.

TABLE 3–1
Comparisons of Possible Advantages (+) and Disadvantages (−) of Batch, CSTR, and PFTR Reactors

	Batch	CSTR	PFTR
reactor size for given conversion	+	−	+
simplicity and cost	+	+	−
continuous operation	−	+	+
large throughput	−	+	+
cleanout	+	+	−
on-line analysis	−	+	+
product certification	+	−	−

TABLE 3–2
Ratio of Residence Times and Reactor Volumes in CSTR and PFTR Versus Conversion for a First-Order Irreversible Reaction

$X = (C_{Ao} - C_A)/C_{Ao}$	τ_{CSTR}/τ_{PFTR}
0.0	1.0
0.5	1.44
0.9	3.91
0.95	6.34
0.99	21.5
0.999	145

It is clear that the CSTR quickly becomes extremely large (large volume or large residence time) compared to the PFTR for high conversions for these kinetics. It is instructive to plot C_A/C_{Ao} versus τ to see how C_A decreases in the two ideal reactors (Figure 3–3).

The example in Figure 3–3 is for a first order irreversible reaction. We can generalize this to say that the PFTR requires a smaller reactor volume for given conversion for any positive order kinetics, and the difference becomes larger as the order of the reaction increases. For zeroth order kinetics the sizes required for a given conversion are exactly equal, and for negative order kinetics the CSTR requires a smaller volume than a PFTR (or batch reactor). We will show later that this is not necessarily true in a nonisothermal reactor, where the CSTR can "win" over the PFTR in both simplicity and residence time.

Figure 3–3 Plots of $C_A(\tau)$ and $C_B(\tau)$ in PFTR and CSTR for a first-order irreversible reaction $A \rightarrow B$, $r = kC_A$. By plotting versus $k\tau$, the graphs appear identical for any value of k.

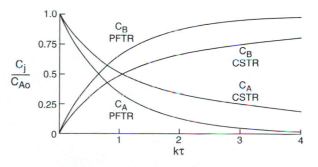

3.7   THE $1/r$ PLOT

There is a graphical construction that shows the difference of residence times in different types and combinations of chemical reactors. We write the mass balance equations as

$$\tau_{CSTR} = \frac{C_{Ao} - C_A}{r(C_A)} = \frac{1}{r(C_A)}(C_{Ao} - C_A) \tag{3.55}$$

and

$$\tau_{PFTR} = -\int_{C_{Ao}}^{C_A} \frac{1}{r(C_A)} dC_A \tag{3.56}$$

The time τ in a CSTR is the area under the rectangle of width $(C_{Ao} - C_A)$ and height $1/r(C_A)$, while the time τ in a PFTR is the area under the curve $1/r(C_A)$ from C_{Ao} to C_A. Shown in Figure 3–4 are plots of r versus $(C_{Ao} - C_A)$ for $r = kC_A^n$ for several values of the order n. It is obvious that r is a monotonically decreasing function of $C_{Ao} - C_A$ if $n > 0$, a horizontal line if $n = 0$, and increases with $C_{Ao} - C_A$ if $n < 0$. The $1/r$ plots obviously have reversed slopes.

Note that $1/r$ goes to infinity as $C_A \to 0$ for any kinetics because reaction rates must go to zero when reactants have been consumed. This is equivalent to saying that the kinetics of all reactions must become positive order in the limit of any reactant disappearing.

From these graphs of $1/r$ versus $C_{Ao} - C_A$ we can construct the residence times in PFTR and CSTR, as shown in Figure 3–5.

In the CSTR the rectangle is drawn with a height equal to $1/r$ *evaluated at the product conversion*, while in the PFTR the height varies from inlet to product conversion. The CSTR rectangle obviously has a larger area as long as $1/r$ is a monotonically increasing function of $C_{Ao} - C_A$ (r monotonically decreasing), while areas are equal for $n = 0$, and the rectangle has a smaller area for $n < 0$.

This construction also shows why the CSTR becomes much less efficient (requires much larger volume) at high conversions. The $1/r$ curve increases rapidly to ∞ as $C_A \to 0$, and therefore τ in a CSTR becomes very large compared to τ_{PFTR}.

Thus it is evident that a PFTR is always the reactor of choice (smaller V) for greater than zero-order kinetics in an isothermal reactor. The CSTR may still be favored for $n > 0$

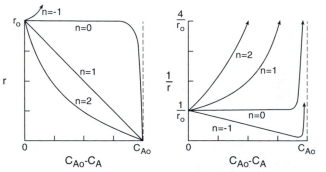

Figure 3–4   Plots of r and $1/r$ versus $C_{Ao} - C_A$ for the nth-order irreversible reactions.

Figure 3–5   Residence times in CSTR (shaded rectangle) and PFTR (area under curve) from the $1/r$ plot.

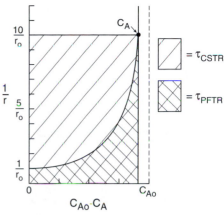

for cost reasons as long as the conversion is not too high, but the isothermal PFTR is much superior at high conversions whenever $n > 0$.

The question of choosing a PFTR or a CSTR will occur throughout this book. From the preceding arguments it is clear that the PFTR usually requires a smaller reactor volume for a given conversion, but even here the CSTR may be preferred because it may have lower material cost (pipe is more expensive than a pot). We will later see other situations where a CSTR is clearly preferred, for example, in some situations to maximize reaction selectivity, in most nonisothermal reactors, and in polymerization processes where plugging a tube with overpolymerized solid polymer could be disastrous.

3.8   SEMIBATCH REACTORS

Semibatch reactors are commonly used for small-volume chemical production. This reactor type is frequently used for biological reactions and for polymerization. In the batch reactor, we required that the system be *closed*, meaning that we neither add nor remove fluids after $t = 0$, where we assumed we started the process. It is of course possible to add feed or withdraw product continuously in a "batch" process, and we call this a *semibatch reactor*. If the reactor is spatially uniform, the mass balance on species j can be written as

$$\frac{dN_j}{dt} = F_{jo} - F_j + V\nu_j r \qquad (3.57)$$

and for reactant A this becomes

$$\frac{dN_A}{dt} = F_{Ao} - F_A - Vr \qquad (3.58)$$

For a constant-density, constant volume reactor we can write

$$V\frac{dC_A}{dt} = v_o C_{Ao} - v C_A - Vr \qquad (3.59)$$

Note that this is exactly the transient CSTR equation we derived previously, and elimination of the flow terms yields the batch reactor. Keeping all these terms and allowing v_o, v, V, and C_{Ao} to vary with time yields the semibatch reactor.

In this reactor species A may be added continuously but nothing removed to give

$$V \frac{dC_A}{dt} = v_o C_{Ao} - Vr \qquad (3.60)$$

but now the volume V of the reactor contents increases linearly with time

$$V = V_o + v_o t \qquad (3.61)$$

We therefore obtain

$$(V_o + v_o t) \frac{dC_A}{dt} = v_o C_{Ao} - (V_o + v_o t)r(C_A) \qquad (3.62)$$

This is a simple first-order differential equation in C_A and t, but the equation is not separable and must be solved numerically.

Semibatch reactors are especially important for bioreactions, where one wants to add an enzyme continuously, and for multiple-reaction systems, where one wants to maximize the selectivity to a specific product. For these processes we may want to place one reactant (say, A) in the reactor initially and add another reactant (say, B) continuously. This makes C_A large at all times but keeps C_B small. We will see the value of these concentrations on selectivity and yield in multiple-reaction systems in the next chapter.

Semibatch processes are also important in solids processing such as in foods and microelectronics, where it is more practical to load the reactor with a batch of solids (bread dough or silicon wafers) and subject the solids to heat and gas or liquid reactants. The processed solids are then withdrawn after a suitable time, and the reactor is reloaded.

3.9 VARIABLE-DENSITY REACTORS

Consider the reaction

$$A \rightarrow 3B, \qquad r = kC_A^2 \qquad (3.63)$$

in a constant-pressure reactor with A and B ideal gases and no diluents present. The density of the fluid in the reactor clearly changes as the reaction proceeds. At completion in a continuous reactor, we make three moles of product for each mole of reactant fed into the reactor so that the volumetric flow rate out of the reactor v would be three times the flow rate into the reactor v_o,

$$v = 3v_o \qquad (3.64)$$

In a tubular reactor of constant diameter the velocity u when the reaction has gone to completion is three times the inlet velocity u_o,

$$u = 3u_o \qquad (3.65)$$

We call this situation *variable density*, and the previous equations clearly do not describe this situation because v, τ, and u in these equations depend on the conversion.

For ideal gases the mole densities do not change if the reaction produces more or less moles of product than the reactants. However, the *mass density* of the gas mixture changes, and this causes the volumetric flow rate and velocity of gases through the reactor to change as the conversion increases.

For the batch reactor we saw in the previous chapter that by switching from C_A as the composition variable to fractional conversion X, we could easily write the differential equation to be solved for compositions versus time. We prefer to use concentration units whenever possible, but, if the density is a function of composition, concentrations become cumbersome variables, and we must switch to another designation of density such as the fractional conversion X.

3.9.1 CSTR

For the CSTR we begin with the mass-balance equation we derived before we substitute for concentration

$$\frac{dN_j}{dt} = F_{jo} - F_j + V\nu_j r \tag{3.66}$$

which for reactant A in steady state becomes

$$F_{Ao} - F_A - Vr = 0 \tag{3.67}$$

Now since the molar flow rate of A varies with conversion as

$$F_A = F_{Ao}(1 - X) \tag{3.68}$$

we can write this mass-balance equation as

$$F_{Ao} - F_A = F_{Ao}X = Vr(X) \tag{3.69}$$

In steady state this equation can be rearranged to become

$$\boxed{V = F_{Ao}\frac{X}{r(X)}} \tag{3.70}$$

While this equation is correct for the steady-state CSTR with variable density, it does not give a correct description of the transient CSTR for variable density because N_A (meaning the number of moles of species A in the reactor) is not given by the preceding expression unless the density is constant.

Note that in the preceding 3 equations we defined the conversion X by the above equation, and this requires that we mean the conversion of reactant A. If there are multiple reactants, we must be careful to define X in a consistent way, and for multiple reactions we use X to signify the conversion of reactant A (we will use X_A to make this clear), but we would need additional Xs for each reactant.

Since the CSTR operates with the same conversion throughout the reactor, we can describe the volumetric flow rate by the equations

$$v = v_o(1 + \nu X) \tag{3.71}$$

where $\nu = \sum_j \nu_j$. Therefore, the residence time in a variable-density reactor is given by the expression

$$\tau = \frac{V}{v} = \frac{V}{v_o(1 + \nu X)} = \frac{C_{Ao} - C_A}{r(C_A)} \tag{3.72}$$

3.9.2 PFTR

For the PFTR we return to the original mass-balance equation

$$F_j(z) - F_j(z + dz) = -A_t\, dz\, \nu_j r \tag{3.73}$$

Since $A_t\, dz = dV$, the element of reactor volume between z and $z + dz$, this equation for reactant A can be written

$$\frac{dF_A}{dV} = -r \tag{3.74}$$

Substituting $dF_A = -F_{Ao}\, dX$, this becomes

$$F_{Ao}\frac{dX}{dV} = r(X) \tag{3.75}$$

which can be integrated to yield

$$\boxed{V = F_{Ao}\int_0^X \frac{dX}{r(X)}} \tag{3.76}$$

These expressions are appropriate whether or not the density of the fluid varies with conversion. The latter is also valid if the cross section of the tube varies with z. Note the similarity (and difference) between the CSTR and PFTR expressions, the first being F_{Ao} times $X/r(X)$ and the second F_{Ao} times the *integral* of $dX/r(X)$.

We will work out the solution to variable-density reactors for a simple example where the number of moles varies with conversion.

Example 3–6   Find expressions the reactor volume V for specified and feed flow rate F_{Ao} for the reaction

$$A \rightarrow 2B + C, \qquad r = kC_A^2$$

among ideal gases with no diluent in a CSTR and in a PFTR.

This is most conveniently set up by considering a basis of N_{Ao} moles of A and writing expressions for N_j in terms of X,

$$N_A = N_{Ao}(1 - X)$$

$$N_B = 2N_{Ao}X$$

$$\underline{N_C = N_{Ao}X}$$

$$\sum N = N_{Ao}(1 + 2X)$$

This shows that one mole of reactant is converted into three moles of product when the reaction goes to completion, $X = 1$. We write the volume change of the fluid (not the volume of the reactor) as $V = V_o(1 + 2X)$ and the number of moles of A in that volume as $N_A = N_{Ao}(1 - X)$. [Note that in the above N_A is not the number of moles in the reactor and V is not the reactor volume but the volume per mole.] Therefore, from these relations we obtain C_A as a function of the conversion as

$$C_A = \frac{N_{Ao}(1 - X)}{V_o(1 + 2X)} = C_{Ao}\frac{1 - X}{1 + 2X}$$

The reaction rate in terms of X is therefore

$$r = kC_A^2 = kC_{Ao}^2\frac{(1 - X)^2}{(1 + 2X)^2}$$

This expression can be inserted into the CSTR and PFTR mass-balance equations to yield

$$V_{CSTR} = F_{Ao}\frac{X}{r(X)} = \frac{F_{Ao}}{kC_{Ao}^2}\frac{X(1 + 2X)^2}{(1 - X)^2}$$

and

$$V_{PFTR} = F_{Ao}\int_{X=0}^{X}\frac{dX}{r(X)} = \frac{F_{Ao}}{kC_{Ao}^2}\int_{X=0}^{X}\frac{(1 + 2X)^2}{(1 - X)^2}\,dX$$

$$= \frac{F_{Ao}}{kC_{Ao}^2}\left(4X - 9 - 12\ln(1 - X) + \frac{9}{1 - X}\right)$$

These equations are significantly more complicated to solve than those for constant density. If we specify the reactor volume and must calculate the conversion, for second-order kinetics we have to solve a cubic polynomial for the CSTR and a transcendental equation for the PFTR. In principle, the problems are similar to the same problems with constant density, but the algebra is more complicated. Because we want to illustrate the principles of chemical reactors in this book without becoming lost in the calculations, we will usually assume constant density in most of our development and in problems.

Example 3–7   Find the reactor volume V required to obtain 90% conversion in the reaction

$$A \rightarrow n_B B, \qquad r = kC_A$$

among ideal gases in a CSTR and in a PFTR with no diluent for $n_B = 2$, 1, and $\frac{1}{2}$ with $C_{Ao} = 2$ moles/liter, $k = 0.5$ min^{-1}, and $v_o = 4$ liters/min. [Note that this would involve a high-pressure reactor because a gas at 2 moles/liter gives $P_A = C_A RT = 2 \times 0.082 \times 300 = 49$ atm. We use these values because we have been using variations of this problem for all examples.]

For $n_B = 2$ on a basis of N_{Ao} moles of A, the expressions for N_A and N_B in terms of X are

$$N_A = N_{Ao}(1 - X)$$

$$\underline{ N_B = 2N_{Ao}X \phantom{N_{Ao}}}$$

$$N_A + N_B = N_{Ao}(1 + X)$$

The volume occupied by these moles is $V = V_0(1 + X)$ and therefore

$$C_A = \frac{N_{Ao}(1 - X)}{V_0(1 + X)} = C_{Ao}\frac{1 - X}{1 + X}$$

The reaction rate in terms of X is

$$r = kC_A = kC_{Ao}\frac{1 - X}{1 + X}$$

This rate expression can be inserted into the CSTR and PFTR mass-balance equations to yield

$$V_{CSTR} = F_{Ao}\frac{X}{r(X)} = \frac{F_{Ao}}{kC_{Ao}}\frac{X(1 + X)}{1 - X} = \frac{v_0}{k}\frac{X(1 + X)}{1 - X}$$

$$= \frac{4}{0.5}\frac{0.9(1 + 0.9)}{1 - 0.9} = 8\frac{1.9}{0.1} = 136.8 \text{ liters}$$

and

$$V_{PFTR} = F_{Ao}\int_{X=0}^{X}\frac{dX}{r(X)} = \frac{F_{Ao}}{kC_{Ao}}\int_{X=0}^{X}\frac{1 + X}{1 - X}dX$$

$$= \frac{v_0}{k}[-X - 2\ln(1 - X)] = 29.6 \text{ liters}$$

For $n_B = 1$, the density does not change with conversion so that $C_A = C_{Ao}(1 - X)$ and

$$V_{CSTR} = F_{Ao}\frac{X}{r(X)} = \frac{F_{Ao}}{kC_{Ao}}\frac{X}{1 - X} = \frac{v_0}{k}\frac{X}{1 - X}$$

$$= \frac{4}{0.5}\frac{0.9}{0.1} = 72 \text{ liters}$$

and

$$V_{PFTR} = F_{Ao}\int_{X=0}^{X}\frac{dX}{r(X)} = \frac{F_{Ao}}{kC_{Ao}}\int_{X=0}^{X}\frac{1}{1 - X}dX = \frac{v_0}{k}\ln\frac{1}{1 - X}$$

$$= 8\ln 10 = 18.4 \text{ liters}$$

For $n_B = 1/2$, the expressions for N_A, N_B, and V give

$$C_A = \frac{N_{Ao}(1 - X)}{V_0(1 - \frac{1}{2}X)} = C_{Ao}\frac{1 - X}{1 - \frac{1}{2}X}$$

Therefore, we obtain

$$V_{CSTR} = F_{Ao}\frac{X}{r(X)} = \frac{F_{Ao}}{kC_{Ao}}\frac{X(1 - \frac{1}{2}X)}{1 - X} = \frac{v_0}{k}\frac{X(1 - \frac{1}{2}X)}{1 - X}$$

$$= \frac{4}{0.5}\frac{0.9(1 - 0.45)}{1 - 0.9} = 39.6 \text{ liters}$$

and

$$V_{\text{PFTR}} = F_{Ao} \int_{X=0}^{X} \frac{dX}{r(X)} = \frac{F_{Ao}}{kC_{Ao}} \int_{X=0}^{X} \frac{1 - \frac{1}{2}X}{1 - X} \, dX$$

$$= \frac{v_o}{2k}[-\ln(1 - X) + X] = 8(\ln 10 + 0.9) = 12.8 \text{ liters}$$

It is interesting to compare these reactor volumes:

n_B	V_{CSTR}	V_{PFTR}
2	136.8	29.6
1	72	18.4
1/2	39.6	12.8

Note that V is larger than with no density change if the reaction produces more moles ($n_B = 2$) because this dilutes the reactant, while V is smaller if the reaction reduces the number of moles ($n_B = \frac{1}{2}$).

It is also interesting to compare the conversions versus reactor volumes for these stoichiometries, and these are shown in Figure 3–6.

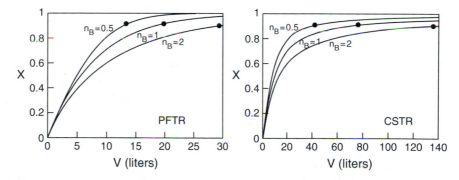

Figure 3–6 Plot of conversion versus reactor volume V for the reaction $A \rightarrow n_B B, r = kC_B$, with ideal gases for $n_B = 2$, 1, and $\frac{1}{2}$. The times are all close until the conversion becomes large, when the product dilutes the reactant for $n_B = 2$ and slows the reaction.

When the density varies with conversion, the analogy between the batch reactor and the PFTR ($dt \rightarrow d\tau$) is no longer appropriate. In the batch reactor with ideal gases, the density varies with conversion in a constant-pressure reactor but not in a constant-volume reactor. In a flow reactor, the reactor volume is fixed, no matter what the density. In a flow reactor the volumetric flow rate changes with conversion if there is a mole number change with ideal gases.

The constant-density approximation is frequently used even when it does not apply exactly because it is much simpler to solve the equations, and the errors are usually not large. The constant-density approximation can be used for the following situations:

1. equal moles of reactants and products,

2. liquids,

3. diluted gases, and

4. estimations.

The errors in assuming constant density are frequently not large, and preliminary calculations with constant density are useful to check for numerical errors in more complex calculations.

We note finally that none of these formulations gives an accurate description of any reactor in which there is a *pressure drop* in gases flowing through the reactor, and we must solve the continuity equation simultaneously with the species mass-balance equation to describe that situation. Another rather complex type of problem involves *nonideal gases and gas mixtures* because then only numerical solutions are possible. Since our primary goal in this course is to find simple approximate solutions to very complex problems, we will not consider reactors with pressure drop or among nonideal gases here.

3.10 SPACE VELOCITY AND SPACE TIME

In a variable-density reactor the residence time depends on the conversion (and on the selectivity in a multiple-reaction system). Also, in any reactor involving gases, the density is also a function of reactor pressure and temperature, even if there is no change in number of moles in the reaction. Therefore, we frequently base reactor performance on the number of moles or mass of reactants processed per unit time, based on the molar or mass flow rates of the feed *into* the reactor. These feed variables can be kept constant as reactor parameters such as conversion, T, and P are varied.

Whenever the density of the fluid in the reactor varies as the reaction proceeds, the reactor residence time τ is not a simple independent variable to describe reactor performance. Typically, we still know the *inlet variables* such as v_o, T_o, F_{jo}, and C_o, and these are independent of conversion.

Since the volumetric flow rate is a function of X, T, and P, the residence time V/v depends on these variables. Instead of using the reactor residence time τ to describe performance, an analogous quantity called the *space time* ST, defined as

$$ST = \frac{V}{v_o} \qquad (3.77)$$

is commonly used. Obviously, ST is equal to τ in a constant-density reactor whenever $v = v_o$ (see Figure 3–7).

Another commonly used designation of this quantity is its inverse, called the *space velocity* SV,

$$SV = \frac{v_o}{V} = \frac{1}{ST} \qquad (3.78)$$

which can be regarded as the number of reactor volumes of feed processed per unit time at the feed conditions. The use of the words *space* and *velocity* in these quantities is obscure, but this nomenclature has become common throughout the chemical and petroleum industry.

Figure 3–7 Plot of nominal space times (or reactor residence times) required for several important industrial reactors versus the nominal reactor temperatures. Times go from days (for fermentation) down to milliseconds (for ammonia oxidation to form nitric acid). The low-temperature, long-time processes involve liquids, while the high-temperature, short-time processes involve gases, usually at high pressures.

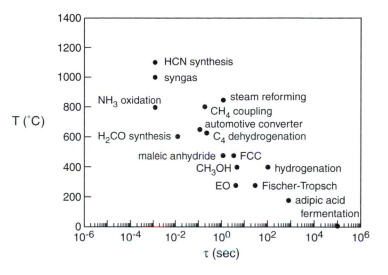

Other complications are that the reactor feed may be preheated and the feed pressure may vary, and thus the volumetric flow rate of gases will be functions of the reactor temperature and pressure at fixed mass flow rate. Therefore, the space velocity of gases is frequently defined at standard conditions: $T = 25°C$ and $P = 1$ atm.

The most common descriptions of these quantities in industry are the gas hourly space velocity

$$\text{GHSV} = \frac{[\text{volume of gaseous feed processed per hour}]}{[\text{volume of reactor}]} \qquad (3.79)$$

and the liquid hourly space velocity

$$\text{LHSV} = \frac{[\text{volume of liquid feed processed per hour}]}{[\text{volume of reactor}]} \qquad (3.80)$$

Both have units of h^{-1}, and they can be regarded as the number of reactor volumes of reactant processed per hour. They are usually evaluated for the feed at 25°C and 1 atm.

3.11 CHEMICAL REACTORS IN SERIES

Next we consider several mixed chemical reactors in series. As sketched in Figure 3–8, the feed to the first is C_{Ao}, the effluent from the first is C_{A1}, which is the feed to the second, the effluent from the second is C_{A2}, which is the feed to the third, etc. The concentrations C_A from the nth reactor are obtained by solving each reactor mass balance successively.

Figure 3–8 Sketch of ideal chemical reactors in series with C_{An}, the product from reactor n, which is also the feed into reactor $n+1$.

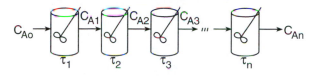

3.11.1 CSTRs in series

For first-order kinetics with equal-volume CSTR reactors (and therefore for all τs equal), the mass balances on species A become

$$C_{Ao} - C_{A1} = \tau_1 r(C_{A1}) \tag{3.81}$$

$$C_{A1} - C_{A2} = \tau_2 r(C_{A2}) \tag{3.82}$$

$$C_{A2} - C_{A3} = \tau_3 r(C_{A3}) \tag{3.83}$$

$$\cdots$$

$$C_{A,n-1} - C_{An} = \tau_n r(C_{An}) \tag{3.84}$$

because each reactor operates at its product concentration. For first-order irreversible reactions, these equations can be solved sequentially to yield

$$C_{A1} = \frac{C_{Ao}}{(1 + k\tau_1)} \tag{3.85}$$

$$C_{A2} = \frac{C_{A1}}{(1 + k\tau_2)} = \frac{C_{Ao}}{(1 + k\tau_1)(1 + k\tau_2)} \tag{3.86}$$

$$C_{A3} = \frac{C_{A2}}{(1 + k\tau_3)} = \frac{C_{Ao}}{(1 + k\tau_1)(1 + k\tau_2)(1 + k\tau_3)} \tag{3.87}$$

or for the nth reactor

$$C_{An} = \frac{C_{A(n-1)}}{(1 + k\tau_n)} = \frac{C_{Ao}}{\prod_\kappa (1 + k\tau_\kappa)} \tag{3.88}$$

Now if each reactor has the same residence time τ (all reactors have the same volume), then the total residence time $\boldsymbol{\tau}$ (bold) in the series of n equal-residence-time CSTRs is

$$\boldsymbol{\tau} = \sum_\kappa \tau_\kappa = n\tau \tag{3.89}$$

and C_{An}, the concentration from the nth reactor is given in terms of C_{Ao} by the simple expression

$$\boxed{C_{An} = \frac{C_{Ao}}{(1 + k\tau_n)^n}} \tag{3.90}$$

Let us examine the total volume and total residence time τ to be expected for a given conversion in CSTRs in series compared to a PFTR or a single CSTR. On our graph of $1/r$ versus $C_{Ao} - C_A$ (Figure 3–9), we see that for n CSTRs in series with equal residence times, we have n equal-area rectangles whose height is given by the intersection of the $1/r$ curve at the product concentration from that reactor, intersecting at the upper right corner of each rectangle. Making all areas equal, we see that for a large number of CSTRs the area (total τ) approaches that of a single PFTR with the same τ.

This is generally true: a series of mixed reactors has a performance closer to that of the PFTR. Thus by using several mixed reactors connected in series, we can gain some of the desired characteristics of the CSTR (cheaper and easier to maintain) while approaching

Figure 3–9 Total residence time from $1/r$ plot for a series of CSTR reactors for 1, 2, 3, 4, and n equal-volume reactors. The total residence time (shaded area) approaches that of a single PFTR for large n.

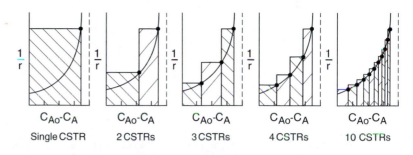

the performance of the PFTR without the volume penalty by using several smaller CSTRs in series. Of course the cost of several smaller reactors is usually greater than one large reactor, and this is another design situation where one must trade off total reactor volume and cost.

Example 3–8 The reaction $A \rightarrow B$, $r = kC_A$ occurs in n equal volume CSTRs in series, each with residence time τ, with 90% overall conversion. If $k = 0.5$ min^{-1}, $C_{Ao} = 2$ moles/liter, and $v = 4$ liters/min, what residence times and reactor volumes will be required for $n = 1, 2, 3,$ and 4?

We rearrange the equation for n equal volume CSTRs in series,

$$C_{An} = \frac{C_{Ao}}{(1 + k\tau_n)^n}$$

to yield

$$\frac{C_A}{C_{Ao}} = (1 + k\tau_n)^{-n}$$

or

$$\tau_n = \frac{n}{k}\left[\left(\frac{C_{Ao}}{C_A}\right)^{1/n} - 1\right]$$

Therefore, we have

$$\tau_1 = \tau = \frac{1}{k}\left[\left(\frac{C_{Ao}}{C_A}\right) - 1\right] = \frac{1}{0.5}(10 - 1) = \frac{C_{Ao} - C_A}{kC_A} = \frac{2.0 - 0.2}{0.5 \times 0.2} = 18 \text{ min}$$

$$\tau_2 = 2\tau = \frac{2}{k}\left[\left(\frac{C_{Ao}}{C_A}\right)^{1/2} - 1\right] = \frac{1}{0.5}(10^{1/2} - 1) = 8.65 \text{ min}$$

$$\tau_3 = 3\tau = \frac{3}{k}\left[\left(\frac{C_{Ao}}{C_A}\right)^{1/3} - 1\right] = \frac{1}{0.5}(10^{1/3} - 1) = 6.92 \text{ min}$$

$$\tau_4 = 4\tau = \frac{4}{k}\left[\left(\frac{C_{Ao}}{C_A}\right)^{1/4} - 1\right] = \frac{1}{0.5}(10^{1/4} - 1) = 6.22 \text{ min}$$

For $n = 1$ we have the solution for a single CSTR, and τ decreases with n to approach the PFTR for which $\tau = 4.61$ min.

The reactor volumes are $V = v\tau_n$, which are 72, 34.6, 27.7, and 24.9 liters, respectively. An infinite number of CSTRs in series would require the total volume of a PFTR, which is $4 \times 4.61 = 18.4$ liters to run this reaction to this conversion.

3.11.2 PFTR + CSTR

We have just seen that a combination of CSTRs gives a total residence time that approaches that of a PFTR. Next we consider a combination of reactors involving both CSTRs and PFTRs as sketched in Figure 3–10.

For a PFTR followed by a CSTR we solve each reactor mass balance sequentially to find $C_{A2}(C_{Ao}, \tau)$. For first-order kinetics this gives

$$C_{A2} = C_{A1}/(1 + k\tau_2) = \frac{C_{Ao}}{1 + k\tau_2} e^{-k\tau_1} \tag{3.91}$$

If we reverse the reactors for a CSTR followed by a PFTR, we obtain

$$C_{A2} = C_{A1}e^{-k\tau_2} = \frac{C_{Ao}}{1 + k\tau_1} e^{-k\tau_2} \tag{3.92}$$

Thus, for two equal-volume reactors ($\tau_1 = \tau_2$) with first-order kinetics the expressions are identical for both configurations.

With two equal-volume reactors, the overall conversion in these two reactors is independent of which reactor is first, but in general it is a common strategy to use a CSTR first where the conversion is low and then switch to a PFTR as the conversion becomes high to minimize total reactor volume. The total residence time from CSTR+PFTR in series is indicated by the $1/r$ plots in Figure 3–11.

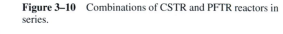

Figure 3–10 Combinations of CSTR and PFTR reactors in series.

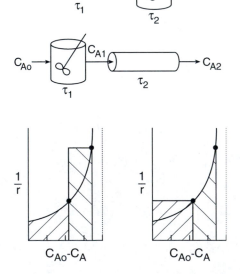

Figure 3–11 Residence times from $1/r$ plots for PFTR and CSTR in series.

3.12 AUTOCATALYTIC REACTIONS

Consider next the reaction

$$A \rightarrow B, \qquad r = kC_A C_B \qquad (3.93)$$

This doesn't look much different than the rate expressions we have been using previously (maybe they are all starting to look the same), but this is a very different reaction because the rate of the forward reaction is enhanced by the concentration of a product. This reaction could also be written

$$A + B \rightarrow 2B, \qquad r = kC_A C_B \qquad (3.94)$$

which correctly indicates the kinetics from the stoichiometry.

Note that the expression $r = kC_A C_B$ requires that the rate is zero if $C_B = 0$. If we try to run this reaction in a batch or PFTR with no B in the feed, there will be no reaction, while a CSTR operating at the product composition gives no problem. We call this type of reaction autocatalytic in that the product B acts as a catalyst to promote the reaction.

Writing this rate as

$$r(C_A) = kC_A C_B = kC_A(C_{Ao} + C_{Bo} - C_A) \qquad (3.95)$$

we see that the rate is a parabola when r is plotted versus C_A or X. Therefore, a $1/r$ versus $C_{Ao} - C_A$ plot is a hyperbola, as sketched in Figure 3–12. If the feed into a reactor is pure A, $C_{Bo} = 0$, then

$$r = kC_A(C_{Ao} - C_A) \qquad (3.96)$$

so we see that $r = 0$ when $C_A = C_{Ao}$ and also when $C_A = 0$. Therefore $1/r \rightarrow \infty$ at $C_A = C_{Ao}$ and also at $C_A = 0$. Thus the PFTR will require an infinite residence time if there is no B in the feed to the reactor, while a CSTR will give the usual rectangle for τ. These τs are shown in the $1/r$ plots in Figure 3–12.

We can solve for autocatalytic reactions fairly simply. Consider the reaction

$$A \rightarrow B, \qquad r = kC_A C_B \qquad (3.97)$$

Figure 3–12 Plots of r and $1/r$ versus $C_{Ao} - C_A$ for the autocatalytic reaction $A \rightarrow B$, $r = kC_A$. The PFTR requires infinite residence time if no B is added to the feed.

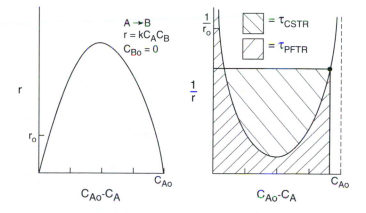

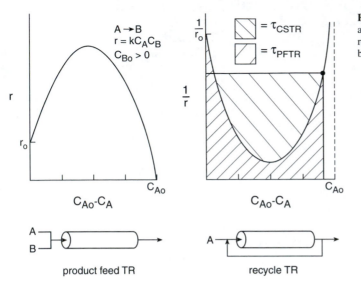

$A \to B$
$r = kC_A C_B$
$C_{Bo} > 0$

$\boxed{\diagdown} = \tau_{CSTR}$

$\boxed{\diagup} = \tau_{PFTR}$

product feed TR

recycle TR

Figure 3–13   Possible tubular reactor configurations for autocatalytic reactions. With large recycle, the tubular reactor performance approaches that of a CSTR, as will be considered in Chapter 8.

In a PFTR the solution is

$$\tau = \int_{C_{Ao}}^{C_A} \frac{dC_A}{kC_A(C_{Ao} + C_{Bo} - C_A)} = -\frac{1}{k(C_{Ao} + C_{Bo})} \left(\int_{C_{Ao}}^{C_A} \frac{dC_A}{C_A} + \int_{C_{Ao}}^{C_A} \frac{dC_A}{C_{Ao} + C_{Bo} - C_A} \right)$$

$$= \frac{1}{k(C_{Ao} + C_{Bo})} \ln \left(\frac{C_A}{C_{Ao} + C_{Bo} - C_A} \frac{C_{Bo}}{C_{Ao}} \right) \tag{3.98}$$

which will be infinite if $C_{Bo} = 0$.

In a CSTR the solution for this autocatalytic reaction is

$$\tau = \frac{C_{Ao} - C_A}{r(C_A)} = \frac{C_{Ao} - C_A}{kC_A(C_{Ao} + C_{Bo} - C_A)} \tag{3.99}$$

which gives no problems when $C_{Bo} = 0$.

We could still use a PFTR reactor for this reaction if we *seeded* the reactant with product or if we used a *recycle tubular reactor* to feed a portion of the product back to the feed (Figure 3–13).

3.12.1   Fermentation

This type of autocatalytic reaction is a simplification of many biological reactions such as fermentation, where the reaction produces products (species B in the previous example), which accelerates the rate. In fermentation, yeast cells in the solution produce enzymes that catalyze the decomposition of sugar to produce ethanol as a byproduct of yeast reproduction. Since the yeast population increases as the reaction proceeds, the enzyme concentration increases, and the process appears to be autocatalytic. A highly simplified description of fermentation might be

$$\text{sugar} + \text{enzyme} \rightarrow \text{alcohol} + 2\,\text{enzyme}, \qquad r = k\,[\text{sugar}]\,[\text{enzyme}] \qquad (3.100)$$

which fits the preceding autocatalytic reaction scheme, although the simple stoichiometry written here is incorrect. This reaction roughly fits our ideal autocatalytic reaction

$$A + B \rightarrow 2B, \qquad r = kC_A C_B \qquad (3.101)$$

with A sugar and B enzyme.

Fermentation reactors are usually CSTRs for continuous operation or batch reactors seeded with yeast (continuous seeding yields the semibatch reactor of the previous section). The single-celled yeast plants reproduce themselves by generating enzymes that break down the sugar into smaller molecules that the yeast cells can metabolize. Ethanol is a byproduct of this digestion process, and it becomes toxic to the yeast cells if its concentration exceeds $\sim$12% by volume; so at high sugar concentrations the reaction must slow down and eventually cease before going to completion.

3.12.2   Combustion processes

Combustion and flame reactions are also highly autocatalytic processes. Here the formation of products and intermediates such as free radicals act to promote or accelerate the reaction in reaction steps, which can be simplified as

$$A + R \rightarrow 2R \qquad (3.102)$$

where A is a fuel molecule and R is a radical species.

Heat generation by exothermic combustion reaction can also accelerates combustion processes (as it also does slightly for fermentation), and we call this thermal autocatalysis which will be discussed more in Chapters 5, 6, and 10.

3.13   REVERSIBLE REACTIONS

For a reversible reaction the rate goes to zero before the reaction reaches completion, and $1/r$ therefore goes to infinity. A plot of $1/r$ versus $C_{Ao} - C_A$, with positive-order kinetics will look as shown in Figure 3–14. The residence time for a given conversion obviously approaches infinity as the conversion approaches equilibrium in either a PFTR or CSTR. Just as for irreversible reactions, the CSTR requires a longer τ for a given conversion.

Figure 3–14   The r and $1/r$ plots for a reversible reaction where r goes to zero and therefore $1/r$ goes to infinity at the equilibrium conversion.

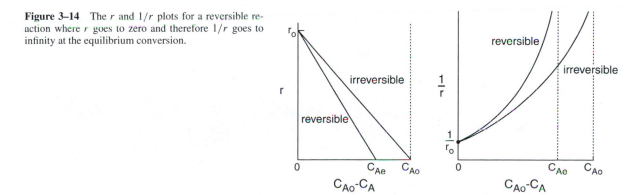

Example 3–9   For the reaction $A \rightleftarrows B$, $r = k_f C_A - k_b C_B$, find the residence times for 50% conversion in a CSTR and in a PFTR if $k_f = 0.5$ min^{-1}, $k_b = 0.1$ min^{-1}, $C_{Ao} = 2$ moles/liter, $v = 4$ liters/min, and $C_{Bo} = 0$.

For the CSTR we solve the equation

$$\tau = \frac{C_{Ao} - C_A}{k_f C_A - k_b C_B} = \frac{C_{Ao} - C_A}{k_f C_A - k_b (C_{Ao} - C_A)}$$

$$= \frac{2.0 - 1.0}{0.5 \times 1.0 - 0.1 - 0.1 \times 1.0} = \frac{1}{0.4} = 2.5 \text{ min}$$

For the PFTR we solve the equation

$$\tau = -\int_{C_{Ao}}^{C_A} \frac{dC_A}{k_f C_A - k_b(C_{Ao} - C_A)} = -\frac{1}{k_f + k_b} \ln \frac{k_f C_{Ao}}{(k_f + k_b)C_A - k_b C_{Ao}}$$

$$= \frac{1}{0.6} \ln \frac{0.5 \times 2}{0.6 \times 1 - 0.1 \times 2} = 1.53 \text{ min}$$

For this reaction calculate the residence time for 90% conversion in a CSTR and in a PFTR.

Answer: $\tau = \infty$ *in either reactor*. If we inserted $C_A = 0.2$ moles/liter in the preceding equations, we would obtain a negative residence time, which is clearly nonsense. We would be trying to go *beyond the equilibrium conversion*, which is

$$\frac{k_f}{k_b} = \frac{C_{B,eq}}{C_{A,eq}} = \frac{C_{Ao} - C_{A,eq}}{C_{A,eq}} = 5.0$$

so that $C_{A,eq} = 0.33$ moles/liter, $C_{B,eq} = 1.67$ moles/liter, and the equilibrium conversion is 83%. This is the maximum conversion obtainable for these kinetics in any single reactor.

3.14   TRANSIENTS IN CONTINUOUS REACTORS

We have thus far considered only steady-state operation of the CSTR and the PFTR. This is the situation some time after the process was started when all transients have died out, and no parameters vary with time. However, all continuous reactors must be started, and parameters such as feed composition, flow rate, and temperature may vary because feed composition and conditions change with time. We therefore need to consider transient operation of the CSTR and the PFTR. Transients are a major cause of problems in reactor operation because they can cause poor performance. Even more important, problems during startup and shutdown are a major cause of accidents and explosions.

Returning to the CSTR mass-balance equation for species A, we obtain

$$\frac{dN_A}{dt} = F_{Ao} - F_A - Vr(C_A) \tag{3.103}$$

If the reactor volume and flow rate are constant and the density is unchanged, this becomes

$$V\frac{dC_A}{dt} = v(C_{Ao} - C_A) - Vr(C_A) \tag{3.104}$$

or dividing by v,

$$\tau\frac{dC_A}{dt} = (C_{Ao} - C_A) - \tau r(C_A) \tag{3.105}$$

as the working equation for transients in a constant-density CSTR.

3.14.1   Solvent replacement

As a simple example, consider the concentration versus time when a pure solvent initially in a tank $(C_{Ai} = 0)$ is replaced by a solute at concentration C_{Ao}, such as replacing pure water in a tank by a brine solution. Since there is no reaction, the mass-balance equation is

$$\tau\frac{dC_A}{dt} = (C_{Ao} - C_A) \tag{3.106}$$

The concentration in the tank at $t = 0$ is C_{Ai}, which is the initial condition in solving this first-order differential equation.

[Note carefully here the difference between C_{Ai} (initial concentration within the tank) and C_{Ao} (feed concentration into the tank). Note also the difference between the reactor residence time τ and the time t after the switch in the feed is initiated.]

The solution to this differential equation is

$$\int_{C_{Ai}}^{C_A} \frac{dC_A}{C_{Ao} - C_A} = \int_0^t \frac{dt}{\tau} \tag{3.107}$$

which gives

$$C_A(t) = C_{Ai} + (C_{Ao} - C_{Ai})e^{-t/\tau} \tag{3.108}$$

which predicts a tank and effluent concentration that varies from C_{Ao} to C_{Ai} with a time constant τ. This is plotted in Figure 3–15.

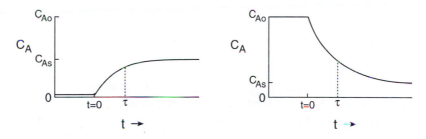

Figure 3–15   Possible transients in CSTR reactors. The left panel shows the situation starting with pure solvent $(C_A = 0)$ in the reactor initially, $C_A = C_{Ao}$ at $t = 0$ with no reaction. The concentration in the reactor in steady state approaches C_{As}. The right panel shows the situation where the reactor initially contains pure reactant at C_{Ao}, and at $t = 0$ reactant flow at C_{Ao} is begun and eventually approaches a steady-state concentration C_{As}.

3.14.2 Reaction

Consider next the situation in which a tank initially filled with pure solvent is switched to reactant at feed concentration C_{Ao} that reacts with $r = kC_A$. As before, $C_A = 0$ at $t = 0$, but now we must include the reaction term to yield

$$\tau \frac{dC_A}{dt} = (C_{Ao} - C_A) - k\tau C_A \tag{3.109}$$

This is solved as before to yield

$$C_A(t) = \frac{C_{Ao} - [C_{Ao} - (1 + k\tau)C_{Ai}]\exp[-(k + 1/\tau)t]}{1 + k\tau} \tag{3.110}$$

Note from this solution that $C_A = C_{Ai}$ at $t = 0$, and at long times the solution approaches

$$C_A(t) = \frac{C_{Ao}}{1 + k\tau} \tag{3.111}$$

the steady state for a first-order irreversible reaction in a CSTR.

Similar transient problems in the CSTR can be easily formulated. If the reactor volume is not constant (such as starting with an empty reactor and beginning to fill it at $t = 0$) or if the volumetric flow rate v into the reactor is changed), then the variable-density version of this reactor involving F_A must be used.

3.14.3 PFTR

Transients in a PFTR require solution of a partial differential equation. The transient version of the derivation of the mass balance in a PFTR is

$$\frac{\partial C_A}{\partial t} + u\frac{\partial C_A}{\partial z} = -r(C_A) \tag{3.112}$$

Note that setting one of the terms on the left side of the equation equal to zero yields either the batch reactor equation or the steady-state PFTR equation. However, in general we must solve the partial differential equation because the concentration is a function of both position and time in the reactor. We will consider transients in tubular reactors in more detail in Chapter 8 in connection with the effects of axial dispersion in altering the perfect plug-flow approximation.

3.15 SOME IMPORTANT SINGLE-REACTION PROCESSES: ALKANE ACTIVATION

Most interesting chemical reaction systems involve multiple reactions, which we will consider in the next chapter. In this section we examine several reaction systems that involve nearly a single reaction and are both industrially important and involve some interesting reaction engineering issues.

3.15.1 Light alkane reactions

Some of the most important reactions outside the petroleum refining industry involve the conversion of light alkanes, CH_4 to C_4H_{10}, into useful chemicals. We will examine some

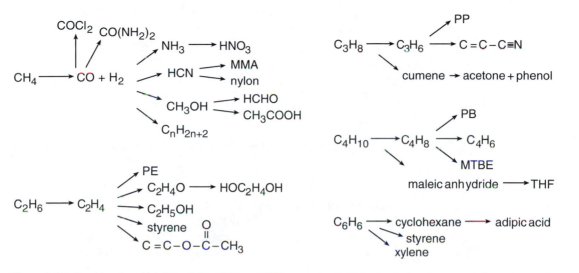

Figure 3–16 Reactions by which CH_4, C_2H_6, C_3H_8, and C_4H_{10} are converted into chemicals and monomers for common polymers. Most of these are major commercial processes, and most will be expanding considerably as production of natural gas and natural gas liquids expands.

of these throughout this book. In Figure 3–16 is shown the "flowsheet" of their conversion into chemicals.

3.16 SYNTHESIS GAS REACTIONS

A mixture of H_2 and CO is an important chemical intermediate as are pure H_2 and pure CO, and this mixture is called *syngas* or *synthesis gas*. These intermediates are used for production of gasoline, ammonia, methanol, and acetic acid.

The process starts by the reaction of natural gas or naphtha (a mixture of larger alkanes) and steam over a nickel catalyst in a tube furnace in a process called *steam reforming*. There are many reactions in this system, but the major products are a mixture of CO and H_2 in the reaction

$$CH_4 + H_2O \rightleftarrows CO + 3H_2 \tag{3.113}$$

This reaction is strongly endothermic and proceeds to near-equilibrium conversion at $\sim 850°C$ at pressures up to ~ 20 atm. The products also contain H_2O and CO_2. If syngas is desired, the CO_2 and H_2O are removed from this mixture (how?).

3.16.1 Water gas shift

To adjust the CO:H_2 ratio from the reactor for a desired downstream process such as methanol synthesis or to form pure H_2 or pure CO, these gases are further reacted in the water gas shift (WGS) reaction,

$$CO + H_2O \rightleftarrows CO_2 + H_2 \tag{3.114}$$

This reaction takes place either at high temperatures ($T > 500°C$) over an Fe or Ni catalyst or at lower temperatures ($\sim250°C$) over a Cu/ZnO catalyst. [Why is it much easier to recover pure H_2 by separating a mixture of H_2 and CO_2 than a mixture of H_2 and CO?]

Thus steam reforming is typically done in two or three staged chemical reactors, the first a reforming reactor operated at $\sim850°C$, and then one or two WGS reactors operated at temperatures as low as 200°C, where the equilibrium conversion to H_2 is higher because the WGS reaction is exothermic.

These reactions can be written as

$$CH_4 + H_2O \rightleftarrows CO + 3H_2, \qquad \Delta H_R = +49 \text{ kcal/mole} \qquad (3.115)$$

$$CO + H_2O \rightleftarrows CO_2 + H_2, \qquad \Delta H_R = -10 \text{ kcal/mole} \qquad (3.116)$$

a system involving at least two simultaneous reactions, the subject of the next chapter.

These reactors operate *near equilibrium*, and therefore the first reactor must be heated to high temperatures because this reaction is endothermic, and the second must be cooled to fairly low temperatures. The kinetics of these reactions are very important if one is designing a reactor in detail, but the major features of the process are governed by *equilibrium limitations* and *heat effects*.

It is difficult to remove the last traces of CH_4 in conventional steam reforming, and the reactions are fairly slow. Therefore, modern syngas plants frequently employ *autothermal reforming*, in which the major reaction is

$$CH_4 + \tfrac{1}{2}O_2 \rightarrow CO + 2H_2, \qquad \Delta H_R = -8.5 \text{ kcal/mole} \qquad (3.117)$$

This process has several advantages: (1) The reaction is exothermic and no process heat is required; and (2) the equilibrium is much more favorable; so the reaction goes to completion. This simplifies both the reforming reactor and the WGS reactors, and the economics of constructing and operating a steam reformer now favor autothermal reforming for large-scale processes in spite of the extra cost of supplying pure O_2 rather than just H_2O. [Why does one need to use O_2 rather than air in autothermal reforming of CH_4 to produce syngas?]

3.16.2 Uses of CO and H_2

H_2 has many uses in hydrogenation, and CO has many uses in carbonylation. The mixture is used in methanol synthesis and in the Fisher Tropsch synthesis of hydrocarbons.

We have already seen some of the uses of H_2 in petroleum refining. A major process is the hydrogenation of heavy oil, which has a formula of approximately $(CH)_n$, while gasoline has a formula of approximately $(CH_2)_n$. Therefore, we can write this hydroprocessing reaction approximately as

$$(CH)_n + (n/2)H_2 \rightarrow (CH_2)_n \qquad (3.118)$$

From this stoichiometry we see that petroleum refinery needs about $\tfrac{1}{2}$ mole of H_2 for each mole of carbon in gasoline produced by hydroprocessing.

As we shall also see, there are also many uses of CO. Examples are acetic acid production, which is made by reacting methanol with CO

$$CH_3OH + CO \rightarrow CH_3COOH \qquad (3.119)$$

and polyurethanes, which contain two CO molecules per monomer molecule. We consider two reactions that use H_2 and a CO–H_2 mixture, the production of ammonia and methanol, respectively.

3.16.3   Ammonia synthesis

All proteins contain one N atom per amino acid. While N_2 is all around us, the N–N bond strength is 225 kcal/mole, and the transformation of N_2 into organic nitrogen is very difficult because the rates of its reactions are small at ambient temperature and the equilibrium conversion is small at higher temperatures. Fortunately, nature accomplishes this readily by nitrogen-fixing bacteria in the soil and in the ocean operating at ambient temperature, and this source of nitrogen is adequate for protein production. However, the large-scale production of crops and the waging of wars require large amounts of fertilizer and explosives, respectively.

We now obtain essentially all this fixed nitrogen from NH_3, which we learned to produce early in the twentieth century, in a method discovered by Fritz Haber and still called the *Haber process*.

The reaction that produces NH_3 is

$$N_2 + 3H_2 \rightleftarrows 2NH_3 \tag{3.120}$$

and the H_2 is produced by steam reforming of CH_4 while the N_2 is produced by separation from air. This reaction has a very favorable equilibrium constant at room temperature (the conditions where bacteria operate using enzyme catalysts), but we do not yet know how to create synthetic catalysts in the laboratory that are effective except at much higher temperatures than the catalysts employed by nature. Since this reaction is exothermic, the equilibrium constant for this reaction becomes very unfavorable at temperatures where the rate is favorable. These equilibrium considerations are illustrated in Table 3–3 and Figure 3–17.

Fritz Haber and Walter Nernst worked on this problem early in the twentieth century, and they had to develop the ideas of thermodynamic equilibrium to figure out how to accomplish this in a practical reactor. Equilibrium in the preceding reaction can be written as

$$K_{eq} = \frac{P_{NH_3}^2}{P_{N_2} P_{H_2}^3} = \frac{y_{NH_3}^2}{y_{N_2} y_{H_2}^3} \frac{1}{P^2} = \exp\left(-\frac{\Delta G_R^\circ}{RT}\right) \tag{3.121}$$

TABLE 3–3
Equilibrium Constant and Equilibrium P_{NH_3} Versus T and P

$T(°C)$	K_{eq}	P_{NH_3}		
		1 atm	10 atm	100 atm
25	5.27×10^5	0.937	9.80	99.3
100	0.275×10^3	0.660	8.76	95.9
200	0.382	0.172	5.37	81.9
300	0.452×10^{-2}	0.031	2.07	58.0
400	0.182×10^{-3}	0.00781	0.682	34.1
500	0.160×10^{-4}	0.00271	0.259	18.2

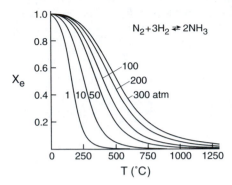

Figure 3–17 Plot of equilibrium conversion X_e versus temperature for ammonia synthesis starting with stoichiometric feed. While the equilibrium is favorable at ambient temperature (where bacteria fix N_2), the conversion rapidly falls off at elevated temperature, and commercial ammonia synthesis reactors operate with a Fe catalyst at pressures as high as 300 atm to attain a high equilibrium conversion.

From the equations developed in the previous chapter, we can find K_{eq} at any temperature from standard state information on reactants and product. These are indicated in Table 3–3, along with the equilibrium partial pressures of NH_3 starting with a stoichiometric mixture of H_2 and N_2 at pressures of 1, 10, and 100 atm.

The equilibrium conversion of NH_3 was calculated for a stoichiometric feed of N_2 and H_2 at the total pressures shown. These conversions were calculated by writing the number of moles of each species versus conversion X, taking a basis of 1 mole of N_2.

$$N_j \qquad\qquad\qquad y_j$$

$$N_{H_2} = 3 - 3X \qquad \frac{3 - 3X}{4 - 2X} = \frac{3}{2}\frac{1 - X}{2 - X}$$

$$N_{N_2} = 1 - X \qquad \frac{1 - X}{4 - 2X} = \frac{1 - X}{2(2 - X)}$$

$$N_{NH_3} = 2X \qquad \frac{X}{2(2 - X)}$$

$$\sum N_j = 4 - 2X \qquad\qquad \sum y_j = 1$$

These expressions can be inserted into the expression for the equilibrium constant to obtain

$$K_{eq} = \frac{P_{NH_3}^2}{P_{N_2} P_{H_2}^3} = \frac{y_{NH_3}^2}{y_{N_2} y_{H_2}^3} \frac{1}{P^2} = \frac{4}{27} \frac{X^2(2 - X)^2}{(1 - X)^4 P^2} \qquad (3.122)$$

The values of P_{NH_3} in the preceding table indicate the conditions for which NH_3 production is possible. It is seen that one must use either low temperatures or very high pressures to attain a favorable equilibrium. At 25°C, where bacteria operate, the equilibrium constant is very large and the conversion is very high, while at higher temperatures K_{eq} and P_{NH_3} fall rapidly. The best catalysts that have been developed for NH_3 synthesis use Fe or Ru with promoters, but they attain adequate rates only above ∼300°C, where the equilibrium conversion is ∼3% at 1 atm.

Modern ammonia synthesis reactors operate at ∼200 atm at ∼350°C and produce nearly the equilibrium conversion (∼70%) in each pass. The NH_3 is separated from unreacted H_2 and N_2, which are recycled back to the reactor, such that the overall process of a tubular reactor plus separation and recycle produces essentially 100% NH_3 conversion. The NH_3 synthesis reactor is fairly small, and the largest components (and the most expensive) are the compressors, which must compress the N_2 and H_2 to 200 atm. [How can NH_3 easily be separated from N_2 and H_2?]

The equilibrium conversion X_e versus temperature and pressure for stoichiometric feed is shown in Figure 3–17.

3.16.4   Nitric acid synthesis

Nitric acid is one of the largest volume commodity chemicals. It is used in fertilizers, in nitrate salts, and in nitrating organic molecules. It is also an important oxidizing agent for organic chemicals because it is a liquid whose products are only N_2 and water; so disposal problems of byproducts are minimal. Nitric acid was produced only from animal wastes before the twentieth century, but Wilhelm Ostwald showed that it could be produced by the direct oxidation of NH_3 over a Pt catalyst, and essentially all industrial HNO_3 is now made by the *Ostwald process*.

Nitric acid is produced by the oxidation of NH_3 in air at pressures up to 10 atm over a Pt catalyst to produce NO, followed by further air oxidation and hydration to produce HNO_3. This is a two-step process with water scrubbing to remove HNO_3 from N_2 and excess air. Since over 90% of the NH_3 fed into the process is converted into HNO_3, we can describe this process qualitatively as the overall single reaction

$$NH_3 + 2O_2 \rightarrow HNO_3 + H_2O \tag{3.123}$$

although the kinetics and reactors are far from simple, and the process is carried out in several distinct reactors,

$$NH_3 + \tfrac{5}{4}O_2 \rightarrow NO + \tfrac{3}{2}H_2O \tag{3.124}$$

$$NO + \tfrac{1}{2}O_2 \rightarrow NO_2 \tag{3.125}$$

$$3NO_2 + H_2O \rightarrow 2HNO_3 + NO \tag{3.126}$$

the sum of which is HNO_3 formation from NH_3. We will leave for a homework problem the sketching of a flow diagram of a nitric acid plant.

3.16.5   Methanol synthesis

Here the reaction is

$$CO + 2H_2 \rightarrow CH_3OH, \qquad \Delta H_R = -22 \text{ kcal/mole} \tag{3.127}$$

This reaction is also strongly exothermic, and the equilibrium yield of CH_3OH decreases as the temperature is increased.

$$K_{eq} = \frac{P_{CH_3OH}}{P_{CO}P_{H_2}^2} = \exp\left(-\frac{\Delta G_R^o}{RT}\right) \tag{3.128}$$

We will leave for a homework problem the calculation of the equilibrium conversion of methanol versus temperature and pressure. Figure 3–18 is a plot of the equilibrium conversion versus temperature.

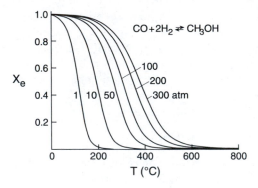

Figure 3–18 Plot of equilibrium conversion X_e versus temperature for methanol synthesis starting with stoichiometric feed. While the equilibrium is favorable at ambient temperature, the conversion rapidly decreases at higher temperature, and industrial reactors operate with a Cu/ZnO catalyst at pressures as high as 100 atm.

This process has many similarities to NH_3 synthesis. The pressure is not as high for acceptable conversions, and modern methanol plants operate at ~250°C at 30–100 atm and produce nearly equilibrium conversions using Cu/ZnO catalysts with unreacted CO and H_2 recycled back into the reactor.

Methanol is a major chemical intermediate for production of acetic acid and the gasoline additive methyl-t-butylether (MTBE). It is an ideal fuel for several types of fuel cells, and there is discussion of using pure methanol as a replacement for gasoline. Therefore, methanol may soon become an even more important chemical.

3.16.6 Toluene nitration

If toluene is heated in HNO_3 (made by the oxidation and hydration of NH_3, which is produced by the Haber process as discussed previously), the toluene is oxidized and nitrated in the reaction

$$CH_3C_6H_5 + 2HNO_3 \rightarrow CH_3C_6H_4NO_2 + H_2O + NO_2 \tag{3.129}$$

This is a fairly simple reaction to run in a flask, but the reaction is far from simple because there are in fact many minor products. [This reaction is not balanced because there are several other products.] First, there are three isomers of nitrotoluene, called *ortho*, *meta*, and *para*. In this reaction there is very little *meta* isomer, and, because of steric considerations, mostly the *para* nitro isomer is formed; so in some sense this is a single reaction.

The next problem with this reaction is that the mononitro compound can add successive nitro groups to produce the di- and trinitro toluenes in the reactions. Fortunately, the rate coefficients for adding the second and especially the third nitro groups are much smaller than for adding the first; so mononitrotoluene can be made with good efficiency by simply heating toluene in nitric acid (Figure 3–19).

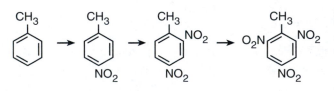

Figure 3–19 Successive reactions in the nitration of toluene. These reactions can be described as series reactions $A \rightarrow B \rightarrow C \rightarrow D$, which are a type of multiple reactions that will be considered in the next chapter.

We will return to these reactions later in connection with *polyurethanes* because one monomer in polyurethanes is toluene diisocyanate, and the first step in its synthesis is the production of dinitrotoluene.

Another use of these reactions is in the production of *amines*, for example, the formation of aniline from benzene. The benzene is first nitrated in nitric acid, and then in another reactor the nitrobenzene is reduced to aniline with H_2 (another use of H_2) in the reactions

$$C_6H_6 + 2HNO_3 \rightarrow C_6H_5NO_2 + H_2O + NO_2 \tag{3.130}$$

followed by

$$C_6H_5NO_2 + 3H_2 \rightarrow C_6H_5NH_2 + 2H_2O \tag{3.131}$$

Note that 2 moles of HNO_3 and 3 moles of H_2 are needed for each mole of aniline produced from benzene. [Why are $4\frac{1}{2}$ moles of H_2 required for the entire process?]

Finally, we note that the product of adding 3 nitro groups to toluene is something one *never* wants to make unintentionally. [Why?] We will consider these reactions later in Chapter 10 in connection with accidents and disasters.

The reactions discussed here are all examples of nearly *single reactions* that have considerable industrial importance. However, the chemical reaction engineering of them is not simple because the pressure, temperature, and reactor type must be carefully designed for successful operation. We must be aware of equilibrium considerations, heat release, and the possibility of undesired reactions, or we can have very unpleasant results. We note finally that most of these reactions in fact use *catalysts* to provide a high rate at low temperatures. The nitration of toluene and its reduction with H_2 can be carried out without catalysts, but the first reaction involves aqueous and organic phases, while the second involves gaseous H_2 and liquid organic. Thus these reactions involve *multiple phases*. We never promised that interesting processes would be simple.

3.17 STAGED REACTORS

These examples all involve multiple reactors with separations between reactors and new feeds added in each stage. As an example of the nitration of toluene, starting with CH_4, we write the steps as

$$CH_4 \rightarrow CO + H_2 \rightarrow H_2 \rightarrow NH_3 \rightarrow NO \rightarrow NO_2 \rightarrow HNO_3 \rightarrow DNT \tag{3.132}$$

Each arrow indicates a single reaction step (frequently with several reactors), and after every reactor stage the products are separated, reactants are recycled, and new feed is added for entry into the next reactor stage. In Table 3–4 we summarize each of these processes.

Thus in this example, the product indicated is fed into the next stage, where it is reacted with other species (H_2O, N_2, O_2, toluene). Different temperatures and pressures are usually used in each stage to attain optimum performance of that reactor. For example, NH_3 synthesis requires very high pressure (200 atm) and low temperature (250°C) because it is an exothermic reversible reaction, while NH_3 oxidation operates at lower pressurse (~10 atm) and the reaction spontaneously heats the reactor to ~800°C because it is strongly exothermic but irreversible. Formation of liquid HNO_3 requires a temperature and pressure where liquid is stable.

TABLE 3–4
Typical Operating Conditions of the Reactions in This Section

Reaction	Name	T (°C)	P (atm)	Catalyst
$CH_4 + H_2O \rightarrow CO + 3H_2$	steam reforming	850	20	Ni
$CO + H_2O \rightarrow CO_2 + H_2$	water gas shift	250	20	Cu/ZnO
$3H_2 + N_2 \rightarrow 2NH_3$	Haber process	250	250	Fe
$NH_3 + \frac{5}{4}O_2 \rightarrow NO + \frac{3}{2}H_2O$	Ostwald process	800	10	Pt/10% Rh
$NO + \frac{1}{2}O_2 \rightarrow NO_2$	NO oxidation	200	10	none
$3NO_2 + H_2O \rightarrow 2HNO_3 + NO$	nitric acid formation	200	10	none
$2HNO_3 + \text{toluene} \rightarrow DNT$	nitration of toluene	100	2	Ni
$CO + H_2 \rightarrow CH_3OH$	methanol synthesis	250	80	Cu/ZnO

For single and multiple reactor stages we sketch reactors and separation units as shown in Figure 3–20. We assume a first reactor running the reaction

$$A \rightarrow B \qquad (3.133)$$

in a first chemical reactor followed by a second reactor running the reaction

$$B + C \rightarrow D \qquad (3.134)$$

in a second reactor so that the overall process would be

$$A + B \rightarrow D \qquad (3.135)$$

with C an intermediate in the staged process.

As a final example of staged reactors, we show in Figure 3–21 the reactions in the production of polyurethane from CH_4 and toluene.

In considering these complex multistage processes, the ideas discussed early in this chapter are still applicable. We need to consider each reactor using the principles described in this chapter, but we need also to consider the separation and recycle components of the process. These aspects can in fact dominate the economics of a chemical process. We also must consider multiple reactions because none of the above reactors produces exclusively the desired product, and this will be the subject of the next chapter. All these reactor stages require different heat management strategies, and most of them use catalysts. Therefore, we cannot design these reactors in detail until we consider these aspects in the following chapters.

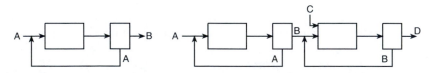

Figure 3–20 Sketch of a single-stage reactor with product separation and recycle and also two stages where reactants from the first stage are separated and recycled into the first reactor while products are mixed with other reactants and fed into a second stage.

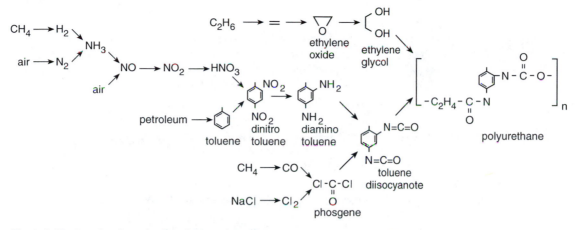

Figure 3–21   Reactions in production of polyurethane. These processes involve many of the individual reaction steps discussed previously.

3.18   THE MAJOR CHEMICAL COMPANIES

The major petroleum and chemical companies have basically *defined* chemical engineering, as their origins and growth have coincided with this discipline. Table 3–5 lists the top chemical companies in the world in 1995.

Some of these companies are mostly or exclusively chemical companies, and several are primarily petroleum companies with a small percentage of their sales from chemicals. Some are very old (DuPont), and others are fairly new (Huntsman Chemical).

TABLE 3–5
Top 20 U.S. Chemical Companies Listed by 1995 Sales (in Millions of Dollars)[a]

1. Dow Chemical	$19,234	95%[b]
2. DuPont	18,433	44
3. Exxon	11,737	9.5
4. Hoechst Celanese	7,395	100
5. Monsanto	7,251	81
6. General Electric	6,628	9.5
7. Mobil	6,155	8.2
8. Union Carbide	5,888	100
9. Amoco	5,655	18
10. Occidental Petroleum	5,410	51
11. Eastman Chemical	5,040	100
12. BASF	4,847	76
13. Shell Oil	4,841	20
14. Huntsman Chemical	4,300	100
15. ARCO Chemicals	4,282	100
16. Rohm and Haas	3,884	100
17. ICI Americas	3,824	100
18. Chevron	3,758	10
19. Allied Signal	3,713	26
20. W. R. Grace	3,665	100

[a]Data from *Chemical and Engineering News*, June 24, 1996. It is interesting to note that 8 of these companies no longer exist.
[b]The last column indicates percentage of sales in chemicals.

By many accounts, this situation is changing as petroleum companies become oil commodity traders rather than refiners and chemical companies attempt to produce commodities with as little research and development as possible.

Nevertheless, it is important for the chemical engineer to see how these industries have evolved in order to see how the discipline might or should continue to evolve in the future. Every company strives to find a "market niche" in which it can make a product cheaper than its competitors. This can be through finding cheaper raw materials, making more valuable products, or keeping competitors out by patent protection and by other means such as buying out a competitor who has a better process.

The major themes we consider are the *evolution of raw materials* from coal to petroleum to gas, and the *evolution of chemical intermediates* in chemical processing. The key organic chemicals have been *acetylene* from 1900 to 1940, *ethylene* from 1940 to the present, and probably *ethane* in the near future. Simultaneously, H_2SO_4, HNO_3, Cl_2, and NaOH have been the major chemicals used in processing organic molecules, and there is a great effort and need to replace them with chemicals that do not leave solid residues such as $CaSO_4$ and salt, liquid residues such as organic chlorides and sulfates and sulfites, and vapor residues such as HCl, SO_x, and NO_x. There is now a trend to replace the oxidizing agents HNO_3 and Cl_2 by direct oxidation with O_2.

We will briefly trace the evolution of chemicals production through three of the largest U.S. chemical companies and the German cartel. We will see that each of these companies can be basically defined through a single chemical: DuPont by HNO_3, Union Carbide by acetylene, Dow Chemical by Cl_2, and BASF by the dye industry.

3.18.1   DuPont Started as an Explosives Company

DuPont is the oldest chemical company in the United States and perhaps in the world. It was started in the eighteenth century when E. I. DuPont came to the United States from France to set up an explosives business to sell gunpowder to the Army. The founders were searching for a site for a plant where reactors could blow up without offending neighbors, and they found a spot on the bank of the Brandywine river near Wilmington, Delaware.

Gunpowder was discovered by the Chinese many centuries ago, but they used it primarily for fireworks until it was imported to the West and used to blow up castles and then to propel projectiles. Explosives are solids that are endothermic with respect to their decomposition products. [We will consider the explosions of solids more in Chapter 10.] Typically explosives contain fuel atoms (C, H, and S) in close proximity to oxygen atoms (as nitrates and perchlorates) so that they will react to form CO_2, H_2O, and SO_2, but only (one hopes) when intentionally ignited. Another class of exothermically decomposing solids is azides, and they are frequently used as igniters or fuses to heat up conventional explosives.

Gunpowder is simply a mixture of charcoal, sulfur, and potassium nitrate. Carbon and S react with O in the nitrate, and the N in the nitrate also releases heat in converting to N_2 (the K combines with something to form the smoke). The reactions are approximately

$$S + C + 2KNO_3 \rightarrow CO_2 + SO_2 + K_2O + N_2 + \tfrac{1}{2}O_2$$

Charcoal is made from wood ashes, sulfur is mined, and potassium nitrate (called Chilean saltpeter) was mined from dry cliffs on the coast of Chile, where fish-eating seabirds had their nests and restroom facilities. Over many centuries, this source accumulated in layers many feet thick, and this was adequate for all nitrate needs until the end of the nineteenth century when deposits began to deplete faster than birds could replenish them and transportation and purification (odor is just part of the problem) kept costs high.

Other explosives, discovered in the nineteenth century, were nitroglycerine, a liquid that is absorbed in a solid to make dynamite, and nitrocellulose, a solid that produces less smoke (smokeless powder). They are made by heating glycerine and cellulose with nitric acid, a process that adds nitro ($-NO_2$) groups. Another important explosive is trinitrotoluene, made by heating (very carefully) toluene in nitric acid,

$$CH_3C_6H_5 + HNO_3 \rightarrow CH_3C_6H_2(NO_2)_3 + H_2O + NO_x$$

Alfred Nobel in Sweden developed this process for manufacturing and selling TNT, became extremely wealthy in the process, and donated some of his wealth to establish prizes for important inventions in science and peace (but that is another story).

In manufacturing these explosives potassium nitrate rapidly became the most difficult and expensive chemical to obtain. Seabird droppings and subsequent leaching and bacterial reactions on the coast of Chile concentrated nitrogen compounds that were originally created by plants that fix N_2 to form amino acids using special enzymes, but this production rate is difficult to increase. Fritz Haber in Germany discovered the process to fix nitrogen to form ammonia in the first decade of the twentieth century, and about the same time Wilhelm Ostwald, also in Germany, found how to convert it to NO and then to HNO_3. (These are other important and interesting stories.)

The DuPont company rapidly saw that HNO_3 was the key in making cheap explosives, all of which need nitrates and nitro compounds, and they rapidly became the largest producers of HNO_3. World War I was a good period for DuPont, but the explosives business is inherently cyclical (fortunately), and dynamite and fireworks have limited markets. Therefore, DuPont began to explore other uses of cellulose, a raw material for smokeless powder. They found that it could be hydrolyzed to form a gel that could then be used to make celluloid, a solid polymer used in shirt collars and corset staves, which were important fashion items in the late nineteenth century. Then they found that it could be spun into fibers. A related natural polymer is cellulose acetate, which they called rayon, and DuPont rapidly evolved from explosives into primarily a textile company.

The major textiles before the 1920s were wool (animal hair), cotton (a seed pod), and silk (a protein used for making cocoons). The silk spider also had a clever device in its abdomen for expelling a gel in a sac through a spinneret where reactions with air made a solid fiber with a uniform cross section. DuPont took this idea in spinning hydrolyzed cellulose into rayon fibers and scaling the process up far beyond the needs of spiders.

In the 1920s and 1930s DuPont research and engineering was supporting these processes and trying to make *synthetic fibers* that had properties similar to wool, cotton, and silk so they could make money by replacing these natural products. A theoretical chemist at the DuPont Experimental Station in Wilmington, Delaware named Wallace Carruthers was examining the flow properties of these large molecules used in products like rayon and wanted to make some molecules that were more uniform than natural hydrolysis products. Since he knew organic chemistry, he was aware that organic acids, alcohols, and amines can be combined to form larger molecules in esters and amides. If he used molecules with *two functional groups*, then the product formed would continue to react with other molecules, and a *linear polymer* with very large molecular weight could be formed. The discovery of linear organic polymers was undoubtedly the most important invention in chemistry in the first half of the twentieth century.

The first product that DuPont commercialized was polyamide, to which they gave the trademark Nylon. These polymers can be made from a diamine and a diacid or from an amino acid (the silk spider). (Remember that proteins are just polypeptides, which are polyamides formed from amino acids; so the DuPont scientists were only adapting and scaling up nature's process.)

DuPont needed large amounts of amines and organic acids, and again their experience with HNO_3 was crucial. HNO_3 contains the needed N atom that can be added to hydrocarbons to make the nitro compound and then reduced to the amine. Nitric acid is also a good oxidizing acid that will selectively oxidize organic molecules to carboxylic acids. The major polyamide of DuPont was made from hexamethylene diamine $[H_2N(CH_2)_6NH_2]$ and adipic acid $[HOOC(CH_2)_4COOH]$. Each of these contains 6 carbon atoms, and the product is called Nylon 66. Another is an amino acid made from caprolactam, called Nylon 6. [We will discuss later how these molecules can be made cheaply.] Cloth made from these products was very strong, and DuPont introduced it in World War II for use in parachutes and tents. After the war they introduced it as Nylon stockings, and the importance and acceptance of synthetic organic polymers by consumers was established.

DuPont had trouble in finding a suitable polyester, because esters of most diacids and dialcohols hydrolyze too easily. Clothing that dissolves in the rain or washing machine is clearly not very satisfactory. However, Imperial Chemical Industries (ICI) in England, a company almost as old and large as DuPont, found the best one, polyethylene terephthalate, which they sold to DuPont, and PET is still the most important polyester.

DuPont no longer sells explosives, and they produce nitric acid only for internal consumption in making other chemicals. However, they learned these technologies from their origins in explosives.

3.18.2   Union Carbide started as a calcium carbide company

Union Carbide had its origin in South Charleston, West Virginia, early in the twentieth century. The company was an amalgamation of several companies, one that made carbon electrodes and another that made organic chemicals. This relationship started because one of the best ways to make many chemicals was to start with acetylene, and carbon was initially crucial in its production.

Acetylene has a triple bond that makes it quite reactive, and therefore it can be reacted in many ways to make many small organic chemicals. For example, by adding two water molecules one can make ethylene glycol. The goal of much of chemicals synthesis is to find a single raw material and a selective process to make one product from one reactant without having to separate the "dog's breakfast" of products that result from most chemical processes. Acetylene was the first of these chemicals. It was discovered in the nineteenth century that acetylene is produced with high yield by a surprising route: by adding water to the product of fusing limestone and charcoal. When these are heated, they form a stable solid, calcium carbide

$$2CaCO_3 + 4C \rightarrow Ca_2C + 3CO_2$$

When water is dripped over this solid, it forms acetylene with almost quantitative yield,

$$2Ca_2C + 8H_2O \rightarrow C_2H_2 + 4Ca(OH)_2 + 3H_2$$

This process had its first application in lamps because both water and Ca_2C can be stored easily and the lamp power can be varied by the water drip rate. It became natural therefore that a carbon company and an organic chemical producer should combine resources.

While acetylene is a good chemical intermediate, Ca_2C must be formed by heating to high temperatures or in an electric arc, and its synthesis is very energy intensive. Also, acetylene is a very dangerous molecule to store because its decomposition

$$C_2H_2 \rightarrow 2C_s + H_2$$

liberates 54 kcal/mole of heat; so, once ignited, acetylene is violently explosive. (Union Carbide did not want to compete with DuPont in the explosives business, especially with this chemical.)

In the 1930s it was discovered that the pyrolysis of alkanes produced large quantities of olefins. This pyrolysis process is not very selective, but the costs of separation were cheaper, and scaleup was simpler and safer in making ethylene rather than acetylene; so during the 1940s ethylene and other small olefins replaced acetylene as the major building block in chemical synthesis. We will consider the reactions and reactors used in olefin synthesis from alkanes in the next chapter.

The Union Carbide Corporation has always been a leader in production of small organic chemicals, and now it does so largely by using olefins as its starting material rather than acetylene. Still the Carbide in the name remains from its acetylene origins.

3.18.3   Dow Chemical started as a chloralkali company

Herbert H. Dow founded the Dow Chemical Company in 1897 as a chloralkali company. He found that there were extensive brine wells near Midland, Michigan, and started a small company there to produce bromine, chlorine, and caustic soda (NaOH) there.

Chlorine has always been produced by electrolysis, and it is still the only major chemical process using electrolysis besides electroplating. It is produced by the stoichiometric reaction

$$NaCl + H_2O \rightarrow NaOH + \tfrac{1}{2}Cl_2 + \tfrac{1}{2}H_2$$

This reaction has a very small equilibrium constant, but by dipping carbon electrodes in concentrated solutions and withdrawing the gaseous Cl_2 and purifying the NaOH, two valuable products could be made rather cheaply because of the large ΔG between anode and cathode in an electrochemical cell. *Electrochemical reactors* will be discussed in Chapter 9.

A chloralkali business should be located either near cheap salt or near cheap electricity, and much of the chemical industry in this country started near Niagara Falls because of the cheap hydroelectric power there. Most of these industries have left this area, and the polluted Love Canal is one of the few remnants of this industry.

Dow Chemical found that the NaOH and Cl_2 they produced were essential intermediates in synthesis of many chemicals, and Dow rapidly expanded into production of many organic chemicals. However, they remain the largest user of chlorine in chemical synthesis.

Dow Chemical is now a major producer of vinyl chloride. This was initially made by adding HCl to acetylene (the Union Carbide intermediate),

$$C_2H_2 + HCl \rightarrow C_2H_3Cl$$

but now it is made by adding Cl_2 to ethylene

$$C_2H_4 + Cl_2 \rightarrow C_2H_3Cl + HCl$$

Another chemical intermediate made from Cl_2 and NaOH is hypochlorous acid,

$$NaOH + Cl_2 \rightarrow HOCl + HCl$$

Hypochlorous acid is a strong oxidizing agent, and it is the dominant bleaching agent, for example, in Chlorox and in the paper industry.

HOCl has another application in adding Cl and OH to olefins,

$$C_2H_4 + HOCl \rightarrow HOC_2H_4Cl$$

this is called ethylene chlorohydrin. This molecule can easily be heated and made to eliminate HCl to form ethylene oxide,

$$HOC_2H_4Cl \rightarrow C_2H_4O + HCl$$

and this molecule readily reacts with water to form ethylene glycol,

$$C_2H_4O + H_2O \rightarrow HOC_2H_4OH$$

used in antifreeze and polyester.

The direct oxidation of ethylene to EO by O_2 has now replaced the chlorohydrin process entirely because it is cheaper and involves less byproducts, but propylene oxide (a monomer in polyurethanes) is still made by the chlorohydrin route.

Chlorine has been a major intermediate in the chemical industry throughout the twentieth century, both because of its oxidizing power and ability to add Cl to molecules. The major drawback in chlorine is the need to dispose of it after the process, and in EO production more than 2 pounds of salt must be disposed of for every pound of ethylene glycol produced. Polyvinyl chloride (PVC) is an important chemical that contains Cl, and there is considerable movement among environmentalists to eliminate PVC completely as a plastic so that no HCl will be emitted in the incineration of plastics.

Three chemicals that contain chlorine and have received much attention for the past few decades are *PCBs*, *dioxin*, and *Freon*. Polychlorinated biphenyls are inert molecules formerly used as heat exchangers such as transformer coolants. Dioxin is a byproduct of the defoliant 2,4-D (Agent Orange). Freons are mainly used as working fluids in refrigerators. Dioxin is toxic (although its toxicity to humans is under debate), as are PCBs, and Freons escape into the stratosphere, where they catalyze the destruction of ozone (discussed in Chapter 8). Serious concerns with these chlorine-containing chemicals have led to strong incentives to replace chlorine as an intermediate in chemical synthesis and as a chemical product.

3.18.4   The English and German chemical industries started as dye companies

The origins of the chemical industries as we know them came first from England and then from Germany. The Industrial Revolution originated in England, and coal and steel were its fuels. As coal was turned into coke by heating it in retorts (large closed pots), the gases driven off were rich in chemicals, originally regarded as just smelly off-gases, but gradually regarded as molecules that could be collected and separated into useful chemicals.

One of the most valuable classes of these chemicals was dyes. While the idea seems quaint to us today, dyes have always been much desired and expensive chemicals for our ancestors. (Salt has an even older and richer history.) Chemicals that turn clothing from brown and gray into bright colors have always been attractive to humans, and we are willing to pay dearly for colored clothes. Purple has been the symbol of royalty in the West, and red-orange is a comparable symbol in China. In the West this goes back to the Bible, where Lydia was a "seller of purple" in Thyatira in Asia Minor whom Paul encountered in the book of Acts and who became the first European convert to Christianity. This was the dye called indigo, which was obtained from the ink produced by a sea animal. The value of dyes was comparable to that of gold in the ancient and into the fairly modern world.

The products of coal tar distillation and extraction contained fractions that were brightly colored. These were mostly polyaromatic molecules containing N=N bonds that contained chromophores that absorbed light in the visible. Several companies were soon started to produce and sell these products. Very quickly organic chemists in England and Germany began to develop techniques to synthesize these compounds, and this market was important in the development of synthetic organic chemistry in the late nineteenth and early twentieth century.

One of these companies was Badische Anilin und Soda Fabrik or BASF (translated as Baden-Baden Aniline and Soda Ash Company) in the town of Ludwigshaven in the Rhine valley of Germany, which combined the synthesis of aniline dyes and alkalis in early products and in their name.

The Rhine and the Ruhr valleys (the Ruhr was a coal producing center in nineteenth-century Germany) rapidly became the center of chemical production, as many companies joined in the production of many chemical products. The synthesis of NH_3 was begun here by Fritz Haber at Karlsruhe and Bosch at BASF in 1910. The large-scale production of acetylene was initiated there. The production of synthetic fuels was also developed in this region by this company in the production of synthesis gas and its subsequent conversion to methanol and to diesel fuel (the Fisher Tropsch process).

In the 1930s Adolph Hitler combined these companies into a cartel called I G Farben (Farben means colors) that rapidly dominated prewar chemical production and then produced Germany's chemicals, munitions, and synthetic fuels during World War II. After the War, in an effort to control the power of this conglomerate, the Allies split I G Farben into three separate companies: BASF, Bayer, and Hoechst, which specialized in commodity chemicals, pharmaceuticals, and fine chemicals, respectively. These companies have continued to grow since the war so that now the three of them are the top three chemical companies in the world.

3.19 REACTOR DESIGN FOR A SINGLE REACTION

We have now developed the basic ideas of continuous chemical reactors, and we have solved these equations for a single reaction in an isothermal reactor. Given $r(C_j)$, we solve the equations

$$C_{jo} - C_j + \tau \nu_j r(C_j) = 0 \qquad (3.136)$$

for the CSTR or

$$u\frac{dC_j}{dz} = v_j r(C_j) \tag{3.137}$$

for the PFTR to obtain $C_A(\tau)$ in one of these ideal reactors.

Since we can also write the species balances in terms of a single reactant species A, we actually usually solve the equations

$$C_{Ao} - C_A = \tau r(C_j) \tag{3.138}$$

or

$$u\frac{dC_A}{dz} = -r(C_A) \tag{3.139}$$

The solution of one of these two equations is all there is to this chapter. The subject will become more complicated, but if you can work problems such as these, you should have little problem as the applications become more interesting. If you can't, you will be in trouble. We suggest that you should not simply memorize these two equations but that you should rather be able to derive them or at least "feel" their meaning.

Most industrial processes are in fact staged operations where separation units between reactors are used to separate the desired intermediate product from reactants and from other products before feeding it to the next stage, where other reactants are added to form another product.

The next chapter on multiple reactions in fact uses almost the same equations, but now we have to replace $v_j r$ in these equations by $\sum v_{ij} r_i$; so, if you understand these equations, the next chapter will be simple (sort of).

3.20 NOTATION

The other aspect of this chapter with which all students need to be familiar is the notation used. We intentionally keep this as simple as possible so that the structure of the problems we deal with are not confused by extensive symbols. We summarize here the quantities that we will use throughout the book.

3.20.1 Concentration

We always use C_j in moles per liter (or in moles per cubic decimeter or 1 kilomole/m^3 for the SI purist) as the only unit of concentration. The subscript j always signifies species, while the subscript i always signifies reaction. We use j as the species designation and species A as the key reactant. For gases the natural concentration unit is partial pressure P_j, but we always convert this to concentration, $C_j = P_j RT$, before writing the mass-balance equations. Conversion X means the fraction of this reactant that is consumed in the reactor, $C_A = C_{Ao}(1 - X)$, but we prefer to use C_A rather than X and find the conversion after we have solved the equation in terms of C_A. We cannot use this unit of density of a species when the density of the fluid varies with conversion, but we prefer to do so whenever possible because the equations are simpler to write and solve.

3.20.2   Reaction

We always use r to signify the rate of a single reaction $\sum_j \nu_j A_j$ and r_i to signify the rate of the ith reaction in the reaction set $\sum_j \nu_{ij} A_j, i = 1, 2, \ldots, R$. The rate r is a positive quantity, and the reaction rate must be defined along with a specified reaction stoichiometry.

3.20.3   Reactor

Reactors have volume V. Continuous-flow reactors have volumetric flow rate v, and constant-density reactors have *residence time* $\tau = V/v$. Until Chapter 8 all continuous reactors are either completely mixed (the CSTR) or completely unmixed (the PFTR).

3.20.4   Flow rates

The molar flow rate of a species in a flow reactor is $F_j = vC_j$. The batch reactor is a closed system in which $v = 0$. The *volumetric flow rate* is v, while the linear velocity in a tubular reactor is u. We usually assume that the density of the fluid in the reactor does not change with conversion or position in the reactor (the constant-density reactor) because the equations for a constant-density reactor are easier to solve.

The definitions of the most used quantities in this book are therefore as follows:

C_j concentration of species j, usually in moles/liter

ν_j stoichiometric coefficient of species j in the reaction $\sum \nu_j A_j$

C_A concentration of key reactant species with stoichiometric coefficient $\nu_A = -1$

P_j partial pressure of species j, $P_j/RT = C_j$ for ideal gases

X fractional conversion of key species A, defined as $C_A = C_{A0}(1-X)$ if density constant

V reactor volume, usually in liters

v volumetric flow rate, usually in liters/time

F_j molar flow rate of species j

o   subscript o always signifies reactor feed parameters

L length of tubular reactor

u average velocity in plug-flow tubular reactor

τ residence time in a reactor, defined as $\tau = V/v$ for a constant-density reactor and also as $\tau = L/u$ for PFTR

r reaction rate of a single reaction in moles/volume time

r_i reaction rate of the ith reaction in a multiple-reaction system

k reaction rate coefficient

k_f reaction rate coefficient for the forward reaction in a single-reaction system (k_f if reversible)

k_b reaction rate coefficient for the back or reverse reaction in a single-reaction system

k_i reaction rate coefficient for the ith reaction

n order of an irreversible reaction

m_j order of a reaction with respect to species j

r_f rate of forward reaction

r_b rate of back reaction

3.21 REFERENCES

Aris, Rutherford, *Introduction to the Analysis of Chemical Reactors*, Prentice-Hall, 1965.
Aris, Rutherford, *Elementary Chemical Reactor Analysis*, Prentice-Hall, 1969.
Boudart, Michel, *The Kinetics of Chemical Processes,* Prentice-Hall, 1968.
Butt, John B., *Reaction Kinetics and Reactor Design*, Prentice-Hall, 1980.
Carberry, James J., *Chemical and Catalytic Reaction Engineering*, McGraw-Hill, 1976.
Denbigh, Kenneth G., *Chemical Reactor Theory*, Cambridge University Press, 1965.
Fogler, H. Scott, *Elements of Chemical Reaction Engineering*, 2nd ed., Prentice-Hall, 1992.
Froment, Gilbert F., and Bischoff, Kenneth B., *Chemical Reactor Analysis and Design*, 2nd ed.,
    Wiley, 1990.
Hill, Charles G., Jr., *An Introduction to Chemical Engineering Kinetics and Reactor Design*, Wiley,
    1977.
Hougen, O. A., and Watson, K. M., *Chemical Process Principles, Volume III: Kinetics and Catalysis*,
    Wiley, 1947.
Laidler, Keith J., *Chemical Kinetics*, 3rd ed., Harper and Row, 1987.
Levenspiel, Octave, *Chemical Reaction Engineering*, Wiley, 1962.
Levenspiel, Octave, *The Chemical Reactor Minibook*, OSU Book Stores, 1979.
Levenspiel, Octave, *The Chemical Reactor Omnibook*, OSU Book Stores, 1984.
Peterson, E. E., *Chemical Reaction Analysis*, Prentice-Hall, 1965.
Pilling, Michael J., and Seakins, Paul W., *Reaction Kinetics*, Oxford University Press, 1995.
Ramachandran, P. A., and Chaudari, R. V., *Three-Phase Catalytic Reactors*, Gordon and Breach,
    1983.
Westerterp, K. R., van Swaaij, W. P. M., and Beenackers, A. A. C. M., *Chemical Reactor Design and
    Operation*, Wiley, 1993.

PROBLEMS

3.1 At your favorite fast food joint the french fries are made by filling a basket with potatoes and dipping them in hot animal or vegetable fat for 4 min and then draining them for 4 min. Every hour the small pieces that fell out of the basket are scooped out because they burn and give a bad taste. At the end of the 16 h day the fat is drained and sent out for disposal because at longer times the oil has decomposed sufficiently to give a bad taste. Approximately 2 pounds of potatoes are used in 10 gallons of oil.

(a) Why is a batch process usually preferred in a restaurant?

(b) Design a continuous process to make 1 ton/day of french fries, keeping exactly the same conditions as above so they will taste the same. Describe the residence times and desired flow patterns in solid and liquid phases. Include the oil recycling loop.

(c) How might you modify the process to double the production rate from that specified for the same apparatus? What experiments would you have to do to test its feasibility?

(d) How would you design this continuous process to handle varying load demands?

[Note: Creation of a continuous process like this would eliminate jobs for a million teenagers.]

3.2 An irreversible first-order reaction gave 95% conversion in a batch reactor in 20 min.

(a) What would be the conversion of this reaction in a CSTR with a 20 min residence time?

(b) What residence time would be required for 95% conversion in a CSTR?

(c) What residence time would be required for 99% conversion in a CSTR?

(d) What residence time would be required for a 95% conversion in a PFTR?

(e) What residence time would be required for 99% conversion in a PFTR?

3.3 Calculate the ratio of residence times in CSTR and PFTR for the nth-order irreversible reaction for conversions of 50, 90, 99, and 99.9% for $n = 0, \frac{1}{2}, 1, 2$, and -1 for $C_{Ao} = 1.0$.

3.4 An aqueous reaction gave the following data in a batch experiment:

t	$T(°C)$	X
10	50	0.5
20	50	0.666
1000	50	0.999
10	80	0.8

What residence time would be required for 95% conversion at 80°C in a pressurized CSTR starting with the same initial composition? In a PFTR? In 3 CSTRs in series?

3.5 Calculate the reactor volumes required to process 100 liter/min of 3 molar A in the aqueous reaction $A \rightarrow 2B$ for PFTR and CSTR reactors.

(a) 90% conversion, $k = 2$ min^{-1}

(b) 99% conversion, $k = 2$ mole liter^{-1} min^{-1}

(c) 99.9% conversion, $k = 2$ mole liter^{-1} min^{-1}

(d) 90% conversion, $k = 2$ min^{-1} liter mole^{-1}

(e) 99.9% conversion, $k = 2$ min^{-1} liter mole^{-1}

3.6 Set up the above problems if the feed is pure A, A and B are ideal gases, and the reactors operate at 100 atm.

3.7 An aqueous feed containing reactant A (1 mole/liter) enters a 2 liter plug-flow reactor and reacts with the reaction $2A \rightarrow B$, r (mole/liter min) $= 0.2C_A$.

(a) What feed rate (liter/min) will give an outlet concentration of $C_A = 0.1$ mole/liter?

(b) Find the outlet concentration of A (mole/liter) for a feed rate of 0.1 liter/min.

(c) Repeat this problem for a stirred reactor.

3.8 An aqueous feed containing reactant A at $C_{Ao} = 2$ moles/liter at a feed rate $F_{Ao} = 100$ mole/min decomposes in a CSTR to give a variety of products. The kinetics of the conversion are represented by

$$A \rightarrow 2.5 \, B, \qquad r = 10 \, C_A \text{ (moles/liter min)}$$

(a) Find the volume CSTR needed for 80% decomposition of reactant A.

(b) Find the conversion in a 30 liter CSTR.

(c) What flow rate will be required to produce 100 moles/min of B at 80% conversion?

3.9 An ester in aqueous solution is to be saponified in a continuous reactor system. Batch experiments showed that the reaction is first order and irreversible, and 50% reaction occurred in 8 min at the temperature required. We need to process 100 moles/h of 4 molar feed to 95% conversion. Calculate the reactor volumes required for this process in

(a) a PFTR,

(b) a CSTR,

(c) two equal-volume CSTRs,

(d) four equal-volume CSTRs.

3.10 The aqueous reaction $A \rightarrow$ products has a rate $r = 2C_A/(1 + C_A)^2$ (rate in moles/liter min), and is to be carried out in a continuous reactor system. We need to process 100 moles/h of 2 molar feed to 90% conversion. Calculate the reactor volumes required in

(a) a PFTR,

(b) a CSTR,

(c) two equal-volume CSTRs.

(d) Use plots of $1/r$ versus conversion to show these results.

(e) What is the ideal combination of reactors for total minimum volume for this reaction?

(f) Show how you would solve problem (e) analytically. Set up equations and indicate the method of solution.

3.11 We have a 100-gallon tank of a product that now contains small particles that must be removed before it can be sold. The product now has 10^4 particles cm^{-3}, and it is only acceptable for sale if the particle concentration is less than 10^2 particles cm^{-3}. We have a filter that will effectively remove all particles.

(a) At what rate must the product be pumped through the filter to make the product acceptable within 2 days if the tank is mixed and the filtered product is continuously fed back into the tank?

(b) At the pumping rate of part (a) how long will be required if the filtered product is placed in a separate tank?

(c) Repeat parts (a) and (b) assuming the filter only removes 90% of the particles on each pass.

3.12 (a) Solve for $\tau(C_A)$ and $C_A(\tau)$ for the reaction $A \rightarrow B + 2C, r = kC_A$ in 1 atm constant-pressure CSTR and PFTR reactors if the feed is pure A and A, B, and C are ideal gases at 100°C.

(b) Repeat with a feed at the same C_{Ao} but with 20 moles of an inert ideal gas solvent per mole of A.

(c) Repeat with an aqueous feed at the same C_{Ao}.

3.13 Find $C_A(\tau)$, $C_B(\tau)$, and $C_C(\tau)$ for the liquid reaction $A + B \rightarrow C$ with $r = kC_A C_B$ in a PFTR and in a CSTR with $C_{CO} = 0$ for $C_{AO} = C_{BO}$ and for $C_{AO} = 2C_{BO}$. Why do the solutions look so different?

3.14 (a) Find $C_A(\tau)$ and $C_B(\tau)$ for the liquid reaction $A \rightarrow B$ with $r = kC_A C_B$ in a PFTR and a CSTR with a feed of $C_{Ao} = 2$ moles/liter and $C_{Bo} = 0$ with $k = 0.2$ liter/mole min..

(b) Why cannot thus reaction be carried out in a PFTR with this composition?

(c) What modification would be required in a PFTR to give finite conversion?

3.15 Sketch $1/r$ versus $C_{Ao} - C_A$ or X for $A \rightarrow$ products for the reactions $r = kC_A, kC_A^2, k$, and k/C_A. Show the relative residence times of PFTR and CSTR graphically.

3.16 Sketch $1/r$ versus $C_{Ao} - C_A$ or X for $A \rightarrow$ products for the reactions $r = kC_A/(1 + KC_A)$ for $C_{A0} = 2, k = 0.1$ min^{-1}, and $K = 1$ liter/mole. From this graph find τ for 90% conversion for

  (a) a PFTR,

  (b) a CSTR,

  (c) 2 equal-volume CSTRs,

  (d) the optimal combination of reactors in series.

3.17 An irreversible first-order reaction gave 80% conversion in a batch reactor in 20 min.

  (a) Calculate the total residence time for this conversion for CSTRs in series for 1, 2, 3, and 4 equal-volume reactors.

  (b) What residence time will be required for a very large number of equal-volume CSTRs? What is the limit of $(1 + k\tau)^{-n}$ in the limit $n \rightarrow \infty$?

3.18 (a) What is the optimum combination of ideal reactors for the reaction $A \rightarrow B$ if it is autocatalytic with $r = kC_A C_B$ and $C_A = C_{Ao}, C_{Bo} = 0$? What is the intermediate concentration between reactors?

  (b) Solve for 90% conversion if $C_{Ao} = 1$ mole/liter and $k = 0.25$ liter/mole min.

3.19 [Computer] In fermentation processes sugar (A) is converted to ethanol (C) as a byproduct of yeast (B) reproduction. In a simple model we can represent this process as

$$A \rightarrow B + 3C, \qquad r = kC_A C_B$$

Starting with 10 weight percent sucrose ($C_{12}H_{22}O_{11}$) in water and assuming that half of the carbon atoms in the sucrose are converted into ethanol (the above stoichiometry), find the times required to produce 3.5 weight percent alcohol (1% sucrose remaining) for initial concentrations of yeast of 0.00001, 0.0001, 0.001, and 0.01 molar. It is found that 2 h are required for this conversion if the initial yeast concentration is maintained at 1 molar. Assume that the density is that of water.

3.20 Consider the reaction $A \rightarrow 4B, r = kC_A^2$, with A and B ideal gases in a constant-pressure reactor and a feed of pure A with $v = 10$ liter/min, $k = 0.1$ liter/mole min, and $C_{Ao} = 0.05$ moles/liter.

  (a) Find an expression for $C_A(X)$.

  (b) Find V for 80% conversion in a CSTR.

  (c) Find V for 80% conversion in a PFTR.

3.21 The gas-phase reaction $A \rightarrow 3B$ obeys zeroth-order kinetics with $r = 0.25$ moles/liter h at 200°C. Starting with pure A at 1 atm calculate the time for 95% of the A to be reacted away in

  (a) a constant-volume batch reactor,

  (b) a constant-pressure batch reactor.

  Calculate the reactor volume to process 10 moles/h of A to this conversion in

  (c) a constant-pressure PFTR,

  (d) a constant-pressure CSTR.

  (e) Calculate the residence times and space times in these reactors.

3.22 One hundred moles of A per hour are available in concentration of 0.1 mole/ liter by a previous process. This stream is to be reacted with B to produce C and D. The reaction proceeds by the aqueous-phase reaction,

$$A + B \rightarrow C + D, \qquad k = 5 \text{ liters/mole h}$$

The amount of C required is 95 moles/h. In extracting C from the reacted mixture A and B are destroyed; hence recycling of unused reactants is not possible. Calculate the optimum reactor size and type as well as feed composition for this process.

Data: B costs \$1.25/mole in crystalline form. It is highly soluble in the aqueous solution and even when present in large amounts does not change the concentration of A in solution. Capital and operating costs are \$0.015/h liter for mixed-flow reactors.

3.23 One hundred gram moles of B are to be produced hourly from a feed consisting of a saturated solution of $A(C_{Ao} = 0.1$ mole/liter) in a mixed-flow reactor. The reaction is

$$A \rightarrow B, \qquad r = 0.2(\text{h}^{-1})C_A$$

The cost of reactant at $C_{Ao} = 0.1$ mole/liter is $\$A = \$0.50$/mole A

The cost of reactor including installation, auxiliary equipment, instrumentation, overhead, labor, depreciation, etc., is $\$m = \$0.01$/h liter.

What reactor size, feed rate, and conversion should be used for optimum operations? What is the unit cost of B for these conditions if unreacted A is discarded?

3.24 Suppose all unreacted A of the product stream of the previous problem can be reclaimed and brought up to the initial concentration $C_{Ao} = 0.1$ mole/liter at a total cost of \$0.125/mole A processed. With this reclaimed A as a recycle stream, find the new optimum operating conditions and unit cost of producing B.

3.25 [Computer] Plot $C_A(t)$ and $C_B(t)$ in the reaction $A \rightarrow B$ for $r = 2C_A/(1 + 4C_A)^2$ for $C_{Ao} = 0.1$, 1, and 10, $C_{Bo} = 0$ for times displaying up to 99% conversion.

3.26 Consider the reaction $A \rightarrow B$, $r = 0.15$ (min^{-1}) C_A in a CSTR. A costs \$2/mole, and B sells for \$5/mole. The cost of the operating the reactor is \$0.03 per liter hour. We need to produce 100 moles of B/hour using $C_{Ao} = 2$ moles/liter. Assume no value or cost of disposal of unreacted A.

(a) What is the optimum conversion and reactor size?

(b) What is the cash flow from the process?

(c) What is the cash flow ignoring operating cost?

(d) At what operating cost do we break even?

3.27 Figure 3–22 shows possible r versus $C_{Ao} - C_A$ curves. What reactor or combination of reactors will give the shortest total residence time for a high conversion?

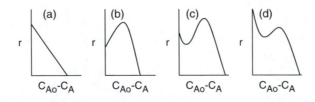

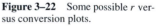

Figure 3–22 Some possible r versus conversion plots.

3.28 Find an expression for the conversion in a dimerization reaction

$$2A \rightarrow B, \qquad r = kC_A^2$$

with A and B ideal gases starting with pure A at 1 atm in

(a) a CSTR,

(b) a PFTR.

3.29 Find an expression for the conversion in the reaction

$$A \rightarrow 3B, \qquad r = k$$

with A and B ideal gases starting with pure A at 1 atm in

(a) a CSTR,

(b) a PFTR.

3.30 Find an expression for the conversion in a dimerization reaction

$$2A \rightarrow B, \qquad r = kC_A^2$$

in a CSTR with A and B ideal gases starting with A at 1 atm for

(a) no diluent in the feed,

(b) 1 atm of inert diluent,

(c) 9 atm of diluent,

(d) 99 atm of diluent.

(e) Compare the volumes required for 90% conversion in these situations with those predicted using the constant-density approximation.

3.31 Find an expression for the conversion in a dimerization reaction

$$A \rightarrow 2B, \qquad r = kC_A^2$$

in a CSTR with A and B ideal gases starting with A at 1 atm for

(a) no diluent in the feed,

(b) 1 atm of inert diluent,

(c) 9 atm of diluent,

(d) 99 atm of diluent.

(e) Compare the volumes required for 90% conversion in these situations with those predicted using the constant-density approximation.

3.32 The aqueous reversible reaction

$$A \rightleftarrows B, \qquad r = kC_A - k'C_B$$

is to be carried out in a series of reactors with separation of unreacted B between stages. At each stage the reaction goes to 50% of the equilibrium composition. How many stages are required for 90% conversion of the initial A for $k = 0.2$ min^{-1}, $K_{eq} = 1$ if

(a) all B is extracted between stages,

(b) 50% of B is extracted between stages.

(c) Find τ with complete extraction if the reactors are CSTRs.

(d) Find τ if the reactors are PFTRs.

3.33 (a) Find an expression for V_1 and V_2 and the intermediate concentration C_{A1} for the reaction

$$A \to B, \qquad r = kC_A$$

in 2 CSTRs to produce a minimum total volume for specified C_{Ao} and C_{A2}.

(b) Repeat for n CSTRs in series.

3.34 Methanol is made by the reaction

$$CO + 2H_2 \to CH_3OH$$

Plot the equilibrium conversion versus temperature at pressures of 1, 10, 30, 50, 100, and 200 atm for temperatures from 300 to 800 K using the following data.

$$\Delta G_{298}^\circ = -24.32 \text{ kJ/mole}$$

$$\Delta H_{298}^\circ = -90.64 \text{ kJ/mole}$$

Compare with the figure in the text.

3.35 A reaction $A \to$ products is known to obey the rate expression $r = kC_A^n$. Using two 50-liter CSTRs in series, it is found that with a feed of 2.0 moles/liter, it is found that after the first reactor $C_{Ao} = 1.0$ mole/liter and after the second $C_{Ao} = 0.3$ mole/liter.

(a) Find n.

(b) What volume PFTR will be required to obtain 90% conversion for this reaction at the same feed rate?

3.36 A certain equipment catalog lists CSTRs as costing

$$\$_{CSTR} = 1000 + 100V^{1/2}$$

and PFTRs as

$$\$_{PFTR} = 500 + 100V$$

where $ is in dollars and V is in liters.

(a) Why might the cost of a chemical reactor be roughly proportional to its surface area? For what reactor geometries might the costs of chemical reactors have these dependences on their volumes? How should the cost of a spherical CSTR depend on its volume?

(b) At what volume will the costs of a CSTR and PFTR be equal in this catalog?

(c) We find that a 1000-liter PFTR will process 500 moles/h of a reactant in a first-order reaction to 90% conversion. How does the cost of this reactor compare with a CSTR for the same conversion?

(d) Repeat this calculation for processing 1000 moles/h to 90% conversion.

3.37 A 10,000-gallon holding tank receives an aqueous byproduct effluent stream from a continuous chemical process. The tank is well mixed and drains into a river. The tank receives 2400 gallons/day of a certain byproduct that decomposes in the tank with a rate coefficient of 0.2 h^{-1}.

(a) What fraction of the byproduct from the process enters the river?

(b) At this flow rate what size tank would be required to react 99% of the byproduct before entering the river?

3.38 A 10,000-gallon holding tank receives an aqueous byproduct effluent stream from a batch chemical process. The constant-volume tank is well mixed and drains into a river. A batch process is recycled every 8 h, and in the cleanup of the reactor 1000 gallons are rapidly drained into the tank at the end of each batch run. The byproduct decomposes in the tank with a rate coefficient of 0.2 h^{-1}.

(a) What fraction of the byproduct from the process enters the river?

(b) At this discharge rate what size tank would be required to react 99% of the byproduct before entering the river?

3.39 Yogurt is produced by adding a small amount of finished yogurt to fresh milk in a process whose kinetics can be described approximately as

$$A \to 2A, \qquad r = kC_A$$

It is found that with a certain culture the concentration of product in fresh milk doubles in 8 h. If the process is started by adding 5% of finished yogurt to milk, how long is required to prepare a batch of yogurt?

3.40 A certain drug is metabolized such that its concentration in the bloodstream halves every 4 h. If the drug loses its effectiveness when the concentration falls to 10% of its initial concentration, how frequently should the drug be taken?

3.41 We wish to produce 90% conversion in the reaction

$$A \to B, \qquad r = kC_A^n$$

where $k = 1/4$ (in units of moles, liters, and minutes), $C_{Ao} = 2$ moles/liter, and n may be 1, 2, or 0.

(a) Find τ required in single PFTR and CSTR reactors for these values of n.

(b) Find the conversions in single PFTR and CSTR reactors for $\tau = 10$ min for these values of n.

3.42 We wish to produce 90% conversion in the reaction

$$A \to B, \qquad r = kC_A^n$$

where $k = 1/4$ (in units of moles, liters, and minutes), $C_{Ao} = 2$ moles/liter, and n may be 1, 2, or 0.

(a) Find τ required and the intermediate conversion in two equal-volume CSTR reactors for these values of n.

(b) From a l/r plot solve these problems graphically by counting squares.

(c) Find τ and the intermediate conversion in an equal-volume PFTR+CSTR.

(d) Find τ and the intermediate conversion in an equal-volume CSTR+PFTR.

3.43 We wish to produce 90% conversion in the reaction

$$A \to B, \qquad r = kC_A C_B$$

with $k = 1/4$ (units of minutes, moles, and liters).

(a) Find τ in the best single reactor for $C_{Ao} = 2$, $C_{Bo} = 0$.

(b) Find τ in 2 equal-volume CSTRs.

(c) Find the reactor volumes for minimum τ in two reactors in series.

3.44 We wish to produce 90% conversion in the reaction

$$A \to B, \quad r = kC_A C_B$$

with $k = 1/4$ (units of minutes, moles, and liters).

(a) Find τ and C_{Bo} where PFTR and CSTR give equal τ for $C_{Ao} = 2 - C_{Bo}$.

(b) What are the best two-reactor combination and optimum C_{Bo} for minimum τ?

3.45 The reaction $A \rightarrow B, r = kC_A$ is to be run in continuous reactors with $C_{Ao} = 3$ moles/liter, $k = 2$ min^{-1}, $v = 5$ liters/min.

(a) At 50% conversion what volume CSTR is required?

(b) At 50% conversion what volume PFTR is required?

(c) We need to produce 100 lb/h of B. If B sells for \$2.00/lb, A costs \$0.50/lb, and we discard unused A at no cost, what is the cash flow of the process?

(d) If the effluent from the CSTR is fed into a second equal-volume CSTR, what is the overall conversion?

3.46 Find an expression for C_{An} for a reaction $r = kC_A^2$ in n equal-volume CSTRs in series for n up to 4. Repeat for zeroth-order kinetics.

3.47 Calculate the total reactor volumes to process 4 liters/min to 90% conversion for the reaction $A \rightarrow B, r = kC_A^2$ for $C_{Ao} = 2$ moles/liter with $k = 0.5$ liter/mole min for

(a) a single CSTR,

(b) a single PFTR,

(c) two equal volume CSTRs in series,

(d) a CSTR followed by a PFTR, with equal volumes,

(e) a PFTR followed by a CSTR, with equal volumes.

3.48 Calculate the total reactor volumes to process 4 liters/min to 90% conversion in two equal-volume CSTRs for the reaction $A \rightarrow B, r = kC_A^n$ for $C_{Ao} = 2$ moles/liter with $k = 0.5$ (in units of moles. liters, and min) for $n = 0, 1$, and 2.

3.49 An empty CSTR of volume V is started up by filling it at flow rate v_o with aqueous reactant at C_{Ao} that undergoes the reaction $A \rightarrow B, r = kC_A$. The exit pipe from the reactor is at the top of the tank, so that no fluid flows out of the tank until the reactor is full.

(a) Describe $C_A(t)$ in the tank before it fills, assuming continuous mixing within the reactor.

(b) Describe $C_A(t)$ as the reactor approaches steady state.

3.50 Pure water is flowing through a PFTR with residence time τ, and at time $t = 0$ flow is switched to a reactant at C_{Ao} at the same flow rate with the reaction $A \rightarrow B, r = kC_A$.

(a) Find $C_A(t)$ at the reactor exit. [This problem is quite simple.]

(b) Find $C_A(t)$ in a CSTR. [This problem is quite difficult.]

3.51 We have run the reaction $A \rightarrow B$ in a 10 liter bench-scale steady-state CSTR and find that with a flow of 0.5 liter/min the concentration of A goes from 0.2 to 0.1 mole/liter.

(a) What size CSTR will be required to produce 100 lb/h of B from this feed composition at this conversion if its molecular weight is 80?

(b) Assuming a first-order irreversible reaction, what size CSTR will be required to produce 100 lb/h of B if the reaction is run to $C_A = 0.01$ mole/liter?

3.52 What rate expressions will give the following conversion versus time data (C_{Ao}, C_A, t) in a batch reactor for a reaction $A \rightleftarrows$ products?

(a) 1.0	0.5	10 min
1.0	0.25	20
1.0	0.00	1000
(b) 1.0	0.5	10
1.0	0.33	20
1.0	0.0	1000
(c) 1.0	0.9	1.0
1.0	0.5	10
1.0	0.33	20
1.0	0.25	1000
(d) 1.0	0.5	10
2.0	1.0	10
2.0	0.0	1000
(e) 1.0	1.0	10
2.0	0.66	10
2.0	0.0	1000

3.53 Find the times (in minutes) required for 10, 50, and 90% conversion for a reaction

$$A \rightarrow \text{products}$$

for the following reaction rates in a batch reactor for $C_{Ao} = 2.0$ moles/liter:

(a) $r = 2C_A$

(b) $r = 2C_A(2 - C_A)$

(c) $r = 2C_A(1.95 - C_A)$

(d) $r = 2C_A(1 - C_A)$

(e) $r = 2C_A(1 + C_A)^{-2}$

(f) $r = 2C_A(1 + C_A)^{-1}$

(g) $r = 2C_A - 0.5(C_{Ao} - C_A)$

(h) $r = 2C_A^4$

(i) $r = 2C_A^{-1}$

3.54 [Computer] Plot C_A versus t for the reaction

$$A \rightarrow \text{products}$$

for the following reaction rates in a batch reactor for $C_{Ao} = 2.0$ moles/liter:

(a) $r = 2C_A$

(b) $r = 2C_A(2 - C_A)$

(c) $r = 2C_A(1.95 - C_A)$

(d) $r = 2C_A(1 - C_A)$

(e) $r = 2C_A(1 + C_A)^{-2}$

(f) $r = 2C_A(1 + C_A)^{-1}$

(g) $r = 2C_A - 0.5(C_{Ao} - C_A)$

(h) $r = 2C_A^4$

(i) $r = 2C_A^{-1}$

(j) $r = 2C_A - 2(C_{Ao} - C_A)$

3.55 We have a reaction we know to be irreversible, and we find that it gives a conversion of 0.50 in 20 min in a batch reactor and 0.75 in 40 min.

(a) What is its order?

(b) We find that the data have error bars of 10% in that both conversions may be 10% lower or higher than those given. Are these data consistent with zeroth-order kinetics? Second-order kinetics?

3.56 We can obtain 50% conversion in an irreversible reaction which obeys first-order kinetics in a batch reactor at 50°C with a reaction time of 20 min. However, for a commercial process we will need 95% conversion in a CSTR.

(a) What must be the residence time at this temperature?

(b) We decide that the residence time in the CSTR must be no more than 30 minutes to develop an economical process. If the activation energy for the reaction is 15 kcal/mole, what reactor temperature will we have to use?

(c) The equilibrium conversion in this reaction at 50°C is found to be 99.5%, so the approximation of irreversibility is good at this temperature. Will reversibility be a problem at the operating temperature if $\Delta H_R = +20$ kcal/mole? If $\Delta H_R = -20$ kcal/mole?

3.57 Our company has a continuous stirred fermentation process that occurs in an open tank in which the liquid is kept at constant height. The tank has a catwalk around it so that the sediment can be removed. The process has been run for many years by a loyal employee who is a bit overweight and has been rumored to have a drinking problem. Yesterday he was missing for the first time in many years. The tank has a volume of 1000 liters. The data log shows that the conversion from the process decreased from its normal 75% to 72%, but the flow rate did not change.

(a) How overweight was he?

(b) If you were the design engineer for this process, you could be in considerable trouble. How should you have modified the process to avoid such accidents?

3.58 Suppose that a bimolecular reaction occurs between A and B with a very large rate coefficient and that both are in dilute solution. What will limit the rate of reaction for the following situations assuming negligible heat effects?

(a) a single phase

(b) a gas–liquid reaction with A and B in different phases

(c) a gas–solid reaction with A and B in different phases

(d) a liquid–liquid reaction with A and B in different phases

(e) a solid–solid reaction with A and B in different phases

3.59 Every hour we need 100 moles of species B to feed a batch process. We produce B in the first-order irreversible reaction $A \rightarrow B + C$ which we now run in a CSTR which gives 80% conversion with a 30-min residence time.

(a) What reactor volume and feed rate are required?

(b) We decide to produce B just as needed by switching our CSTR to transient operation in which we add A to the batch reactor every hour. Assume that the feed addition and product outflow are instantaneous to maintain constant volume and that there is no mixing between reactant and product in the addition. Calculate the reactor size, C_B, and the amount of 2 molar A that must be added each hour if we discard unreacted A.

3.60 A CSTR operates to give 90% conversion in the first-order irreversible reaction $A \rightarrow B + C$ with a residence time of 1 h and a feed rate of 200 moles/h.

(a) Calculate the conversion and the total amount of product formed if this reactor is switched to semibatch with 400 moles of fresh feed added every 30 min. Assume that feed addition and product outflow are instantaneous with no mixing of reactant and product in the addition.

(b) What size reactor would be required to produce the same amount of B as in the CSTR operating in steady state?

3.61 We need to run the reaction

$$A \rightarrow B, \qquad r = kC_A^{-1}$$

with $C_{Ao} = 2$ moles/liter and $k = 0.1$ mole^2 liter^{-2} min^{-1}.

(a) Find an expression for $\tau(C_A)$ in a CSTR.

(b) What residence time will give 20, 50, and 80% conversion in a CSTR?

(c) What residence time will give 20, 50, and 80% conversion in a PFTR?

(d) Sketch τ in a CSTR and in a PFTR for these kinetics. Show the data points from parts b–d on this graph.

3.62 A reaction $A \rightarrow$ products obeys a rate expression

$$r = kC_A/(1 + KC_A^2)$$

with $k = 0.1$ min^{-1}, $K = 10$ liters/mole, and $C_{Ao} = 4$ moles/liter.

(a) Find τ in a CSTR for 80% conversion.

(b) Find τ in a PFTR for 80% conversion.

(c) Sketch qualitatively $C_A(\tau)$ expected for a CSTR and PFTR. Put the above data points on this graph.

3.63 The reaction

$$3A \rightarrow B, \qquad r = kC_A$$

is to be carried out to 90% completion with $k = 0.5$ min^{-1} and $C_{Ao} = 2$ moles/liter with A and B ideal gases.

Find the reactor volumes required in

(a) a constant volume batch reactor,

(b) a constant pressure batch,

(c) a constant pressure PFTR,

(d) a constant pressure CSTR.

(e) Repeat these calculations if the feed has 90% diluent.

Chapter 4

MULTIPLE REACTIONS IN CONTINUOUS REACTORS

W e noted earlier that chemical engineers are seldom concerned with single-reaction systems because they can always be optimized simply by heating to increase the rate or by finding a suitable catalyst. [You don't need to hire a chemical engineer to solve the problems in Chapter 3.] Essentially all important processes involve multiple reactions where the problem is not to increase the rate but to create a reactor configuration that will maximize the production of desired products while minimizing the production of undesired ones.

In this chapter we consider the performance of isothermal batch and continuous reactors with multiple reactions. Recall that for a single reaction the single differential equation describing the mass balance for batch or PFTR was always separable and the algebraic equation for the CSTR was a simple polynomial. In contrast to single-reaction systems, the mathematics of solving for performance rapidly becomes so complex that analytical solutions are not possible. We will first consider simple multiple-reaction systems where analytical solutions are possible. Then we will discuss more complex systems where we can only obtain numerical solutions.

4.1 SOME IMPORTANT INDUSTRIAL CHEMICAL PROCESSES

Table 4.1 lists some of the important industrial reaction processes in the chemical and petroleum processing industries. Indicated are the typical times τ required for reaction, the typical pressure P used for the process, the reactor temperature T, the conversion of the major feed X, and the selectivity to the desired product X usually obtained. Most of these reactions are carried out in continuous reactors rather than batch, and many of the processes use multiple chemical reactors rather than just a single reactor.

The selectivity S is the fraction of the reactant that is converted into the desired product. Almost all practical processes involve many chemical reactions, and it is usually more difficult to obtain a high selectivity to the desired product than to obtain a high conversion of the reactant.

148

TABLE 4–1
Some Important Chemical Processes

Process	Reaction	τ	P	T	X	S	Catalyst
fermentation	sugar $\rightarrow C_2H_5OH + 2CO_2$	1 week	1 atm	35°C	100%	50%	yeast
methanol synthesis	$CO + 2H_2 \rightarrow CH_3OH$	1 min	50	250	90	99	Cu/ZnO
steam cracking	$C_2H_6 \rightarrow C_2H_4 + H_2$	1 sec	2	850	70	85	none
ethylene oxidation	$C_2H_4 + \frac{1}{2}O_2 \rightarrow C_2H_4O$	5 sec	2	280	10	60–90	Ag
FCC reactor	$C_{16} \rightarrow C_8 + C_8^=$	10 sec	2	450			zeolite
FCC regenerator	$C + O_2 \rightarrow CO_2$	1 min	2	550	100	100	none
hydrotreating	$C_{24} + H_2 \rightarrow 2C_{12}$	1 min	30				Co–Mo
steam reforming	$CH_4 + H_2O \rightarrow CO + 3H_2$	1 sec	—	250		99	Ni
NH$_3$ synthesis	$N_2 + 3H_2 \rightarrow 2NH_3$	10 sec	200	250	50	100	Fe
auto catalytic converter	$CO \rightarrow CO_2, NO \rightarrow N_2$	0.1 sec	1	400	95		Pt–Rh–Pd
maleic anhydride	$C_4H_{10} + O_2 \rightarrow C_4H_4O_3$	10 sec	2	400			VPO
NH$_3$ oxidation	$NH_3 + \frac{5}{4}O_2 \rightarrow NO + H_2O$	10^{-3} sec	10	900	90	90	Pt–Rh
HCN synthesis	$CH_4 + NH_3 + O_2 \rightarrow HCN$	10^{-3} sec	2	1100	70	70	Pt–Rh
steam reforming	$CH_4 + H_2O \rightarrow CO + 3H_2$	1 sec	—	800	90	100	Ni
cyclohexane oxidation	$c\text{-}C_6H_{12} \rightarrow$ adipic acid	1 h	1	150	5	80	Co homog.
p-xylene oxidation	xylene $\rightarrow$ terephthalic acid	20 min	1	150	80	99	
water–gas shift	$CO + H_2O \rightarrow CO_2 + H_2$	1 sec	—	250	80	100	Cu/ZnO
autothermal reform	$CH_4 + \frac{1}{2}O_2 \rightarrow CO + 2H_2$	10^{-3} sec	—	1000	95	95	Rh
formaldehyde	$CH_3OH + \frac{1}{2}O_2 \rightarrow HCHO$	10^{-2} sec	—	400	98	98	Fe–Mo

Notes: All values are very approximate. Selectivities are based on carbon atoms in the major reactant.

It is the subject of this chapter to learn how to calculate selectivities and yields in a multiple reaction process.

4.2 THE PETROCHEMICAL INDUSTRY

In Chapter 2 we sketched the processes by which petroleum is refined and polyester is made from petroleum fractions and natural gas liquids. In Chapter 3 we sketched the history of the major petrochemical companies and looked at several important single-reaction systems. In this chapter we will consider the evolution of feedstocks and intermediates in the petroleum and chemical industries.

Recall that the petroleum refining industry involves multiple reactions because crude oil contains an almost infinite number of molecules, and almost the same number of product molecules is produced by each of the reactors we described in Chapter 2. However, in a sense the petroleum refining industry is simply a sophisticated "cooking" of large molecules to crack them down to smaller molecules.

The chemical industry typically involves much more high technology but smaller reactors because one usually desires to produce a single molecule as an intermediate to make a particular product. These molecules usually can be sold for a much greater price than gasoline; so the extra value added in the petrochemical processing industry justifies the increased sophistication and cost of these reactors. The costs of separating the desired product from reactants and undesired products can dominate the economics of petrochemical processes.

Even more than the petroleum industry, the petrochemical industry defines the chemical engineering profession. The evolution of this industry has produced the changes in the country's needs for engineers and continues to evolve and employ most chemical engineers. The raw materials used in chemical synthesis have made the transitions

$$coal \rightarrow petroleum \rightarrow natural\ gas$$

and the major chemical intermediates have made the transitions

$$acetylene \rightarrow olefins \rightarrow alkanes$$

The transition from petroleum to light alkanes as direct chemical feedstocks is still in progress.

In the previous chapters we described the history of the petroleum and chemical industries through the evolution of the companies that produce these feedstocks and intermediates. In this chapter we describe the highlights of these transitions and the chemical reactors that have implemented them.

4.2.1 Coal

As with fuels, the feedstocks for chemicals in the nineteenth century were primarily derived from coal, and coal was important in the refining of ores.

Copper and bronze (Cu, Sn, and Zn) can all be recovered from their sulfide and oxide ores by simply heating to sufficiently high temperatures,

$$CuO \rightarrow Cu + \tfrac{1}{2}O_2$$

and the Copper and Bronze Ages began when our ancestors learned to make fires sufficiently hot to run these reaction processes by constructing kilns and blowing air through them. Iron oxide is too stable to be reduced by heating alone, and the Iron Age needed the discovery of carbon as a reducing agent. In the eighteenth and nineteenth centuries the iron and steel industries needed large quantities of coke, which is made by pyrolyzing coal to remove volatile components and leave nearly pure carbon (coke). This is oxidized to CO, which is used to reduce iron oxide in the reaction

$$Fe_2O_3 + 3CO \rightarrow 2Fe + 3CO_2$$

These smelting reactions are gas–solid reactions whose principles will be considered in Chapters 9 and 14.

The volatile components from coal pyrolysis are primarily small hydrocarbons and oxygen-containing molecules. By adding H_2O and limited O_2 and while heating coal, these products incorporate considerable O atoms into the volatile products to form alcohols, aldehydes, ketones, and acids. However, these products consist of many molecules that must be separated into pure components by distillation or extraction, and these separations are expensive for commodity chemical production.

Alumina (Al_2O_3) cannot be reduced to aluminum metal by any direct chemical reaction, and Al was almost an unknown metal until the twentieth century. [How is aluminum (widely used in cans, foil, and automobile and airplane parts) produced from bauxite, its oxide ore?]

Petroleum became the primary source of hydrocarbons for chemical feedstocks, beginning in about 1850 with the discovery of easily extracted crude oil in eastern Pennsylvania and in the Ural mountains of Russia. The gases from the primary distillation of crude oil and the light products from FCC, catalytic reforming, and hydroprocessing are ideal mixtures of C_2 to C_8 alkanes that can be used to make many chemicals. Petroleum products are also cleaner than those from coal, producing no ash and less sulfur.

Acetylene first became the primary intermediate molecule in chemical synthesis in the early twentieth century. As noted previously, the triple bond in acetylene is ideal to add one or two small molecules. For example, addition of one mole of water produces vinyl alcohol,

$$C_2H_2 + H_2O \rightarrow C_2H_3OH$$

and addition of two water molecules yields ethylene glycol,

$$C_2H_2 + 2H_2O \rightarrow HOC_2H_4OH$$

Addition of HCl yields vinyl chloride

$$C_2H_2 + HCl \rightarrow C_2H_3Cl$$

and addition of Cl_2 yields ethylene dichloride

$$C_2H_2 + Cl_2 \rightarrow ClC_2H_2Cl$$

Addition of a molecule across a triple bond yields a molecule that still contains a double bond and is therefore reactive. Polymerization of vinyl chloride yields polyvinyl chloride (PVC).

Acetylene can be produced by pyrolysis of coal and by fuel-rich combustion of alkanes, a process that also produces soot or carbon black, the major ingredient in automobile tires. Acetylene can also be made with nearly stoichiometric efficiency by decomposing calcium carbide, CaC, which is produced by pyrolyzing limestone and coke,

$$CaCO_3 + C \rightarrow CaC + CO_2$$

Calcium carbide is a stable solid that reacts with water to form acetylene

$$2Ca_2C + 8H_2O \rightarrow C_2H_2 + 4Ca(OH)_2 + 3H_2$$

This was first burned for miners' lamps, but acetylene was soon recognized as a valuable chemical intermediate. [Recall that Union Carbide started as a company that produced acetylene from CaC.]

Acetylene production is a fairly complex and dirty process because it involves processing solids. It was discovered in Germany in the 1920s that acetylene can also be made by pyrolyzing petroleum in an oxygen-deficient flame. Although this process also produced large amounts of soot, it was cleaner and could be scaled up to produce larger quantities of acetylene than from coal.

However, although acetylene is an excellent feedstock because two molecules can be added across the triple bond, it is intrinsically hazardous because acetylene can spontaneously decompose

$$C_2H_2 \rightarrow 2C_s + H_2, \qquad \Delta H_R^o = -54 \text{ kcal/mole}$$

a very exothermic reaction that releases heat and will explode in a confined container. This reaction can occur unpredictably in stored C_2H_2. While stabilizers that scavenge free radicals make acetylene safer to handle (tanks of acetylene used for welding torches contain liquid acetone to make it safer), acetylene is intrinsically a rather dangerous feedstock to use in large quantities.

4.3 OLEFINS

Before we develop the equations for dealing with multiple-reaction systems, we consider a very important reaction system that is the largest and most important petrochemical process outside the petroleum refinery, the olefins industry.

In the 1930s ethylene began to replace acetylene as the primary chemical intermediate, and by the end of the 1940s the transition from acetylene to ethylene as an intermediate was nearly complete. Ethylene is much safer to store and handle, and ethylene now is by far the largest-volume organic chemical produced (80 billion pounds per year worldwide).

4.3.1 Steam cracking

Olefins are now made primarily by a process called *steam cracking* in which alkanes are reacted homogeneously (cracked) at high temperature. With ethane, the overall reactions are primarily

$$C_2H_6 \rightarrow C_2H_4 + H_2 \qquad \text{ethylene}$$

$$C_2H_6 \rightarrow C_2H_2 + 2H_2 \qquad \text{acetylene}$$

$$C_2H_6 + H_2 \rightarrow 2CH_4 \qquad \text{methane}$$

$$3C_2H_6 \rightarrow C_6H_6 + 6H_2 \qquad \text{benzene}$$

$$C_2H_6 \rightarrow 2C_s + 3H_2 \qquad \text{coke}$$

while for *n*-butane we can write

$$C_4H_{10} \rightarrow C_4H_8 + H_2 \qquad \text{butylene}$$

$$C_4H_{10} \rightarrow C_4H_6 + H_2 \qquad \text{butadiene}$$

$$C_4H_{10} \rightarrow 2C_2H_4 + H_2 \qquad \text{ethylene}$$

$$C_4H_{10} \rightarrow C_3H_8 + CH_4 \qquad \text{propylene}$$

$$C_4H_{10} \rightarrow 4C_s + 5H_2 \qquad \text{coke}$$

These are just a few of the many products, and the reactions are by no means as simple as these equations imply. These reactions occur by *homogeneous free-radical chain reactions*, which we will consider later in this chapter and in Chapter 10. Free-radical reactions make many products with little control of selectivity.

Figure 4–1 shows the components in a typical steam cracking olefins plant. The reactor to accomplish these reactions must be operated at high temperature and low pressure because equilibrium in steam cracking is unfavorable at lower temperatures and higher pressures. Since steam cracking is very endothermic, extra process heat must be supplied. Recall that NH_3 and CH_3OH synthesis reactions were exothermic and produced an increasing number

Figure 4–1 Flow sheet of a steam cracking plant to produce olefins from alkanes.

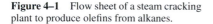

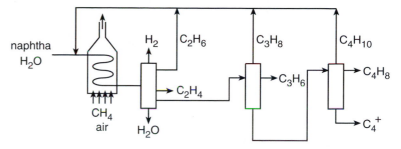

of moles. We therefore wanted to run these processes at as low a temperature as possible to attain a favorable equilibrium yield, and we needed to design a reactor with efficient cooling to avoid overheating. With an endothermic process, the strategy is reversed: We need high temperature and low pressure for maximum olefin yield.

Modern steam cracking reactors use 4-in. steel tubes 100 ft long in a tube furnace heated to ~850°C. Pressures are approximately 2 atm (sufficiently above 1 atm to force reactants through the reactor), and residence times are typically 1 sec. Water (steam) interacts very little with the hydrocarbons by homogeneous reactions, but more water than alkane is typically added to reduce coke formation.

Water reacts with carbon on the walls of the tube by the reaction

$$C_s + H_2O \rightarrow CO + H_2$$

which is called *steam gasification*. This reaction cannot completely eliminate carbon, which inevitably deposits on the reactor walls. Carbon deposition on the tube walls of this reactor eventually requires the shutdown of the reactor so air can be blown through the tubes to remove the coke in the reaction

$$C_s + O_2 \rightarrow CO_2$$

Typically, steam cracking reactors operate from 10 to 40 days between coke removal cycles.

Thus we see from these examples that almost all the important chemical reactions in the petroleum and chemical industries require the skilled processing of feedstocks to produce specific products. These industries are based on the successful handling of simultaneous reaction systems involving an almost infinite set of chemical reactions.

All companies have conducted considerable research and development on these processes, and they have also developed reactor simulation programs to predict the behavior of these complicated plants. The low cost producer will be the company that has the most efficient process. This requires optimizing the reactor, and this in turn requires understanding the reactions and their engineering.

4.4 MASS BALANCES

In this chapter we consider how we should design chemical reactors when we want to produce a specific product while converting most of the reactant and minimizing the production of undesired byproducts. It is clear that in order to design any chemical process, we need to be able to formulate and solve the species mass-balance equations in multiple-reaction systems to determine how we can convert reactants into valuable products efficiently and economically.

As developed in Chapter 2, for any multiple-reaction system we write the reactions as

$$\sum_{j=1}^{s} \nu_{ij} A_j = 0, \qquad r_i, \qquad i = 1, \ldots, R \tag{4.1}$$

where the rate of each reaction is r_i. We now write the mass balances in batch, PFTR, and CSTR just as we did for a single reaction.

For a *batch reactor* with multiple reactions the mass-balance equation on species j is

$$\frac{dN_j}{dt} = V \sum_{i=1}^{R} \nu_{ij} r_i \tag{4.2}$$

and for constant volume it becomes

$$\boxed{\frac{dC_j}{dt} = \sum_{i=1}^{R} \nu_{ij} r_i} \tag{4.3}$$

For a PFTR, the substitution $dt \rightarrow dz/u$ transforms the batch reactor equation into the PFTR mass-balance equation,

$$\boxed{u\frac{dC_j}{dz} = \sum_{i=1}^{R} \nu_{ij} r_i} \tag{4.4}$$

at constant density.

For a CSTR the transient mass-balance equation is

$$\boxed{V\frac{dC_j}{dt} = v(C_{jo} - C_j) + V \sum_{i=1}^{R} \nu_{ij} r_i} \tag{4.5}$$

and for steady state this becomes

$$\boxed{C_j - C_{jo} = \tau \sum_{i=1}^{R} \nu_{ij} r_i} \tag{4.6}$$

again ignoring any mole number changes in the reactions.

Note that for multiple-reaction systems we can simply substitute $\sum_i \nu_{ij} r_i$ for $\nu_j r$ in the mass-balance expressions for a single-reaction mass-balance equation. The difference with multiple reactions is that now we must solve R simultaneous mass-balance equations rather than the single equation we had with a single reaction.

4.5 CONVERSION, SELECTIVITY, AND YIELD

For multiple reactions we are not only interested in the conversion but also the selectivity to form a desired product and the yield of that product. In fact, selectivity is frequently much more important than conversion because we can always increase the conversion by using a larger reactor, a lower flow rate, or a higher temperature, but poor selectivity necessarily requires consumption (loss) of more reactant for a given amount of desired product, and separation of reactants and products and disposal costs increase markedly as the amount of undesired product increases.

4.5.1 Simple reaction systems

In the following sections we will consider in detail only two sets of reactions, the *parallel reaction* system $A \rightarrow B$ and $A \rightarrow C$, and the *series reaction* system $A \rightarrow B \rightarrow C$. These are prototype reactions for most reaction systems, and their characteristics show many of the features of any complex reaction system. The characteristics of these simple reactions are the following:

1. There is a single reactant A.

2. There are two products B and C and we usually want to produce B and minimize C.

3. The stoichiometric coefficients of all species are unity. These stoichiometries require that, if we feed pure A, it must react to form either B or C. Therefore, the loss of A is equal to the gain in B and C,

$$N_{Ao} - N_A = N_B + N_C \qquad (4.7)$$

or with constant density

$$C_{Ao} - C_A = C_B + C_C \qquad (4.8)$$

These relations assume that there is no B or C in the feed, $C_{Bo} = C_{Co} = 0$. With this simple stoichiometry the definitions become simpler compared to more complex sets of reactions.

4.5.2 Conversion

As before, we define conversion as

$$\boxed{\text{conversion} = \frac{\text{loss of reactant}}{\text{feed of reactant}}} \qquad (4.9)$$

For reactant A this can be written as

$$\boxed{X_A = \frac{N_{Ao} - N_A}{N_{Ao}} = \frac{F_{Ao} - F_A}{F_{Ao}}} \qquad (4.10)$$

and in a constant-density situation (V or v constant) this becomes

$$X_A = \frac{C_{Ao} - C_A}{C_{Ao}} \tag{4.11}$$

For a single reaction this was called the fractional conversion X (or X_A), a number between zero and unity, because in a single reaction there is always a single variable that describes the progress of the reaction (we used C_A or X). For multiple reactants and multiple reactions there is not always a single species common to all reactions to designate as A. However, there is frequently a *most valuable reactant* on which to base conversion. We emphasize that by conversion X_j we mean the fractional conversion of reactant species j in all reactions.

4.5.3 Selectivity

There are several common definitions of selectivity, but we will use simply the formation of the desired product (species j) divided by the formation of all products, all based on moles,

$$S_j = \frac{\text{amount of desired product } j \text{ formed}}{\text{amount of all products formed}} \tag{4.12}$$

This is a number that goes from zero to unity as the selectivity improves. We can use the number of moles N_j, chosen on some basis for each species such that we divide each N_j by its stoichiometric coefficient to normalize them. For a steady-state flow system the molar flow rates F_j are appropriate.

　　If many products are formed, we sometimes have many possible definitions such as one based on carbon products, hydrogen products, a particular functional group, etc. It is also common to base these on weights or volumes if these are measured quantities in a particular process. Another and quite different definition is to take the ratio of particular desired and undesired products, a quantity that goes from zero to infinity; this definition becomes ambiguous for other than simple reaction sets, and we will not use it here.

　　Note also that selectivity becomes more complicated in systems where densities and mole numbers change. The above definition is straightforward once a suitable basis is chosen (such as moles of reactant N_{Ao} or molar flow rate F_{Ao}), but in terms of concentrations and partial pressures one must be careful in substituting these quantities in the preceding equation.

　　If we can identify a reactant A on which to base the selectivity, then we can define the selectivity as

$$S_B = \frac{N_B}{N_{Ao} - N_A} = \frac{F_B}{F_{Ao} - F_A} \tag{4.13}$$

as long as loss of one mole of A can form one mole of B.

　　If the system is also at constant density, this can also be written

$$S_B = \frac{C_B}{C_{Ao} - C_A} \tag{4.14}$$

and, for the $A \rightarrow B \rightarrow C$ or $A \rightarrow B, A \rightarrow C$ systems, the selectivity to form B becomes

$$S_B = \frac{C_B}{C_B + C_C} \tag{4.15}$$

4.5.4 Yield

We also need to express the yield Y_j with multiple reactions. This is generally the amount of the desired product formed divided by the amount of the reactant fed,

$$\boxed{Y_j = \frac{\text{desired product } j \text{ formed}}{\text{reactant fed}}} \tag{4.16}$$

while the selectivity is the amount of the desired product formed divided by the amount of the reactant that has been reacted. For reactant A and product B with one mole of B formed per mole of A reacted, the yield is given by the expression

$$Y_B = \frac{N_B}{N_{Ao}} = \frac{F_B}{F_{Ao}} = \frac{C_B}{C_{Ao}} \tag{4.17}$$

where the last equality assumes constant density.

Note finally that the yield is always the selectivity times the conversion,

$$\boxed{Y_B = S_B X_A} \tag{4.18}$$

Since the conversion is based on a reactant species, the yield is based on a specific product and a specific reactant.

For our $A \rightarrow B \rightarrow C$ and $A \rightarrow B, A \rightarrow C$ reaction systems, species A is the reactant, species B is the desired product, and C the undesired product, and there is no change in number of moles ($-\nu_{1A} = \nu_{1B} = -\nu_{2B} = \nu_{2C} = 1$); so these expressions become particularly simple,

$$X_A = \frac{C_{Ao} - C_A}{C_{Ao}} \tag{4.19}$$

$$S_B = \frac{C_B}{C_B + C_C} = \frac{C_B}{C_{Ao} - C_A} \tag{4.20}$$

$$Y_B = X_A S_B = \frac{C_B}{C_{Ao}} \tag{4.21}$$

4.5.5 Feed recycle

We need to complicate these definitions further by noting that the above definitions are the *per pass yield* of a reactor. If unreacted A can be recovered and recycled back into the feed, then the overall yield of the reactor plus the separation system becomes the single-pass selectivity of the reactor because no unreacted A leaves the reactor system.

Note that with separation of reactants from products and recycle of reactant back to the feed, the conversion of the reactant approaches completion, $X_A \rightarrow 1$ (Figure 4–2). The

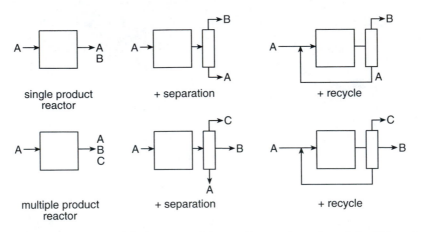

Figure 4–2   Sketches of single reactor and reactor with separation and recycle in which reactant can be fed back into the reactor to increase the overall yield of a process.

yield of a reactor with complete feed recycle therefore approaches the selectivity, $Y_B \rightarrow S_B$. Again, note that in this situation we are describing not just the reactor but the reactor plus separation combination.

   In addition to recycling reactants back into the reactor, there are several other "tricks" to keep reactants in the reactor and thus attain higher conversions and yields. Among these are formation of vapor-phase products from liquid or solid feeds and the use of membranes that pass products but retain reactants. We will discuss the integration of reactor–separation systems in Chapter 14.

4.5.6 Production rate

Another quantity we need to define is the rate of production of a given product. By this we mean the amount of the desired product produced by the reactor per unit time F_j, usually in moles/time. If species A is the key reactant and species B is the desired product, then

$$\boxed{F_B = Y_B F_{Ao} = S_B X_A F_{Ao}}\qquad (4.22)$$

   We frequently want to know the production rate of a process in mass per time of a product (tons/day, pounds/year, etc.), and these are obtained simply by multiplying the molar production rate F_B by the molecular weight of that species.

4.5.7 Profit

The final (and ultimately the most important) quantity we need to specify in any chemical reactor is the profit from a process. The cost (\$/time) of reactant A is $\$_A F_{Ao}$, while the price of product B is $\$_B F_B$, where $\$_j$ is the price per mole of species j, such as listed in prices taken from the Chemical Marketing Reporter in Chapter 2.

The rate of profit from a process is

$$\text{profit} = \sum_{\text{products}} \$_j F_j - \sum_{\text{reactants}} \$_j F_{jo} - \text{fixed costs} - \text{operating costs} \qquad (4.23)$$

Detailed economic considerations are the subject of process design courses, where you will learn about net present value, cash flow, and rate of return. Here we are concerned only with technical aspects of design such as achieving maximum production of a particular product with minimum reactor volume or designing a reactor with minimum heat input required. The costs associated with these options and separation costs before and after the reactor obviously must be included to achieve maximum profit in a chemical process.

4.6 COMPLEX REACTION NETWORKS

The definitions in the previous section are simple for simple stoichiometry, but they become more complicated for complex reaction networks. In fact, one frequently does not know the reactions or the kinetics by which reactants decompose and particular product form. The stoichiometric coefficients (the ν_{ij}) in Section 4.4 are complicated to write in general, but they are usually easy to figure out for given reaction stoichiometry.

Consider the reactions

$$A \rightarrow 2B + C \qquad (4.24)$$

$$2A \rightarrow C + D \qquad (4.25)$$

$$A + 2C \rightarrow E \qquad (4.26)$$

For this reaction network we can define selectivities of B and D simply as

$$S_B = \frac{\frac{1}{2}C_B}{C_{Ao} - C_A} \qquad (4.27)$$

$$S_D = \frac{C_D}{\frac{1}{2}(C_{Ao} - C_A)} \qquad (4.28)$$

because A is a reactant in all reactions; however, species C is formed in the first two reactions and lost in the third; so the definition of S_C is ambiguous but would probably be defined as

$$S_C = \frac{C_C}{C_{Ao} - C_A} \qquad (4.29)$$

although S_C could perhaps be negative if C were fed into the reactor and the third reaction dominated.

A simple example of an important reaction network mentioned is the oxidation of methane to make syngas,

$$CH_4 + \frac{1}{2}O_2 \rightarrow CO + 2H_2 \qquad (4.30)$$

The major reactions might be

$$CH_4 + 2O_2 \rightarrow CO_2 + H_2O \qquad \text{combustion} \qquad (4.31)$$

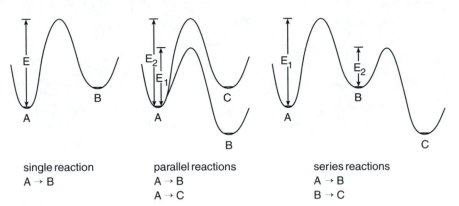

single reaction
A → B

parallel reactions
A → B
A → C

series reactions
A → B
B → C

Figure 4–3 Energy diagrams for a single reaction (left) and for parallel reactions (center) and series reactions (right). For parallel and series reactions there are several activation energies that must be considered in describing selectivity and yield of a desired product.

$$CO + H_2O \rightarrow CO_2 + H_2 \qquad \text{water gas shift} \qquad (4.32)$$

$$CH_4 + H_2O \rightarrow CO + 3H_2 \qquad \text{steam reforming} \qquad (4.33)$$

$$CH_4 + CO_2 \rightarrow 2CO + 2H_2 \qquad \text{CO}_2 \text{ reforming} \qquad (4.34)$$

We can write these as

$$A + 2B \rightarrow E + F \qquad (4.35)$$

$$C + F \rightarrow E + D \qquad (4.36)$$

$$A + F \rightarrow C + 3D \qquad (4.37)$$

$$A + E \rightarrow 2C + 2D \qquad (4.38)$$

as reaction steps in the overall desired reaction

$$A + \tfrac{1}{2}B \rightarrow C + 2D \qquad (4.39)$$

It is seen that A and B are written as reactants only, but all reactions are in fact reversible; so all species in this reaction network are both products and reactants.

We could write species mass-balance equations ($S = 6$ in this example) on any such reaction sequence and solve these ($R = 4$ are inseparable) to find $C_j(\tau)$, and in most practical examples we must do this. However, there are two simple reaction networks that provide insight into these more complex networks, and we will next consider them, namely, series and parallel reaction networks (Figure 4–3).

4.7 SERIES REACTIONS

The simplest sets of reactions involve series or parallel first-order irreversible reactions. We will first consider these cases because they have simple analytical solutions and are useful prototypes of more complicated reaction sets. These can be considered in the energy diagrams similar to those we discussed in the previous chapter for single reactions.

The reactants must cross energy barriers to form products. For series reactions an intermediate is formed that can then further react, while for parallel reactions the reactant can react by alternate pathways. The activation energy of the first reaction is E_1, while the activation energy of the second reaction is E_2. A smaller energy barrier can be expected to result in a higher reaction rate, although different preexponential factors can also affect rates.

4.7.1 Irreversible first-order reactions in a PFTR

For the reaction system $A \rightarrow B \rightarrow C$ we write the reactions in the standard form

$$A \rightarrow B, \qquad r_1 = k_1 C_A \tag{4.40}$$

$$B \rightarrow C, \qquad r_2 = k_2 C_B \tag{4.41}$$

For a batch reactor or PFTR the mass balances on A, B, and C are

$$\frac{dC_A}{d\tau} = -r_1 = -k_1 C_A \tag{4.42}$$

$$\frac{dC_B}{d\tau} = +r_1 - r_2 = +k_1 C_A - k_2 C_B \tag{4.43}$$

$$\frac{dC_C}{d\tau} = +r_2 = +k_2 C_B \tag{4.44}$$

These come from simple application of the mass-balance equation $dC_j/d\tau = \sum_i \nu_{ij} r_i$, which you should verify. Differential equations always need initial or boundary conditions, and for the batch reactor these are the initial concentrations of A, B, and C. For this system, the feed may be expected to be pure A, so $C_A = C_{Ao}$ and $C_{Bo} = C_{Co} = 0$.

We need to solve these equations to obtain $C_A(\tau)$, $C_B(\tau)$, and $C_C(\tau)$. It can be seen by inspection that the first equation is independent of the others, so it can be solved by separating variables to yield

$$\boxed{C_A(\tau) = C_{Ao} e^{-k_1 \tau}} \tag{4.45}$$

We next solve the second equation by substituting $C_A(\tau)$ to obtain

$$\frac{dC_B}{d\tau} = +k_1 C_A - k_2 C_B \tag{4.46}$$

$$\frac{dC_B}{d\tau} = +k_1 C_{Ao} e^{-k_1 \tau} - k_2 C_B \tag{4.47}$$

This is a function of C_B and τ only, but it is *not separable*. However, it can be written as

$$\frac{dC_B}{d\tau} + k_2 C_B = k_1 C_{Ao} e^{-k_1 \tau} \tag{4.48}$$

This equation is *linear* in C_B.

The *linear first-order ordinary differential equation* has the standard form of

$$\frac{dy}{dx} + p(x)y = q(x) \tag{4.49}$$

which has an integrating factor

$$P(x) = \int_{x_0}^{x} p(x')\, dx' \tag{4.50}$$

to yield the solution

$$y(x) = y_0 \exp[-P(x)] + \int_{x_0}^{x} q(x')\, \exp\{-[P(x) - P(x')]\}\, dx' \tag{4.51}$$

Our equation is a simple form of this expression because $P(x) = k_2\tau$, and the solution becomes

$$\boxed{C_B(\tau) = C_{Ao}\frac{k_1}{k_2 - k_1}(e^{-k_1\tau} - e^{-k_2\tau})} \tag{4.52}$$

We could substitute this $C_B(\tau)$ into the differential equation for C_C and separate and solve. However, this is not necessary since the concentrations are related by

$$C_{Ao} - C_A = C_B + C_C \tag{4.53}$$

to yield

$$\begin{aligned}
C_C(\tau) &= C_{Ao} - C_A - C_B \\[4pt]
&= C_{Ao} - C_{Ao}e^{-k_1\tau} - C_{Ao}\frac{k_1}{k_2 - k_1}(e^{-k_1\tau} - e^{-k_2\tau}) \\[4pt]
&= C_{Ao}\left(1 - e^{-k_1\tau} - \frac{k_1}{k_2 - k_1}(e^{-k_1\tau} - e^{-k_2\tau})\right)
\end{aligned} \tag{4.54}$$

$$\boxed{C_C(\tau) = C_{Ao}\left(1 - \frac{k_2}{k_2 - k_1}e^{-k_1\tau} + \frac{k_1}{k_2 - k_1}e^{-k_2\tau}\right)} \tag{4.55}$$

These expressions give the concentrations C_A, C_B, and C_C versus residence time τ. These are plotted in Figure 4–4 for different values of k_2/k_1. It is convenient to plot a

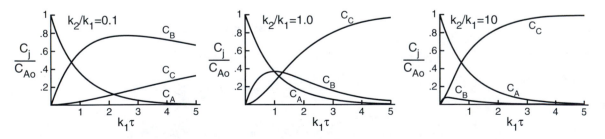

Figure 4–4 C_A, C_B, and C_C versus $k_1\tau$ for series reactions $A \to B \to C$ in a PFTR. Plots shown are for $k_2/k_1 = 0.1$, 1.0, and 10. The yield to form B exhibits a maximum at a particular residence time τ_{max}.

dimensionless time $k_1\tau$ because this gives a single curve for C_A for any k_1, and the only variable in the plot is k_2/k_1.

For $k_1 = k_2$ the preceding expressions cannot be used because they become indeterminate (zeros in numerator and denominator). We will leave the derivation of the relevant equations as a homework problem.

When $k_2/k_1 \rightarrow \infty$, then C_B remains small, and the reaction system $A \rightarrow B \rightarrow C$ actually looks like the single reaction $A \rightarrow C$, $r = k_1 C_A$, and there is no evidence of the intermediate B in the reaction because it immediately reacts to form C. We will discuss series reactions with large rate coefficients later in this chapter in connection with mechnisms of reactions.

4.7.2 Reversible series reactions in a PFTR

Consider next the reaction system

$$A \rightleftarrows B, \qquad r_1 = k_{f1}C_A - k_{b1}C_B \tag{4.56}$$

$$B \rightarrow C, \qquad r_2 = k_2 C_B \tag{4.57}$$

These are almost identical to the irreversible series reactions solved previously except that now the first reaction is assumed to be *reversible*. The mass-balance equations for A, B, and C are

$$\frac{dC_A}{d\tau} = -k_{f1}C_A + k_{b1}C_B \tag{4.58}$$

$$\frac{dC_B}{d\tau} = +k_{f1}C_A - k_{b1}C_B - k_2 C_B = +k_{f1}C_A - (k_{b1} + k_2)C_B \tag{4.59}$$

$$\frac{dC_C}{d\tau} = +k_2 C_B \tag{4.60}$$

The first two equations are functions of C_A and C_B only, but now the first equation is coupled to the second because it contains C_B and the second contains C_A. Thus we must solve the two equations simultaneously. This is generally a more complicated problem than for a single first-order differential equation. However, note that both of these equations are *linear* in C_A and C_B.

Elementary courses in differential equations show that the equations

$$\frac{dx}{dt} = \alpha x + \beta y \tag{4.61}$$

$$\frac{dy}{dt} = \gamma x + \delta y \tag{4.62}$$

can be solved by differentiating the first equation with respect to t again and substituting for y and its derivative to yield the single second-order differential equation

$$\frac{d^2 x}{dt^2} - (\alpha + \delta)\frac{dx}{dt} + (\alpha\delta - \beta\gamma)x = 0 \tag{4.63}$$

This is a second-order differential equation with constant coefficients. A solution of the form

$$x(t) = C_1 e^{\lambda_1 t} + C_2 e^{\lambda_2 \tau} \tag{4.64}$$

satisfies this differential equation with coefficients C_1 and C_2, which must be determined from initial conditions. The coefficients λ_1 and λ_2, called the *eigenvalues* of the equation, are the solutions of the quadratic of the coefficients

$$\lambda^2 - (\alpha + \delta)\lambda + (\alpha\delta - \beta\gamma) = 0 \tag{4.65}$$

which has two roots

$$\lambda_1, \lambda_2 = \tfrac{1}{2}\{(\alpha + \delta) \pm [(\alpha - \delta)^2 + 4\beta\gamma]^{1/2}\} \tag{4.66}$$

[These derivations are simple applications of material that was covered in lower-division math courses, although you may not remember it (or want to remember it). We have derived everything you will need from this material so that you can solve any problems like these without referring to your math text.]

Returning to our problem, we let $C_A = x$, $C_B = y$, $-k_{f1} = \alpha$, $k_{b1} = \beta$, $k_{f1} = \gamma$, and $\delta = -k_{b1} - k_2$. Thus the general solution becomes

$$C_A = C_1 e^{\lambda_1 t} + C_2 e^{\lambda_2 t} \tag{4.67}$$

where

$$\lambda_1, \lambda_2 = \tfrac{1}{2}\{-(k_{f1} + k_{b1} + k_2) \pm [(k_{f1} + k_{b1} + k_2)^2 - 4k_{f1}k_2]^{1/2}\} \tag{4.68}$$

Initial conditions are $C_A = C_{Ao}$ and $C_B = C_C = 0$ at $\tau = 0$. Also at $\tau = 0$, $dC_A/d\tau = -k_{f1}C_{Ao}$

Substitution and elimination of the coefficients C_1 and C_2 yields

$$C_A = \frac{C_{Ao}}{\lambda_2 - \lambda_1}[(\lambda_2 + k_{f1})e^{\lambda_1 t} - (\lambda_1 + k_{f1})e^{\lambda_2 t}] \tag{4.69}$$

$$C_B = \frac{C_{Ao}k_{f1}}{\lambda_2 - \lambda_1}(e^{\lambda_2 t} - e^{\lambda_1 t}) \tag{4.70}$$

$$C_C = C_{Ao} - C_A - C_B = \frac{C_{Ao}}{\lambda_2 - \lambda_1}(1 - e^{\lambda_1 t} - e^{\lambda_2 t}) \tag{4.71}$$

assuming $C_{Bo} = C_{Co} = 0$. Note that if we set $k_{b1} = 0$, we obtain the solution we derived for the irreversible series reactions.

We can next solve the series reactions where both reactions are reversible

$$\frac{dC_A}{d\tau} = -k_{f1}C_A + k_{b1}C_B \tag{4.72}$$

$$\frac{dC_B}{d\tau} = +k_{f1}C_A - (k_{b1} + k_{f2})C_B + k_{b2}C_C \tag{4.73}$$

$$\frac{dC_C}{d\tau} = +k_{f2}C_B - k_{b2}C_C \tag{4.74}$$

Now we see that we appear to have three coupled first-order ordinary differential equations with C_C apparently coupled in the second and third equations. However, we can eliminate C_C by the stoichiometric relation $C_C = C_{Ao} - C_A - C_B$ to form two simultaneous linear equations, which can be solved as above.

For zeroth-order reaction steps, we still have linear equations, but for other orders of kinetics we have *nonlinear* simultaneous equations, which generally have no closed-form analytical solutions. We must solve these sets of equations *numerically* to find $C_A(\tau)$, $C_B(\tau)$, and $C_C(\tau)$.

4.7.3 Series reactions in a batch reactor

We do not need to repeat these derivations for the batch reactor, because all expressions derived are simply translated from PFTR to batch by replacing the residence time τ by the reaction time t, although these times have quite different meanings. Students have probably worked out the previous derivations for the batch reactor in courses in chemical kinetics.

4.7.4 Series reactions in a CSTR

For the reaction system $A \rightarrow B \rightarrow C$ we again write the reactions in our standard form

$$A \rightarrow B, \qquad r_1 = k_1 C_A \tag{4.75}$$

$$B \rightarrow C, \qquad r_2 = k_2 C_B \tag{4.76}$$

For the CSTR the mass balances on A, B, and C are

$$C_{Ao} - C_A = \tau k_1 C_A \tag{4.77}$$

$$C_B - C_{Bo} = \tau (k_1 C_A - k_2 C_B) \tag{4.78}$$

$$C_C - C_{Co} = +\tau k_2 C_B \tag{4.79}$$

assuming $C_{Bo} = C_{Co} = 0$. These expressions come from simple application of the CSTR mass-balance equations $C_j - C_{jo} = \sum_i \nu_{ij} r_i$ for species A, B, and C. These algebraic equations can be simply solved to yield

$$\boxed{C_A = \frac{C_{Ao}}{1 + k_1 \tau}} \tag{4.80}$$

$$\boxed{C_B = \frac{k_1 C_{Ao} \tau}{(1 + k_1 \tau)(1 + k_2 \tau)}} \tag{4.81}$$

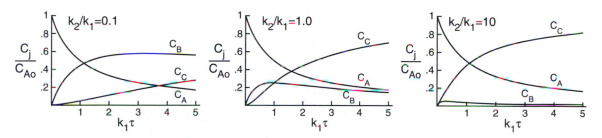

Figure 4–5 C_A, C_B, and C_C versus $k_1 \tau$ for series reactions $A \rightarrow B \rightarrow C$ in a CSTR. Plots shown are for $k_2/k_1 = 0.1$, 1.0, and 10. The yield to form B exhibits a maximum at a particular residence time τ_{max}, but this maximum is lower than in a PFTR.

$$C_C = \frac{k_1 k_2 C_{Ao} \tau^2}{(1 + k_1 \tau)(1 + k_2 \tau)} \qquad (4.82)$$

These solutions are plotted in Figure 4–5 for the CSTR. These curves exhibit a monotonic decrease for C_A, a maximum for C_B, and a monotonic increase for C_C, just as in a PFTR.

Example 4–1   We wish to produce B in the reaction $A \rightarrow B$ in a continuous reactor at $v = 4$ liter/min with $C_{Ao} = 2$ moles/liter. However, we find that there is a second reaction $B \rightarrow C$ that can also occur. We find that both reactions are first order and irreversible with $k_1 = 0.5$ min^{-1} and $k_2 = 0.1$ min^{-1}. Find τ, V, C_B, S_B, and Y_B for 90% conversion of A.

This is the same example $A \rightarrow B$ of the previous chapter, but now there is a second reaction $B \rightarrow C$, so these are *series* reactions.

PFTR. Application of this equation yields

$$\tau = \frac{1}{k_1} \ln \frac{C_{Ao}}{C_A} = \frac{1}{0.5} \ln 10 = 4.61 \text{ min}$$

so that

$$V = v\tau = 4 \times 4.61 = 18.4 \text{ liters}$$

Therefore

$$C_B(\tau) = C_{Ao} \frac{k_1}{k_2 - k_1} (e^{-k_1 \tau} - e^{-k_2 \tau})$$

$$= 2 \times \frac{0.5}{-0.4} \times (e^{-0.5 \times 4.61} - e^{-0.1 \times 4.61}) = 1.33 \text{ moles/liter}$$

so that

$$S_B = \frac{C_B}{C_{Ao} - C_A} = \frac{1.33}{1.8} = 74\%$$

$$Y_B = \frac{C_B}{C_{Ao}} = \frac{1.33}{2} = 66\%$$

CSTR. Application of these equations yields

$$\tau = \frac{C_{Ao} - C_A}{(k_1) C_A} = \frac{2.0 - 0.2}{0.5 \times 0.2} = 18 \text{ min}$$

so that

$$V = 4 \times 18 = 72 \text{ liters}$$

Substitution in the other equations yields

$$C_B = \frac{k_1 C_{Ao} \tau}{(1 + k_1 \tau)(1 + k_2 \tau)} = \frac{0.5 \times 2 \times 18}{(1 + 9)(1 + 1.8)} = 0.64$$

Therefore $S_B = 0.64/1.8 = 36\%$ and $Y_B = 0.64/2.0 = 32\%$.

As expected, the CSTR requires a longer residence time for this conversion. It also gives much lower selectivity and yield to the intermediate product because we have chosen a residence time far too long where much of the B formed has reacted on to form C.

4.7.5 Maximum yield of B in PFTR

The selectivity of formation of B in this reaction system is defined as

$$S_B = \frac{C_B}{C_B + C_C} = \frac{C_B}{C_{Ao} - C_A} \tag{4.83}$$

and the yield of B is

$$Y_B = \frac{C_B}{C_{Ao}} \tag{4.84}$$

It is clear from the plots of Figure 4–5 that C_B increases to a maximum and then decreases as τ increases. Thus, if species B is desired, there is an optimum residence time to maximize C_B in a PFTR. We can find this time and the maximum value of C_B, $C_{B\,max}$, by differentiating $C_B(\tau)$ with respect to τ and setting the derivative equal to zero,

$$\frac{dC_B}{d\tau} = 0 = \frac{k_1 C_{Ao}}{k_2 - k_1}(-k_1 e^{-k_1 \tau} + k_2 e^{-k_2 \tau}) \tag{4.85}$$

[This seems like we have just differentiated the equation we previously integrated, but we are now trying to find the maximum of the integrated function.]

This algebraic equation can be solved to yield

$$k_1 e^{-k_1 \tau} = k_2 e^{-k_2 \tau} \tag{4.86}$$

or

$$\boxed{\tau_{max} = \frac{ln\,(k_2/k_1)}{k_2 - k_1}} \tag{4.87}$$

Substituting this residence time back into the expression for C_B yields

$$C_{B\,max} = C_{Ao}\left(\frac{k_1}{k_2}\right)^{k_2/(k_2-k_1)} \tag{4.88}$$

so that the maximum yield is

$$\boxed{Y_{B\,max} = \frac{C_{B\,max}}{C_{Ao}} = \left(\frac{k_1}{k_2}\right)^{k_2/(k_2-k_1)}} \tag{4.89}$$

The maximum value of the selectivity S_B is 1 at a conversion of zero, because initially no C is being formed, but the conversion of A and the yield of B are zero. If C is desired, then the reactor should be operated at long residence time to attain complete conversion.

4.7.6 Maximum yield of B in series reactions in CSTR

We can calculate the value of τ_{max} and $C_{B\,max}$ just as we did previously for the PFTR and batch reactors, by setting $dC_B/d\tau = 0$ and solving for C_B and τ,

$$\frac{dC_B}{d\tau} = 0 = \frac{k_1 C_{Ao}}{(1 + k_1\tau)(1 + k_2\tau)} - \frac{k_1^2 C_{Ao}\tau}{(1 + k_1\tau)^2(1 + k_2\tau)} - \frac{k_1 k_2 C_{Ao}\tau}{(1 + k_1\tau)(1 + k_2\tau)^2}$$

$$= \frac{k_1 C_{Ao}}{(1 + k_1\tau)(1 + k_2\tau)} \left(1 - \frac{k_1\tau}{(1 + k_1\tau)} - \frac{k_2\tau}{(1 + k_2\tau)}\right) = 0 \qquad (4.90)$$

The terms in the large parentheses must be equal to zero, and after clearing of fractions this yields the surprisingly simple expression

$$k_1 k_2 \tau_{max}^2 = 1 \qquad (4.91)$$

or

$$\boxed{\tau_{max} = \frac{1}{(k_1 k_2)^{1/2}}} \qquad (4.92)$$

Substitution back into the expression for C_B yields

$$C_{B\,max} = \frac{k_1 C_{Ao}}{(k_1^{1/2} + k_2^{1/2})^2} \qquad (4.93)$$

so the maximum yield of B is given by

$$\boxed{Y_{B\,max} = \frac{C_{B\,max}}{C_{Ao}} = \frac{k_1}{(k_1^{1/2} + k_2^{1/2})^2}} \qquad (4.94)$$

[Note how much simpler this problem is to solve in the CSTR than in the PFTR (Figure 4–6), where we had to solve simultaneous differential equations. The CSTR involves only simultaneous algebraic equations; so we just need to find roots of polynomials.]

It can be easily shown that the maximum of C_B is greater in the PFTR than in the CSTR. Here we see an important principle. In any series process with positive-order kinetics where we want maximum selectivity to an intermediate, the PFTR will give a higher selectivity with the same kinetics than will the CSTR. We can see the reason for this quite simply. In the PFTR or batch reactor there is only A initially; so initially only B is being created. Only as B builds up does it react further to form C. In the CSTR at a particular τ there is only a single composition of A, B, and C; so B is always reacting to form C, and therefore B does not build up to as high a maximum as in the PFTR or batch. [Does this argument make sense?]

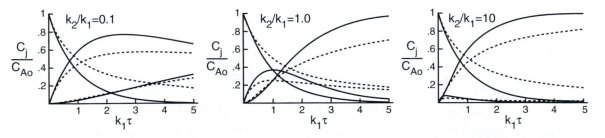

Figure 4–6   Comparison of concentrations in the reactions $A \rightarrow B \rightarrow C$ in a PFTR (solid curves) and CSTR (dashed curves). The PFTR is seen to give a larger Y_B and shorter τ for maximum yield of B.

Example 4–2   We wish to produce B in the reaction $A \to B$ in a continuous reactor at $v = 4$ liters/min with $C_{Ao} = 2$ moles/liter. However, we find that there is a second reaction $B \to C$ that can also occur. We find that both reactions are first order and irreversible with $k_1 = 0.5$ min^{-1} and $k_2 = 0.1$ min^{-1}. Find τ, V, S_B, and Y_B for *maximum yield* of B.

This is the same example $A \to B$ of the previous example, but now we want to choose τ for a maximum yield of B.

PFTR. Application of the equation given yields

$$\tau_{max} = \frac{\ln (k_2/k_1)}{k_2 - k_1} = \frac{\ln 5}{0.4} = 4.0 \text{ min}$$

$$C_{B\,max} = C_{Ao} \left(\frac{k_1}{k_2}\right)^{k_2/(k_2-k_1)} = 2 \times 5^{-1/4} = 1.34 \text{ moles/liter}$$

$$C_A = C_{Ao} e^{-k_1 \tau} = 2e^{-2} = 0.27 \text{ mole/liter}$$

so that the maximum yield is

$$Y_{B\,max} = \frac{C_{B\,max}}{C_{Ao}} = \left(\frac{k_1}{k_2}\right)^{k_2/(k_2-k_1)} = 67\%$$

The other quantities are

$$V = v\tau = 4 \times 4 = 16 \text{ liters}$$

$$S_B = \frac{C_B}{C_{Ao} - C_A} = \frac{1.34}{2.0 - 0.27} = 77\%$$

$$Y_B = \frac{C_B}{C_{Ao}} = \frac{1.33}{2} = 67\%$$

CSTR. Application of these equations yields

$$\tau_{max} = \frac{1}{(k_1 k_2)^{1/2}} = \frac{1}{0.05^{1/2}} = 4.47 \text{ min}$$

$$C_{B\,max} = \frac{k_1 C_{Ao}}{(k_1^{1/2} + k_2^{1/2})^2}$$

$$= \frac{0.5 \times 2}{(0.5^{1/2} + 0.1^{1/2})^2} = 0.95 \text{ mole/liter}$$

$$C_{A\,max} = \frac{C_{Ao}}{1 + k_1 \tau} = \frac{2}{1 + 2.23} = 0.62 \text{ mole/liter}$$

Therefore, we obtain

$$V = 4 \times 4.47 = 17.9 \text{ min}$$

$$Y_B = \frac{C_{B\,max}}{C_{Ao}} = \frac{0.955}{2.0} = 48\%$$

$$S_B = \frac{.955}{4 - .955} = 69\%$$

As expected, the CSTR requires a longer residence time for maximum yield of B. It also gives lower maximum selectivity and yield to the intermediate product than a PFTR.

4.7.7 Series reactions with other kinetics

We have solved $A \to B \to C$ for first-order irreversible kinetics, and for more complex kinetics the procedure is identical except that the expressions become more difficult to solve except for zeroth-order kinetics.

To see selectivities for other rate expressions, let us consider the same reactions with zeroth-order kinetics.

$$A \to B, \qquad r_1 = k_1 \tag{4.95}$$

$$B \to C, \qquad r_2 = k_2 \tag{4.96}$$

For a PFTR or batch reactor the mass balances on A, B, and C are

$$\frac{dC_A}{dt} = -k_1, \qquad C_A > 0 \tag{4.97}$$

$$\frac{dC_A}{dt} = 0, \qquad C_A = 0 \tag{4.98}$$

$$\frac{dC_B}{dt} = +k_1 - k_2, \qquad C_A > 0, C_B > 0 \tag{4.99}$$

$$\frac{dC_B}{dt} = -k_2, \qquad C_A = 0, C_B > 0 \tag{4.100}$$

$$\frac{dC_B}{dt} = 0, \qquad C_B = 0 \tag{4.101}$$

$$\frac{dC_C}{dt} = +k_2, \qquad C_B > 0 \tag{4.102}$$

These equations can be simply integrated to yield

$$C_{Ao} - C_A = \tau k_1, \qquad C_A > 0 \tag{4.103}$$

$$C_B = \tau(k_1 - k_2), \qquad C_A > 0 \tag{4.104}$$

$$C_B = C_{B,C_A \to 0} - \tau k_2, \qquad C_A = 0 \tag{4.105}$$

$$C_C = +\tau k_2, \qquad C_B > 0 \tag{4.106}$$

Note that with zeroth-order kinetics the rate of formation of a species goes to zero when the concentration of the reactant forming it goes to zero. We write $C_{B,C_A \to 0}$ to indicate the value of C_B when C_A goes to zero because after that time no more B is being formed. You should plot these functions and compare them with curves for first-order kinetics.

For the CSTR the mass-balance equations for these kinetics for $C_{Bo} = C_{Co} = 0$ are

$$C_{Ao} - C_A = \tau k_1, \qquad C_A > 0 \tag{4.107}$$

$$C_B = \tau(k_1 - k_2), \qquad C_A > 0, C_B > 0 \tag{4.108}$$

$$C_B = C_{B,C_A \to 0} - \tau k_2, \qquad C_A = 0 \tag{4.109}$$

$$C_C = +\tau k_2, \qquad C_B > 0 \tag{4.110}$$

These are immediately explicit expressions for C_A, C_B, and C_C. However, note that, in contrast to first-order kinetics, they are *identical* to the expressions we obtained for the PFTR, as expected for any zeroth-order processes.

We can now begin to formulate the general principles of the choice of PFTR and CSTR for *optimum yield* of an intermediate in irreversible series reactions.

1. There is an *optimum residence time* in a continuous reactor or an optimum reaction time in a batch reactor to maximize yield of an intermediate.

2. The PFTR will always give a higher maximum yield of an intermediate if all reactions obey positive-order kinetics.

For zeroth-order kinetics the maximum selectivities are identical, and for negative-order kinetics the CSTR will give a higher maximum selectivity. [What type of reactor will be better if one reaction is positive order and the other negative order?]

4.8 PARALLEL REACTIONS

4.8.1 Irreversible first-order reactions in a PFTR

Here we consider the reaction set $A \to B$, $A \to C$, which we write in our standard form

$$A \to B, \qquad r_1 = k_1 C_A \tag{4.111}$$

$$A \to C, \qquad r_2 = k_2 C_A \tag{4.112}$$

These look similar to series reactions, but the solution is quite different. In a PFTR or batch reactor the mass-balance equations are

$$\frac{dC_A}{d\tau} = -k_1 C_A - k_2 C_A \tag{4.113}$$

$$\frac{dC_B}{d\tau} = +k_1 C_A \tag{4.114}$$

$$\frac{dC_C}{d\tau} = +k_2 C_A \tag{4.115}$$

The first is uncoupled and separable to yield

$$\boxed{C_A(\tau) = C_{Ao} e^{-(k_1+k_2)\tau}} \tag{4.116}$$

Thus $C_A(\tau)$ can be substituted into the second equation

$$\frac{dC_B}{d\tau} = +k_1 C_{Ao} e^{-(k_1+k_2)\tau} \tag{4.117}$$

which can be separated and integrated to yield

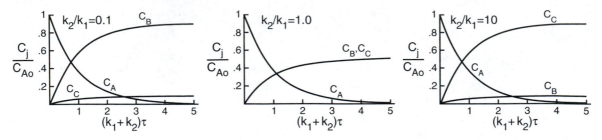

Figure 4–7 C_A, C_B, and C_C versus $(k_1 + k_2)\tau$ for parallel reactions $A \to B$, $A \to C$ in a PFTR. Plots shown are for $k_2/k_1 = 0.1$, 1.0, and 10. The selectivity to B and yield of B are independent of conversion for equal-order kinetics.

$$C_B(\tau) = C_{Ao} \frac{k_1}{k_1 + k_2}(1 - e^{-(k_1+k_2)\tau})$$
(4.118)

Similarly, the third equation can be integrated to yield

$$C_C(\tau) = C_{Ao} \frac{k_2}{k_1 + k_2}(1 - e^{-(k_1+k_2)\tau})$$
(4.119)

when $C_{Bo} = C_{Co} = 0$

These are plotted in Figure 4–7 versus a dimensionless time $(k_1 + k_2)\tau$ for different values of k_2/k_1. Note here that it is convenient to plot the dimensionless time as $(k_1 + k_2)\tau$, which gives a single curve for C_A, irrespective of k_1 and k_2.

The selectivity to form B is defined as

$$S_B = \frac{C_B}{C_B + C_C}$$
(4.120)

and substituting for these concentrations we obtain

$$S_B = \frac{k_1}{k_1 + k_2}$$
(4.121)

so that S_B is independent of conversion if both reactions are first order.

4.8.2   Parallel reactions in a batch reactor

We do not need to repeat the derivation of $A \to B \to C$ for the batch reactor because the transformation $d\tau = dz/u \to t$ immediately transform the results predicted for the PFTR to the batch reactor.

4.8.3   Parallel reactions in a CSTR

Next we consider parallel first-order irreversible reactions in a CSTR. The mass-balance equations with $C_{Bo} = C_{Co}$ are

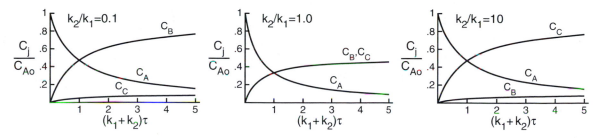

Figure 4–8 C_A, C_B, and C_C versus $(k_1 + k_2)\tau$ for parallel reactions $A \to B$, $A \to C$ in a CSTR. Plots shown are for $k_2/k_1 = 0.1$, 1.0, and 10. The selectivity to B and yield of B are independent of conversion for equal-order kinetics. The selectivities and yields are the same for both PFTR and CSTR, but the CSTR requires a longer τ for a given conversion.

$$C_{Ao} - C_A = \tau(k_1 + k_2)C_A \tag{4.122}$$

$$C_B = \tau k_1 C_A \tag{4.123}$$

$$C_C = \tau k_2 C_A \tag{4.124}$$

Since there are two reactions, only two of these equations are independent. In fact, the first is a function of C_A alone and can be solved to yield

$$\tau = \frac{C_{Ao} - C_A}{(k_1 + k_2)C_A} \tag{4.125}$$

or, solving for C_A, we obtain

$$\boxed{C_A = \frac{C_{Ao}}{1 + (k_1 + k_2)\tau}} \tag{4.126}$$

Substitution in the other equations yields

$$\boxed{C_B = \frac{\tau k_1 C_{Ao}}{1 + (k_1 + k_2)\tau}} \tag{4.127}$$

$$\boxed{C_C = \frac{\tau k_2 C_{Ao}}{1 + (k_1 + k_2)\tau}} \tag{4.128}$$

These concentrations versus τ are plotted in Figure 4–8. The selectivity to B is

$$S_B = \frac{C_B}{C_B + C_C} = \frac{k_1}{k_1 + k_2} \tag{4.129}$$

which is identical to the expression for the PFTR. Thus we see that for parallel irreversible first-order reactions the residence time required for a given conversion is smaller in a PFTR, but the selectivity is identical in the PFTR and the CSTR. These are compared in the PFTR and CSTR in Figure 4–9.

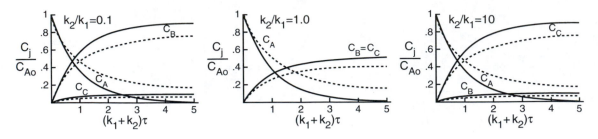

Figure 4–9 Comparison of C_A, C_B, and C_C for parallel first-order reactions in a PFTR (solid curves) and CSTR (dashed curves).

Example 4–3 We wish to produce B in the reaction $A \rightarrow B$ in a continuous reactor at $v = 4$ liter/min with $C_{Ao} = 2$ moles/liter. However, we find that there is a second reaction $A \rightarrow C$, which can also occur. We find that both reactions are first order and irreversible with $k_1 = 0.5$ min^{-1} and $k_2 = 0.1$ min^{-1}. Find τ, V, C_B, S_B, and Y_B for 90% conversion of A.

This is the same example $A \rightarrow B$ of the previous chapter, but now there is a second reaction $A \rightarrow C$, so these are *parallel* reactions.

PFTR. Application of these equations yields

$$\tau = \frac{1}{k_1 + k_2} \ln \frac{C_{Ao}}{C_A} = \frac{1}{0.5 + 0.1} \ln 10 = 3.84 \text{ min}$$

so that

$$V = v\tau = 4 \times 3.84 = 15.4 \text{ liters}$$

Therefore

$$C_B(\tau) = C_{Ao} \frac{k_1}{k_1 + k_2} (1 - e^{-(k_1 + k_2)\tau})$$

$$= 2 \times \frac{0.5}{0.6}[1 - \exp(-0.6 \times 3.84)] = 1.50 \text{ moles/liter}$$

so that

$$S_B = \frac{C_B}{C_{Ao} - C_A} = \frac{k_1}{k_1 + k_2} = \frac{0.5}{0.6} = \frac{1.5}{1.6} = 83\%$$

$$Y_B = \frac{C_B}{C_{Ao}} = \frac{1.5}{2} = 75\%$$

CSTR. Application of the preceding equations yields

$$\tau = \frac{C_{Ao} - C_A}{(k_1 + k_2)C_A} = \frac{2.0 - 0.2}{0.6 \times 0.2} = 15 \text{ min}$$

so that

$$V = 4 \times 15 = 60 \text{ liters}$$

Substitution in the other equations yields

$$C_B = \frac{\tau k_1 C_{Ao}}{1 + (k_1 + k_2)\tau} = \frac{15 \times 0.5 \times 2}{1 + 0.6 \times 15} = 1.50 \text{ moles/liter}$$

Therefore, $S_B = 1.5/1.8 = 83\%$ and $Y_B = 1.5/2.0 = 75\%$. As expected, we obtain the same selectivity and yield for the CSTR and PFTR for these kinetics, but the CSTR requires a longer residence time.

4.8.4 Parallel reactions with other kinetics

For other kinetics and for reversible reactions the expressions of course become more complex to solve. Let us first examine the situation with different order kinetics for two irreversible reactions such as

$$A \rightarrow B, \qquad r_1 = k_1 C_A^{m_1} \tag{4.130}$$

$$A \rightarrow C, \qquad r_2 = k_2 C_A^{m_2} \tag{4.131}$$

which is identical to the previous example except the orders are not unity but are rather m_1 and m_2.

For the PFTR or batch reactor the equation for A is

$$\frac{dC_A}{d\tau} = -k_1 C_A^{m_1} - k_2 C_A^{m_2} \tag{4.132}$$

which is still separable. For the CSTR the equation for A is

$$C_{Ao} - C_A = \tau(k_1 C_A^{m_1} + k_2 C_A^{m_2}) \tag{4.133}$$

which is a polynomial in C_A.

Let us consider a simple example of this with $m_1 = 1$ and $m_2 = 2$. We must solve the PFTR equation using partial fractions. The solution for the CSTR is a quadratic,

$$C_{Ao} - C_A = \tau(k_1 C_A + k_2 C_A^2) \tag{4.134}$$

whose solution is

$$\tau = \frac{C_{Ao} - C_A}{k_1 C_A + k_2 C_A^2} \tag{4.135}$$

This is a quadratic equation in C_A, which can be solved to yield the expression

$$C_A(\tau) = \frac{-(1 + k_1\tau) \pm [(1 + k_1\tau)^2 + 4k_2\tau C_{Ao}]^{1/2}}{2k_2\tau} \tag{4.136}$$

For these reactions in a PFTR the mass-balance equation becomes

$$\tau = \int_{C_{Ao}}^{C_A} \frac{dC_A}{k_1 C_A + k_2 C_A^2} \tag{4.137}$$

which must be solved by partial fractions or numerically.

Example 4–4   Calculate the selectivities of forming B in CSTR and PFTR reactors at 90% conversion of A starting with $C_{Ao} = 4$ moles/liter in the parallel reactions

$$A \to B, \qquad r_1 = k_1 = 2$$

$$A \to C, \qquad r_2 = k_2 C_A = C_A$$

where rates are in moles/liter min. The first reaction that produces B is zeroth-order in C_A and is the desired one, while the reaction that produces C is first order in C_A and is undesired.

For the PFTR the mass-balance equation for A is

$$\frac{dC_A}{d\tau} = -k_1 - k_2 C_A = -(2 + C_A)$$

which can be integrated to yield

$$\tau = \int_{C_{Ao}}^{C_A} \frac{dC_A}{2 + C_A} = \ln \frac{2 + C_{Ao}}{2 + C_A} = \ln \frac{2 + 4}{2 + 0.4} = \ln \frac{6}{2.4} = 0.92 \text{ min}$$

The mass-balance equation for B is

$$\frac{dC_B}{d\tau} = 2$$

$$C_B = 2\tau = 1.83 \text{ moles/liter}, \qquad C_A > 0$$

Thus the selectivity to B in a PFTR for this conversion is

$$S_B = \frac{C_B}{C_{Ao} - C_A} = \frac{1.83}{4.0 - 0.4} = 0.51$$

For the CSTR we write the mass-balance equations as

$$C_{Ao} - C_A = \tau(2 + C_A), \qquad C_A > 0$$

$$C_B = 2\tau, \qquad C_A > 0$$

$$C_C = \tau C_A, \qquad C_A > 0$$

This gives

$$C_A = \frac{C_{Ao} - 2\tau}{1 + \tau}$$

$$C_B = 2\tau$$

$$C_C = \frac{C_{Ao}\tau - 2\tau^2}{1 + \tau}$$

until all A is consumed. The residence time is

$$\tau = \frac{C_{Ao} - C_A}{2 + C_A} = \frac{4 - 0.4}{2 + 0.4} = \frac{3.6}{2.4} = 1.5 \text{ min}$$

This gives

$$C_B = 2\tau = 3 \text{ moles/liter}$$

Thus in a CSTR we obtain the selectivity

$$S_B = \frac{C_B}{C_{Ao} - C_A} = \frac{3.0}{4.0 - 0.4} = 0.83$$

From this example we see that the CSTR requires a longer residence time for the required 90% conversion (it must since the kinetics are positive order), but the CSTR gives a higher selectivity to B. One can always design for a larger reactor, but the A that was converted to C is a continual loss. Thus the CSTR is the clear choice of reactor in selectivity for this example, where we wanted to favor a lower-order reaction rather than a high-order reaction.

4.8.5 Differential and total selectivity

The calculation of selectivities becomes more complicated for other than these simple examples. Some of the features controlling selectivity can be seen by considering the differential selectivity.

If there is a single reactant A on which we can base the conversion and selectivity, then we can define a *differential selectivity* s_j and integrate it to obtain the total selectivity. Consider the following relation where B is the desired product

$$s_B = \frac{dC_B}{dC_B + dC_C} = -\frac{dC_B}{dC_A} \tag{4.138}$$

which is the rate of forming B divided by the rate of loss of A. The total selectivity in a PFTR can be found by writing this as

$$dC_B = -s_B\, dC_A \tag{4.139}$$

and integrating along the reactor to obtain

$$C_B = -\int_{C_{Ao}}^{C_A} s_B\, dC_A \tag{4.140}$$

Since $s_B = k_1/(k_1 + C_A\, m_2 - m_1)$, we obtain

$$S_B = \frac{C_B}{C_{Ao} - C_A} = \frac{-1}{C_{Ao} - C_A} \int_{C_{Ao}}^{C_A} s_B\, dC_A \tag{4.141}$$

For two parallel irreversible reactions with orders m_1 and m_2 this becomes

$$C_B = -\int_{C_{Ao}}^{C_A} \frac{r_1}{r_1 + r_2}\, dC_A = -\int_{C_{Ao}}^{C_A} \frac{k_1 C_A^{m_1}}{k_1 C_A^{m_1} + k_2 C_A^{m_2}}\, dC_A \tag{4.142}$$

If $m_1 = m_2$, then the integral becomes

$$C_B = -\int_{C_{Ao}}^{C_A} \frac{k_1}{k_1 + k_2}\, dC_A = \frac{k_1}{k_1 + k_2}(C_{Ao} - C_A) \tag{4.143}$$

and $S_B = k_1/(k_1 + k_2)$, as we derived previously for first-order parallel reactions.

If $m_1 = 0$ and $m_2 = 1$, this expression becomes

$$C_B = -\int_{C_{Ao}}^{C_A} \frac{k_1}{k_1 + k_2 C_A} \, dC_A = \frac{k_1}{k_2} \ln \frac{k_1 + k_2 C_{Ao}}{k_1 + k_2 C_A} \tag{4.144}$$

and the selectivity is

$$S_B = \frac{C_B}{C_{Ao} - C_A} = \frac{k_1}{k_2(C_{Ao} - C_A)} \ln \frac{k_1 + k_2 C_{Ao}}{k_1 + k_2 C_A} \tag{4.145}$$

We can generalize this by plotting the s_B versus $C_{Ao} - C_A$.

Example 4–5   Consider the parallel reactions

$$A \rightarrow B, \qquad r_1 = 4C_A$$

$$A \rightarrow C, \qquad r_2 = 2C_A^2$$

for $C_{Ao} = 2$ moles/liter. Determine the selectivities in PFTR and CSTR versus the conversion.

We could solve the equations directly, but it is easier to use the differential selectivity,

$$s_B = \frac{r_1}{r_1 + r_2} = \frac{4C_A}{4C_A + 2C_A^2} = \frac{2}{2 + C_A}$$

This gives $s_B = 0.5$ at $C_A = 2$ (no conversion), $s_B = 0.67$ at $C_A = 1$ (50% conversion), and $s_B = 1$ at $C_A = 0$ (complete conversion). Since s_B increases with conversion, the CSTR will give a higher selectivity ($S_B = s_B$) than a PFTR (S_B is the average of s_B). These are plotted in Figure 4–10.

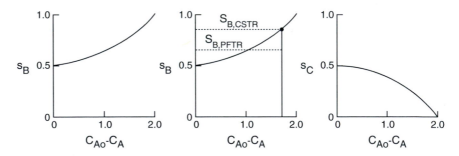

Figure 4–10   Differential selectivity s_B versus conversion $C_{Ao} - C_A$ for situations where the desired reaction has a lower order and where the desired reaction has a higher order. The overall selectivity S_B increases or decreases for these situations. The overall selectivity in the CSTR will be higher than in the PFTR if the desired reaction has a lower order.

S_B in a CSTR is s_B evaluated at the exit conversion. In a PFTR S_B is the average of s_B between C_{Ao} and C_A, which is the area under the s_B versus $C_{Ao} - C_A$ curve divided by $C_{Ao} - C_A$. The CSTR always requires a longer τ for a given conversion.

This graphical construction illustrates the comparison in selectivity of CSTR and PFTR. If s_B is an increasing function of conversion, then the CSTR gives a higher selectivity for a given conversion than a PFTR (but with a larger reactor required).

For our example, S_B is favored more in the CSTR as the conversion increases. Again this is seen simply because the entire process in the CSTR occurs at the final conversion where C_A is smallest; so less C is produced compared to the PFTR.

All these arguments require a single reactant A on which to base the calculation of selectivity. For more complex situations we can still determine how the selectivity varies with conversion in PFTR and CSTR, but calculation of the selectivity requires complete solution of the mass-balance equations.

Example 4–6   Consider the reactions

$$A \rightarrow B, \qquad r_1 = k_1$$

$$A \rightarrow C, \qquad r_2 = k_2 C_A$$

$$A \rightarrow D, \qquad r_3 = k_3 C_A^2$$

What is the best single ideal PFTR or CSTR to produce B, C, and D? Do S_B, S_C, and S_D increase or decrease with conversion?

If we want to produce B, which has the lowest order, then a CSTR is preferred and a high conversion will give the highest yield.

If we want to produce D, which has the highest order, then a PFTR is preferred and a low conversion will give the highest yield.

If we want to optimize the reaction to produce C, which has intermediate reaction order between B and D, then we have to solve the problem completely.

Solve these equations for all species versus τ in a PFTR and a CSTR for $k_1 = 2$, $k_2 = 4$, and $k_3 = 2$. Which would be the best reactor to produce C for different C_{Ao} and conversion?

Shown in Figure 4–11 is a plot of s_C versus $C_{Ao} - C_A$. It is seen that s_B has a maximum at $C_A = 1$ and decreases to zero at complete conversion.

Thus for this example the CSTR will give the higher selectivity to C for up to ~70% conversion, but if the conversion is above 70%, the PFTR gives a higher selectivity, as well as requiring a shorter τ.

Figure 4–11 Plot of differential selectivity s_C for the example of three reactions in parallel. If C is desired, the CSTR gives a larger S_C, and this occurs when the reactor is operated at $C_A = 1$ mole/liter.

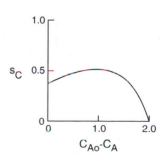

4.9 MULTIPLE REACTIONS WITH VARIABLE DENSITY

Previously in this chapter we have implicitly assumed that the density of the fluid did not change as the reaction proceeded. In our parallel and series reactions, $A \rightarrow B$, $A \rightarrow C$, and $A \rightarrow B \rightarrow C$, the number of moles is obviously constant; so the problems can always be worked in terms of concentrations.

When the density varies, we need to find another variable to express the progress of a reaction. Earlier we defined the fractional conversion X for a single reaction, and in this chapter we defined the conversion of a reactant species X_A for reactant A and X_j for reaction j. For the conversion in a reaction we need a different variable, and we shall use $\mathbf{X_i}$ (bold type), with the index i describing the reaction. We will first work our series and parallel reactions with these variables and then consider a variable-density problem.

We define the conversion of reactant A as

$$X_A = \frac{C_{Ao} - C_A}{C_{Ao}} \qquad (4.146)$$

assuming that $\nu_{iA} = -1$, while in conversion of a reaction we choose a basis N_{Ao} and define

$$N_j = N_{jo} + N_{Ao} \sum_{i=1}^{R} \nu_{ij} \mathbf{X_i} \qquad (4.147)$$

For a constant-density system where the volume occupied by the number of moles chosen for the basis is constant, this can be written as

$$C_j = C_{jo} + C_{Ao} \sum_{i=1}^{R} \nu_{ij} \mathbf{X_i} \qquad (4.148)$$

4.9.1 Parallel reactions

Consider the reactions

$$A \rightarrow B, \qquad r_1 = k_1 C_A \qquad (4.149)$$

$$A \rightarrow C, \qquad r_2 = k_2 C_A \qquad (4.150)$$

From stoichiometry with pure A at C_{Ao} initially we can write these species in fractional conversions $\mathbf{X_1}$ and $\mathbf{X_2}$ as

$$C_A = C_{Ao}(1 - \mathbf{X_1} - \mathbf{X_2}) \qquad (4.151)$$

$$C_B = C_{Ao}\mathbf{X_1} \qquad (4.152)$$

$$C_C = C_{Ao}\mathbf{X_2} \qquad (4.153)$$

Note that for these parallel reactions $X_A = \mathbf{X_1} + \mathbf{X_2}$ because A is lost through both reactions. Since C_A must be greater than zero, X_A must be less than 1, and therefore $0 < \mathbf{X_1} + \mathbf{X_2} < 1$. This is a consequence of the fact that A can react to form either B or C.

In a CSTR the mass balances on the three species become

$$C_{Ao} - C_A = C_{Ao}(\mathbf{X_1} + \mathbf{X_2}) = \tau(r_1 + r_2) = \tau(k_1 + k_2)C_{Ao}(1 - \mathbf{X_1} - \mathbf{X_2}) \qquad (4.154)$$

$$C_B = C_{Ao}\mathbf{X_1} = \tau r_1 = \tau k_1 C_{Ao}(1 - \mathbf{X_1} - \mathbf{X_2}) \qquad (4.155)$$

$$C_C = C_{Ao}\mathbf{X_2} = \tau r_2 = \tau k_2 C_{Ao}(1 - \mathbf{X_1} - \mathbf{X_2}) \qquad (4.156)$$

Eliminating C_{Ao} from these equations, we obtain

$$\mathbf{X_1} + \mathbf{X_2} = \tau(k_1 + k_2)(1 - \mathbf{X_1} - \mathbf{X_2}) \tag{4.157}$$

$$\mathbf{X_1} = \tau r_1 = \tau k_1(1 - \mathbf{X_1} - \mathbf{X_2}) \tag{4.158}$$

$$\mathbf{X_2} = \tau r_2 = \tau k_2(1 - \mathbf{X_1} - \mathbf{X_2}) \tag{4.159}$$

which can be solved to obtain

$$\mathbf{X_1} = \frac{\tau k_1}{1 + \tau k_1 + \tau k_2} \tag{4.160}$$

$$\mathbf{X_2} = \frac{\tau k_2}{1 + \tau k_1 + \tau k_2} \tag{4.161}$$

These values of $\mathbf{X_1}$ and $\mathbf{X_2}$ can be inserted into the expressions for C_A, C_B, and C_C to give the expressions

$$C_A = C_{Ao}(1 - \mathbf{X_1} - \mathbf{X_2}) = \frac{C_{Ao}}{1 + \tau k_1 + \tau k_2} \tag{4.162}$$

$$C_B = C_{Ao}\mathbf{X_1} = \frac{C_{Ao}\tau k_1}{1 + \tau k_1 + \tau k_2} \tag{4.163}$$

$$C_C = C_{Ao}\mathbf{X_2} = \frac{C_{Ao}\tau k_2}{1 + \tau k_1 + \tau k_2} \tag{4.164}$$

These are the same expressions we derived previously (they had better be or we made an error somewhere) using concentrations directly. We note that the method just used, where we solved in terms of $\mathbf{X_1}$ and $\mathbf{X_2}$, is more complicated and subject to mathematical mistakes than from just solving the original equations in terms of C_A, C_B, and C_C directly

4.9.2   Series reactions

Consider next the series reactions

$$A \rightarrow B, \qquad r_1 = k_1 C_A \tag{4.165}$$

$$B \rightarrow C, \qquad r_2 = k_2 C_B \tag{4.166}$$

Now the stoichiometric relations among the concentration in terms of $\mathbf{X_1}$ and $\mathbf{X_2}$ starting with $C_{Bo} = C_{Co} = 0$ are

$$C_A = C_{Ao}(1 - \mathbf{X_1}) \tag{4.167}$$

$$C_B = C_{Ao}(\mathbf{X_1} - \mathbf{X_2}) \tag{4.168}$$

$$C_C = C_{Ao}\mathbf{X_2} \tag{4.169}$$

Note that these are quite different than for parallel reactions. In this reaction system $X_A = \mathbf{X_1}$ because A is lost only through the first reaction. Also, in this reaction system $\mathbf{X_1} \rightarrow 1$ and $\mathbf{X_2} \rightarrow 1$ when the reaction goes to completion.

In a CSTR the mass-balance equations are

$$C_{Ao} - C_A = C_{Ao}\mathbf{X_1} = \tau_1 = \tau k_1 C_{Ao}(1 - \mathbf{X_1}) \tag{4.170}$$

$$-C_B = -C_{Ao}(X_1 - X_2) = \tau(r_1 - r_2)$$

$$= \tau[(k_1 C_{Ao}(1 - X_1) - k_2 C_{Ao}(X_1 - X_2)] \tag{4.171}$$

$$-C_C = -C_{Ao}X_2 = \tau r_2 = \tau k_2 C_{Ao}(X_1 - X_2) \tag{4.172}$$

C_{Ao} can be eliminated from these equations to yield

$$X_1 = \tau k_1(1 - X_1) \tag{4.173}$$

$$X_2 - X_1 = \tau[(k_1(1 - X_1) - k_2(X_1 - X_2)] \tag{4.174}$$

$$-X_2 = \tau k_2(X_1 - X_2) \tag{4.175}$$

Again, these can be solved for X_1 and X_2 and substituted back into C_A, C_B, and C_C to obtain the same results we obtained previously.

For both parallel and series reactions we have not derived these equations in the PFTR. This involves writing equations such as $dX_1/d\tau = \cdots$ and solving the differential equations for X_1 and X_2. This would be uninstructive because the answers are identical to those derived without using these variables (assuming no mistakes).

4.9.3 Variable-density series reactions

Consider next the reactions

$$A \rightarrow 2B, \qquad r_1 = k_1 C_A \tag{4.176}$$

$$2B \rightarrow 2C, \qquad r_2 = k_2 C_B \tag{4.177}$$

where all species are ideal gases with no diluents and $C_B = C_C = 0$.

Now the mass density varies with conversion because 1 mole of A is converted into 2 moles of C when the reaction system goes to completion. Therefore, we cannot use the CSTR mass-balance equations, $C_{jo} - C_j = \tau \sum v_{ij} r_i$, and we need to use the variable-density versions of the reactor equations,

$$F_{jo} - F_j = \tau \sum_{i=1}^{R} v_{ij} r_i(C_j) \tag{4.178}$$

We therefore need to write the mass balances in terms of a basis N_{Ao} moles initially. These give

$$N_A = N_{Ao}(1 - X_1) \tag{4.179}$$

$$N_B = 2N_{Ao}(X_1 - X_2) \tag{4.180}$$

$$\underline{N_C = 2N_{Ao}X_2} \tag{4.181}$$

$$\sum N = N_{Ao}(1 + X_1) \tag{4.182}$$

Note that here the N_j are quite different than for series reactions because of the $v_{1B} = +2$, $v_{2B} = -2$, and $v_{2C} = +2$. The volume occupied by these species is given by the expression

$$V = V_o(1 + X_1) \tag{4.183}$$

(Recall that in this equation V is the volume containing N moles of fluid, not the reactor volume.)

We can now write the concentrations of these species as

$$C_A = \frac{N_A}{V} = C_{Ao}\frac{1-X_1}{1+X_1} \tag{4.184}$$

$$C_B = \frac{N_B}{V} = C_{Ao}\frac{X_1-X_2}{1+X_1} \tag{4.185}$$

$$C_C = \frac{N_C}{V} = C_{Ao}\frac{X_2}{1+X_1} \tag{4.186}$$

In a CSTR the mass-balance equations are

$$F_{Ao} - F_A = F_{Ao}X_1 = Vk_1C_A = Vk_1C_{Ao}\frac{1-X_1}{1+X_1} \tag{4.187}$$

$$-F_B = -F_{Ao}(X_1-X_2) = V(k_1C_A - k_2C_B) \tag{4.188}$$

$$-F_B = V\left(k_1C_{Ao}\frac{1-X_1}{1+X_1} - Vk_2C_{Ao}\frac{X_1-X_2}{1+X_1}\right) \tag{4.189}$$

$$-F_C = -F_{Ao}X_2 = Vk_2C_B = Vk_2C_{Ao}\frac{X_1-X_2}{1+X_1} \tag{4.190}$$

Solving these equations for the reactor volume V, we obtain the expressions

$$V = F_{Ao}\frac{X_1(1+X_1)}{C_{Ao}(1-X_1)} = v_o\frac{X_1(1+X_1)}{1-X_1} \tag{4.191}$$

$$V = v_o\frac{(X_1-X_2)(1+X_1)}{1-X_1} \tag{4.192}$$

$$V = v_o\frac{X_2(1+X_1)}{X_1-X_2} \tag{4.193}$$

These equations can be solved for X_1 and X_2, and these can be inserted back into the expressions for C_A, C_B, and C_C, such as

$$C_A = C_{Ao}\frac{1-X_1}{1+X_1} \tag{4.194}$$

In these equations X_1 and X_2 are found by solving a quadratic equation; so the solutions are not simple, even for these simple kinetics.

It is evident that for multiple reactions with variable density, we rapidly arrive at rather complex expressions that require considerable manipulation even to formulate the expressions, which can be used to calculate numerical values of the reactor volume required for a given conversion and selectivity to a desired product.

4.10 REAL REACTION SYSTEMS AND MODELING

We have considered several simple examples of multiple reaction systems. While the simplest of these systems have analytical solutions, we rapidly come to situations where only numerical solutions are possible.

However, the ideas in the examples given are still very useful in dealing with more complex problems because simple examples with analytical solutions allow us to estimate reactor performance quickly without resorting to complex numerical solutions. Further, the kinetics of the complete reactions are frequently unknown, or, if rate expressions are available, there may be large errors in these rate parameters that will make even the complete solution approximate. This is especially true for surface-catalyzed reactions and biological reactions, where the rate expressions are complex with fractional orders and activation energies that vary between catalyst or enzyme batches. Also, activation and deactivation processes can cause time-dependent reactor performance.

Therefore, complex processes are frequently simplified to assume (1) a single reaction in which the major reactant is converted into the major product or for a more accurate estimate; (2) simple series or parallel processes in which there is a major desired and a single major undesired product. The first approximation sets the approximate size of the reactor, while the second begins to examine different reactor types, operating conditions, feed composition, conversion, separation systems required, etc.

Thus there are enough uncertainties in the kinetics in most chemical reaction processes that we almost always need to resort to a simplified model from which we can estimate performance. Then, from more refined data and pilot plant experiments, we begin to refine the design of the process to specify the details of the equipment needed.

In almost all examples of designing a process for which detailed experience is not available from a previous operating process, the reactor and separation systems are overdesigned with 10 to 100% margins of error compared to actual predicted performance. This allows the engineer a margin of safety in engineering predictions used for economic evaluations. The added benefit of conservative design is that, if the process operates at or above estimations, the process can deliver added capacity, which really makes the engineer look good. The clever engineer needs to have a design that is conservative enough to work at predicted performance with high probability but bold enough to beat the competition.

At some point in most processes, a detailed model of performance is needed to evaluate the effects of changing feedstocks, added capacity needs, changing costs of materials and operations, etc. For this, we need to solve the complete equations with detailed chemistry and reactor flow patterns. This is a problem of solving the R simultaneous equations for S chemical species, as we have discussed. However, the real process is seldom isothermal, and the flow pattern involves partial mixing. Therefore, in formulating a complete simulation, we need to add many additional complexities to the ideas developed thus far. We will consider each of these complexities in successive chapters: temperature variations in Chapters 5 and 6, catalytic processes in Chapter 7, and nonideal flow patterns in Chapter 8. In Chapter 8 we will return to the issue of detailed modeling of chemical reactors, which include all these effects.

4.11 APPROXIMATE RATE EXPRESSIONS FOR MULTIPLE-REACTION SYSTEMS

We have gone about as far as is useful in finding closed-form analytical solutions to mass-balance equations in batch or continuous reactors, described by the set of reactions

$$\sum_{j=1}^{s} v_{ij} A_j = 0, \quad i = 1, \ldots, R, \qquad r_i = k_{\mathrm{f}i} \prod_{j=1}^{S} C_j^{m_{\mathrm{f}ij}} - k_{\mathrm{b}i} \prod_{j=1}^{S} C_j^{m_{\mathrm{b}ij}} \tag{4.195}$$

Let us summarize the general procedure we would follow for an arbitrary reaction system. For the batch or PFTR we write S first-order differential equations describing each of the S species. We then eliminate as many equations as possible by finding $S - R$ suitable stoichiometric relations among species to obtain R irreducible equations. Sometimes some of these are uncoupled from others so that we may need to solve smaller sets simultaneously.

For the CSTR we have S algebraic species mass-balance equations. We can eliminate $S - R$ of these to obtain R irreducible algebraic polynomials in R of the S species which we must solve for C_j, and again we frequently do not need so solve them all simultaneously.

However, for more than a few reactions and with reversible reactions, we rapidly reach a situation where only numerical solutions are possible. While numerical solutions can easily be found with differential equation solvers or computer programs that find roots of simultaneous polynomials, note that these are only *simulations* for a given set of ks and C_{jo}s: We have to specify all rate coefficients and initial compositions and let the computer plot out the values of $C_j(t)$ or $C_j(\tau)$. This is quite unsatisfactory compared to an analytical solution because we have to reevaluate the solution for each set of ks and C_{jo}s, whereas with an analytical solution we just need to plug in the values of ks and C_{jo}s.

We repeat the argument that, even if we are going to obtain numerical solutions to a problem, we still need to pose a proper mathematical problem for the computer program to solve. A problem involving two equations with three unknowns cannot be solved with the most sophisticated computer problem. Even worse, with an ill-posed problem, the computer may give you answers, but they will be nonsense.

4.12 SIMPLIFIED REACTIONS

There is another difficulty with exact solutions of complex sets of reactions. We frequently don't want to know all the $C_j(t)$, just the important reactants and products. In many chemical reaction systems there are many intermediates and minor species whose concentrations are very small and unmeasured.

Therefore, we need to find approximate methods for simultaneous reaction systems that will permit finding analytical solutions for reactants and products in simple and usable form. We use two approximations that were developed by chemists to simplify simultaneous reaction systems: (1) the equilibrium step approximation and (2) the pseudo-steady-state approximation.

4.12.1 Equilibrium step assumption

Here we assume simply that some reaction steps remain in thermodynamic chemical equilibrium throughout the process. The validity of the approximation relies on the fact that both the forward and the reverse reaction steps for the reaction assumed to be in equilibrium are *very fast* compared to others.

We will work out these approximations for the series reactions

$$A \rightleftarrows B, \qquad r_1 = k_{\mathrm{f}1} C_A - k_{\mathrm{b}1} C_B \tag{4.196}$$

$$B \rightarrow C, \qquad r_2 = k_2 C_B \tag{4.197}$$

which is simply the first-order reversible reaction we worked out previously. However, now we consider this the exact solution to the approximate single reaction

$$A \rightarrow C, \qquad r(C_A) \tag{4.198}$$

Thus we search for the approximate solution for the single nonelementary reaction $A \rightarrow C$, which proceeds through the reaction intermediate B in the two preceding reactions. We of course know the exact solution in batch, PFTR, and CSTR. This will allow us to test the approximate solutions. As before, we write the mass-balance equations for batch or PFTR as

$$\frac{dC_A}{dt} = -k_{f1}C_A + k_{b1}C_B \tag{4.199}$$

$$\frac{dC_B}{dt} = +k_{f1}C_A - k_{b1}C_B - k_2 C_B = +k_{f1}C_A - (k_{b1} + k_2)C_B \tag{4.200}$$

$$\frac{dC_C}{dt} = +k_2 C_B \tag{4.201}$$

which has the solution derived previously.

In the equilibrium step assumption, we assume that the first reaction remains in thermodynamic equilibrium, which gives

$$r_1 = k_{f1}C_A - k_{b1}C_B = 0 \tag{4.202}$$

This equation can easily be solved for C_B,

$$C_B = \frac{k_{f1}}{k_{b1}}C_A \tag{4.203}$$

We now write the approximate reaction as the rate of forming species C, the product,

$$\frac{dC_C}{dt} = r = r_2 = k_2 C_B = \frac{k_{f1}k_2}{k_{b1}}C_A = k_{eff}C_A \tag{4.204}$$

Therefore, the equilibrium step assumption gives an approximation for the rate,

$$A \rightarrow C, \qquad r = k_{eff}C_A \tag{4.205}$$

which is a first-order irreversible reaction, but now the "effective rate coefficient" k_{eff} is given by

$$k_{eff} = \frac{k_{f1}k_2}{k_{b1}} \tag{4.206}$$

a combination of rate coefficients of the elementary steps.

Note carefully the logic of this very simple derivation. We want an overall rate r for the single reaction in terms of C_A by eliminating the intermediate C_B in the two-step reaction. We did this by assuming the first reaction in the exact two-step process to be in equilibrium, and we then solved the algebraic expression for C_B in terms of C_A and rate coefficients. We then put this relation into the second reaction and obtained an expression for the overall approximate in terms of the reactant species alone. We eliminated the intermediate from the overall expression by assuming an equilibrium step.

We can use this approximation for many complex reaction systems by assuming many steps to be in thermodynamic equilibrium. Whether or not the expression is a good approximation to the exact expression of course depends on the values of particular rate coefficients.

4.12.2 Pseudo-steady-state approximation

Next we consider the same reaction system, but now we do not assume that the first reaction is in thermodynamic equilibrium but rather that the concentration of the intermediate species B does not change during the reaction or that $dC_B/dt = 0$,

$$\frac{dC_B}{dt} = +k_{f1}C_A - k_{b1}C_B - k_2 C_B = 0 \tag{4.207}$$

We set this rate equal to zero, which is what we would do if this reaction were in steady state (no B being formed or reacted), which emphatically is not the case, since B is the intermediate in forming C, the major product.

The steady state in the preceding equation gives an algebraic equation relating C_A and C_B,

$$C_B = \frac{k_{f1}}{k_{b1} + k_2} C_A \tag{4.208}$$

Proceeding as before, the approximate rate of the overall reaction to form C from A is

$$\frac{dC_C}{dt} = r = r_2 = k_2 C_B = \frac{k_{f1} k_2}{k_{b1} + k_2} C_A \tag{4.209}$$

which looks almost identical to the expression assuming thermodynamic equilibrium in step 1 except for the extra term k_2 in the denominator. Thus the pseudo-steady-state approximation on this reaction yields the approximate solution

$$A \rightarrow C, \qquad r = k_{\text{eff}} C_A \tag{4.210}$$

with

$$k_{\text{eff}} = \frac{k_{f1} k_2}{k_{b1} + k_2} \tag{4.211}$$

4.12.3 Nitric oxide oxidation

The first example we will use as an application of the equilibrium step approximation and pseudo-steady-state approximation is the reaction

$$NO + \tfrac{1}{2}O_2 \rightarrow NO_2 \tag{4.212}$$

This is a very important reaction because it converts NO from engine exhaust into the brown NO_2, which oxidizes hydrocarbons in the atmosphere into aldehydes, peroxides, and other noxious pollutants. The reaction occurs homogeneously in air. This reaction obeys the empirical rate expression

$$r = k C_{NO}^2 C_{O_2} \tag{4.213}$$

One might expect that this is an elementary reaction because the reaction agrees with the stoichiometry (after multiplying all coefficients in the preceding expression by two). However, one peculiarity of this reaction is that the rate actually decreases with increasing temperature or has a negative activation energy, an unreasonable situation for an elementary process.

The reaction actually occurs by a succession of bimolecular reaction steps, one possible set being

$$NO + NO \rightleftarrows N_2O_2, \qquad r_1 = k_{f1}C_{NO}^2 - k_{b1}C_{N_2O_2} \qquad (4.214)$$

$$N_2O_2 + O_2 \rightarrow 2NO_2, \qquad r_2 = k_2 C_{N_2O_2} C_{O_2} \qquad (4.215)$$

These are two elementary bimolecular reaction steps, the sum of which is the oxidation of NO to NO_2,

$$NO + NO + O_2 \rightarrow 2NO_2 \qquad (4.216)$$

Let us begin by assuming that the first reaction is in thermodynamic equilibrium,

$$r_1 = k_{f1}C_{NO}^2 - k_{b1}CC_{N_2O_2} = 0 \qquad (4.217)$$

Solving this expression for the intermediate N_2O_2, we obtain

$$C_{N_2O_2} = \frac{k_{f1}}{k_{b1}} C_{NO}^2 \qquad (4.218)$$

The rate of forming NO_2 is the rate of the second reaction

$$r = 2r_2 = 2k_2 C_{N_2O_2} C_{O_2} = 2k_2 \frac{k_{f1}}{k_{b1}} C_{NO}^2 C_{O_2} \qquad (4.219)$$

which agrees with the experimentally observed rate expression.

Let us next apply the pseudo-steady-state approximation to see if it agrees with the experimentally observed rate. We write the mass balance on $C_{N_2O_2}$,

$$\frac{dC_{N_2O_2}}{dt} = +r_1 - r_2 = k_{f1}C_{NO}^2 - k_{b1}C_{N_2O_2} - k_2 C_{N_2O_2} C_{O_2} = 0 \qquad (4.220)$$

We now solve this expression for $C_{N_2O_2}$ to yield

$$C_{N_2O_2} = \frac{k_{f1}}{k_{b1} + K_2 C_{O_2}} C_{NO}^2 \qquad (4.221)$$

Now the rate of formation of NO_2 is given by

$$r = r_2 = k_2 C_{N_2O_2} C_{O_2} = k_2 \frac{k_{f1}}{k_{b1} + k_2 C_{O_2}} C_{NO}^2 C_{O_2} \qquad (4.222)$$

Thus we see that the pseudo-steady-state approximation gives orders of the reaction as the thermodynamic equilibrium approximation, the only difference being the definition of the rate constant

$$k_{eff} = k_2 \frac{k_{f1}}{k_{b1}} \qquad (4.223)$$

in the equilibrium step approximation and

$$k_{eff} = k_2 \frac{k_{f1}}{k_{b1} + k_2 C_{O_2}} \qquad (4.224)$$

in the pseudo-steady-state approximation. There is an extra term in the denominator of the pseudo-steady-state approximation expression. If $k_{b1} \gg k_2 C_{O_2}$, then the expressions are identical by the two approximations. However, if $k_{b1} \ll k_2 C_{O_2}$, then the pseudo-steady-state approximation gives the rate expression

$$r = k_{f1} C_{NO}^2 \qquad (4.225)$$

which predicts different orders (zeroth order in O_2) and a rate constant of k_{f1}.

Thus both approximations predict rate expressions that agree with experimental data as long as the second term in the denominator of the pseudo-steady-state approximation is small.

4.12.4 Decomposition of acetaldehyde

Acetaldehyde decomposes homogeneously at temperatures of several hundred degrees Celsius to produce methane and carbon monoxide,

$$CH_3CHO \rightarrow CH_4 + CO \qquad (4.226)$$

and the rate of this reaction is empirically found to be $\frac{3}{2}$ order in acetaldehyde,

$$r = k[CH_3CHO]^{3/2} \qquad (4.227)$$

This is a reaction of the form $A \rightarrow B + C$, but it is clearly not an elementary reaction because it is not first order in the reactant.

This reaction is actually a multiple-reaction system that involves the major steps

$$A \rightarrow CH_3 + CHO, \qquad r_1 = k_1[A] \qquad (4.228)$$

$$CH_3 + A \rightarrow CH_4 + CH_3CO, \qquad r_2 = k_2[CH_3][A] \qquad (4.229)$$

$$CH_3CO \rightarrow CH_3 + CO, \qquad r_3 = k_3[CH_3CO] \qquad (4.230)$$

$$CH_3 + CH_3 \rightarrow C_2H_6, \qquad r_4 = k_4[CH_3]^2 \qquad (4.231)$$

In words, we describe the process as initiated by the decomposition of acetaldehyde to form the methyl radical CH_3 and the formyl radical CHO. Then methyl attacks the parent molecule acetaldehyde and abstracts an H atom to form methane and leave the acetyl radical CH_3CO, which dissociates to form another methyl radical and CO. Finally, two methyl radicals combine to form the stable molecule ethane.

This is a set of four elementary reactions (unimolecular or bimolecular processes obeying first- or second-order kinetics). Examination of these four steps shows that they do involve the reaction of acetaldehyde (called $[A]$ for convenience) and that methane is formed by step 2 and CO by step 3. If we add steps 2 and 3, we get

$$A \rightarrow CH_4 + CO \qquad (4.232)$$

but this occurs by a sequence of two steps.

Note also that we form products CHO and C_2H_6, which are not the products we said were formed. These species are *minor products* (no more than 1% of the amounts of CH_4 and CO), which can be ignored and perhaps not even measured.

This reaction system is an example of a *chain reaction*, which we will consider in more detail in Chapter 10. However, here we will just use the pseudo-steady-state approximation to find the approximate rate expression above. The only reactant is acetaldehyde, and there are six products listed: CH_4, CO, CH_3, CH_3O, CHO, and C_2H_6. The first two, CH_4 and CO, are the major products; C_2H_6 is a stable but minor product. The other species, CH_3, CH_3CO, and CHO, are free radicals that are very reactive and never build up to high concentrations.

To show that the above rate expression can yield this rate expression, let us apply the pseudo-steady-state approximation on $[CH_3]$ and $[CH_3CO]$ and see what it predicts. Mass balances on all species yield,

$$\frac{d[A]}{dt} = -k_1[A] - k_2[A][CH_3] \tag{4.233}$$

$$\frac{d[CH_3]}{dt} = k_1[A] - k_2[CH_3][A] + k_3[CH_3CO] - 2k_4[CH_3]^2 \tag{4.234}$$

$$\frac{d[CH_3CO]}{dt} = +k_2[CH_3][A] - k_3[CH_3CO] \tag{4.235}$$

$$\frac{d[CO]}{dt} = +k_3[CH_3CO] \tag{4.236}$$

$$\frac{d[CH_4]}{dt} = +k_2[CH_3][A] \tag{4.237}$$

$$\frac{d[CHO]}{dt} = +k_1[A] \tag{4.238}$$

$$\frac{d[C_2H_6]}{dt} = +k_4[CH_3]^2 \tag{4.239}$$

These are batch reactor or PFTR mass-balance equations on all seven species we identified in our mechanism. The reader should verify each term, particularly the signs and the factor of 2 in some terms.

We will use the pseudo-steady-state approximation on $[CH_3]$ and on $[CH_3CO]$. These yield

$$\frac{d[CH_3]}{dt} = k_1[A] - k_2[CH_3][A] + k_3[CH_3CO] - 2k_4[CH_3]^2 = 0 \tag{4.240}$$

$$\frac{d[CH_3CO]}{dt} = +k_2[CH_3][A] - k_3[CH_3CO] = 0 \tag{4.241}$$

We want to solve explicitly for these species concentrations to obtain an expression in terms of $[A]$ alone. Examining this equation, we see that we can *add them* to obtain

$$k_1[A] - 2k_4[CH_3]^2 = 0 \tag{4.242}$$

Now solving for $[CH_3]$, we obtain

$$[CH_3] = \left(\frac{k_1}{2k_4}\right)^{1/2} [A]^{1/2} \tag{4.243}$$

Next, recognizing that the overall rate of reaction is the rate of loss of $[A]$,

$$\frac{d[A]}{dt} = -k_1[A] - k_2[A][CH_3] = -k_1[A] - k_2\left(\frac{k_1}{2k_4}\right)^{1/2}[A]^{3/2} \qquad (4.244)$$

Finally, assuming that the second term is much larger than the first (we will consider why later), we obtain for the overall rate

$$r = k_2\left(\frac{k_1}{2k_4}\right)^{1/2}[A]^{3/2} \qquad (4.245)$$

so that we predict $\frac{3}{2}$ order kinetics with an effective rate coefficient $k = k_2(k_1/2k_4)^{1/2}$.

4.12.5 Phosgene synthesis

Phosgene is a key intermediate used in synthesis of many chemicals, a major one being toluene diisocyanate, a monomer in polyurethanes. Phosgene is made from chlorine and carbon monoxide through the overall reaction

$$CO + Cl_2 \rightarrow COCl_2 \qquad (4.246)$$

However, the rate of the reaction is empirically found to be

$$r = k[CO][Cl_2]^{3/2} \qquad (4.247)$$

so the reaction is clearly not elementary and must consist of several steps. The mechanism is thought to be

$$Cl_2 \rightleftarrows 2Cl, \qquad r_1 = k_{f1}[Cl_2] - k_{b1}[Cl]^2 \qquad (4.248)$$

$$CO + Cl \rightleftarrows COCl, \qquad r_2 = k_{f2}[CO][Cl] - k_{b2}[COCl] \qquad (4.249)$$

$$COCl + Cl_2 \rightleftarrows COCl_2 + Cl, \qquad r_3 = k_{f3}[COCl][Cl_2] - k_{b3}[COCl_2][Cl] \qquad (4.250)$$

In words, we describe the process as initiated by the dissociation of the chlorine molecule to form two chlorine atoms. The reactive chlorine atom then adds to CO to form another radical carbonyl chloride. This species is reactive enough to abstract a Cl atom from the chlorine molecule to form phosgene and produce another chlorine atom.

Phosgene is produced in the third reaction, and the rate of its formation is given by

$$r = r_3 = k_{f3}[COCl][Cl_2] - k_{b3}[COCl_2][Cl] \qquad (4.251)$$

However, this expression is not very useful because it is written as functions of two radical intermediates, $[COCl]$ and $[Cl]$, as well as the product $[COCl_2]$ and one reactant $\{Cl_2\}$. We need a rate expression in terms of the reactants and product only

$$r = r([CO], [Cl_2], [COCl_2]) \qquad (4.252)$$

To do this, we will use the above approximate methods to eliminate the intermediates in terms of the major species.

We first write the mass-balance equations for all five species,

$$\frac{d[CO]}{dt} = -r_2 = -k_{f2}[CO][Cl] + k_{b2}[COCl] \tag{4.253}$$

$$\frac{d[Cl_2]}{dt} = -r_1 - r_3 = -k_{f1}[Cl_2] + k_{b1}[Cl]^2 - k_{f3}[COCl][Cl_2]$$
$$+ k_{b3}[COCl_2][Cl] \tag{4.254}$$

$$\frac{d[Cl]}{dt} = +2r_1 - r_2 + r_3 = 2k_{f1}[Cl_2] - 2k_{b1}[Cl]^2 - k_{f2}[CO][Cl] + k_{b2}[COCl]$$
$$+ k_{f3}[COCl][Cl_2] - k_{b3}[COCl_2][Cl] \tag{4.255}$$

$$\frac{d[COCl]}{dt} = +r_2 - r_3 = k_{f2}[CO][Cl] - k_{b2}[COCl] - k_{f3}[COCl][Cl_2]$$
$$+ k_{b3}[COCl_2][Cl] \tag{4.256}$$

$$\frac{d[COCl_2]}{dt} = r_3 = k_{f3}[COCl][Cl_2] - k_{b3}[COCl_2][Cl] \tag{4.257}$$

We could solve these equations numerically for specific values of the ks and initial conditions. Note that we have written the mass balances as in a batch reactor. Replacing t by τ, we have the relevant equations in a PFTR, and we could easily write them for a CSTR. However, the solutions would not be very instructive because they are just numerical solutions, which we would have to solve separately for each set of initial conditions and assumed ks. We want a solution in terms of reactants and products alone.

To do this we assume that steps 1 and 2 are in thermodynamic equilibrium or

$$r_1 = k_{f1}[Cl_2] - k_{b1}[Cl]^2 = 0 \tag{4.258}$$

$$r_2 = k_{f2}[CO][Cl] - k_{b2}[COCl] = 0 \tag{4.259}$$

We can easily solve these algebraic equations for the concentrations of the intermediate species

$$[Cl] = \left(\frac{k_{f1}}{k_{b1}}\right)^{1/2} [Cl_2]^{1/2} \tag{4.260}$$

$$[COCl] = \frac{k_{f2}}{k_{b2}}[CO][Cl] = \frac{k_{f2}}{k_{b2}}\left(\frac{k_{f1}}{k_{b1}}\right)^{1/2} [CO][Cl_2]^{1/2} \tag{4.261}$$

Thus we have used the equilibrium step approximation to find expressions for the intermediates in terms of reactants and products.

We can now write the rate of the forward reaction in terms of these species alone,

$$r = r_3 = k_{f3}[COCl][Cl_2] \tag{4.262}$$

$$= k_{f3}\frac{k_{f2}}{k_{b2}}\left(\frac{k_{f1}}{k_{b1}}\right)^{1/2} [CO][Cl_2]^{3/2} \tag{4.263}$$

which agrees with the experimentally determined rate expression. We will leave the reverse rate expression for a homework problem.

4.13 REACTION MECHANISMS

All reactions occur by collisions between molecules or by collisions of molecules with surfaces. We will consider reactions at surfaces later, but here we consider the theory

Figure 4–12 Elementary reaction steps involving collisions of two reactants A and B, of A with an inert M, or with A with a surface S.

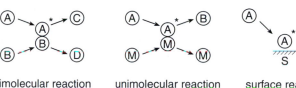

bimolecular reaction   unimolecular reaction   surface reaction
enzyme reaction

of homogeneous reactions. We will not attempt a quantitative or thorough description of reaction mechanisms but will only describe them in enough detail to be able to see how the engineer can control them. These collisions occur as sketched in Figure 4–12.

4.13.1 Theory of unimolecular reactions

An isolated molecule at zero pressure cannot react. Thus processes such as isomerization

$$A \to B \qquad (4.264)$$

or dissociation

$$A \to B + C \qquad (4.265)$$

do not involve simply the molecule rearranging its structure by itself. Rather, *collisions* of molecule A with other molecules (either another A molecule, a product, or an "inert") are necessary for reaction.

This is because there is a potential energy barrier for isomerization or dissociation that can only be overcome by adding energy to the appropriate bonds. Let us consider as prototypes the isomerization of cyclopropane to propylene

$$\text{cyclopropane} \to \text{propylene} \qquad (4.266)$$

and the hydrogenolysis of propane to methane and ethylene

$$C_3H_8 \to CH_4 + C_2H_4 \qquad (4.267)$$

In both reactions a C–C bond must be broken and C–H bonds must be broken and H atoms moved to other C atoms.

This requires sufficient energy inserted into the relevant bond vibration for the bond to break or for bonding locations to move. C–C and C–H bond energies in stable alkanes are greater than 80 kcal/mole, and these processes are very infrequent. As we will see later, hydrocarbon decomposition, isomerization, and oxidation reactions occur primarily by chain reactions initiated by bond breaking but are propagated by much faster abstraction reactions of molecules with parent molecules.

Consider the dissociation of a molecule $A–B$, a process whose heat of reaction ΔH_R is D_{AB}, the dissociation energy of AB. We can regard this as a chemical reaction

$$AB \to A + B \qquad (4.268)$$

or in our usual notation as the reaction $A \to B + C$. On a diagram of energy versus the $A–B$ distance, d_{AB}, this might look as shown in Figure 4–13. The activation energy for bond breaking will be equal to ΔH_R if there is no maximum in the $E(d_{AB})$ curve. However, if there is a maximum, then $E > \Delta H_R$. We expect that this reaction is probably endothermic,

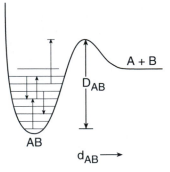

Figure 4–13   Energy diagram for the dissociation of a molecule into two products.

but it is possible that the dissociation reaction is exothermic if the dissociated state $A + B$ has a lower energy than AB.

Any energy well between atoms or fragments has vibrational levels because vibrations are quantized. The excitation of the $A–B$ bond therefore occurs by excitation of vibrational states to sufficiently high energies that the bond is broken. The AB molecule is colliding with other molecules in gas or liquid states and exchanging energy with colliding neighbors. If there are two states E_1 and E_2, then the ratio of populations (concentrations) of these two states on average is given by the Boltzmann factor

$$\frac{C_1}{C_2} = \exp\left(-\frac{E_2 - E_1}{RT}\right) \tag{4.269}$$

and if there is a ladder of states, the probability of excitation from the ground vibrational state to the top of the barrier is

$$p(E) = \exp\left(-\frac{E}{RT}\right) \tag{4.270}$$

This argument shows simply where the *Arrhenius temperature dependence* of reaction rates originates. Whenever there is an energy barrier that must be crossed for reaction, the probability (or rate) of doing so is proportional to a Boltzmann factor. We will consider the value of the pre-exponential factor and the complete rate expression later.

Returning to our $A–B$ dissociation process, the vibrational levels of the molecule can only be excited by collisions with other molecules. Consider the energy transfer of molecule A by collision with other A molecules as a chemical reaction

$$A + A \rightarrow A + A^*, \qquad r = kC_A^2 \tag{4.271}$$

We have transformed species A into species A^* by colliding with other A molecules that remained unchanged except that they must have lost the energy gained by forming A^*.

We write the reaction $A \rightarrow B$ as the sequence

$$A + A \rightleftarrows A^* + A, \qquad r_1 = k_{f1}C_A^2 - k_{b1}C_A C_{A^*} \tag{4.272}$$

$$A^* \rightarrow B, \qquad r_2 = k_2 C_{A^*} \tag{4.273}$$

with the first being reversible and the second irreversible.

Thus our unimolecular isomerization reaction actually occurs by a sequence of steps and is therefore a multiple-reaction system! We need a simplified expression for the overall rate of this rate in terms of C_A alone because A^* is an intermediate whose concentration is always very small just as for the free-radical intermediates and dimers in the previous examples.

Applying the pseudo-steady-state approximation to A^*, we obtain

$$\frac{dC_{A^*}}{dt} = r_1 - r_2 = k_{f1}C_A^2 - k_{b1}C_A C_{A^*} - k_2 C_{A^*} = 0 \qquad (4.274)$$

which can be solved for C_{A^*} to yield

$$C_{A^*} = \frac{k_{f1}C_A^2}{k_{b1}C_A + k_2} \qquad (4.275)$$

Now the rate of forming B is r_2; so we can write

$$r_2 = k_2 C_{A^*} = k_2 \frac{k_{f1}C_A^2}{k_{b1}C_A + k_2} \qquad (4.276)$$

This looks quite complicated because it does not even seem to obey first-order kinetics. However, we quickly recover a simple expression by simply assuming that $k_{b1}C_A \gg k_2$ to yield

$$r_2 = \frac{k_{f1}k_2}{k_{b1}}C_A \qquad (4.277)$$

Thus we have the mechanism of a first-order elementary reaction as being proportional to C_A and having a rate coefficient

$$k = \frac{k_{f1}k_2}{k_{b1}} \qquad (4.278)$$

The activation energy of this process is predicted to be $E_{f1} + E_2 - E_{b1}$. Although we call this an elementary reaction, it still proceeds through intermediate states of different vibrational and electronic energy.

4.14 COLLISION THEORY OF BIMOLECULAR REACTIONS

We consider the rate of the bimolecular reaction

$$A + B \rightarrow \text{ products}, \qquad r = kC_A C_B \qquad (4.279)$$

and we want to find an expression for k and to assure ourselves that the process should be first order in C_A and C_B. We write this reaction rate as

$$r = [\text{rate of collision of } A \text{ and } B] \times [\text{probability of reaction upon collision}]$$

The rate of collision of gas molecules is given by gas kinetic theory. Molecules have an average kinetic energy given by the expression

$$E_{\text{kin}} = \tfrac{3}{2}RT = \tfrac{1}{2}M\bar{u}^2 \qquad (4.280)$$

where M is the molecular weight, R is the gas constant per mole, and $\bar{u}^2$ is the square of the average thermal velocity of the molecule. The average velocity from kinetic theory of gases is

$$\bar{u} = \left(\frac{8RT}{\pi M}\right)^{1/2} \simeq 50,000 \text{ cm/sec} \qquad (4.281)$$

for N_2 at 300k. [Do not confuse the speed of thermal motion of a molecule $\bar{u}$ with u the speed that a molecule travel down a PFTR.]

From the kinetic theory of gases we find that the frequency ν_{AB} of an A molecule colliding with a B molecule is

$$\nu_{AB} = \tfrac{1}{3}\bar{u}\sigma_{AB}C_B \qquad (4.282)$$

in units of collisions per time. The quantity σ_{AB} is called the *collision cross section*, an area that is approximately the area of a molecule πd_{AB}^2, where d_{AB} is roughly the average radius of the molecules (assuming them spheres). The density of A molecules is given by the ideal gas law

$$C_A = \frac{P_A}{RT} \qquad (4.283)$$

Now putting all these together, we obtain a rate of collision of A with B

$$[\text{rate of collision}] = \nu_{AB}C_B$$

$$= \frac{\bar{u}\sigma_{AB}}{(RT)^2}P_A P_B$$

$$= \frac{(8RT/\pi M)^{1/2}\pi d_{AB}^2}{(RT)^2}P_A P_B$$

$$= \left(\frac{8RT}{\pi M}\right)^{1/2} \pi d_{AB}^2 C_A C_B \qquad (4.284)$$

This gives a collision rate constant for a second-order reaction as

$$k^{(2)} = \left(\frac{8RT}{\pi M}\right)^{1/2} \pi d_{AB}^2 = \bar{u}\sigma_{AB} \qquad (4.285)$$

This is the *maximum possible rate of bimolecular reaction*, the collision rate of the molecules that can react. We must multiply this by a probability of reaction in the collision; so actual rates must be less than this. We know that the units of k should be liters/mole time, and, since velocity is in length/time and cross section is in area/time, the units are correct if we make sure that we use volume in liters, and compute the area of a mole of molecules. If the molecular weight is 28 (air) and the temperature is 300 K, then we have

$$k^{(2)} \simeq 10^{11} \text{ liters/mole sec} \qquad (4.286)$$

By collision theory this is the maximum rate of a bimolecular reaction, because we must multiply this value by the probability that reaction will occur during the collision (a number less than unity). Thus we predict from collision theory that the pre-exponential factor k_o in a bimolecular reaction should be no larger than 10^{11} liters/mole sec.

4.15 ACTIVATED COMPLEX THEORY

The next more sophisticated theory of bimolecular reactions is called *activated complex theory*, which assumes that the collision of A and B forms a complex $(AB)^*$ and that the rate of the reaction depends on the rate of decomposition of this complex. We write this as

Figure 4–14 Energy diagram illustrating the reactants A and B and the activated complex $(AB)^*$ in the activated complex theory of bimolecular reactions.

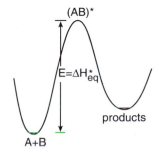

$$A + B \rightleftarrows (AB)^* \rightarrow \text{products} \qquad (4.287)$$

The intermediate $(AB)^*$ exists at the transition between reactants A and B and the products. It exists at the top of the energy barrier between these stable species as sketched in Figure 4–14. In activated complex theory one assumes that the reactants are in thermodynamic equilibrium with the activated complex,

$$A + B \rightleftarrows (AB)^*, \qquad K_{eq} = \frac{[(AB)^*]}{[(A)][(B)]} = e^{-\Delta G_o^*/RT} \qquad (4.288)$$

Since

$$\Delta G_o^* = \Delta H_R - T\Delta S^* \qquad (4.289)$$

this equilibrium composition is given by the expression

$$K_{eq} = \frac{[(AB)^*]}{[(A)][(B)]} = e^{-\Delta G_o^*/RT} = e^{\Delta S/R}e^{-\Delta H_R/RT} = K_o e^{-\Delta H_R/RT} \qquad (4.290)$$

The quantity ΔH_R in this expression is the energy barrier E in Figure 4–14.

The rate of the overall reaction is therefore

$$r = k'[(AB)^*] = k'K_{eq}[(A)][(B)] = k'K_o e^{-E/RT}[(A)][(B)] \qquad (4.291)$$

This gives a rate coefficient $k = k'K_o$, where these quantities can be calculated from statistical thermodynamics (which we shall not do here). [It can be shown from statistical mechanics that $k' = kT/h$, where k is Boltzmann's constant and h is Planck's constant.]

4.15.1 Statistical-mechanical theories

The most accurate theories of reaction rates come from statistical mechanics. These theories allow one to write the *partition function* for molecules and thus to formulate a quantitative description of rates. Rate expressions for many homogeneous elementary reaction steps come from these calculations, which use quantum mechanics to calculate the energy levels of molecules and potential energy surfaces over which molecules travel in the transition between reactants and products. These theories give

$$r = \frac{kT}{h}\frac{Q^*}{Q_A Q_B}e^{-E/RT} \qquad (4.292)$$

where the Qs are partition functions for the species.

These theories may have been covered (or at least mentioned) in your physical chemistry courses in statistical mechanics or kinetic theory of gases, but (mercifully) we will not go through them here because they involve a rather complex notation and are not necessary to describe chemical reactors. If you need reaction rate data very badly for some process, you will probably want to find the assistance of a chemist or physicist in calculating reaction rates of elementary reaction steps in order to formulate an accurate description of processes.

4.15.2   Reactions in liquids, on surfaces, and by enzymes

Here the problem of formulating a reaction rate expression is much more difficult because there are many atoms involved, and consequently the statistical mechanics and quantum mechanics are much more complex. We will consider the forms of rate expressions for surface- and enzyme-catalyzed processes in Chapter 7, but fundamental theories are usually not obtainable.

For all these types of reactions the rates must be obtained empirically under conditions as close to conditions of the industrial process as possible, but caution must be used in interpreting these rates.

4.16   DESIGNING REACTORS FOR MULTIPLE REACTIONS

For a single reaction in an isothermal reactor the design principles involved primarily the reactor configuration for minimum residence time. This generally favors the PFTR over the CSTR for positive-order kinetics. However, the CSTR frequently is less costly and easier to maintain, and one or more CSTRs can frequently be preferred over a PFTR.

When selectivity and yield of a given product need to be maximized, the design issues become more complicated. While minimum τ is frequently desired, it is usually more important to obtain maximum selectivity to a desired product and minimum selectivity to undesired products. For simple series and parallel reaction systems, we can fairly easily summarize the choices.

For series reactions with an intermediate desired, there is always an optimum τ for maximum yield, and the PFTR gives a higher maximum yield if both reactions have positive order, while the CSTR gives a higher maximum yield if the reactions are negative order (a rather rare occurrence). For series reactions with the final product desired, the PFTR requires the shorter time and gives less intermediate for positive-order kinetics.

For parallel reactions both reactor types give equal selectivities if the orders of both reactions are the same, and the CSTR can be preferred even though its τ would be longer because it may be less expensive. The CSTR is preferred for parallel reactions whenever the desired reaction has a lower order than the undesired reaction, while the PFTR is preferred (for both yield and τ) whenever the desired reaction has a higher order.

Most multiple-reaction systems are more complicated series–parallel sequences with multiple reactants, some species being both reactant and product in different reactions. These simple rules obviously will not work in those situations, and one must usually solve the mass-balance equations to determine the best reactor configuration.

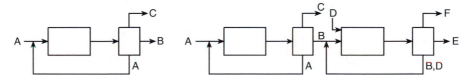

Figure 4–15 Sketch of a two-stage process where the reactant A from the first reactor is separated from the desired product B and the undesired product C. The desired product B is then fed into a second reactor, where another reactant D may be added to make a desired product E and an undesired product F, which are further separated.

4.16.1   Staged reactors

In practice, most industrial processes are staged with multiple reaction processes and separation units as sketched in Figure 4–15. If A is the key raw material and E is the key product, it is clear that many factors must be included in designing the process to maximize the yield of E. The effectiveness of the separations are obviously critical as well as the kinetics of the reactions and the choice of reactor type and conversion in each reactor. If separations are perfect, then the yields are equal to the selectivities, so that the overall yield in the process of making E from A is the product of the selectivity on making B in the first stage and the selectivity in making E from B in the second stage.

These considerations are only valid for isothermal reactors, and we shall see in the next two chapters that the possibility of temperature variations in the reactor can lead to much more interesting behavior. We will also see in Chapter 7 that with catalytic reactors the situation becomes even more complicated. However, these simple ideas are useful guides in the choice of a chemical reactor type to carry out multiple-reaction systems. We will still use these principles as the chemical reactors become more complicated and additional factors need to be included.

PROBLEMS

4.1 Write the reactor mass-balance equations that must be solved for the methane oxidation reaction system,

$$CH_4 + 2O_2 \rightarrow CO_2 + H_2O, \qquad r_1 = k_1 C_{CH_4} C_{O_2}$$

$$CO + H_2O \rightarrow CO_2 + H_2, \qquad r_2 = k_{f2} C_{CO} C_{H_2O} - k_{b2} C_{CO_2} C_{H_2}$$

$$CH_4 + H_2O \rightarrow CO + 3H_2, \qquad r_3 = k_3 C_{CH_4} C_{H_2O}$$

$$CH_4 + CO_2 \rightarrow 2CO + 2H_2, \qquad r_4 = k_4 C_{CH_4} C_{CO_2}$$

described in the text for a feed ot CH_4 and O_2

(a) in a batch reactor,

(b) in a CSTR,

(c) in a PFTR.

These reaction rates are written in very simple form. In all industrial processes catalysts are used to increase rates and selectivities, and the rate expressions are more complex than those indicated.

4.2 Write the equations to be solved for the reaction set

$$A \;\rightarrow\; B \;\rightarrow\; D$$
$$\downarrow \qquad\quad \downarrow$$
$$C \qquad E \;\rightarrow\; F$$

assuming all reactions are first order and irreversible

(a) in a PFTR,

(b) in a CSTR.

(c) Solve these equations for $C_j(\tau)$. [Partial credit will be given for setting up the equations in some situations.]

4.3 Find τ, S_B, and Y_B for 90% conversion in the reactions in a PFTR for all ks in min^{-1}.

(a) $A \rightarrow B,$ $r_1 = 0.5 C_A$

    $A \rightarrow C,$ $r_2 = 0.1 C_A$

(b) $A \rightarrow B,$ $r_1 = 0.5 C_A$

    $B \rightarrow C,$ $r_2 = 0.1 C_B$

4.4 Find τ, S_B, and Y_B for 90% conversion in the reactions in a CSTR for all ks in min^{-1}.

(a) $A \rightarrow B,$ $r_1 = 0.5 C_A$

    $A \rightarrow C,$ $r_2 = 0.1 C_A$

(b) $A \rightarrow B,$ $r_1 = 0.5 C_A$

    $B \rightarrow C,$ $r_2 = 0.1 C_B$

4.5 Solve for C_A, C_B, and C_C versus τ for the reactions

$$A \rightarrow B, \qquad r_1 = k_1$$
$$A \rightarrow C, \qquad r_2 = k_2 C_A$$

in a PFTR and in a CSTR for $k_1 = 2$ moles liter^{-1} min^{-1} and $k_2 = 1$ min^{-1}.

4.6 Find τ and S_B at 90% conversion in a PFTR and in a CSTR for $C_{A_o} = 4$ in the reactions of the previous problem. Which reactor type is better?

4.7 For the reactions

$$A \rightarrow B, \qquad r_1 = k_1 C_A$$
$$A \rightarrow C, \qquad r_2 = k_2 C_A^n$$

will a PFTR or a CSTR give the better selectivity to B for

(a) $n = 0$

(b) $n = 1$

(c) $n = 2$?

4.8 Ethanol is to be oxidized to acetaldehyde in aqueous solution. The aldehyde can further oxidize to acetic acid, which can decarboxylate to methane. There is an excess of O_2 in the solution; so all reactions are first order in the organics and irreversible.

(a) Write out the species mass-balance equations to be solved in PFTR and CSTR reactors.

(b) Find $C_j(\tau)$ in PFTR and CSTR.

(c) For $C_{Ao} = 0.1$ mole/liter, find the optimum residence times, conversions, and selectivities to acetaldehyde in PFTR and CSTR. The rate coefficients with a certain catalyst are 1.0, 0.1, and 0.2 min^{-1}, respectively.

(d) Sketch a reactor–separator configuration that will separate products and recycle unused ethanol. What is the best reactor?

(e) What reactor size is required if we need to produce 100 moles/h of acetaldehyde?

(f) How would your answer change if the second reaction step were second order?

4.9 Consider the reactions

$$A \rightarrow B, \qquad r_1 = k_1$$
$$A \rightarrow C, \qquad r_2 = k_2 C_A$$
$$A \rightarrow D, \qquad r_3 = k_3 C_A^2$$

(a) What is the best reactor to produce B, C, and D? Do S_B, S_C, and S_D increase or decrease with conversion?

(b) Solve these equations for all species versus τ in PFTR and CSTR for $k_1 = 8$, $k_2 = 8$, and $k_3 = 2$ in appropriate units with time in minutes and $C_{Ao} = 4$ moles/liter.

(c) Plot the differential selectivity s_C versus C_A and use this plot to determine S_C in PFTR and CSTR versus conversion.

(d) What would be the best single reactor to produce C for different C_{Ao} and conversion? For what situations would multiple reactors give the best selectivity to C?

4.10 Find X_A, S_B, Y_B, and F_B in a PFTR and in a CSTR for the reactions

$$A \rightarrow B, \qquad r_1 = k_1 C_A^2$$
$$A \rightarrow C, \qquad r_2 = k_2 C_A$$

with $k_1 = 2$ liter mole^{-1} min^{-1}, $k_2 = 1$ min^{-1}, $v = 20$ liters/min, $V = 100$ liters, and $C_{Ao} = 4$ moles/liter.

4.11 (a) Develop expressions for X_A, S_B, Y_B, and F_B in a CSTR and in a PFTR for the reactions

$$A \rightarrow B, \qquad r_1 = k_1 C_A$$
$$B \rightarrow C, \qquad r_2 = k_2 C_B^2$$

(b) For the CSTR find numerical values for these quantities for $k_1 = 2$ min^{-1}, $k_2 = 0.2$ liter mole^{-1} min^{-1}, $v = 20$ liters/min, $V = 100$ liters, and $C_{Ao} = 4$ moles/liter.

4.12 We want to isomerize a species A to form B in liquid solution. The rate for this process is $r_1 = k_1 C_A$. However, A can also dimerize to form a worthless and unreactive species C with a rate $r_1 = k_1 C_A^2$. Calculate the yield of B in a CSTR and in a PFTR if 90% of A is reacted if $C_{Ao} = 2$ molar, $k_2 = 0.2$ liter/mole min, and $k_1 = 0.1$ min^{-1}. How might conditions be altered to improve the performance of a tubular reaction?

4.13 For the parallel reactions $A \rightarrow B, r_1 = k_1 C_A$ and $A \rightarrow C, r_2 = k_2$, calculate the residence time and selectivity and yield of B in PFTR and CSTR for $C_{Ao} = 2$ moles/liter, $k_1 = 5$ min^{-1}, and $k_2 = 1$ mole liter^{-1} min^{-1} at 50, 90, 95, and 99% conversion of A. Note the tradeoff between conversion and selectivity in the two reactors.

4.14 You find that a process and reactor that you don't know much about gives a higher selectivity to the desired product at higher conversion.

(a) If the process involves parallel reactions, what can you say about the kinetics?

(b) If the process involves series reactions, what can you say about the kinetics?

(c) Would the above processes give better selectivity in a mixed or an unmixed reactor?

4.15 [Computer] Plot $C_j(\tau)$ versus $k_1\tau$ for the reaction $A \rightarrow B \rightarrow C$ in a PFTR for

(a) First-order irreversible reactions with $k_2/k_1 = 0.1, 0.3, 1, 3, 10$.

(b) First-order reversible reactions with $k_2/k_1 = 0.1, 0.3, 1, 3, 10$ and equilibrium constants $= 2.0$.

(c) Repeat (a) for second-order kinetics for all steps.

(d) Plot (by hand is OK) $C_{B\,max}$ and τ_{max} versus k_2, k_1 for part (a).

4.16 [Computer] Plot $C_j(\tau)$ versus $k\tau$ for the polymerization process $A_1 \rightarrow A_2 \rightarrow A_3 \rightarrow A_4 \rightarrow \cdots \rightarrow A_j \rightarrow \cdots$ to $j = 20$ for all ks equal if all reactions are first order and irreversible with rate coefficient k_i

(a) in a CSTR,

(b) in a PFTR.

[Note: This is a prototype of polymerization reactions and reactors that will be considered further in Chapter 11.]

4.17 Ethylene oxide (EO) is made by passing ethylene and oxygen over a promoted silver catalyst in a multitube reactor with each tube 20 ft long and 1 in. in diameter. The feed is 0.2 atm of O_2, 0.6 atm of C_2H_4, and 2.0 atm of CO_2. At 50% O_2 conversion the selectivity to EO is 80% [based on carbon, $S_{EO} = C_{EO}/(C_{EO} + 2C_{CO_2})$] at 270°C at a reactor residence time of 1.2 sec with the rest of the products CO_2 and H_2O. After the reactor the EO is condensed in H_2O and enough CO_2 is stripped in caustic to maintain the recycled product at this composition after adding makeup ethylene and oxygen.

(a) Sketch the flow diagram of the process and calculate the partial pressures before the reactor and before and after the EO stripper (which also removes all water) assuming no pressure drops.

(b) What are the makeup flow rates of ethylene and oxygen?

(c) The reaction to form EO is 1/2 order in P_{O_2} and independent of ethylene. Ignoring mole number changes, calculate the rate constant for this reaction from these data.

(d) How many tubes must be used to produce 200 tons/day of EO?

(e) Experiments show that at 300°C the O_2 conversion is 60% and the selectivity is 70%. Calculate the number of tubes required to produce 200 tons/day of EO at this temperature.

4.18 The reaction $NO + O_2 \rightarrow NO_2$ could proceed through the elementary steps

$$NO + O_2 \rightleftarrows NO_3$$

$$NO_3 + NO \rightarrow 2NO_2$$

rather than the mechanism described in the text. Show that this mechanism can predict $r = kC_{NO}^2 C_{O_2}$, which is identical to the expression derived in the text. How might you determine experimentally which mechanism is correct?

4.19 (a) Show that a unimolecular reaction should obey second-order kinetics as the pressure goes to zero.

(b) Show that a termolecular reaction such as NO oxidation to NO_2 should obey second-order kinetics as the pressures go to zero.

4.20 The parallel first-order reactions $A \rightarrow B, A \rightarrow C$ have activation energies of 8 and 10 kcal/mole, respectively. In a 1 liter batch reactor at 100°C the selectivity to B is 50% and the conversion is 50% in a reaction time of 10 min with $C_{Ao} = 1$ mole/liter. The solvent is water and the reactor can be pressurized as needed to maintain liquids at any temperature.

(a) What temperature and reactor volume are required to produce 90% selectivity to B at 90% conversion in a CSTR using a feed of 2 molar A at a flow rate of 10 liters/min?

(b) What temperature and reactor volume are required to produce 90% selectivity to C at 90% conversion in a CSTR using a feed of 2 molar A at a flow rate of 10 liters/min?

(c) What temperature and reactor volume are required to produce 90% selectivity to B at 90% conversion in a PFTR using a feed of 2 molar A at a flow rate of 10 liters/min?

(d) What temperature and reactor volume are required to produce 90% selectivity to C at 90% conversion in a PFTR using a feed of 2 molar A at a flow rate of 10 liters/min?

4.21 The first-order series reactions $A \rightarrow B, B \rightarrow C$ have activation energies of 8 and 10 kcal/mole, respectively. In a 1-liter batch reactor at 100°C the selectivity to B is 50% and the conversion is 50% in a reaction time of 10 min with $C_{Ao} = 1$ mole/liter. The solvent is water and the reactor can be pressurized as needed to maintain liquids at any temperature.

(a) What temperature and reactor volume are required to produce 90% selectivity to B at 90% conversion in a CSTR using a feed of 2 molar A at a flow rate of 10 liters/min?

(b) What temperature and reactor volume are required to produce 90% selectivity to C at 90% conversion in a CSTR using a feed of 2 molar A at a flow rate of 10 liters/min?

(c) What temperature and reactor volume are required to produce 90% selectivity to B at 90% conversion in a PFTR using a feed of 2 molar A at a flow rate of 10 liters/min?

(d) What temperature and reactor volume are required to produce 90% selectivity to C at 90% conversion in a PFTR using a feed of 2 molar A at a flow rate of 10 liters/min?

4.22 Consider the reaction sequence

$$A \xrightarrow{k_1} B \xrightarrow{k_2} C \xrightarrow{k_3} D$$
$$k_4 \downarrow \qquad k_5 \downarrow \qquad k_6 \downarrow$$
$$E \qquad\qquad F \qquad\qquad G$$

with all reactions first order and irreversible.

(a) Find $C_j(\tau)$ for all species in a CSTR.

(b) Find $C_j(\tau)$ for all species in a PFTR.

(c) Find S_E in each of these reactors.

(d) Find S_D in each of these reactors for very long residence times.

4.23 (a) What is the overall yield of product in a sequence of n steps if each of the n reactors has the same selectivity S and conversion X if unconverted reactant after each stage is discarded?

(b) What is the overall yield of product in a sequence of n reactor steps if each reactor has the same selectivity S and conversion X if unconverted reactant after each stage is recycled?

(c) What is the overall yield of a six-step process in which each step has a yield of 0.8? If the raw materials cost $0.10/lb, what must the products sell for to obtain a positive cash flow?

(d) Suppose that this six-step process can be replaced by a three-step process. What per step yield would be necessary to achieve the same overall yield? Why might the three-step process be preferred even if its overall yield is less than the six-step process?

4.24 We wish to produce ethylene by the oxidative dehydrogenation of ethane. Over a suitable catalyst the reactions and rates are found to be

$$C_2H_6 + \tfrac{1}{2}O_2 \rightarrow C_2H_4 + H_2O, \qquad r_1 = k_1 C_{C_2H_6} C_{O_2}$$

$$C_2H_6 + \tfrac{7}{2}O_2 \rightarrow 2CO_2 + 3H_2O, \qquad r_2 = k_2 C_{C_2H_6} C_{O_2}^2$$

The reaction is to be run at 1000 K, where it is found that $k_1 = 9$ l/mole sec and $k_2 = 1670$ l^2 /mole^2 sec. The feed pressures are $P_{C_2H_6} = 2$ atm, $P_{O_2} = 1$ atm, and $P_{N_2} = 4$ atm. The pressure is constant and the reaction is to have a residence time such that the product contains 0.05 atm of O_2. Assume that there is sufficient diluent that the density remains constant.

(a) Will a PFTR or a CSTR give a higher ethylene selectivity?

(b) Will a PFTR or a CSTR require the longer residence time?

(c) Calculate the selectivities to ethylene on a carbon atom basis in PFTR and CSTR for this O_2 conversion.

(d) Set up the problem to calculate the residence times required in PFTR and CSTR. [This illustrates how complex multiple-reaction systems can become and why the selectivity formula is useful.]

4.25 Yttrium 90, which has a half-life of 64.2 h, is produced from the fission product strontium 90, which has a half-life of 27.5 years. How much Y-90 will be present after 1 week from a sample of pure Sr-90 in a closed container?

4.26 We wish to produce B in the reaction system

$$A \rightarrow B, \qquad r_1 = k_1 C_A$$

$$A \rightarrow C, \qquad r_2 = k_2 C_A$$

A costs $1.20/lb, B sells for $2.50/lb, and C costs $0.50/lb to dispose. Assume that unreacted A can be recycled into the reactor at no cost.

(a) What relation between the ks will give a positive cash flow if the reactor and operating costs are ignored?

(b) What relation between the ks will give a 30% profit on sales if the reactor and operating costs are ignored?

4.27 We have a ten-step process to produce a product from a reactant, and each step has an 80% yield. What must be the yield of each step in a three-step process to obtain improved economics before we worry about lower reactor and separation process costs?

4.28 On a triangular diagram of C_A, C_B, and C_C sketch trajectories (not detailed plots) in the reactions $A \rightarrow B \rightarrow C$ and also in the reactions $A \rightarrow B, A \rightarrow C$ for the following situations:

(a) $k_1 = 2k_2$, irreversible, series, PFTR,

(b) $k_1 = 2k_2$, irreversible, series, CSTR,

(c) $k_1 = 2k_2, k_1' = 0.5k_1, k_2' = 0.5k_2$, series, PFTR,

(d) $k_1 = 2k_2$, $k_1' = 0.5k_1$, $k_2' = 0.5k_2$, series, CSTR,

(e) $k_1 = 2k_2$, irreversible, parallel, PFTR,

(f) $k_1 = 2k_2$, irreversible, parallel, CSTR,

(g) $k_1 = 2k_2$, $k_1' = 0.5k_1$, $k_2' = 0.5k_2$, parallel, PFTR,

(h) $k_1 = 2k_2$, $k_1' = 0.5k_1$, $k_2' = 0.5k_2$, parallel, CSTR.

4.29 On a triangular diagram of C_A, C_B, and C_C sketch trajectories (not detailed plots) in the reactions $A \rightarrow B$, $A \rightarrow C$ for $r_1 = k_1$, $r_2 = 0.5k_1 C_A$ in PFTR and CSTR.

4.30 In the cross-flow reactor a species B is introduced at several places in a reactor while species A is introduced at the entrance as sketched in Figure 4–16. What might be the benefits on selectivity and in autocatalytic reactions by this flow situation?

Figure 4–16 Cross-flow reactor in which a species B is introduced at several points in the reactor.

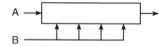

4.31 The reactions

$$A \rightarrow B, \qquad r_1 = k_1$$
$$A \rightarrow C, \qquad r_2 = k_2 C_A$$

give 67% selectivity to the desired product B at 75% conversion of A in a CSTR with a reaction time of 1 min with $C_{Ao} = 2$ moles/liter.

(a) What are the selectivity and conversion for this reaction and feed in a PFTR with $\tau = 1$ min?

(b) What are the conversion and selectivity in this reactor if $C_{Ao} = 1$ mole/liter?

(c) What τ and conversion will give 99% selectivity in a CSTR?

(d) Give short reasons why the trends in each of the above situations should be expected.

4.32 The reactions

$$A \rightarrow B, \qquad r_1 = k_1$$
$$A \rightarrow C, \qquad r_2 = k_2 C_A$$

give 67% selectivity to the desired product B at 75% conversion of A in a CSTR with a reaction time of 1 min with $C_{Ao} = 2$ moles/liter at a temperature of 400 K. It is found that $E_1 = 5$ kcal/mole and $E_2 = 10$ kcal/mole. At what temperature will S_B be 95% for $\tau = 1$ min and $C_{Ao} = 2$ moles/liter in a CSTR?

4.33 Consider the reactions

$$A \rightarrow B, \qquad r_1 = k_1 C_A$$
$$B \rightarrow C, \qquad r_2 = k_2 C_B$$
$$A \rightarrow D, \qquad r_3 = k_3 C_A^2$$

with B the desired product.

(a) How would you describe this reaction set?

(b) Should one use a PFTR or CSTR, and is there an optimum residence time? [The answer depends on the relative ks.]

(c) Find $C_j(\tau)$ in PFTR and CSTR if $r_3 = k_3 C_A$.

4.34 Consider the liquid-phase reaction system

$$A \rightarrow B, \qquad r_1 = k_1 C_A$$

$$A \rightarrow C, \qquad r_2 = k_2 C_A^2$$

(a) Will a PFTR or CSTR give a shorter residence rime for the same conversion?

(b) Find an expression for τ in a CSTR in terms of k_1, k_2, and C_{Ao}, and C_A.

(c) Find S_B and Y_B in a CSTR with rates in moles/liter h, $k_1 = 4$ and $k_2 = 1$ (in appropriate units) and $C_{Ao} = 2$ moles/liter if the reaction is to be run to 90% conversion of A.

4.35 As discussed in the text, the reaction of CO and Cl_2 to form phosgene obeys the rate expression $r = k C_{CO} C_{Cl_2}^{3/2}$.

(a) Find an expression for C_{COCl_2} versus τ in a CSTR for stoichiometric feed assuming constant density.

(b) Find an expression for C_{COCl_2} versus τ in a PFTR for stoichiometric feed assuming constant density.

(c) Repeat these calculations (or at least set them up) for pure feeds with no diluent assuming all species are ideal gases.

4.36 Some of the raw materials in manufacturing polyurethanes (from TDIC and EG monomer) are Cl_2 and HNO_3, which are used to prepare the TDIC monomer.

(a) Assuming all stages in processing the monomer are 100% efficient, how many tons of salt are used to prepare a ton of PU?

(b) Assuming that the HNO_3 is derived from ammonia, which is prepared by steam reforming of methane, how many tons of CH_4 are required per ton of PU if all stages are 100% efficient?

(c) Repeat the above calculations assuming that each stage is 90% efficient.

4.37 In nitric acid synthesis from methane the synthesis of H_2 has a yield of 90%, NH_3 90%, NO 90%, N_2O_5 90%, HNO_3 90%. What is the yield of HNO_3 synthesis from CH_4, and how many pounds of CH_4 are required to produce a pound of HNO_3?

4.38 It is desired to produce B in the series reactions $A \rightarrow B \rightarrow C$, $r_1 = k_1 C_A$, $r_2 = k_2 C_B$, in two equal-volume CSTRs in series. Find expressions for the optimum residence time and maximum yield of B. Find expressions for the optimum residence time and maximum yield of B for a large number of CSTRs in series.

4.39 We wish to optimize the selectivity to B from A in the reactions

$$A \rightarrow B, \qquad r = k C_A$$

$$A \rightarrow C, \qquad r = k C_A^2$$

with $k = 1/4$ (units of moles, liters, and minutes and $C_{Ao} = 2$ moles/liter.

Calculate S_B and Y_B in a CSTR for $X_A = 0.99, 0.90$, and 0.5.

4.40 We wish to optimize the selectivity to B from A in the reactions

$$A \rightarrow B, \qquad r = k C_A$$

$$A \rightarrow C, \qquad r = k C_A^2$$

with $k = 0.25$ (units of moles, liters, and minutes and $C_{Ao} = 2$ moles/liter.

(a) Calculate S_B and Y_B in a PFTR for $X_A = 0.99, 0.90$, and 0.5.

(b) Calculate S_B and Y_B in a CSTR for $X_A = 0.99, 0.90$, and 0.5.

4.41 We wish to optimize the selectivity to B from A in the reactions

$$A \rightarrow B, \qquad r = kC_A^2$$
$$A \rightarrow C, \qquad r = kC_A$$

with $k = 1/4$ (units of moles, liters, and minutes and $C_{Ao} = 2$ moles/liter.

(a) Calculate S_B and Y_B in the optimum single reactor for $X_A = 0.99, 0.90$, and 0.5.

(b) What combination of two CSTRs will give the optimum S_B for these conversions?

4.42 The reaction to form NO

$$N_2 + O_2 \rightarrow 2NO$$

in high-temperature combustion processes is thought to proceed by the elementary steps

$$O_2 \rightleftarrows 2O, \qquad r_{i,t} = k_i C_{O_2} - k_t C_O^2$$
$$N_2 + O \rightarrow NO + N, \qquad r_{p1} = k_{p1} C_{N_2} C_O$$
$$O_2 + N \rightarrow NO + O, \qquad r_{p2} = k_{p2} C_{O_2} C_N$$

Find a reasonable reaction rate for $r(N_2, O_2)$.

4.43 The gas-phase dehydrogenation of ethane to ethylene

$$C_2H_6 \rightarrow C_2H_4 + H_2$$

proceeds through the elementary reaction steps

$$C_2H_6 + H \rightarrow C_2H_5 + H_2$$
$$C_2H_6 \rightleftarrows C_2H_5 + H$$
$$C_2H_5 \rightarrow C_2H_4 + H$$

(a) Identify the initiation, propagation, and termination steps.

(b) Find a reasonable expression for $r(C_{C_2H_6})$.

(c) The bond energy of ethane is 104 kcal/mole, and the propagation steps have very low activation energies. What is the approximate activation energy of this reaction?

4.44 Using the mechanism for phosgene synthesis

$$CO + Cl_2 \rightleftarrows COCl_2$$

discussed in the text, find the rate expression predicted for the reverse reaction.

Show that this expression consistent with thermodynamic equilibrium in this reaction.

4.45 Consider the reaction system

$$A \rightleftarrows B \rightleftarrows C$$

with $k_{1f} = 100, k_{1b} = 100, k_{2f} = 1, k_{2b} = 1$, where we start with pure A, and B is the desired product.

(a) What are S_B and Y_B for a very long residence time?

(b) Estimate S_B and Y_B in a PFTR for $\tau = 0.02$. Do not solve the equations.

(c) What reaction–separation scheme will give 100% yield of B?

(d) Reactant A can also react directly to form C with rate coefficient $k_{3f} = 0.1$. What is k_{3b}?

4.46 Consider the reaction system

$$A \rightleftarrows B \rightleftarrows C \rightleftarrows D \rightleftarrows A$$

with $k_1 = 1$, $k_1' = 0.1$, $k_2 = 1$, $k_2' = 1$, $k_3 = 0.1$, $k_3' = 0.1$, $k_4 = 10$ in a continuous reactor.

(a) What is k_4'?

(b) Estimate the optimum S_D and Y_D obtainable, starting with pure A.

(c) Estimate the optimum S_C and Y_C obtainable, starting with pure A.

4.47 Suppose you need to produce B in the reactions

$$A \rightarrow B \rightarrow C$$

with $k_1 = 1\,\mathrm{h}^{-1}$ and $k_2 = 0.1\,\mathrm{h}^{-1}$. Find S_B, Y_B, and τ in CSTR and PFTR reactors for maximum yield of B. Calculate these from the formulas in the text and check your answers with the figures in the text.

4.48 The expressions in the text for the series reactions $A \rightarrow B \rightarrow C$ cannot be used if $k_1 = k_2$ because the expressions become degenerate (zeros in numerator and denominator). Solve for $C_A(\tau)$, $C_B(\tau)$, and $C_C(\tau)$ for these reactions in PFTR and CSTR for $k_1 = k_2$. Note that these expressions are simpler than the general expressions.

4.49 Our plant produces a chemical A, which is not toxic, but which slowly hydrolyzes in H_2O to form a highly toxic substance B. This chemical B is slowly oxidizes in air to form a harmless chemical C.

(a) Describe the concentrations in a river downstream from our plant for a steady leak of A of 2 ppm in the river if $k_1 = 0.002$ and $k_2 = 1000$ with both ks in liter/mole h. The river flows in plug flow at 5 mph, and the saturation concentration of O_2 in water is $C_{O_2} = 1.5 \times 10^{-3}$. Assume that O_2 is replenished as fast as it is consumed.

(b) How far downstream will the toxic chemical reach a maximum?

(c) The intermediate B is highly toxic to fire ants and boll weevils, but only moderately toxic to humans. How might we market this product as a pesticide with interesting properties? How and under what conditions should this chemical A be applied? Monsanto makes an herbicide Roundup with similar properties.

4.50 Our plant produces a chemical A that is not toxic but slowly hydrolyzes in H_2O to form a highly toxic substance B. This chemical A also slowly oxidizes in air to form a harmless chemical C.

(a) Describe the concentrations in a river downstream from our plant for a steady leak of A of 2 ppm in the river if $k_1 = 0.002$ and $k_2 = 1000$ with both ks in liter/mole h. The river flows in plug flow at 5 mph, and the saturation concentration of O_2 in water is $C_{O_2} = 1.5 \times 10^{-3}$. Assume that O_2 is replenished as fast as it is consumed.

(b) The intermediate B is highly toxic to fire ants and boll weevils, but only slightly toxic to humans. How might we market this product as a pesticide with interesting properties? How and under what conditions should this chemical A be applied?

4.51 Consider the reactions

$$A \rightarrow B, \qquad r_1 = 9$$
$$A \rightarrow C, \qquad r_2 = 6C_A$$
$$A \rightarrow D, \qquad r_3 = C_A^2$$

with rates in moles/liter min and concentrations in mole/liter.

(a) What type of reactor and what conversion will give the maximum selectivity to C for $C_{Ao} = 2$ moles/liter?

(b) Calculate the maximum selectivity and yield obtainable to form C.

(c) Repeat these calculations if B or D are the desired products.

4.52 Consider the parallel reactions

$$A \rightarrow 2B, \qquad r_1 = k_1 C_A$$
$$A \rightarrow 3C, \qquad r_2 = k_2 C_A$$

where all species are ideal gases with no diluent.

(a) Write expressions for $C_j(\mathbf{X_i})$.

(b) Find $C_j(\tau)$ in a CSTR.

(c) Set up expressions for $\tau(C_j)$ in a PFTR.

4.53 We wish to produce B in the reaction $A \rightarrow B$ in a continuous reactor at $v = 5$ liters/min with $C_{Ao} = 4$ moles/liter. However, we find that there is a second reaction $B \rightarrow C$ that can also occur. We find that both reactions are first order and irreversible with $k_1 = 1.0$ min^{-1} and $k_2 = 0.1$ min^{-1}. Find τ, V, C_B, S_B, and Y_B for 95% conversion of A in a PFTR and in a CSTR.

4.54 We wish to produce B in the reaction $A \rightarrow B$ in a continuous reactor at $v = 5$ liters/min with $C_{Ao} = 4$ moles/liter. However, we find that there is a second reaction $B \rightarrow C$ that can also occur. We find that both reactions are first order and irreversible with $k_1 = 1.0$ min^{-1} and $k_2 = 0.1$ min^{-1}. Find τ, V, S_B, and Y_B for maximum yield of B in a PFTR and in a CSTR.

4.55 We wish to produce B in the reaction $A \rightarrow B$ in a continuous reactor at $v = 5$ liters/min with $C_{Ao} = 4$ moles/liter. However, we find that there is a second reaction $A \rightarrow C$ that can also occur. We find that both reactions are first order and irreversible with $k_1 = 1.0$ min^{-1} and $k_2 = 0.1$ min^{-1}. Find τ, V, C_B, S_B, and Y_B for 95% conversion of A in a PFTR and in a CSTR.

4.56 We wish to produce B in the reaction $A \rightarrow B$ in a continuous reactor at $v = 5$ liters/min with $C_{Ao} = 4$ moles/liter. However, we find that there is a second reaction $B \rightarrow C$ that can also occur. We find that both reactions are second order and irreversible with $k_1 = 1.0$ liter mole^{-1} min^{-1} and $k_2 = 0.1$ liter mole^{-1} min^{-1}. Find τ, V, C_B, S_B, and Y_B for 95% conversion of A in a PFTR and in a CSTR.

4.57 We wish to produce B in the reaction $A \rightarrow B$ in a continuous reactor at $v = 5$ liters/min with $C_{Ao} = 4$ moles/liter. However, we find that there is a second reaction $B \rightarrow C$ that can also occur. We find that both reactions are second order and irreversible with $k_1 = 1.0$ liter mole^{-1} min^{-1} and $k_2 = 0.1$ liter mole^{-1} min^{-1}. Find τ, V, S_B, and Y_B for *maximum yield* of B in a PFTR and in a CSTR.

Chapter 5

NONISOTHERMAL REACTORS

We have thus far considered only isothermal reactors in which we specified the temperature. However, real reactors are almost always and unavoidably operated nonisothermally. This is because (1) reactions generate or absorb large amounts of heat and (2) reaction rates vary strongly with temperature.

In fact, we usually *want* to operate exothemic reactions nonisothermally to take advantage of the heat release in the reaction to heat the reactor to a temperature where the rates are higher and reactor volumes can be smaller. However, if the temperature is too high, equilibrium limitations can limit the conversion, as we saw previously for NH_3 and CH_3OH synthesis reactions.

Small lab reactors are frequently thermostatted, either by contact with the surroundings (a small beaker on a bench) or by placing them on a hot plate or in an oven. This is because in a lab reactor the wall area (where heat transfer occurs) is usually large compared to the reactor volume (where reaction heat is generated).

However, the large reactors encountered in industry are usually so large that it is impossible to thermostat the reactor at a fixed temperature that would be independent of the reactions and conversions. Further, it is frequently desired to use the heat release from exothermic reactions to heat the reactor above the feed temperature to attain a higher rate. On the other hand, one of the greatest problems in exothemic reactions is that of overheating if the heat generated is too large.

The temperature dependence of reactions comes from dependences in properties such as concentration ($C_j = P_j/RT$ for ideal gases) but especially because of the temperature dependence of rate coefficients. As noted previously, the rate coefficient usually has the Arrhenius form

$$k(T) = k_o e^{-E/RT} \tag{5.1}$$

if we ignore the small temperature dependence in the pre-exponential factor.

One important reason to consider the nonisothermal reactor is because it is the major cause of accidents in chemical plants. Thermal runaway and consequent pressure buildup and release of chemicals is an ever-present danger in any chemical reactor. Engineers must be intimately acquainted with these characteristics in designing and operating chemical reactors.

Chemical reactors may look similar to other units of chemical processing and sometimes they behave similarly, but the nonisothermal chemical reactor has nonlinearities that never occur in nonreacting systems.

5.1 HEAT GENERATION AND REMOVAL

We first need to develop energy-balance equations for these reactors. Shown in Figure 5–1 is a sketch of a generalized reactor with heat flows indicated. Heat flows into the reactor with the reactants and out with the products. Reaction generates or absorbs heat, depending on whether the reactions are exothermic or endothermic. Any stirring and friction generate heat. Finally, heat is transferred through walls in heat exchange. In a CSTR the temperature in the reactor is equal to the temperature at the exit, but in any other reactor the temperature and conversion are functions of position in the reactor.

A reactor will be isothermal at the feed inlet temperature T_0 if (1) reactions do not generate or absorb significant heat or (2) the reactor is thermostatted by contact with a temperature bath at coolant temperature T_0. For any other situation we will have to solve the energy-balance equation along with the mass balance to find the temperature in the reactor. We therefore must set up these equations for our mixed and unmixed reactors.

5.1.1 Thermodynamics of a flow system

Before we develop the energy balances for our reactors, it is worthwhile to define some quantities from thermodynamics because the energy balances we need are *thermal energy balances*. We begin with the First Law of Thermodynamics,

$$d\mathbf{U} = d\mathbf{q} - d\mathbf{w} \tag{5.2}$$

where $\mathbf{U}$ is the internal energy per mole, $\mathbf{q}$ is the amount of heat *added* to the system per mole, and $\mathbf{w}$ is the work done *by* the system per mole. For a flow system, pressure–volume work is done in flowing a fluid through the system,

$$d\mathbf{w} = d(\mathbf{PV}) + d\mathbf{w_s} \tag{5.3}$$

where $\mathbf{w_s}$ is any shaft work that may be done *by* the system. Since the enthalpy is defined through the relation

$$d\mathbf{H} = d\mathbf{U} + d(\mathbf{PV}) \tag{5.4}$$

Figure 5-1 Energy balance in chemical reactors. The volume shown could be the total reactor volume in a CSTR or a differential volume in a PFTR.

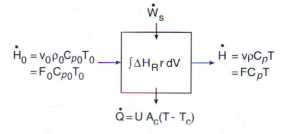

an energy balance on a flow system becomes

$$d\mathbf{H} = d\mathbf{q} - d\mathbf{w}_s \tag{5.5}$$

These are the fundamental thermodynamic equations from which we can develop our energy balances in batch, stirred, and tubular reactors.

In thermodynamics these quantities are usually expressed in *energy per mole* (quantities in bold), while we are interested in the rate of energy change or energy flow in *energy per time*. For these we replace the molar enthalpy $\mathbf{H}$ (cal/mole) of the fluid by the rate of enthalpy flow $\dot{H}$ in a flow system.

For the energy balance in a flow system we therefore assume that we can make an *enthalpy balance* on the contents of the reactor. We can write the rate of enthalpy generation $\Delta \dot{H}$ in any flowing or closed system as

$$\Delta \dot{H} = \dot{H}_o - \dot{H} \tag{5.6}$$

$$\Delta \dot{H} = \sum_j F_{oj} \mathbf{H}_{oj} - \sum_j F_j \mathbf{H}_j \tag{5.7}$$

where the first term on the right is the rate of enthalpy flow into the system and the second is the enthalpy flow out of the system.

5.1.2   Heat generated by reaction

Our major concern is the heat generated or absorbed by the chemical reactions that occur in the reactor. If $\mathbf{H}_j$ is the molar enthalpy of a chemical species, then the heat of a chemical reaction is defined as

$$\sum_j \nu_j \mathbf{H}_j = -\Delta H_R \tag{5.8}$$

If a reaction is exothermic, then the reaction generates heat, which tends to increase the reactor temperature, while if a reaction is endothermic, the reaction absorbs heat, which tends to cool the reactor.

5.1.3   Heat removal and addition through reactor walls

Heat can be removed from or added to a reactor through heat exchange across the walls. For an integral reactor we write the rate of heat removal from the reactor $\dot{Q}$ as

$$\dot{Q} = U A_c (T - T_c) \tag{5.9}$$

In this equation, U is the heat transfer coefficient in energy per area A_c per temperature difference (don't confuse this U with the internal energy $\mathbf{U}$), and A_c is the area across which heat exchange occurs between the reactor at temperature T and coolant at temperature T_c. We want to define $\dot{Q}$ as the rate of heat removal as a positive quantity; so $\dot{Q}$ will be positive if $T > T_c$. If $T < T_c$, heat flows into the reactor; so the reactor is being heated. We could use T_h as the heating temperature, but since the cooled reactor is the more interesting situation, we shall use T_c as the temperature of the fluid, which is exchanging heat with the reactor.

For a differential reactor we write

$$d\dot{Q} = U(T - T_c)\, dA_c \tag{5.10}$$

as the heat flow through an element of reactor wall area dA_c between the reactor at T and coolant at T_c.

5.1.4   Shaft work

Any stirring within the reactor will generate heat, and we call this term $\dot{W}_s$. We omit gravity and kinetic energy terms in the energy balance because these are usually very small compared to the other terms in the energy balance in a chemical reactor.

Note that we call this term positive when work is done on the system, just as we make $\dot{Q}$ positive when heat is removed from the reactor. Thus the signs on $\dot{W}_s$ and $\dot{Q}$ are opposite to those of $d\mathbf{q}$ and $d\mathbf{w}_s$ in the thermodynamic equations. We are always interested in shaft work done by a stirrer on the reactor fluid, and we are usually interested in cooling rather than heating. Therefore, we carry these sign conventions so that we can be sure of the signs of these terms.

5.1.5   Heat flow with fluid

The other mode of heat interchange with the surroundings in a flow reactor is by inflow and outflow of heat with the fluid entering and leaving the reactor. If the fluid is flowing into the reactor with volumetric flow rate v_o and has a heat capacity per reactor volume ρC_{po} or with total molar flow rate F_o with a heat capacity per mole $\mathbf{C}_{po}$ (written in bold), then the rate of enthalpy flow into the reactor is

$$\dot{H} = v_o \rho C_{po} T_o = F_o \mathbf{C}_{po} T_o \tag{5.11}$$

and the enthalpy carried out is similarly

$$\dot{H} = v \rho C_p T = F \mathbf{C}_p T \tag{5.12}$$

where the quantities with subscript o represent inflow and without the subscripts those in the outflow. The total molar flow rate F is the sum of the molar flow rates of all species $F = \sum F_j$. The enthalpy per mole of fluid $\mathbf{H} = \mathbf{C}_p T$, so that multiplication of $\mathbf{H}$ by the molar flow F rate gives $\dot{H}$, or $\dot{H} = F\mathbf{H}$. The reference state of enthalpy where $\mathbf{H} = 0$ is arbitrary, and we always write enthalpy changes; so the above expression is adequate (assuming that C_p is independent of T).

For liquid water and for aqueous solutions we will assume $C_p = 1$ cal/g K, and, since the density ρ of water is ~ 1 g/cm^3, we have $\rho C_p = 1$ cal/cm^3 K or $\rho C_p = 1000$ cal/liter K. To estimate the heat capacity of gases, we will usually assume that the molar heat capacity $\mathbf{C}_p$ is $\frac{7}{2}R$ cal/mole K. There are thus three types of heat capacity, the heat capacity per unit mass C_p, the heat capacity per unit volume ρC_p, and the heat capacity per mole $\mathbf{C}_p$. However, we will use heat capacity per unit volume for much of the next two chapters, and we use the symbol ρC_p for most of the equations.

Another approximation we frequently make in simplifying the energy balance is that the molar heat capacity, which might be approximated as

$$\mathbf{C}_p(T) = \sum_{j=1}^{S} y_j \mathbf{C}_{pj}(T) \tag{5.13}$$

is independent of temperature and composition, so that we can assume $\mathbf{C}_{po} = \mathbf{C}_p$ and $\rho_o \mathbf{C}_{po} = \rho \mathbf{C}_p$. Thus we continue the approximation that all thermal properties of the fluid in a reactor are constants that are independent of composition and temperature. We can easily modify these to be more accurate, but for most preliminary calculations in chemical reactors, these approximations make the equations much simpler.

In the above expressions we used symbols in bold for thermodynamic quantities per mole to show how these thermodynamic relations are used to develop the energy balance. We will not use them elsewhere except for ΔH_R, which is the heat of reaction per mole (but will not be written in bold).

5.2   ENERGY BALANCE IN A CSTR

We first derive the energy balance in a CSTR. For the mass balance in a constant-density reactor we wrote an integral balance on the rate of change of the number of moles N_j of species j in the reactor to obtain

$$\frac{dN_j}{dt} = V\frac{dC_j}{dt} = v(C_{jo} - C_j) + Vv_j r \tag{5.14}$$

$$\frac{dN_j}{dt} = V\frac{dC_j}{dt} = v(C_{jo} - C_j) + V\sum_{i=1}^{R} v_{ij} r_i \tag{5.15}$$

for single and multiple reactions, respectively. (We follow our custom of ignoring density changes so that V is a constant and C_j is an appropriate species density unit.) For the steady-state CSTR we set the time derivatives equal to zero and obtained mass balances on species j as in the previous chapters

$$C_j - C_{jo} = \tau v_j r \tag{5.16}$$

$$C_j - C_{jo} = \tau \sum_i v_{ij} r_i \tag{5.17}$$

For the corresponding energy balance in a CSTR we write an analogous expression

[accumulation of heat] = [heat flow in] − [heat flow out]

+ [heat generation by reaction] − [heat removal to surroundings]

The last term is not present in the mass balance (unless the reactor leaks), but heat can be carried in and out not only with flow but also by heat transfer through the walls. An enthalpy balance on the contents of this CSTR gives

$$\frac{dH}{dt} = \rho C_p V\frac{dT}{dt} = v\rho C_p(T_o - T) + V(-\Delta H_R)r - UA_c(T - T_c) + \dot{W}_s \tag{5.18}$$

$$\frac{dH}{dt} = \rho C_p V\frac{dT}{dt} = v\rho C_p(T_o - T) + V\sum_{i=1}^{R}(-\Delta H_{Ri})r_i - UA_c(T - T_c) + \dot{W}_s \tag{5.19}$$

for single and multiple reactions, respectively. Heat transfer occurs across the walls of the reactor with A_c the wall area across which heat transfer occurs (the subscript c stands for cooling).

Each term in the preceding equations has units of energy/time. Note the signs on each term indicating that heat is removed or added to the reactor. We preserve the minus sign on ΔH_R because we are more interested in exothermic reactions for which $\Delta H_R < 0$. The student can recognize each term on the right side from the steady-state enthalpy balance we derived in the previous section from the thermodynamics of a steady-state flow system.

For steady state in the CSTR we set $d\dot{H}/dt = 0$ and divide each term by $v\rho C_p$ to yield steady-state energy balance

$$T - T_o = \tau \frac{-\Delta H_R}{\rho C_p} r - \frac{U A_c}{v\rho C_p}(T - T_c) \tag{5.20}$$

or

$$T - T_o = \tau \sum_{i=1}^{R} \frac{-\Delta H_{Ri}}{\rho C_p} r_i - \frac{U A_c}{v\rho C_p}(T - T_c) \tag{5.21}$$

for single and multiple reaction systems, respectively. We have omitted the $\dot{W}_s$ term from these expressions because, even for very vigorous stirring, this term is usually small compared to the other terms in the energy balance.

5.3 ENERGY BALANCE IN A PFTR

For the PFTR we wrote the mass balance on species j by writing a shell balance between position z and $z + dz$. After letting $dz \to 0$ we obtained the expressions

$$u \frac{dC_j}{dz} = v_j r \tag{5.22}$$

$$u \frac{dC_j}{dz} = \sum_{i=1}^{R} v_{ij} r_i \tag{5.23}$$

for single and multiple reactions, respectively.

For the energy balance we make a similar differential balance on the enthalpy flow between z and $z + dz$ for a tube of length L and diameter D as sketched in Figure 5–2. In steady state this energy balance is

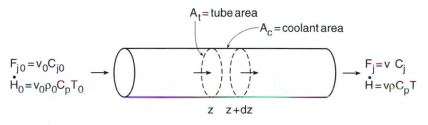

Figure 5–2 Energy balance in a PFTR. A shell balance is made on an element of volume dV between z and $z + dz$. Species flow F_j and enthalpy flow in and out of this element of volume are balanced by species and energy generated by reaction.

$$0 = [uA_t\rho C_p T]_{z+dz} - [uA_t\rho C_p T]_z + A_t(-\Delta H_R)r\,dz - Up_w(T-T_c)\,dz \qquad (5.24)$$

In this expression A_t is the cross-sectional area of the tube ($\pi D^2/4$ for a cylindrical tube); so the volume dV of this element is $A_t\,dz$. Heat transfer occurs across the external wall of the tube, which has area dA_c (subscript c stands for coolant), so that

$$dA_c = p_w\,dz \qquad (5.25)$$

where p_w is the perimeter length of the wall of the tube (πD for a cylindrical tube). The heat transfer rate in this element is

$$d\dot{Q} = Up_w\,dz(T-T_c) \qquad (5.26)$$

Next we assume that the fluid density and the tube diameter are the same at z and $z+dz$ to obtain

$$0 = A_t u\rho C_p(T_{z+dz} - T_z) + A_t(-\Delta H_R)r\,dz - Up_w(T-T_c)\,dz \qquad (5.27)$$

Finally we take the limit as $dz \to 0$,

$$\lim_{dz\to 0}(T_z - T_{z+dz}) = -\frac{dT}{dz}\,dz \qquad (5.28)$$

The energy-balance equation after dividing by dz is

$$uA_t\rho C_p\frac{dT}{dz} = A_t(-\Delta H_R)r - Up_w(T-T_c) \qquad (5.29)$$

$$uA_t\rho C_p\frac{dT}{dz} = A_t\sum_{i=1}^{R}(-\Delta H_{Ri})r_i - Up_w(T-T_c) \qquad (5.30)$$

for single and multiple reactions, respectively. Division by $A_t\rho C_p$ yields

$$u\frac{dT}{dz} = \frac{-\Delta H_R}{\rho C_p}r - \frac{Up_w}{\rho C_p A_t}(T-T_c) \qquad (5.31)$$

$$u\frac{dT}{dz} = \sum_{i=1}^{R}\frac{-\Delta H_{Ri}}{\rho C_p}r_i - \frac{Up_w}{\rho C_p A_t}(T-T_c) \qquad (5.32)$$

for single and multiple reactions, respectively.

5.3.1 Batch reactor

For a single reaction in a nonisothermal batch reactor we can write the species and energy-balance equations

$$V\frac{dC_A}{dt} = -Vr \qquad (5.33)$$

$$V\rho C_p\frac{dT}{dt} = V(-\Delta H_R)r - UA_c(T-T_c) \qquad (5.34)$$

or

$$\frac{dC_A}{dt} = -r \qquad (5.35)$$

$$\frac{dT}{dt} = -\frac{\Delta H_R}{\rho C_p} r - \frac{U A_c}{V \rho C_p}(T - T_c) \tag{5.36}$$

We can also obtain these expressions from the energy-balance equation for the steady-state PFTR by simply transforming $dz/u \to dt$ with A_c/V replacing p_w/A_t. The solutions of these equations for the batch reactor are *mathematically identical* to those in the PFTR, although the physical interpretations are quite different.

5.3.2 Assumptions

In deriving these equations we have made many assumptions to keep them simple. We have assumed *constant density* so that we can use concentration as the composition variable. We have also assumed that the parameters in these systems are independent of temperature and composition. Thus parameters such as ΔH_R, ρC_p, and U are considered to be *constants*, even though we know they all depend at least weakly on temperature. To be exact, we would have to find the heat of reaction, heat capacity, and heat transfer coefficient as functions of temperature and composition, and for the PFTR insert them within the integrals we must solve for temperature and composition. However, in most situations these variations are small, and the equations written will give good approximations to actual performance.

In this chapter and in Chapter 6 we will usually solve these equations assuming a single first-order irreversible reaction, $r = k(T)C_A$. Other orders and multiple reactions could of course be considered, but the equations are much more difficult to solve mathematically, and the solutions are qualitatively the same. We will see that the solutions with these simple kinetics are sufficiently complicated that we do not want to consider more complicated kinetics and energy balances at the same time.

There are many interesting problems in which complex chemistry in nonisothermal reactors interact to produce complex and important behavior. As examples, the autocatalytic reaction, $A \to B$, $r = kC_A C_B$, in a nonisothermal reactor can lead to some quite complicated properties, and polymerization and combustion processes in nonisothermal reactors must be considered very carefully in designing these reactors. These are the subjects of Chapters 10 and 11.

5.4 EQUATIONS TO BE SOLVED

For the nonisothermal reactors we need to solve the mass- and energy-balance equations *simultaneously*.

5.4.1 CSTR

For a single reaction in a steady-state CSTR the mass-balance equation on reactant A and temperature T give the equations

$$\boxed{C_{Ao} - C_A = \tau\, r(C_A, T)} \tag{5.37}$$

and

$$T - T_{\mathrm{o}} = \tau \frac{-\Delta H_R}{\rho C_p} r(C_A, T) - \frac{U A_{\mathrm{c}}}{v \rho C_p}(T - T_{\mathrm{c}}) \qquad (5.38)$$

These are two coupled algebraic equations, which must be solved simultaneously to determine the solutions $C_j(\tau)$ and $T(\tau)$. For multiple reactions the $R + 1$ equations are easily written down, as are the differential equations for the transient situation. However, for these situations the solutions are considerably more difficult to find. We will in fact consider the solutions of the transient CSTR equations in Chapter 6 to describe phase-plane trajectories and the stability of solutions in the nonisothermal CSTR.

5.4.2 PFTR

For a single reaction in a steady-state PFTR the mass-balance equations on species A and the temperature are

$$u \frac{dC_A}{dz} = -r(C_A, T) \qquad (5.39)$$

and

$$u \frac{dT}{dz} = \frac{-\Delta H}{\rho C_p} r(C_A, T) - \frac{U p_{\mathrm{w}}}{\rho C_p A_{\mathrm{t}}}(T - T_{\mathrm{c}}) \qquad (5.40)$$

These are first-order ordinary differential equations that have two initial conditions at the inlet to the reactor,

$$C_A = C_{A\mathrm{o}} \qquad \text{and} \qquad T = T_{\mathrm{o}} \qquad \text{at } z = o \qquad (5.41)$$

We have to solve these two equations simultaneously to find $C_A(z)$ and $T(z)$ versus position z or at the exit L of the reactor, $C_A(L)$ and $T(L)$.

These are two simultaneous differential equations with two initial conditions for a single reaction. For R simultaneous reactions we have to solve $R + 1$ simultaneous differential equations with $R + 1$ initial equations because there are R independent mass-balance equations and one temperature equation.

These equations look innocuous, but they are highly nonlinear equations whose solution is almost always obtainable only numerically. The nonlinear terms are in the rate $r(C_A, T)$, which contains polynomials in C_A, and especially the very nonlinear temperature dependence of the rate coefficient $k(T)$. For first-order kinetics this is

$$r(C_A, T) = k_{\mathrm{o}} e^{-E/RT} C_A \qquad (5.42)$$

and it is the factor $e^{-E/RT}$ that is most nonlinear and creates the most interesting mathematical and physical complexities in the nonisothermal reactor.

We will find that the solutions of the algebraic equations (CSTR) or differential equations (PFTR) can be very complicated. However, we must examine these solutions in some detail because they reveal some of the most important features of chemical reactors.

For a single reaction in the CSTR we solve two algebraic equations

$$F_1(C_A, T) = 0 \qquad (5.43)$$

$$F_2(C_A, T) = 0 \qquad (5.44)$$

which are generally polynomials in C_A and exponentials in T. While algebraic equations are generally simpler to solve than differential equations, these algebraic equations sometimes yield multiple roots, which indicate *multiple steady states* in CSTRs. This will be the subject of the next chapter.

In this chapter we consider mainly the plug-flow reactor. For the PFTR we solve the equations

$$\frac{dC_A}{dz} = F_1(C_A, T) \qquad (5.45)$$

$$\frac{dT}{dz} = F_2(C_A, T) \qquad (5.46)$$

along with the inlet condition

$$C_A = C_{Ao}, \qquad T = T_o, \qquad \text{at } z = 0 \qquad (5.47)$$

Since dC_A/dz and dT/dz are the slopes of the functions $C_A(z)$ and $T(z)$, they must be unique at $z = 0$ and at all z. Therefore, in contrast to the CSTR, the PFTR can exhibit only a single steady state.

5.5   HEAT REMOVAL OR ADDITION TO MAINTAIN A REACTOR ISOTHERMAL

One of the first considerations in a reactor is the heat load that is necessary to control the temperature. It was implicit in all the previous chapters on isothermal reactors that the reactor was somehow thermostatted to maintain its temperature at that specified. For this case we did not have to solve the energy-balance equation, but we do need to consider the heat load

$$\dot{Q} = U A_c (T - T_c) \qquad (5.48)$$

to maintain the reactor temperature at T.

Example 5–1   Consider the reaction $A \rightarrow B$, $r = kC_A$, $k_{300} = 0.05 \text{ min}^{-1}$, $\Delta H_R = -20$ kcal/mole in a 10 liter reactor with $C_{Ao} = 2$ moles/liter and $T_o = 300$ K. At what rate must heat be removed to maintain the reactor isothermal at 300 K for

(a) a batch reactor at 90% conversion?

This is an isothermal reactor with $r = 0.05C_A$. If batch, then

$$C_A = C_{Ao}e^{-kt}$$

$$\dot{Q}(t) = V(-\Delta H_R)kC_A(t)$$

$$= V(-\Delta H_R)kC_{Ao}e^{-kt}$$

$$= 10 \times 20 \times 0.05 \times 2 \times e^{-0.05t}$$

$$= 20\,e^{-0.05t}\ \text{kcal/min}$$

(b) a CSTR at 90% conversion?

In a CSTR

$$\dot{Q}(\tau) = V(-\Delta H_R)kC_A$$

$$= 10 \times 20 \times 0.05 \times 0.2 = 2\ \text{kcal/min (constant)}$$

(c) a PFTR at 90% conversion?

In a PFTR the answer is identical to that in a batch reactor except that t is replaced by τ,

$$\dot{Q}(z) = 20\,e^{-0.05z/u}\ \text{kcal/min}$$

Note, however, that the significance of heat removal or addition (Q positive or negative) is quite different with the batch reactor, requiring a heat removal that varies in time, the CSTR requiring a constant heat removal rate, and the PFTR requiring a heat removal rate that varies with position z in the reactor. These are sketched in Figure 5–3.

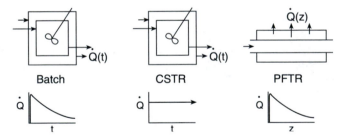

Figure 5–3 Plot of reactor configurations and $\dot{Q}$ necessary to maintain a batch reactor, CSTR, and PFTR isothermal.

If this reaction were endothermic, we would have opposite signs to those shown in the previous example, because we would have to add heat to the reactor to maintain the process isothermal rather than cool it.

Note that the batch reactor requires *programming* $\dot{Q}(t)$, and the PFTR requires either a very high $U\,A_c$ [to remain at $T(z)$] or programming $\dot{Q}(z)$. The CSTR operates with a fixed $\dot{Q}$ because T is uniform throughout the reactor for given conditions, and for steady-state operation $\dot{Q}$ is independent of time. Therefore, the CSTR is usually much simpler to design for stable heat removal. This is another attractive feature of the CSTR: It is much easier to maintain constant temperature in a CSTR reactor than in a batch or PFTR because the heat load required is not a function of either time or position in the reactor. We will see later that the CSTR has even higher heat transfer characteristics when we consider that stirring increases convective heat transfer so that U is frequently higher in a CSTR.

We still must use a PFTR in many chemical processes, and we must then determine how to program the cooling or heating to attain a temperature profile in the reactor close to that desired. The subject of this chapter is the proper temperature management to attain desired operation of a PFTR. In the next chapter the nonisothermal CSTR will be considered specifically.

5.6 ADIABATIC REACTORS

The mass- and energy-balance equations must be solved numerically in the general situation where heat is transferred to or from walls. There are three terms on the right side of the energy equation, heat flow with reactants and products, reaction heat, and heat transfer through walls. However, the adiabatic reactor is a special case where we need to solve only one equation for a single reaction.

The adiabatic reactor is also an important limiting case of real reactors. As we have noted previously, the isothermal reactor (previous chapters) is a limiting case where the energy equation can be ignored. This situation occurs in practice whenever heat transfer

$$\dot{Q} = U A_c (T - T_c) \tag{5.49}$$

is so large that the reactor temperature T is maintained isothermal at $T = T_c$.

The other limit of real reactors is where the heat removal term is not infinite but instead is negligible. We call case where $\dot{Q} = 0$ the *adiabatic reactor*. For wall cooling, A_c is proportional to $V^{1/2}$ for a tube and to $V^{2/3}$ for a spherical reactor. Since heat generation is proportional to V and cooling is proportional to A_c, cooling becomes more difficult to maintain as the reactor becomes larger, and, as noted, large reactors frequently must be operated adiabatically.

Another reason why calculations of the adiabatic reactor is important is for safety. Suppose we have a reactor operating in a stable fashion with cooling. What happens if the cooling is suddenly stopped? The limit of this situation is the adiabatic reactor, and the engineer must always be aware of this mode because it is the worst-case scenario of any exothermic reactor. Note that if $\Delta H_R > 0$, we must supply heat to maintain the reactor temperature, and loss of heat will cause the reactor to cool down and the rate will decrease safely.

5.6.1 Adiabatic CSTR

Here the mass- and energy-balance equations are

$$C_{Ao} - C_A = \tau \, r(C_A, T) \tag{5.50}$$

$$T - T_o = \tau \left(\frac{-\Delta H}{\rho C_p} \right) r(C_A, T) \tag{5.51}$$

We still have the $r(C_A, T)$ factors in both terms, but note that they are identical in both energy- and mass-balance equations. Therefore, $r(C_A, T)$ may be eliminated between them by dividing the second equation by the first to yield

$$T - T_o = \tau \frac{-\Delta H_R}{\rho C_p} r = \left(\frac{-\Delta H_R}{\rho C_p}\right)(C_{Ao} - C_A) = \left(\frac{-\Delta H_R}{\rho C_p}\right) C_{Ao} X \qquad (5.52)$$

or

$$\Delta T = -\left(\frac{-\Delta H_R}{\rho C_p}\right) \Delta C_A \qquad (5.53)$$

Note the sign on terms in these equations. T increases as a reaction proceeds (C_A decreases) if ΔH_R is negative (reaction exothermic), as we know it must be since reaction is then *generating* heat. For an endothermic reaction T decreases as a reaction proceeds.

To simplify the appearance of these equations, we will find it convenient to define a new variable

$$J = \frac{-\Delta H_R C_{Ao}}{\rho C_p} \qquad (5.54)$$

so that the previous equation becomes

$$T - T_o = J X \qquad (5.55)$$

This has exactly the same significance as the preceding equation, but the constants are grouped in J, which has the units of temperature and will be positive if the reaction is exothermic.

5.6.2 Adiabatic PFTR

Here the equations are

$$u \frac{dC_A}{dz} = -r(C_A, T) \qquad (5.56)$$

$$u \frac{dT}{dz} = \left(\frac{-\Delta H_R}{\rho C_p}\right) r(C_A, T) \qquad (5.57)$$

The right-hand sides both contain r; so we can divide the first equation by the second to yield

$$\frac{dT}{dC_A} = -\left(\frac{-\Delta H_R}{\rho C_p}\right) \qquad (5.58)$$

The right-hand side is constant (assuming ΔH_R and ρC_p are independent of T and composition); so we can integrate this single differential equation between $T = T_o$ and $C_A = C_{Ao}$ at $z = 0$ to T and C_A at position z. This gives

$$\boxed{T - T_o = \left(\frac{-\Delta H_R}{\rho C_p}\right)(C_{Ao} - C_A) = \left(\frac{-\Delta H_R}{\rho C_p}\right) C_{Ao} X = J X} \qquad (5.59)$$

which is an identical expression as we obtained for the adiabatic CSTR.

Figure 5–4 C_A and X versus T for an exothermic adiabatic reactor with a single reaction with constant $-\Delta H_R / \rho C_p$.

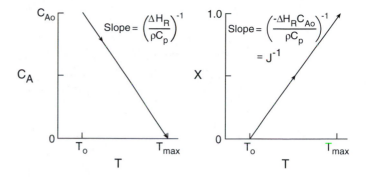

We can generalize this for any reactor since we assumed nothing about $r(C_A, T)$. The preceding relation is true for any single reaction in any adiabatic reactor, no matter what the kinetics or reactor flow pattern. This simple relation between C_A or X versus T is plotted in Figure 5–4.

Example 5–2 Find $C_{Ao} - C_A$ versus T for a reaction with $\Delta H_R = -20$ kcal/mole, $\rho C_p = 1000$ cal/liter K, $T_o = 300$ K, and $C_{Ao} = 4$ moles/liter in an adiabatic reactor. What is the temperature if the reaction goes to completion?

The previous equation

$$T - T_o = \left(\frac{-\Delta H_R}{\rho C_p} \right) (C_{Ao} - C_A)$$

becomes

$$T = 300 + \frac{20000}{1000}(4 - C_A) = 300 + 80 - 20 C_A$$

$$= 380 - 20 C_A$$

Thus T is 300 K initially and 380 K if the reaction goes to completion, and this would be true in *any* adiabatic reactor.

This reaction is in aqueous solution. Why might bad things happen if this reactor were operated with $C_{Ao} = 5$ moles/liter?

Throughout this chapter (and throughout this book) we will make the frequently unstated assumption that ΔH_R and ρC_p are independent of temperature and composition. They are of course dependent on T because C_p of each species depends on T. Each species of course has a different heat capacity, but we will unless stated otherwise assume that these can be ignored. The actual value of ΔH_R is computed through the relation

$$\Delta H_R = \Delta H_{R,298}^\circ + \int_{298}^{T} \Sigma_j \nu_j C_{pj}\, dT \tag{5.60}$$

where $\Delta H^{\circ}_{R,298}$ is the heat of reaction in standard state at 298 K, and $\mathbf{C}_{pj}$ is usually written as a polynomial in T. It is frequently a good approximation to assume that $\Delta H_R = \Delta H^{\circ}_{R,298}$, which is equivalent to having all heat capacities constant. We also assume that the average heat capacity of the fluids is independent of composition, especially when reactants are diluted with solvent. The mathematics becomes more complex if we have to include these factors, although the principles remain the same.

5.6.3 Multiple reactions

For multiple reactions similar arguments hold. For multiple reactions in the CSTR in which A is a reactant ($\nu_A = -1$) in all reactions, we can write

$$C_{Ao} - C_A = C_{Ao} \sum_{i=1}^{R} \mathbf{X}_i = \tau \sum_{i=1}^{R} r_i \tag{5.61}$$

$$\rho C_p (T - T_o) = \tau \sum_{i=1}^{R} (-\Delta H_{Ri}) r_i \tag{5.62}$$

Therefore, division of the second equation by the first equation yields

$$T - T_o = \sum_{i=1}^{R} \frac{-\Delta H_{Ri}}{\rho C_p} C_{Ao} \mathbf{X}_i \tag{5.63}$$

In these expressions we have defined the total conversion of A as a summation of its conversion in all reactions,

$$X_A = \frac{C_{Ao} - C_A}{C_{Ao}} = \sum_{i=1}^{R} \mathbf{X}_i \tag{5.64}$$

Only if A is a reactant in all reactions ($\nu_{iA} = -1$) can this simple relation be written in terms of $\mathbf{X}_i$. In the previous chapter we also used conversion to describe a reactant *species* such as X_A, but this definition of $\mathbf{X}_i$ refers to the conversion of the ith reaction.

5.6.4 Solutions

For the adiabatic reactor we have a unique relation between T and conversion. We can therefore solve for T and eliminate it from the mass-balance equation. For the CSTR the mass-balance equation for a single first-order irreversible reaction

$$A \rightarrow B, \qquad r = k(T)C_A, \qquad \Delta H_R \tag{5.65}$$

is

$$C_{Ao} - C_A = \tau r(C_A, T) = \tau k(T) C_A \tag{5.66}$$

We can eliminate T in $k(T)$ to obtain

$$C_{Ao} - C_A = \tau k_o C_A \exp\left(-\frac{E}{R[T_o - (\Delta H_R/\rho C_p)(C_{Ao} - C_A)]}\right) \tag{5.67}$$

This is a *single equation* in C_A and τ. It is not easy to solve explicitly for $C_A(\tau)$, but it can easily be solved for $\tau(C_A)$,

$$\tau = \frac{C_{Ao} - C_A}{k(T)C_A} = \left[(C_{Ao} - C_A)/C_A k_o \exp\left(-\frac{E}{R[T_o - (\Delta H_R/\rho C_p)(C_{Ao} - C_A)]}\right)\right] \quad (5.68)$$

Therefore, this problem can be solved simply by determining values of τ versus values of C_A for specified values of constants.

In the adiabatic PFTR the differential equation is

$$\frac{dC_A}{d\tau} = -C_A k_o \exp\left(-\frac{E}{R[T_o - (\Delta H_R/\rho C_p)(C_{Ao} - C_A)]}\right) \quad (5.69)$$

which can be separated and integrated to yield

$$\tau = -\int_{C_{Ao}}^{C_A} \frac{dC_A}{k(T)C_A} = -\int_{C_{Ao}}^{C_A} dC_A/\left[C_A k_o \exp\left(\frac{-E}{R[T_o - (\Delta H_R/\rho C_p)(C_{Ao} - C_A)]}\right)\right] \quad (5.70)$$

This expression must be solved numerically, but it is a single integral expression, which can be integrated directly rather than two simultaneous differential equations.

Example 5–3 Plot $C_A(\tau)$ and $T(\tau)$ for the reaction $A \rightarrow B$, $r = k(T)C_A$ in an adiabatic PFTR for $E = 30$ kcal/mole, $k_o = 2.6 \times 10^{20}$ min^{-1}, $\Delta H_R = -20$ kcal/mole, $\rho C_p = 1000$ cal/liter K, $T_o = 300$ K, and $C_{Ao} = 2$ moles/liter.

This is the same reaction we used in the previous example, because $k = 0.05$ min^{-1} at 300 K, but now the temperature varies. The preceding equation becomes

$$\tau = -\int_{C_{Ao}}^{C_A} \frac{dC_A}{k(T)C_A}$$

$$= -\int_2^{C_A} \frac{dC_A}{C_A \times 2.6 \times 10^{20} \exp\{(-15000)/[300 - 20(2 - C_A)]\}}$$

$$+ 300 + 20(2 - C_A)$$

We can easily integrate this equation numerically to find the solutions for the PFTR. We can also solve the algebraic equation for the CSTR. We will solve this problem numerically using a spreadsheet in the next example, but in Figure 5–5 we just plot the solutions $X(\tau)$ and $T(\tau)$.

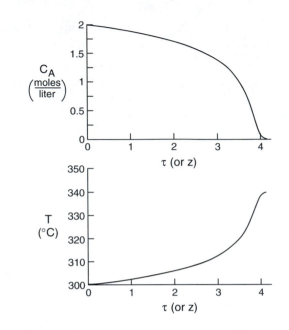

Figure 5–5 C_A and T versus τ or z in an adiabatic PFTR. The time is in minutes.

It is interesting to compare $C_A(\tau)$ for the adiabatic reactor with the solution for the reactor maintained isothermal at 300 K. For the isothermal PFTR, the solution is

$$\tau = \frac{1}{k} \ln \frac{C_{Ao}}{C_A} = \frac{1}{0.05} \ln \frac{2}{C_A} \qquad (5.71)$$

or

$$C_A(\tau) = 2e^{-0.05\tau} \qquad (5.72)$$

These solutions $C_A(\tau)$ are plotted in Figure 5–6. Several features of these curves are worth noting. First, the adiabatic reactor requires a much shorter τ to attain complete conversion than does the isothermal reactor. Second, the shapes of the curves are quite different, with the isothermal reactor exhibiting the standard exponential decay but the adiabatic reactor exhibiting an *acceleration* in rate as the reaction proceeds because the temperature increases.

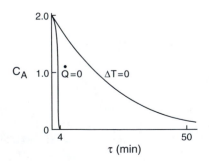

Figure 5–6 C_A versus τ or z in an isothermal PFTR maintained at 300 K and in an adiabatic PFTR starting at $T_o = 300$ K.

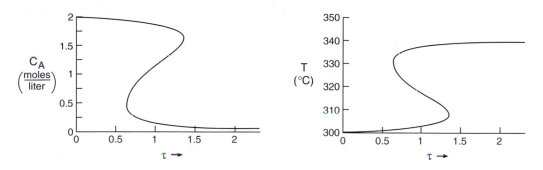

Figure 5–7 C_A and T versus τ in an adiabatic CSTR. The unusual shapes of these curves will require considerable examination in this chapter and in Chapter 6. Residence time is in minutes.

For the adiabatic CSTR we have to solve the equation

$$\tau = \frac{C_{Ao} - C_A}{k(T)C_A} = (C_{Ao} - C_A) / \left[k_o C_A \exp\left(-\frac{E}{R[T_o - (\Delta H_R/\rho C_p)(C_{Ao} - C_A)]} \right) \right]$$

$$= \frac{2.0 - C_A}{C_A \times 2.6 \times 10^{20} \exp\left\{ -\dfrac{15000}{300 - 20(2 - C_A)} \right\}} + 300 + 20(2 - C_A) \qquad (5.73)$$

This is also plotted in Figure 5–7.

Note several features of these solutions. First the $C_A(\tau)$ and $T(\tau)$ solutions have *identical shapes* (but T increases as C_A decreases for the exothermic reaction) for the adiabatic reaction in these reactors or in any adiabatic reactor. If we plot $C_{Ao} - C_A$ versus T from these solutions we obtain Figure 5–8, because, by the previous arguments, this must be a straight line for any single reaction in any reactor as long as parameters do not depend on temperature or composition. Second, note that the CSTR in this example requires a *shorter residence time* for a given conversion than a PFTR. Third, note that the CSTR exhibits *multiple values* of C_A and T for a range of τ. This situation is physically real and will be the subject of the next chapter.

While the straight-line graphs of the adiabatic trajectories in the X–T plane appear similar for CSTR and PFTR (but they are different curves), their significance is quite different. For the PFTR the graphs can be thought of as plots of concentration and

Figure 5–8 Plot of $C_{Ao} - C_A$ versus reactor temperature T for the adiabatic reactors of Figure 5–7. For the CSTR the points indicate the steady states for $\tau = 1.0$ min.

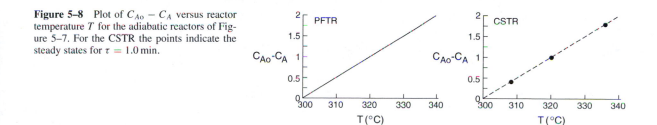

temperature versus position in the reactor, while in the CSTR $\tau = V/v$ is the residence time for a particular set of feed conditions. In the CSTR the solutions $C_A(\tau)$ and $T(\tau)$ are the *locus of solutions* versus residence time. This difference has great significance for the CSTR, and, as we will see, it suggests that sometimes states on these curves are not accessible but are in fact *unstable*.

5.6.5 Other reactions

We have used the first-order irreversible reaction as an example, but this is easy to generalize for any reaction, irreversible or reversible, with any kinetics. In a PFTR the mass-balance equation for an arbitrary reaction becomes

$$\tau = \int_{C_{Ao}}^{C_A} \frac{dC_A}{r(C_A, T)} = \int_{C_{Ao}}^{C_A} \frac{dC_A}{r_{ad}(C_A)} \tag{5.74}$$

where we define $r_{ad}(C_A)$ as the rate versus C_A with the temperature varying as the conversion changes.

5.6.6 The $1/r$ plot

Let us return to the graphical construction we developed in earlier chapters for isothermal reactors, because for nonisothermal reactors τ is still the area under curves of plots of $1/r$ versus $C_{Ao} - C_A$. For the first-order irreversible reaction in an adiabatic reactor $1/r_{ad}$ is given by

$$\frac{1}{r_{ad}(C_A)} = \frac{1}{C_A k_o \exp\{-E/R[T_o + (-\Delta H/\rho C_p)(C_{Ao} - C_A)]\}} \tag{5.75}$$

Example 5–4 For the irreversible first-order reaction with parameters in the previous example, calculate τ in adiabatic PFTR and CSTR from a table of $1/r$ versus C_A.

We compute $1/r$ versus $C_{Ao} - C_A$ from the equation

$$\frac{1}{r_{ad}(C_A)} = \frac{1}{C_A k_o \exp\{-E/R[T_o + (-\Delta H/\rho C_p)(C_{Ao} - C_A)]\}}$$

$$= \frac{1}{C_A \times 2.6 \times 10^{20} \exp\{-15000/[300 + 80(2 - C_A)]\}}$$

We can calculate values of τ using a simple spreadsheet as shown in Table 5–1. The previous figures of C_A and T versus τ were obtained from the first and second columns of this spreadsheet for the τs in the last two columns, because from the previous equations, we see that

$$\tau_{PFTR} = \int_{C_{Ao}}^{C_A} \frac{dC_A}{kC_A} = \sum \frac{\Delta C_A}{kC_A}$$

$$\tau_{CSTR} = \frac{C_{Ao} - C_A}{k(T)C_A}$$

These values of $1/r$ are plotted versus $C_{Ao} - C_A$ in Figure 5–9. The rectangles have areas indicated in Table 5–1. The sum of these areas is an approximation to the residence time in an adiabatic PFTR because

$$\tau_{PFTR} = \int_{C_{Ao}}^{C_A} \frac{dC_A}{r_{ad}(C_A)} = \sum \frac{\Delta C_A}{kC_A}$$

The last two columns show τ_{PFTR} and τ_{CSTR}; so one can simply read $C_A(\tau)$ and $T(\tau)$ from Table 5–1, and, since this is the same problem worked previously, the previous graphs can be plotted simply from this spreadsheet (see Figure 5–10).

TABLE 5–1
Spreadsheet Solution for Adiabatic PFTR and CSTR

C_A	T	k	r kC_A	$1/r$ $1/kC_A$	Area $\Delta C_A/kC_A$	τ_{PFTR} $\sum \Delta C_A/kC_A$	τ_{CSTR} $(C_{Ao} - C_A)/kC_A$
2.0	300	0.05	0.1	10	—	—	—
1.8	304	0.097	0.174	5.74	1.15	1.15	1.15
1.6	308	0.184	0.294	3.40	0.680	1.83	1.36
1.4	312	0.343	0.480	2.08	0.416	2.24	1.25
1.2	316	0.630	0.757	1.32	0.264	2.51	1.06
1.0	320	1.14	1.14	0.876	0.175	2.68	0.876
0.8	324	2.04	1.63	0.614	0.123	2.81	0.737
0.6	328	3.58	2.15	0.465	0.093	2.90	0.651
0.4	332	6.21	2.48	0.402	0.080	2.98	0.64
0.2	336	10.6	2.13	0.470	0.094	3.07	0.84
0.0	340	18.0	0	∞	∞	∞	∞

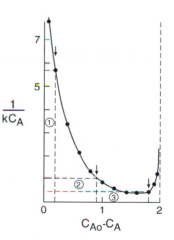

Figure 5–9   Plot of $1/r$ versus $C_{Ao} - C_A$ for the previous example. It is seen that r increases and then decreases so that $1/r$ has a strong minimum. The arrows indicate the three possible steady states in a CSTR for $\tau = 1.0$ min (dashed rectangles).

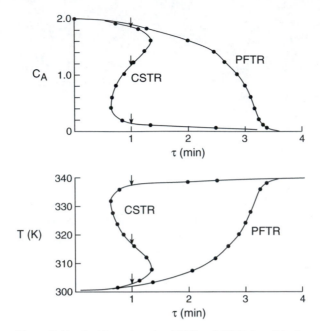

Figure 5–10   Residence times in a PFTR and CSTR for adiabatic reactors. The CSTR can require a much smaller τ than the PFTR and can exhibit multiple steady states for some τ (arrows).

5.6.7   Euler's method solutions of differential equations

The $1/r$ solution is in fact just an Euler's method approximation to the integral for the PFTR, in which one approximates the integral as a summation. The calculation is not very accurate because we used a 0.2 moles/liter step size to keep the spreadsheet small, but it illustrates the method and the identity between Euler's method and a spreadsheet solution.

Thus we see that for nonisothermal reactors this $1/r$ versus $C_{Ao} - C_A$ curve is not always an increasing function of conversion as it was for isothermal reactors even with positive-order kinetics. Since the $1/r$ curve can have a minimum for the nonisothermal reactor, we confirm the possibility that the CSTR requires a smaller volume than the PFTR for positive-order kinetics. This is true even before the multiple steady-state possibilities are accounted for, which we will discuss in the next chapter. This is evident from our $1/r$ plot for the PFTR and CSTR and will occur whenever r has a sufficiently large maximum that the area under the rectangle is less than the area under the curve of $1/r$ versus $C_{Ao} - C_A$.

5.6.8   Reversible reactions

The nonisothermal behavior of reversible reactions is especially interesting because the equilibrium composition is temperature dependent. Recalling our description of reversible reactions, we wrote for the rate expression

$$r = r_f - r_b = k_1(T) \prod_{j=1}^{R} C_j^{m_{fj}} - k_b(T) \prod_{j=1}^{R} C_j^{m_{bj}} \tag{5.76}$$

and we noted that near equilibrium the reaction kinetics become elementary and that the rate coefficients are related to the equilibrium constant by

$$\frac{k(T)}{k'(T)} = \frac{k_{fo}}{k_{bo}} e^{-(E-E')/RT} = K_{eq}(T) = e^{-\Delta G_0/RT} = K_o e^{-\Delta H/RT} = \prod C_j^{v_j} \quad (5.77)$$

We will find it convenient to use X rather than C_A as the composition variable for a single reaction and consider C_{Ae} and X_e as the corresponding quantities at equilibrium. For an endothermic reaction $\Delta H_R > 0$ and X_e increases with T, while for an exothermic reaction $\Delta H_R < 0$ and X_e decreases with T.

Since our goal is frequently to attain a high conversion in a reactor, the relation of equilibrium to reactor temperature is obviously important.

Example 5–5 A chemist obtained the kinetic data in the previous example. [Rumors are circulating that he started out in chemical engineering but failed thermodynamics and never had a course in reaction engineering.] He says that he made a slight mistake and the reaction is not quite irreversible but the equilibrium conversion is actually 0.95 at 300 K. He doesn't think this will be a serious error. Is he correct?

We need to calculate the rate versus temperature and the equilibrium conversion versus temperature. From the requirement that the rate must be consistent with equilibrium, we have

$$\Delta H_R = E - E' = -20 = 30 - 50$$

Therefore, since the conversion is given at 300 K, we can determine k'_o and the complete rate expression,

$$r = 2.6 \times 10^{21} \exp(-30{,}000/RT)C_{Ao}(1 - X)$$
$$- 3.9 \times 10^{34} \exp(-50{,}000/RT)C_{Ao}X$$

and the equilibrium conversion is given by the expression

$$K_{eq} = 6.7 \times 10^{-14} \exp(+20{,}000/RT) = \frac{X_e}{1 - X_e}$$

This gives the equilibrium conversions in the following table:

T (K)	K_{eq}	X_e
300	19	0.95
310	6.8	0.87
320	2.5	0.71
330	0.97	0.49
340	0.40	0.29
350	0.17	0.15
400	0.005	0.005

These data are plotted in Figure 5–11. Thus we have a very serious problem if this reaction is reversible because the adiabatic reactor trajectory intersects the equilibrium curve at a low conversion. For these kinetics, equilibrium limits the process to a very low conversion at high temperatures.

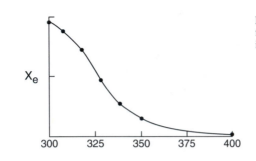

Figure 5–11   Equilibrium conversion versus T for an exothermic reversible reaction in preceding example.

5.7 TRAJECTORIES AND PHASE-PLANE PLOTS

We need to examine the solutions of the reactor mass- and energy-balance equations, starting at C_{Ao}, $X = 0$, $T = T_o$ and proceeding (continuously in a PFTR and discontinuously in a CSTR) to C_A, X, and T at τ, z, or L.

For this we will find it convenient to examine *trajectories* of $C_A(\tau)$ or $X(\tau)$ and $T(\tau)$ in a *phase plane* of X versus T. Thus we do not examine explicitly the time or position dependences but instead examine the trajectory followed as the reaction proceeds.

We need to follow $r(X, T)$, and a convenient way to do this is to map out lines of constant r in the $X–T$ plane. At equilibrium $r = 0$ and we have $X_e(T)$, which is a curve going from $X = 0$ at low temperature to $X_e = 1$ at high temperature for an endothermic reaction and $X = 1$ at low T and $X_e = 0$ at high T for an exothermic reaction.

We can plot lines of constant r for any single reaction, but for simplicity we consider $A \to B$,

$$r(X, T) = k_o e^{-E/RT} C_A - k_o' e^{-E'/RT}(C_{Ao} - C_A)$$
$$= k_o e^{-E/RT} C_{Ao}(1 - X) - k_o' e^{-E'/RT} C_{Ao} X \qquad (5.78)$$

At $r = 0$ this gives

$$K_{eq}(T) = \frac{[C_B]}{[C_A]} = \frac{X_e}{1 - X_e} = \frac{k_o}{k_o'} e^{-(E-E')/RT} = K_o e^{-\Delta H/RT} \qquad (5.79)$$

which can easily be plotted.

Example 5–6   For the example discussed plot lines of constant r for $r = 0, 1, 10, 100$ in the $X(T)$ plane.

We have to solve these equations numerically, and the solutions are shown in Figure 5–12.

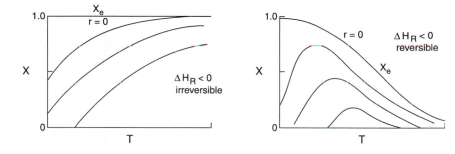

Figure 5–12   Plots of lines of constant rate in the X–T plane for an exothermic reversible reaction.

5.8   TRAJECTORIES OF WALL-COOLED REACTORS

We next examine trajectories on these graphs. Two of these are simple straight lines: a vertical line at temperature T_c for the isothermal $(UA \rightarrow \infty)$ reactor and a straight line starting at T_o with slope $(-C_{Ao}\Delta H/\rho C_p)^{-1}$ for the adiabatic reactor.

5.8.1   Possible trajectories

In Figure 5–13 are plotted the possible trajectories for the exothermic reaction. These are the limiting cases of the trajectory in a wall-cooled reactor, and any wall-cooled reactor will have a trajectory between these two straight lines. The trajectory cannot go above the equilibrium curve $X_e(T)$. For an adiabatic reactor the curve stops there, and for finite UA_w the curve finishes at X_e at T_c.

For an endothermic reaction we assume that the reactor is heated rather than cooled so that $T_c > T_o$. (We are now usually heating so we should write T_h rather than T_c.) Since $\Delta H_R > 0$, the slope of the adiabatic curve is also negative, and the equilibrium conversion is the reverse of that for the exothermic reaction. The corresponding regime of possible trajectories for an endothermic reaction is shown in Figure 5–14. This situation is generally uninteresting because we usually want to heat an endothermic reactor as hot as practical because the rate and the equilibrium conversion both increase with temperature (see Figure 5–15).

Figure 5–13   Possible region of trajectories for exothermic reversible reactions, starting at feed temperature T_o with cooling from the wall at temperature T_c. Trajectories must be in the shaded region between the adiabatic and isothermal curves and below the equilibrium curve.

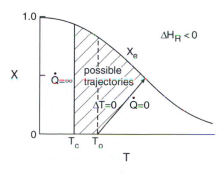

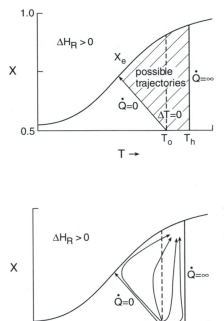

Figure 5–14 Possible region of trajectories for endothermic and for reversible reactions, starting at feed temperature T_o with heating from the wall at temperature T_h. Trajectories must be in the shaded region between the adiabatic and isothermal curves and below the equilibrium curve.

Figure 5–15 Possible trajectories for different amounts of wall heating for an endothermic reversible reaction, starting at feed temperature T_o with heating from the wall at temperature T_h.

However, for exothermic reactions we see that we have a tradeoff between high initial rates at high temperature and high equilibrium conversion at low temperature. We therefore seek an optimum where the initial rate is sufficiently high to require a small size and the equilibrium conversion is acceptable. For an exothermic reaction in a tubular reactor there is therefore some optimum program of T_o, T_c, and geometry for a given reactor with given kinetics and feed.

These arguments do not hold in a CSTR because the conversion and temperature jump discontinuously from $X = 0$, $T = T_o$ to X, T in the reactor and at the exit. Trajectories are continuous curves for the PFTR but are only single points for the CSTR. We will examine this in more detail in the next chapter.

Finally we sketch the possible trajectories that might be followed in an exothermic reaction in a cooled reactor. The trajectory starts at the feed condition of $C_A = C_{Ao}$, $X = 0$, $T = T_o$, and when τ goes to infinity, the trajectory will end up at T_c at the equilibrium composition X_e. Sketched in Figure 5–16 are four trajectories.

We distinguish between four types of trajectories of these curves:

(a) Nearly adiabatic. If UA_c is small, then the reaction first follows the adiabatic trajectory up to X_e at the adiabatic temperature, and then it moves along the equilibrium curve out to T_c. This is not efficient because the rate will be very small near X_e.

(b) Optimal trajectory. If we choose UA_c coptimally, we initially trace the adiabatic curve (maximum T and therefore maximum rate) out to where the rate is a maximum. Then the temperature falls as conversion increases to finally wind up at X_e and T_c.

(c) Moderate cooling. Here the cooling is enough that the temperature does not increase strongly but rises nearly isothermally until the rate slows enough that the temperature cools back to T_c as X approaches 1.

(d) Strong cooling.

(e) Nearly isothermal. If UA_c is large, then the reactants immediately cool down to T_c and the reaction proceeds up to X_e at temperature T_c. This curve gives a high conversion, but the rate should be slow because the temperature in the reactor is low.

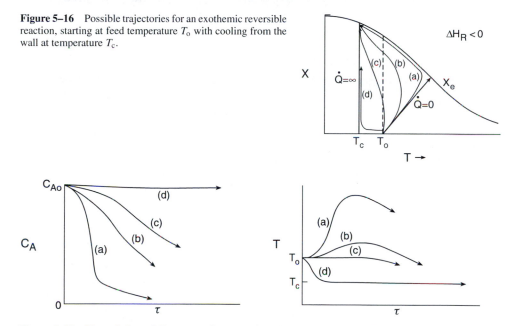

Figure 5–16   Possible trajectories for an exothermic reversible reaction, starting at feed temperature T_o with cooling from the wall at temperature T_c.

Figure 5–17   Plots of C_A and T versus τ for an exothermic reversible reaction for situations sketched in the previous figure.

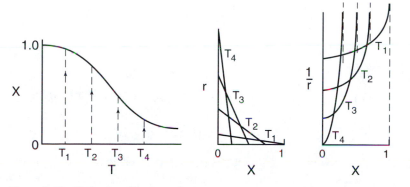

Figure 5–18   Illustration of isothermal trajectories for an exothermic reversible reaction. At the lowest temperature the rate is low but the equilibrium conversion is high, while at the highest temperature the initial rate is high but the equilibrium conversion is low. From the $1/r$ versus X plot at these temperatures, it is evident that T should decrease as X increases to require a minimum residence time.

In designing a wall-cooled tubular reactor, we want to operate such that the trajectory stays near the maximum rate for all temperatures. Thus for an exothermic reversible reaction the temperature should increase initially while the conversion is low and decrease as the conversion increases to stay away from the equilibrium constraint. One can easily program a computer to compute conversion and T versus τ to attain a desired conversion for minimum τ in a PFTR. These curves are shown in Figure 5–17 for the three situations.

Another way of visualizing the optimal trajectory for the exothermic reversible reaction is to consider isothermal reaction rates at increasing temperatures. At T_1 the equilibrium conversion is high but the rate is low, at T_2 the rate is higher but the equilibrium conversion is lower, at T_3 the rate has increased further and the equilibrium conversion is even lower, and at a high temperature T_4 the initial rate is very high but the equilibrium conversion is very low.

These four isothermal trajectories are shown in Figure 5–18, along with plots of r versus X and plots of $1/r$ versus X. It is seen that the minimum residence time in such a reactor is one in which the temperature decreases to maintain the lowest $1/r$ as the conversion increases from 0 to 1.

5.9   EXOTHERMIC VERSUS ENDOTHERMIC REACTIONS

It is interesting to note that the exothermic reversible reaction $A \rightleftarrows B$ is *identical* to the same endothermic reversible reaction $B \rightleftarrows A$. This can be seen by plotting X versus T and noting that the $r = 0$ equilibrium line separates these two reactions. In the reaction $A \rightleftarrows B$ the rate in the upper portion where $r < 0$ is exactly the reaction $B \rightleftarrows A$. We defined all rates as *positive* quantities, but if we write the reaction as its reverse, both r and ΔH_R reverse signs. This is plotted in Figure 5–19.

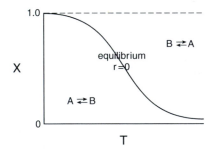

Figure 5–19   Plot of X versus T for an exothermic reversible reaction $A \rightleftarrows B$. The upper portion of this curve where $r < 0$ can be regarded as the endothermic reversible reaction $B \rightleftarrows A$ if the sign of r is reversed or if X is replaced by $1 - X$.

5.10   OTHER TUBULAR REACTOR CONFIGURATIONS

5.10.1   Multiple reactors with interstage cooling

It is sometimes possible to overcome the problem of low X_e with adiabatic operation by using several adiabatic reactors with cooling between reactors. Thus one runs the first reactor up to near the maximum rate, then cools and runs the second reactor up to near the maximum at that temperature, and continues through successive reactors and heat exchangers until a satisfactory conversion is attained.

Figure 5–20   Trajectories in the X versus T plane for three reactors in series with interstage cooling.

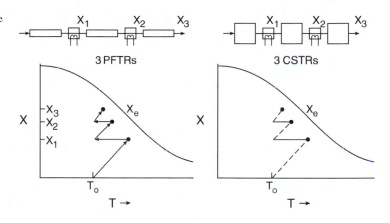

The reactor configurations and possible trajectories for three adiabatic reactors with interstage cooling are shown in Figure 5–20.

There are several examples of exothermic reversible reactors where interstage cooling is practiced. One is methanol synthesis from syngas

$$CO + 2H_2 \rightarrow CH_3OH, \qquad \Delta H_R = -25 \text{ kcal/mole} \qquad (5.80)$$

As discussed in Chapter 3, this reaction is reversible at 250°C, where the commercial reactors are operated, and it is common practice to use interstage cooling and cold feed injection to attain high conversion and extract reaction heat. Another important reaction is the water–gas shift, used to prepare industrial H_2 from syngas,

$$CO + H_2O \rightarrow CO_2 + H_2, \qquad \Delta H_R = -10 \text{ kcal/mole} \qquad (5.81)$$

This process is typically run with two reactors, one at ~800°C, where the rate of the reaction is high, and the second at ~250°C, where the equilibrium conversion is higher.

We have discussed these reactions previously in connection with equilibrium limitations on reactions, and we will discuss them again in Chapter 7, because both use catalysts. These reactions are very important in petrochemicals, because they are used to prepare industrial H_2 and CO as well as methanol, formaldehyde, and acetic acid. As noted previously, these processes can be written as

$$CH_4 \rightarrow CO + H_2 \rightarrow CH_3OH \rightarrow HCHO \qquad (5.82)$$

$$\rightarrow CH_3COOH \qquad (5.83)$$

All steps except the first are exothermic, all reactions except HCHO synthesis are reversible, and all involve essentially one reaction. Energy management and equilibrium considerations are crucial in the design of these processes.

5.10.2   Jacketed PFTR

The cooled tubular reactor is frequently operated with a cooling jacket surrounding the reactor in a tube-and-shell configuration. This looks simply like a tube-in-shell heat exchanger, typically with catalyst in the tube to catalyze the reaction.

A major example of this reactor is ethylene oxidation to ethylene oxide (EO)

$$C_2H_4 + \tfrac{1}{2}O_2 \rightarrow C_2H_4O \tag{5.84}$$

$$C_2H_4 + 3O_2 \rightarrow 2CO_2 + 2H_2O \tag{5.85}$$

As noted in Chapter 2, the heat release in the second reaction is much greater, so that much more heat is released if the selectivity to EO decreases. For world-scale ethylene oxide plants the reaction is run in a reactor consisting of several thousand 1-in.-diameter tubes ~20 ft long in a tube-and-shell heat exchanger. Heat transfer is accomplished using boiling recirculating hydrocarbon on the shell side to absorb the reaction heat. This heat transfer method assures that the temperature of the tubes is constant at the boiling point of the fluid.

If the coolant flow rate is sufficiently high in a jacketed reactor or if boiling liquid is used as the heat transfer fluid, then T_c is constant, as we have assumed implicitly in the previous discussion.

However, with finite flow rate in the jacket, the coolant may heat up, and we must now distinguish between cocurrent and countercurrent flows. While it is true that countercurrent operation gives better heat transfer for an ordinary exchanger, this is not usually desired for a tubular reactor. Heat is generated mostly near the reactor entrance, and the maximum amount of cooling is needed there. In countercurrent flow, the coolant has been heated before it reaches this region, while in cocurrent operation, the coolant is the coldest exactly where the heat load is greatest. Thus cocurrent operation reduces the hot spot, which can plague cooled tubular reactors with exothermic reactions.

Illustrated in Figure 5–21 are possible temperatures of reactor and coolant versus position z for cocurrent and countercurrent operation.

A major goal in wall cooling is to spread out the hot zone and prevent very high peak temperatures. High peak temperatures cause poor reaction selectivity, cause carbon formation, deactivate catalysts, and cause corrosion problems in the reactor walls. Cocurrent

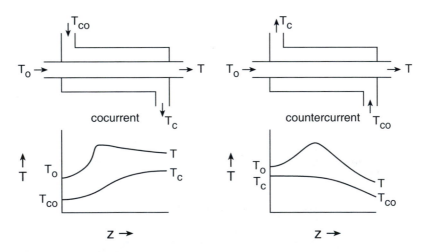

Figure 5–21 Temperature profiles for cocurrent and countercurrent cooled PFTR with feed temperature T_o and coolant feed temperature T_{co}.

flows spread out the hot zone and cause lower peak temperatures, but many additional design features must be considered in designing jacketed reactors.

Temperature profiles in jacketed reactors can be examined rather simply by writing the energy balance on the coolant and on the reactor. In analogy with the energy-balance equation in the PFTR

$$u\frac{dT}{dz} = \frac{-\Delta H}{\rho C_p}r(C_A, T) - \frac{Up_w}{\rho C_p A_t}(T - T_c) \qquad (5.86)$$

we assume that the coolant also obeys plug flow; so the energy balance on the coolant becomes

$$u_c\frac{dT_c}{dz} = +\frac{Up_w}{\rho C_{pc} A_t}(T - T_c) \qquad (5.87)$$

We have of course omitted any heat generation by reaction in the coolant, and we have defined the corresponding quantities for the coolant with the subscript c. Note also that we have reversed the sign of the heat-removal term because whatever heat is transferred from the reactor must be transferred into the coolant.

The student should recognize these problems with the jacketed reactor as similar to those encountered in analysis of heat exchangers. The only difference in chemical reactors is that now we have considerable heat generation by chemical reaction on one side of the heat exchanger. The equipment used for these reactors in fact looks very similar to heat exchangers, but pressures, temperatures, and catalysts must be chosen very carefully, and materials of construction are frequently more difficult to deal with in the reactive environment.

5.10.3   Feed cooling

A final mode of heat transfer in tubular reactors is the feed-cooled reactor, where the hot products from the reactor are cooled by the feed before it enters the reactor. As shown in Figure 5–22, the cold feed in the jacket is preheated by the reaction in the inner tube or

Figure 5–22   Temperature profiles in a wall-cooled reactor with countercurrent feed cooling.

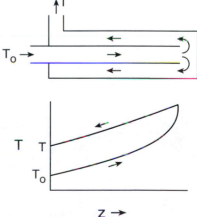

a heat exchanger is used for this purpose before the reactor. Temperature profiles for feed cooling are shown in Figure 5–22.

One interesting characteristic of this type of reactor is that the maximum temperature of the products can be above the adiabatic temperature predicted for reactant temperatures before heat exchange. Heat is retained in the reactor by preheating the feed, and temperatures in some situations can be many hundreds of degrees above adiabatic. This can be useful in combustors for pollution abatement where dilute hydrocarbons need to be heated to high temperatures to cause ignition and attain high conversion with short residence times.

5.11 TEMPERATURE PROFILE IN A PACKED BED

Thus far we have only considered the PFTR with gradients in the axial direction. The heat transfer to the wall at temperature T_c was handled through a heat transfer coefficient U. The complete equations are

$$\frac{\partial C_j}{\partial t} + \mathbf{u} \cdot \nabla C_j = D_j \nabla^2 C_j + v_j r \tag{5.88}$$

$$\rho C_p \left(\frac{\partial T}{\partial t} + \mathbf{u} \cdot \nabla T \right) = k_T \nabla^2 T + (-\Delta H_R) r \tag{5.89}$$

where $\mathbf{u}$ is the velocity vector, ∇C_j and ∇T are the gradients of concentration and temperature, respectively, and k_T is the effective thermal conductivity in the reactor. If we retain the plug-flow approximation, then these gradients are in the z direction only and we lump heat transfer in the radial direction through an overall heat transfer coefficient h. However, heat transfer from the walls must occur by temperature gradients in the radial direction R, and these show up in the term

$$\nabla^2 T = \frac{\partial^2 T}{\partial z^2} + \frac{1}{R} \frac{\partial}{\partial R} \left(R \frac{\partial T}{\partial R} \right) \tag{5.90}$$

If we assume that the reactor is in steady state and that in the z direction the reactor is still in plug flow, then the equations for a first-order irreversible reaction become

$$u \frac{\partial C_A}{\partial z} = -kC_A(z, R) \tag{5.91}$$

$$\rho C_p u \frac{\partial T}{\partial z} = k_T \frac{1}{R} \frac{\partial}{\partial R} \left(R \frac{\partial T}{\partial R} \right) + (-\Delta H_R)kC_A \tag{5.92}$$

We need to solve these equations with inlet conditions

$$C_A = C_{Ao} \tag{5.93}$$

$$T = T_o \tag{5.94}$$

at $z = 0$ as before. However, now we need to add two boundary conditions on temperature in the radial direction,

$$T = T_w, \qquad R = \pm R_o/2 \tag{5.95}$$

$$\frac{\partial T}{\partial R} = 0, \qquad R = 0 \tag{5.96}$$

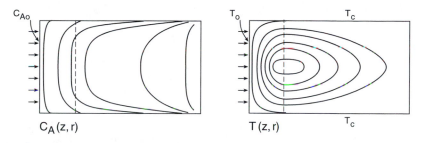

$C_A(z, r)$ $T(z, r)$

Figure 5–23   Plots of reactant and temperature profiles versus axial position z and radial position x in a wall-cooled tubular reactor. The reactor can exhibit a hot spot near the center where the rate is high and cooling is least.

as well as additional boundary conditions on C_A. For a tubular reactor we can solve these equations numerically to yield $C_A(z, R)$ and $T(z, R)$. These solutions are plotted in Figure 5–23.

The notable feature of the wall-cooled tubular reactor is that there can exist a *hot spot* near the center of the reactor and near the entrance. We saw this for the lumped model, which allowed only for variations in the z direction, but when radial variations are allowed, the effect can become even more severe as both temperature and concentration vary radially.

If we examine the temperature and concentration profile in the radial direction for the plot in Figure 5–23, we obtain graphs approximately as those in Figure 5–24.

We used the wall temperature T_w in the boundary condition, and this may be different from the coolant temperature T_c. There may be temperature variations across the wall as well as through the coolant. These are described through the overall heat transfer coefficient U, but in practice all these effects must be considered for a detailed description of the wall-cooled tubular reactor.

Figure 5–24   Plots of C_A and T versus radial position R for a wall-cooled reactor. The temperature is highest and the conversion is lowest near the centerline.

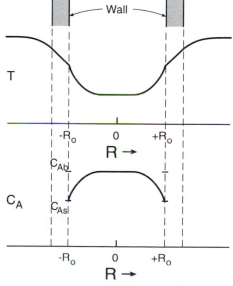

5.1 Find $\dot{Q}(t)$ or $\dot{Q}(z)$ necessary to maintain a 5 liter batch reactor or PFTR isothermal reactor at 300 K for the reaction $A \rightarrow B, r = kC_A$ in aqueous solution with $k = 2.0 \, \text{min}^{-1}, C_{Ao} = 2$ mole/liter, $\Delta H_R = -30$ kcal/mole. What is the average rate of heat removal for 95% conversion?

5.2 Calculate $C_A(\tau)$ and $T(\tau)$ if the above reaction is run in an adiabatic PFTR to 90% conversion if the reaction activation energy is 6000 cal/mole.

5.3 Under what conditions would you have to worry about boiling in the reaction of the previous problem? How might one design a reactor to process 5 molar feed to eliminate boiling?

5.4 What is the ratio of volumes required for adiabatic and isothermal operation for this reaction?

5.5 Compare $T(z)$ and $T_c(z)$ trajectories of a wall-cooled PFTR with cocurrent and countercurrent flows. Which configuration is more likely to produce more problems with a hot spot in the reactor?

5.6 The maximum reactor temperature is much more sensitive to feed temperature for feed-cooled tubular reactors than for tubular reactors with a separate cooling stream. Why?

5.7 A tubular reactor in which the feed is heated by product in a countercurrent adiabatic configuration can attain a maximum temperature much hotter than the adiabatic temperature (and in fact it can be infinite). Sketch $T(z)$ to show why this is so.

5.8 [Computer] Plot $C_A(\tau)$ and $T(\tau)$ for the reaction $A \rightarrow B, r = k(T)C_A$ in an adiabatic PFTR for $E = 30$ kcal/mole, $k_o = 2.6 \times 10^{20} \, \text{min}^{-1}, \Delta H_R = -20$ kcal/mole, $\rho C_p = 1000$ cal/liter K, $T_o = 300$ K, and $C_{Ao} = 2$ moles/liter. [This is the example in the text.] Repeat these calculations for $C_{Ao} = 0.1, 0.5, 1.0, 2.0, 4.0,$ and 10.0. Wherever possible, plot $C_A(\tau)$ and $T(\tau)$ on the same graphs to display the differences with initial concentration.

5.9 [Computer] Plot $C_A(\tau)$ and $T(\tau)$ for the reaction $A + B \rightarrow C, r = k(T)C_A^2$ in an adiabatic PFTR for $E = 30$ kcal/mole, $k_o = 2.6 \times 10^{20}$ liters/mole min, $\Delta H_R = -20$ kcal/mole, $\rho C_p = 1000$ cal/liter K, $T_o = 300$ K, and $C_{Ao} = C_{Bo} = 2$ moles/liter. [This is the example in the text but with second-order kinetics.] Repeat these calculations for $C_{Ao} = C_{Bo} = 0.1, 0.5, 1.0, 2.0, 4.0,$ and 10.0. Wherever possible, plot $C_A(\tau)$ and $T(\tau)$ on the same graphs to display the differences with initial concentration.

5.10 [Computer] Plot $C_A(\tau)$ and $T(\tau)$ for the reaction $A \rightarrow B, r = k(T)C_A$ in an adiabatic PFTR for $E = 30$ kcal/mole, $k_o = 2.6 \times 10^{20} \, \text{min}^{-1}, \Delta H_R = -20$ kcal/mole, $\rho C_p = 1000$ cal/liter K, $T_o = 300$ K, and $C_{Ao} = 2$ moles/liter. [This is the example in the text.] Repeat these calculations for $T_o = 280, 290, 300, 310, 320, 330, 340,$ and 350 K. Plot $C_A(\tau)$ and $T(\tau)$ on the same graphs to display the differences with initial T_o.

5.11 The reaction $A \rightarrow B, r = kC_A$ is carried out in an adiabatic CSTR with $C_{Ao} = 2.0$ moles/liter and $T_o = 300$ K. It is found that 50% conversion is obtained with $\tau = 4$ min, and the reactor temperature is 350 K.

(a) Find k at 350 K.

(b) The reaction is in water with $\rho C_p = 10^3$ cal/liter K. Find the heat of the reaction.

5.12 [Computer] Sketch lines of $r(X, T)$ on an X versus T plot for $A \rightarrow B$ assuming the reaction is (a) irreversible, (b) reversible and endothermic, and (c) reversible and exothermic.

5.13 Show that $r(X, T)$ for $A \rightarrow B$, $r = k_f C_A - k_b C_B$ has a maximum $(\partial X/dT)_r = 0$ if the reaction is exothermic but not if endothermic.

5.14 [Computer] Plot $C_A(\tau)$ and $T(\tau)$ in a wall-cooled tubular reactor for the liquid-phase reaction $A \rightarrow$, $k(T) = 1000 \exp(-3000/T) \min^{-1}$, $\Delta H_R/\rho C_p = 17$ K liter/mole, $C_{Ao} = 1$ mole/liter. The heat transfer coefficient is 5 $\min^{-1}$. Consider a feed temperature of 300 K and different wall temperatures in 50 K increments starting with 250 K.

5.15 (a) Sketch r versus $C_{Ao} - C_A$ for a reversible exothermic reaction if the reactor is operated isothermally at a range of temperatures where the equilibrium conversions go from high to low.

(b) From $1/r$ plots show how several isothermal PFTRs in series could achieve a higher conversion with a smaller total volume than a single isothermal PFTR.

(c) Why is this generally not an economical reactor design?

5.16 Sketch r and $1/r$ versus $C_{Ao} - C_A$ for a reversible exothermic reaction for fixed T_o and a fairly low T_c if the reactor

(a) is adiabatic,

(b) is isothermal at T_o, or

(c) quickly cools to T_c.

5.17 Sketch some possible r and $1/r$ versus $C_{Ao} - C_A$ curves for an irreversible exothermic reaction for nonisothermal operation. Show that a CSTR becomes even more attractive for nonisothermal operation.

5.18 The reaction

$$A \rightarrow B, \qquad r = k(T)C_A, \qquad \Delta H_R = -25{,}000 \text{ cal/mole}$$

is to be run in aqueous solution at atmospheric pressure.

(a) If the reactor is adiabatic and $T_o = 300$ K, what is the highest C_{Ao} for the reactor not to boil?

(b) It is found that k at 300 K is 0.2 $\min^{-1}$ and the activation energy is 30,000 cal/mole. Calculate $X(\tau)$ in an adiabatic PFTR and in an adiabatic CSTR for $C_{Ao} = 2$ moles/liter.

5.19 In the preceding reaction it is found that the equilibrium constant is 0.99 at 300 K.

(a) What is the maximum conversion attainable in an adiabatic reactor?

(b) Calculate $X(\tau)$ in an adiabatic PFTR and in an adiabatic CSTR for $C_{Ao} = 2$ moles/liter.

5.20 The reaction

$$A \rightarrow B, \qquad r = k(T)C_A, \qquad \Delta H_R = -25{,}000 \text{ cal/mole}$$

is to be run in aqueous solution in an adiabatic CSTR with $T_o = 300$ K, $C_{Ao} = 2$ moles/liter, $k = 2.0 \min^{-1}$ at 300 K, and $E = 30{,}000$ cal/mole. Calculate and plot $X(t)$, $T(t)$, and $X(T)$, for initial temperatures varying in 10 K increments starting at 280 K for initial conversions of $X = 0$ and 1. Do calculations for

(a) $\tau = 0.3$ min,

(b) $\tau = 1.5$ min,

(c) $\tau = 3.0$ min.

5.21 Consider methanol synthesis from CO and H_2 with a stoichiometric feed at pressures of 10, 30, and 100 atm.

(a) If the reactor goes to equilibrium, what conversions will be obtained at 250, 300, and 350°C at each of these pressures?

(b) If the reactor operates adiabatically, what T_o will give these conversions? Assume $C_p = \frac{7}{2}R$.

(c) How many stages are necessary to give 90% conversion in adiabatic reactors with interstage cooling to 25°C at each of these pressures?

(d) Repeat with interstage methanol separation.

(e) What are the final temperatures expected to compress these gases isentropically to these pressures starting at 1 atm and 25°C? Assume $C_p = \frac{7}{2}R$.

(f) Assuming that electricity costs $0.02/kwh, what is the cost per pound of compression for each pressure?

5.22 Construct a table of the reactors needed to synthesize methanol from natural gas and water. Include reactions, ΔH_R, reactor type, heat management method, and catalyst.

5.23 Plywood contains approximately 30% binder, which is a urea–formaldehyde polymer. Indicate the reactions and steps in its synthesis from natural gas and water (and trees). Approximately how many SCF of natural gas go into a 4×8 sheet of 3/4-in. plywood?

5.24 Consider an autocatalytic reaction $A \rightarrow B$, $r = k(T)C_A C_B$, in an adiabatic PFTR. This reaction has both chemical autocatalysis, which can produce a lag in the rate as the concentration of product B builds up, and also thermal autocatalysis, in which the rate accelerates at the temperature increases. Solve these equations for the example in the text but replacing the kinetics by the above expression to show how the temperature and C_A vary with τ.

5.25 The oxidative coupling of methane to ethylene

$$2CH_4 + O_2 \rightarrow C_2H_4 + 2H_2O, \qquad \Delta H_R = -50 \text{ kcal/mole}$$

occurs in an adiabatic tube wall reactor. The feed is 90% CH_4 in O_2 at atmospheric pressure at 300 K. In excess CH_4 the reaction can be regarded as $O_2 \rightarrow$ products with a rate first order in P_{O_2}, and the change in moles may be ignored. [Do these arguments make sense?] Assume that the heat capacity of the gases is $\frac{7}{2}R$.

(a) Find the adiabatic temperature if the reaction goes to completion assuming the wall has negligible thermal conductivity. [Answer: ≈ 800°C.]

(b) Find the reactor wall temperature versus oxygen conversion X.

(c) It is found that 90% O_2 conversion occurs in a reactor with a very large wall thermal conductivity such that the wall is isothermal at the exit temperature. For a velocity of 1 m/s, a tube diameter of 0.5 mm, and a length of 1 cm, find the surface reaction rate coefficient in molecules/cm² sec atm at this temperature. Assume that the gases are incompressible (not a very good assumption).

(d) Assuming instead that the wall thermal conductivity is zero, show that the conversion versus position z is given by the expression

$$z(X) = u \int_0^X dX / \left[(1 - X)k_o \exp\left(\frac{-E}{R[T_o + (-\Delta H_R/NC_p)X]} \right) \right]$$

(e) The reactive sticking coefficient of O_2 has an activation energy of 20 kcal/mole. Estimate k_o.

(f) Find L for 90% conversion of O_2.

5.26 Suppose your apartment has a 50 gallon hot water heater and your shower uses 10 gallons per minute. The water in the tank initially is at 120°F and the supply water is at 60°F. The heater has an on–off control and requires 1 h to heat 60°F water to 120°F.

(a) Describe the temperature of your shower versus time assuming a mixed tank, ignoring the effect of the heater.

(b) What is the wattage of the heater?

(c) Describe $T(t)$ assuming that the heater is on continuously.

(d) This design is clearly unsatisfactory, and manufacturers configure hot water tanks to minimize mixing. How can this be done fairly simply? What is $T(t)$ for this system?

(e) Expensive water heaters provide hot water continuously on demand. Sketch the reactor diagram and control system that will accomplish this with no storage reservoir. What heater wattage is required to maintain your shower at 120°F?

5.27 Derive the mass-balance equation on species A and the energy-balance equation for a batch reactor of volume V with rate $r(C_A, T)$ with wall coolant area A_c and heat transfer coefficient U.

5.28 [Computer] We have a catalyst that causes the oxidative dehydrogenation of ethane to ethylene

$$C_2H_6 + \tfrac{1}{2}O_2 \rightarrow C_2H_4 + H_2O, \qquad \Delta H_R = -25,000 \text{ cal/mole ethane}$$

The process produces only ethylene in a reaction that goes to completion with a feed of 5 mole % C_2H_6 in air in an adiabatic CSTR operated at a pressure of 2 atm with $\tau = 10$ min. We find that the reaction rate is given approximately by the expression

$$r = k_o \exp(-E/RT)C_{C_2H_6}$$

with $k_o = 1.4 \times 10^8$ mole/liter min and $E = 30,000$ cal/mole. Assume constant density and that the heat capacity is $\tfrac{7}{2}R$ at all temperatures and compositions. Calculate $C_{C_2H_6}$ and T versus τ in steady state in a CSTR and in a PFTR using a spreadsheet to solve the equations as discussed in the example in the text. Assume $C_{C_2H_6} = P_{C_2H_60}(1-X)/RT_o$.

5.29 We have a catalyst that causes the oxidative dehydrogenation of ethane to ethylene

$$C_2H_6 + \tfrac{1}{2}O_2 \rightarrow C_2H_4 + H_2O, \qquad \Delta H_R = -25,000 \text{ cal/mole ethane.}$$

The process produces only ethylene in a reaction that goes to completion with a feed of 5 mole % C_2H_6 in air in a CSTR operated at a pressure of 2 atm.

(a) If the reaction goes to completion, how many moles are produced for each mole fed into the reactor?

(b) Find an approximate expression for the temperature versus conversion in an adiabatic reactor with $T_o = 350°C$. Assume constant density and that the heat capacity is $\tfrac{7}{2}R$ at all temperatures and compositions. [Answer: $T \sim 500°C$ at completion.]

(c) What is the relationship between the space time and the residence time if the reaction goes to completion?

(d) We find that the reaction rate is given approximately by the expression

$$r = k_o \exp(-E/RT)C_{C_2H_6}$$

with $k_o = 1.4 \times 10^8$ mole/liter min and $E = 30,000$ cal/mole. Plot $X(T)$ for $\tau = 1, 3, 10,$ 30, and 100 min. [This problem can be worked with a calculator and graphed by hand or on a spreadsheet with graphics.]

(e) From this graph determine the conversion and reactor temperature for these residence times. What startup procedure would you use to attain the upper steady state for $\tau = 10$ min?

(f) From these curves plot (by hand is OK) X versus τ.

5.30 Solve (or at least set up) the preceding problem exactly, taking account of the density change caused by mole number changes in the reaction and to the temperature variations.

Chapter 6

MULTIPLE STEADY STATES AND TRANSIENTS

$\quad$I$\quad$n the previous chapter we showed how nonisothermal reactors can exhibit much more complex behavior than isothermal reactors. This occurs basically because $k(T)$ is strongly temperature dependent. Only a single steady state is possible in the PFTR, but the CSTR, although (or because) it is described by algebraic equations, can exhibit even more interesting (and potentially even more dangerous) behavior.

$\quad$This chapter will also be more mathematical than previous chapters. We need to analyze these equations to find these steady states and examine their stability. Some students will find this mathematics more than they are comfortable with or than they want to learn. We assure you that this section will be short and fairly simple. The intent of this chapter is (1) to show how multiple steady states arise and describe their consequences and (2) to show how mathematical analysis can reveal these behaviors. Attempts at nonmathematical discussions of them makes their occurrence and consequences quite mysterious.

$\quad$We will close this chapter by summarizing the principles of designing nonisothermal reactors; so if you want to skim the more mathematical sections, you should still read the last section carefully.

6.1 HEAT GENERATION AND REMOVAL IN A CSTR

We solved the mass- and energy-balance equations for the CSTR in the previous chapter. The mass balance on species A is

$$\tau \frac{dC_A}{dt} = C_{Ao} - C_A - \tau r(C_A, T) \tag{6.1}$$

For the energy balance the terms were

[accumulation] = [heat flow in] − [heat flow out] + [reaction heat] − [cooling]

which for constant density gave

$$\rho C_p V \frac{dT}{dt} = v \rho C_p (T_o - T) + V(-\Delta H_R) r - U A_c (T - T_c) + W_s \tag{6.2}$$

247

The first term on the right side of the equation is the rate of heat removal in the inlet and outlet streams, the middle term is the rate of heat generation by reaction, and the third term is the rate of heat removal by cooling. We usually omit the shaft work term W_s in this equation because it is usually much smaller than the other terms.

We next divided by $v\rho C_p$ to obtain

$$\tau\frac{dT}{dt} = T_0 - T + \tau\frac{-\Delta H_R}{\rho C_p}r(C_A, T) - \frac{U A_c}{v\rho C_p}(T - T_c) \tag{6.3}$$

where each term has the units of temperature.

We want to find solutions of these equations. It will be convenient to make the energy-balance equation dimensionless before solving it. As introduced in the previous chapter, we define an almost dimensionless rate of heat generation as

$$J = \frac{-\Delta H_R C_{Ao}}{\rho C_p} \tag{6.4}$$

where J has the units of temperature. We also define a dimensionless heat transfer coefficient as

$$\kappa = \frac{U A_c}{v\rho C_p} \tag{6.5}$$

to obtain a transient energy-balance equation

$$\boxed{\tau\frac{dT}{dt} = T_0 - T + \frac{J\tau r(X, T)}{C_{Ao}} - \kappa(T - T_c)} \tag{6.6}$$

or in steady state

$$\boxed{T - T_0 = \frac{J\tau r(X, T)}{C_{Ao}} - \kappa(T - T_c)} \tag{6.7}$$

Returning to the mass-balance equation, we write it in terms of fractional conversion (another dimensionless variable) as

$$C_{Ao} - C_A = C_{Ao}X = \tau r(X, T) \tag{6.8}$$

or

$$\boxed{X(T) = \frac{\tau r(X, T)}{C_{Ao}}} \tag{6.9}$$

We can now write the preceding energy-balance equation in terms of temperature alone as

$$\boxed{\frac{T - T_0 + \kappa(T - T_c)}{J} = X(T)} \tag{6.10}$$

We need to examine the nature of these solutions $X(T)$ for given inlet and reactor parameters. These are algebraic equations; so after elimination of X from the mass-balance equation, we only have to solve the single energy-balance equation for the reactor temperature T versus conversion X for specified fluid, reaction, and other reactor parameters.

This seems to be a simpler set of equations than the differential equations of the PFTR, but since the PFTR equations are first-order differential equations, their solutions must be *unique* for specified flow, reaction, and reactor parameters, as we discussed in the previous chapter. This is not necessarily true for the CSTR.

6.1.1 Dimensionless variables

Until now we have tried to keep all quantities in dimensional form so their physical significances can be readily appreciated. Here, however, we define groups of parameters that we define as new parameters (J and κ) so that the equations look simpler and will be easier to manipulate.

Thus, even though each term in the energy-balance equation is now dimensionless, we associate each term with a term in the original energy-balance equation: $(T - T_o)/J$ represents the heat flow out minus the heat flow in, $\kappa(T - T_c)/J$ represents the heat transfer through the walls, and $X(T)$ represents the rate of heat generation by reaction.

6.1.2 First-order irreversible reaction

We begin by considering our old friend the first-order irreversible reaction

$$A \rightarrow B, \qquad r = k(T)C_A \tag{6.11}$$

in the CSTR. For these kinetics the mass-balance equation,

$$C_{Ao} - C_A = C_{Ao}X = \tau k(T)C_{Ao}(1 - X) \tag{6.12}$$

can be solved explicitly for $X(T)$ to yield

$$X(T) = \frac{\tau k(T)}{1 + \tau k(T)} \tag{6.13}$$

This is a sigmoidal function of T as plotted in Figure 6–1. Note that at low temperature, $X = \tau k(T)$; so X increases rapidly with T as $k(T)$, while at high temperature $X \rightarrow 1$ as the reaction goes to completion. Note also that $X = \frac{1}{2}$ when $k\tau = 1$.

Figure 6–1 Plot of conversion X versus temperature for an irreversible first-order reaction in a CSTR.

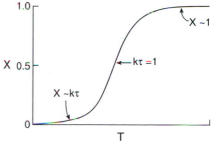

Now we substitute this expression for $X(T)$ into the energy-balance equation to yield a combined energy- and mass-balance equation for a first-order irreversible reaction in a CSTR,

$$\boxed{\frac{T - T_o + \kappa(T - T_c)}{J} = \frac{\tau k(T)}{1 + \tau k(T)}} \tag{6.14}$$

We further write this equation as

$$\mathbf{R}(T) = \mathbf{G}(T) \tag{6.15}$$

as a short-hand description of the energy-balance equation. The right-hand side of the equation $\mathbf{G}(T)$ is a dimensionless description of the rate of energy generation, while the left-hand side is the rate of energy removal. The term $T - T_o$ represents the net rate of heat removal with the feed and exit streams, while the rest of the left-hand side is the rate of heat removal from the coolant.

Thus on the left-hand side the removal term is

$$\mathbf{R}(T) = \frac{T - T_o + \kappa(T - T_c)}{J} = \frac{(1 + \kappa)T - T_o - \kappa T_c}{J} \tag{6.16}$$

while the right-hand side is

$$\mathbf{G}(T) = \frac{\tau k(T)}{1 + \tau k(T)} = X(T) \tag{6.17}$$

Again, we encourage the student to try not to become confused by the fact that all these terms were made dimensionless to keep them simple. However, we still associate each term with a particular term in the original energy balance.

6.1.3 Graphical solution

We search for the solution(s) to this equation. Note that heat removal term $\mathbf{R}(T)$ is *linear* in T, while the heat generation term $\mathbf{G}(T)$ contains $k(T)$ and is *highly nonlinear*. The method we will use to solve this equation is to construct a graph of the left- and right-hand sides and look for intersections that must be the roots of the equation or T solutions to the CSTR. We do this in a plot of $\mathbf{R}(T)$ and $\mathbf{G}(T)$ versus T.

One can always find solutions of algebraic equations of the form $F_1(x) = F_2(x)$ by plotting F_1 and F_2 versus x and determining intersections (Figure 6–2). For the situation on the left there is only one solution x_1, while for the situation on the right there are three solutions $x_1, x_2,$ and x_3.

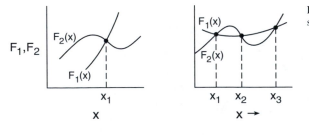

Figure 6–2   Graphical solution to the equation $F_1(x) = F_2(x)$ for situations with 1 and 3 intersections (solutions).

6.2   ADIABATIC CSTR

We first consider the adiabatic reactor for which $Q = 0$ and therefore $\kappa = 0$. The left-hand side or heat removal $\mathbf{R}(T)$ is

$$\mathbf{R}(T) = (T - T_0)/J \qquad (6.18)$$

This is a straight line whose slope is $1/J$ and which intersects the T axis at $X = 0$ where T_0 (Figure 6–3). It is evident that there may be one or three intersections of these curves (the large dots in the figure), and these represent one or three steady states in the adiabatic CSTR.

We can now remove the $\mathbf{R}$ and $\mathbf{G}$ curves and retain only the intersections that are the possible steady states in the reactor (Figure 6–4).

We solved this example for the first-order irreversible reaction $A \rightarrow B$ because it is easy to solve. If the reactor is not adiabatic, the $\mathbf{R}(T)$ curve has a different slope and intercept, but it is *still a straight line*. If the kinetics are not first order but are still irreversible, $X(T)$ still has the same qualitative shape, and if the reaction is reversible and exothermic, $X(T)$ decreases at sufficiently high temperature as the equilibrium X decreases. We can generalize therefore to say that multiple steady states may exist in any exothermic reaction in a nonisothermal CSTR.

The existence and properties of these steady states and their implications in the design of CSTRs is the major subject of this chapter.

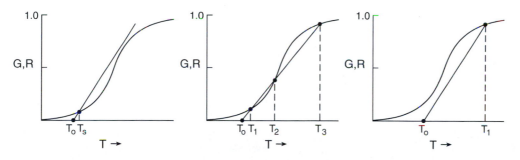

Figure 6–3   Dimensionless heat removal curve $\mathbf{R}(T)$ versus T for the adiabatic reactor plotted along with the heat generation curve $\mathbf{X}(T)$. There can be one or three intersections corresponding to one or three possible temperatures in the adiabatic CSTR, depending on T_0.

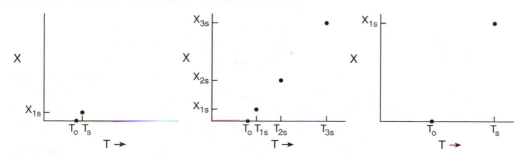

Figure 6–4   Steady-state solutions for the preceding situations, with a single solution X_1, T_1 or three solutions $X_1, T_1, X_2, T_2,$ and $X_3, T_3,$ depending on T_0. These steady states are obtained by removing the $\mathbf{R}$ and $\mathbf{G}$ curves and keeping the intersections.

6.3   STABILITY OF STEADY STATES IN A CSTR

These steady states are within the physically possible range of $T(0 < T < \infty)$ and $X(0 < X < 1)$. This is in contrast to many situations in the physical sciences where equations have multiple roots but only one root is physically acceptable because the other solutions are either outside the bounds of parameters (such as negative concentrations or temperatures) or occur as imaginary or complex numbers and can therefore be ignored.

Multiple solutions to equations occur whenever they have sufficient nonlinearity. A familiar example is equilibrium composition calculations for other than $A \rightleftarrows B$. The reaction composition in the reaction $A \rightleftarrows 3B$ yields a cubic polynomial that has three roots, although all but one give nonphysical concentrations because thermodynamic equilibrium (the solution for a reactor with $t \to \infty$ or $\tau \to \infty$) is unique.

However, even if there are three steady states, there is still the question of whether a steady state is *stable* even if it is in the physically significant range of parameters. We ask the question, starting the reactor at some initial situation, will the reactor eventually approach these solutions? Or, starting the reactor near one of the steady states, will it remain there? To answer this, we must examine the *transient equations*.

For a single reaction in the CSTR, the transient mass- and energy-balance equations are

$$\tau \frac{dX}{dt} = -X + \frac{\tau r(X, T)}{C_{Ao}} \tag{6.19}$$

$$\tau \frac{dT}{dt} = T_o - T + \frac{\tau J r(X, T)}{C_{Ao}} - \kappa(T - T_c) \tag{6.20}$$

Note that the first equation is dimensionless, while each term in the second equation has the dimensions of temperature. [We repeat these equations many times to emphasize their importance; if we can understand their solutions, we will understand multiple steady states.]

Stability would occur if there are initial conditions in the reactor C_{As} and T_s (subscript s for steady state) from which the system will evolve into each of these steady states. Also, if we start the reactor at exactly these steady states, the system will remain at that state for long times if they are stable. We will look at the first case later in connection with transients in the nonisothermal CSTR. Here we examine stability by asking if a steady state, once attained, will persist.

Consider a steady-state solution $(X_s T_s)$. This solution is stable if the system will return to it following a small perturbation away from it. To decide this, we linearize the equations about the steady state and examine the stability of the linear equations. First we subtract the steady-state version of these equations from the transient equations to obtain

$$\tau \frac{d(X - X_s)}{dt} = -(X - X_s) + \frac{\tau(r - r_s)}{C_{Ao}} \tag{6.21}$$

$$\tau \frac{d(T - T_s)}{dt} = (T_s - T) + \frac{J(r - r_s)}{C_{Ao}} - \kappa(T - T_s) \tag{6.22}$$

Note the cancellation of the T_o and T_c terms in the energy balance. For small deviations from steady state, the deviations of T and r from their steady-state values can be expanded in a *Taylor series*. [More math again.]

Consider a function of a single variable $f(x)$ that is not too far from $f(x_s)$. We can expand $f(x)$ about x_s to give

$$f(x) - f(x_s) = \frac{df(x_s)}{dx}\delta x + \frac{1}{2}\frac{d^2 f(x_s)}{dx^2}(\delta x)^2 + \cdots \tag{6.23}$$

and for two variables $f(x, y)$ this expansion becomes

$$f(x, y) - f(x_s, y_s) = \frac{\partial f(x_s, y_s)}{\partial x}\delta x + \frac{\partial f(x_s, y_s)}{\partial y}\delta y + \cdots \tag{6.24}$$

The rate for a first-order irreversible reaction can be written as

$$r(X, T) = k_o \exp(-E/RT)C_{Ao}(1 - X) \tag{6.25}$$

The deviation of r from the steady-state r_s may be expanded to yield

$$r(X, T) - r(X_s, T_s) = \frac{\partial r}{\partial X}(X - X_s) + \frac{\partial r}{\partial T}(T - T_s) + \cdots \tag{6.26}$$

Therefore, for the first-order irreversible reaction these derivatives are

$$\frac{\partial r}{\partial X} = -C_{Ao}k(T_s) \tag{6.27}$$

$$\frac{\partial r}{\partial T} = \frac{-E}{RT_s^2}k(T_s)C_{Ao}(1 - X_s) \tag{6.28}$$

where both derivatives are evaluated at the steady state.

The linearized mass-balance equation therefore becomes

$$\tau\frac{dx}{dt} = -x - \tau k(T_s)x + \frac{-E}{RT_s^2}k(T_s)(1 - X_s)y \tag{6.29}$$

$$\tau\frac{dx}{dt} = -[1 + \tau k(T_s)]x + \left(\frac{-E}{RT_s^2}k(T_s)(1 - X_s)\right)y = \alpha x + \beta y \tag{6.30}$$

where we have defined

$$x = X - X_s \tag{6.31}$$

$$y = T - T_s \tag{6.32}$$

and have given the coefficients of x and y the symbols α and β. Although the coefficients α and β look messy, they are just constants. Similarly, we can linearize the energy-balance equation to obtain an expression of the form

$$\tau\frac{dy}{dt} = \gamma x + \delta y \tag{6.33}$$

where we will not bother to write out the coefficients γ and δ.

Thus we have two simultaneous linear first-order differential equations in x and y,

$$\tau\frac{dx}{dt} = \alpha x + \beta y \tag{6.34}$$

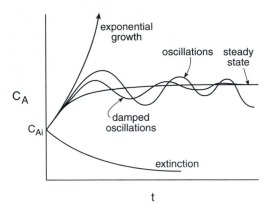

Figure 6–5 Possible transients in a CSTR following small perturbations from the steady state x_s. If λ is real, the solution will return to x_s if $\lambda < 0$ but not if $\lambda > 0$. If λ is complex, the system will exhibit either damped or growing oscillations depending on whether the real part of λ is positive or negative.

$$\tau \frac{dy}{dt} = \gamma x + \delta y \tag{6.35}$$

These simultaneous first-order differential equations can be written as a single second-order differential equation with constant coefficients,

$$\frac{d^2 x}{dt^2} - (\alpha + \delta)\frac{dx}{dt} + (\alpha\delta - \beta\gamma)x = 0 \tag{6.36}$$

The general solution of this equation is

$$x(t) = C_1 e^{\lambda t/\tau} + C_2 e^{-\lambda t/\tau} \tag{6.37}$$

The coefficients C_1 and C_2 are determined from the two initial conditions. The quantities $\pm\lambda$ are called *eigenvalues*. They are related to the coefficients through the relation

$$\lambda = \tfrac{1}{2}(\alpha + \delta) \pm \tfrac{1}{2}[(\alpha - \delta)^2 + 4\beta\gamma]^{1/2} \tag{6.38}$$

[Recall that the same equations were encountered earlier for the reversible $A \rightleftarrows B \rightleftarrows C$ reaction system. However, for that example there was only one solution that had physically possible concentrations, and that solution was always stable.]

The stability of $x(t)$ is determined by whether the solution will return back to x_s, y_s following a perturbation. This will occur if the solutions $x \sim e^{\lambda t/\tau}$ are stable, and this requires that the real part of λ be negative. This can also be regarded as a problem of finding the eigenvalues of the matrix $\begin{bmatrix} \alpha & \beta \\ \gamma & \delta \end{bmatrix}$ from the Jacobean of the original linear equations. Solutions of $x(t)$ can be as shown in Figure 6–5. This pair of equations is stable if $[(\alpha - \delta)^2 + 4\beta\gamma] > 0$ (λ real). When we insert coefficients α, β, γ, and δ as defined, we find that when there is only one steady state, it is stable. However, when there are three steady states, the first and third are stable but the middle steady state is always unstable.

6.4 OBSERVATION OF MULTIPLE STEADY STATES

We next examine the variations of these solutions with respect to parameters of the system, T_o and v. We consider the situations where we vary these parameters slowly and examine how the steady states change.

6.4.1 Variation of feed temperature

Picture a CSTR in which the feed is preheated to different temperatures T_0 while holding all other feed and reactor parameters constant (Figure 6–6). In this "experiment" we assume that we have a preheater of the feed so that we are changing T_0 sufficiently slowly that the reactor moves to its new steady state each time T_0 changes with no significant transients. On the plot of X versus T a variation in T_0 corresponds to moving the heat removal line while holding its slope constant and maintaining the heat generation curve unchanged (Figure 6–7). As T_0 is slowly varied, the steady-state intersections go from a single T at low T_0, through three intersections at intermediate T_0, to a single intersection at high T_0. We can now plot T versus T_0 and track the solutions (Figure 6–8). As T_0 increases, T first follows a line with a slope near unity, then increases discontinuously, and finally increases with a

Figure 6–6 Adiabatic CSTR in which feed temperature T_0 is varied.

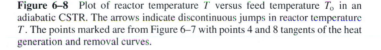

Figure 6–7 Heat removal curves in which T_0 is increased or decreased. One or three intersections with $X(T)$ indicates that one or three steady states occur.

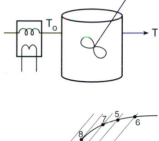

Figure 6–8 Plot of reactor temperature T versus feed temperature T_0 in an adiabatic CSTR. The arrows indicate discontinuous jumps in reactor temperature T. The points marked are from Figure 6–7 with points 4 and 8 tangents of the heat generation and removal curves.

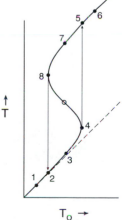

slope near unity. When T_o is decreased from high values, T is first retraced, but then T continues along the upper curve until some point where T decreases discontinuously to the original T versus T_o curve. Thus, from this "experiment" in which we varied T_o, we see both *discontinuities* and *hysteresis* in the plot of T versus T_o.

Example 6–1

(a) Plot $X(T)$ for the reaction $A \rightarrow B$, $r = kC_A$, $\Delta H_R = -20$ kcal/mole, $E = 30$ kcal/mole, $k = 0.05$ at 300 K for $\tau = 1$ min.

(b) Plot $X(T)$ for the adiabatic reactor with these parameters for $C_{Ao} = 2$ moles/liter and T_o in intervals of 20 K, starting at 300 K.

(c) Plot T versus T_o for these parameters.

This is the example from the previous chapter, and $k(T)$ was determined previously. $X(T)$ is shown in Figure 6–9 for $\tau = 1$ min. As T_o increases so that the curves move into the situation where multiple roots emerge, the system does not jump to them because it is already in a stable steady state. Only when T_o is so high that the heat removal line becomes *tangent* to the heat generation curve does the lower intersection disappear. The system then has no alternative but to jump to the upper intersection. A similar argument holds in the decrease in T_o from the upper steady state.

Note that in this "experiment" in which we varied T_o, we somehow never reached the middle intersection but jumped from the lower steady state to the upper one and back again. We said previously that this middle state is unstable, and we see that it is *never attained* in a slow variation of T_o.

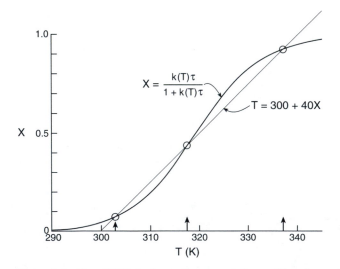

Figure 6–9   Plot of $X(T)$ for the previous example for conditions that produce three steady states for $\tau = 1$ min.

Example 6–2   Plot $C_A(\tau)$ and $T(\tau)$ for the previous reaction for $T_0 = 300$ K by plotting $X(T)$ curves versus τ and finding intersections (Figure 6–10).

It is seen that multiple intersections exist for a limited range of τ. For $\tau > 1.3$ min the reaction goes nearly to completion, while for $\tau < 0.8$ min the conversion is very small. Between these τs the system exhibits two steady states as τ (or v) is varied.

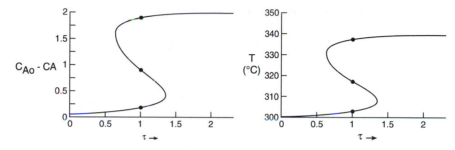

Figure 6–10   Plot of $C_A(\tau)$ and $T(\tau)$ for the preceding reaction for $T_0 = 300$ K from intersections of $X(T)$. This is the same example that was worked in Chapter 5.

6.4.2   Variation of other parameters

The previous examples showed that we can attain multiple steady states in a CSTR by varying the feed temperature T_0. Many other parameters in the reactor such as v and C_{Ao} can also be varied, and these can give multiple steady states.

Figure 6–11 shows a plot of X versus T for this example for $T_0 = 300$ K and $\tau = 1$ min. It is seen that for this situation a single low conversion steady state exists for low C_{Ao}, a single high conversion steady state exists for high C_{Ao}, and multiple steady states can exist for an intermediate range of C_{Ao}. We will discuss these further in a homework problem.

Whenever multiple steady states in a reactor are possible, we must be very concerned that we are operating on the desired steady-state branch. This requires a proper *startup*

Figure 6–11   Possible multiple steady states in an adiabatic CSTR for variations in the feed concentrations C_{Ao}. For low C_{Ao} the conversion is low, while for $C_{Ao} > 2.6$ moles/liter only the high conversion steady state can exist. For a range of C_{Ao}, three steady states are possible.

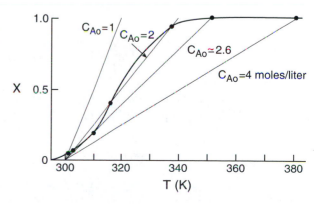

procedure to attain the desired steady state and suitable operation limits to make sure that we never exhibit a sufficiently *large transient* to cause the system to *fall off* the desired conversion branch. We will consider transients in the CSTR in the next section.

6.5 TRANSIENTS IN THE CSTR WITH MULTIPLE STEADY STATES

Next we consider the time dependence of C_A and T when a CSTR is started from different initial conditions. We solve the equations

$$\tau \frac{dX}{dt} = -X + \tau k(T)(1 - X) \tag{6.39}$$

$$\tau \frac{dT}{dt} = T_0 - T + \tau J k(T)(1 - X) \tag{6.40}$$

We choose the reaction $A \rightarrow B$, $r = k(T)C_A$, with $k = 0.05$ min^{-1} at 300 K, $E = 30$ kcal/mole, $\Delta H_R = -40$ kcal/mole, $\rho C_P = 1000$ cal/liter K, and $C_{Ao} = 2$ moles/liter. We choose $\tau = 1$ min, the conditions where calculations in the previous chapter showed that multiple steady states exist in the steady-state CSTR.

Figure 6–12 shows plots of $C_A(t)$ and $T(t)$ for different initial reactor variables. We choose $C_{Ai} = 0$ (starting the reactor with pure solvent) and also $C_{Ai} = 2.0$ (starting the reactor with pure feed). Shown are $C_A(t)$ and $T(t)$ for different values of T_i. The solutions converge on X_{1s}, T_{1s} or X_{3s}, T_{3s} for long times as the reactor approaches steady state. However, it is seen from Figure 6–11 that the time to converge to the steady states varies strongly with initial conditions. Also, there is a *temperature overshoot* for some initial conditions where the reactor temperature has a maximum much higher than the steady-state

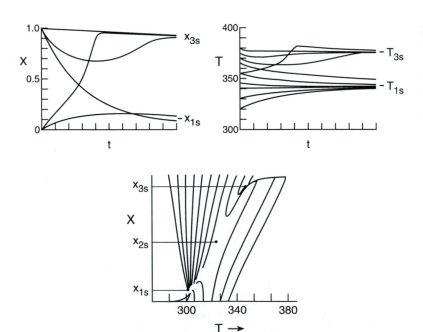

Figure 6–12 Transients in the adiabatic CSTR for the irreversible reaction of the previous example. The upper panels show $X(t)$ and $T(t)$, while the lower panel displays $X(T)$ for the same curves shown in the upper panel. The system converges on one of the stable steady states but never on the middle unstable steady state.

temperature. Note also that, even though the reactor appears to be approaching the unstable steady state at X_{2s}, T_{2s} for some initial conditions, it always moves away from this state to one of the two stable steady states.

Detailed transient studies and plots such as these must be done whenever one is interested in designing a nonisothermal reactor to determine how startup and shutdown transients might cause unacceptable transient conditions in the reactor.

6.6 OTHER REACTIONS IN A CSTR

We have as usual examined an almost trivial example, the first-order irreversible reaction. We did this because the math is simple and we could easily see what can happen. For more complex examples the equations become more difficult to solve.

6.6.1 Other order irreversible reaction

Consider $2A \rightarrow$ products or $A + B \rightarrow$ products with $C_{Ao} = C_{Bo}$ for a rate $r = kC_A^2$. The mass-balance equation now becomes

$$C_{Ao}X = \tau k(T)C_{Ao}^2(1 - X)^2 \tag{6.41}$$

which is a quadratic in X,

$$X(T) = 1 + \frac{1 \pm (1 + 4\kappa\tau C_{Ao})^{1/2}}{2k\tau X_{Ao}} \tag{6.42}$$

Since $X < 1$, we must take the smaller root,

$$X(T) = 1 + \frac{1 - (1 + 4\kappa\tau C_{Ao})^{1/2}}{2k\tau X_{Ao}} \tag{6.43}$$

The heat removal line is of course unchanged no matter what the kinetics, and $X(T)$ has a qualitatively similar sigmoidal shape for second-order kinetics. We could continue this for any order irreversible reaction, and the solutions would be qualitatively similar, although the algebra becomes messier.

Therefore, we can generalize the previous discussion to say that all qualitative features of multiple steady states in the CSTR remain unchanged for the nth-order irreversible reaction as long as is obeys positive-order kinetics. We will consider zeroth-order and negative-order kinetics in problems.

6.6.2 Reversible reactions

For $A \rightleftarrows B$, $r = kC_A - k'C_B$ we have a similar $X(T)$ curve

$$C_{Ao}X = \tau C_{Ao}[k(T)(1 - X) - k'(T)X] \tag{6.44}$$

which has the solution

$$X(T) = \frac{\tau k(T)}{1 + \tau[k(T) + k'(T)]} \tag{6.45}$$

The shapes of these curves is plotted in Figure 6–13 for endothermic and exothermic reactions. If $\Delta H > 0$, then the shape of the $X(T)$ curve is nearly unchanged because

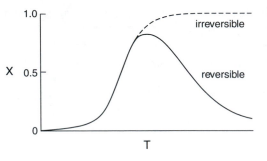

Figure 6–13   Plots of $X(T)$ for the exothermic irreversible (dashed curve) and reversible (solid curve) reactions. If the reaction is reversible, the conversion falls at high temperature, but multiple steady states are still possible.

the equilibrium conversion is lower at low temperatures, but if $\Delta H < 0$, then $X(T)$ increases with T initially but then decreases at high T as the reversibility of the reaction causes X to decrease. However, the multiplicity behavior is essentially unchanged with reversible reactions.

6.6.3 Multiple reactions

For R simultaneous reactions we have R different X_i values and R simultaneous mass-balance equations to solve along with the energy-balance equation. In the energy-balance equation the flow and heat removal terms are identical, but the energy generation has terms for each of the R reactions,

$$T - T_o + \kappa(T - T_c) = \frac{\tau \sum_{i=1}^{R} J_i r_i(X_1, X_2, \ldots, X_R, T)}{C_{Ao}} \tag{6.46}$$

$$= \frac{\tau \sum_{i=1}^{R} J_i r_i(X_1, X_2, \ldots, X_R, T)}{C_{Ao}} \tag{6.47}$$

These $R + 1$ equations must be solved for $X_1, X_2, X_3, \ldots, X_R, T$. For a particular set of reactions there may be many steady states, and in general no simple graphical or analytical simplifications are possible.

For conversions with multiple reactions we encounter a problem we have escaped previously. In a multiple-reaction system we need a symbol for the conversion of each reaction and also a symbol for the conversion of each reactant. In Chapter 4 we defined the conversion of a species as X_A for reactant A and as X_j for the jth reactant. If we are talking about the conversion of each reaction, we need a symbol such as X_i for the ith reaction. We should therefore not use Xs for both of these quantities. However, rather than introducing a new variable, we will simply use X_j to specify the conversion of a species and X_i to indicate the conversion of a reaction. We will not use these quantities enough to make this ambiguity too confusing.

For $A \rightarrow B \rightarrow C$ with $C_{Bo} = C_{Co} = 0$ in a CSTR the two mass-balance equations are

$$C_{Ao}X_1 = \tau r_1 = \tau k_1 C_{Ao}(1 - X_1) \tag{6.48}$$

$$C_{Ao}(X_1 - X_2) = \tau C_{Ao}k_1(1 - X_1) - \tau C_{Ao}k_2(X_1 - X_2) \tag{6.49}$$

Figure 6–14   Plot of possible conversions $(X_1 + X_2)$ versus reactor temperature T in an adiabatic CSTR with series exothermic reactions $A \rightarrow B \rightarrow C$. Up to five steady state intersections are possible for this situation.

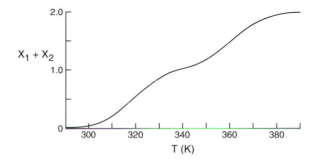

These give

$$X_1 = \frac{\tau r_1}{C_{Ao}} = \frac{\tau k_1}{1 + \tau k_1} \tag{6.50}$$

$$X_2 = \frac{\tau^2 k_1 k_2}{(1 + k_1 \tau)(1 + k_2 \tau)} \tag{6.51}$$

The energy-balance equation is

$$T - T_o + \frac{U A_c}{v \rho C_p}(T - T_c) = \frac{-\tau}{\rho C_p}(\Delta H_1 r_1 + \Delta H_2 r_2)$$

$$= J_1 X_1 + J_2 X_2 \tag{6.52}$$

Solutions can be found by plotting the left-hand side and the right-hand side together on a plot of $X_1 + X_2$ versus T. In this case $X_1 + X_2$ versus T may appear as shown in Figure 6–14. For appropriate values of the parameters, these curves give up to five intersections, corresponding to three stable steady states. This situation will be considered in a homework problem.

6.7   VARIABLE COOLANT TEMPERATURE IN A CSTR

We have thus far considered only the situation where the coolant is at constant temperature T_c. This requires that the heat load from the reactor and the coolant flow rate are adequate to maintain the coolant at constant temperature.

The reactor configuration might look as shown in Figure 6–15 for the jacketed reactor and with an internal cooling coil. If T_c is not constant, we have to solve an energy balance on the coolant along with the mass and energy balance on the reactor, where now the coolant has inlet temperature T_{co} and outlet temperature and the coolant flows with volumetric flow rate v_c and has contact area A_c with the reactor.

These configurations are essentially two reactors connected by the heat exchange area A_c, and we will consider this in more detail in a later chapter in connection with multiphase reactors. We do not need to write a species balance on the coolant, but the energy balance in the jacket or cooling coil is exactly the same as for the reactor except that we omit the reaction generation term.

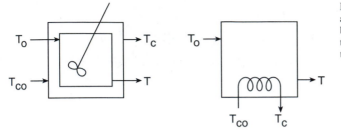

Figure 6–15 Sketches of a CSTR with cooling either through a jacket (left) or a cooling coil (right). A separate energy balance must be solved for the coolant temperature if the temperature of the cooling fluid T_c is not constant and equal to the coolant feed temperature T_{co}.

6.7.1 The jacketed CSTR

The jacketed reactor may be assumed to have mixed flow (either by stirring the jacket or by mixing through natural convection); so the jacket also obeys the CSTR equation. Therefore, for the jacketed reactor with a single reaction we write the three equations

$$C_{Ao} - C_A = C_{Ao}X = \tau r(X, T) \tag{6.53}$$

$$v\rho C_p(T_o - T) + V(-\Delta H_R)r - UA_c(T - T_c) = 0 \tag{6.54}$$

$$v_c\rho C_{pc}(T_{co} - T_c) + UA_c(T - T_c) = 0 \tag{6.55}$$

Note that the heat transfer rates across the wall are the same except for the sign change. The residence time of the coolant in the jacket is $\tau_c = V_c/v_c$.

We now have to solve the three algebraic equations to find C_A, T, and T_c for given values of parameters. It is evident that the possible heating of the coolant by exothermic reactions provides added instability to the CSTR, and makes multiple steady states even more likely. This occurs because the reaction heats the coolant and thus lowers its effectiveness in controlling the reactor temperature.

6.7.2 Cooling coil

If there is an internal cooling coil in a CSTR, the coil can exhibit a coolant temperature that is a function of position in the coil. Thus we should write a mass balance on the coolant, which is that of a PFTR without chemical reaction,

$$u_cC_{pc}\frac{dT_c}{dz_c} = +UA_c(T - T_c) \tag{6.56}$$

Note that now T_c is a variable that is a function of position z_c in the cooling coil, while T, the reactor temperature in the CSTR reactor, is a constant. We can solve this differential equation separately to obtain an average coolant temperature to insert in the reactor energy-balance equation. However, the heat load on the cooling coil can be complicated to calculate because the heat transfer coefficient may not be constant.

6.8 DESIGNING REACTORS FOR ENERGY MANAGEMENT

In this and the previous chapters we considered the effects of nonisothermal operation on reactor behavior. The effects of nonisothermal operation can be dramatic, especially for exothermic reactions, often leading to reactor volumes many times smaller than if isothermal and often leading to the possibility of multiple steady states. Further, in nonisothermal operation, the CSTR can require a smaller volume for a given conversion than a PFTR. In this section we summarize some of these characteristics and modes of operation. For endothermic reactions, nonisothermal operation cools the reactor, and this reduces the rate, so that these reactors are inherently stable. The modes of operation can be classified as follows:

6.8.1   Isothermal

This mode is seldom used except for extremely slow processes such as fermentation or for very small reactors where the surface area for heat transfer is large enough to maintain the reactor thermostatted at the temperature of the surroundings. In fact, we seldom want to operate a reactor isothermally, because we want to optimize the temperature and temperature profile in the reactor to optimize the rate and selectivity, and this is most efficiently achieved by allowing the heat released by the reaction to supply the energy to maintain the desired temperature.

6.8.2   Adiabatic

This mode is easiest to achieve in large reactors, and it will be attained if heat removal or addition is stopped. For endothermic reactions the reactor simply shuts itself off as it cools, but for exothermic reactions the situation becomes much more complex and interesting.

This is frequently the required mode of operation for fast oxidation reactions because the heat release is too fast to provide efficient heat exchange. Most combustion processes are nearly adiabatic (your home furnace and your automobile engine), and many catalytic oxidation processes such as NH_3 oxidation in HNO_3 synthesis are nearly adiabatic.

Some endothermic reactors are operated nearly adiabatically. Fluidized catalytic cracking (FCC), which was described in Chapter 2, is adiabatic in the sense that, when the hot catalyst is mixed with the cold gas oil in the riser reactor, the reactions are sufficiently fast (a few seconds residence time in the riser section) that little heat is lost and the fluid and catalyst cool strongly as they flow up the reactor pipe. The inlet temperature of the regenerated catalyst determines the temperature profile in the reactor, and the endothermicities of the different cracking reactions also will have a strong effect on the temperature profile within the pipe. The regenerator of the FCC is also nearly adiabatic, and the exothermic burning of carbon from the spent catalyst heats the catalyst back up to the desired temperature to enter the reactor. A key feature of FCC is that the entire reactor–regenerator combination is nearly adiabatic. The heat needed for the endothermic cracking reactions in the reactor is supplied by the exothermic combustion reactions in the regenerator. The heat is carried from reactor to regenerator, not by conventional heat exchangers through reactor walls but by heat carried in the flowing catalyst. It is clear that the reactor engineers who designed and continue to improve FCC must be very concerned with the mass and heat flows throughout the reactor. We repeat our comment that FCC is the most important and probably the most complex chemical reactor in our discipline.

6.8.3   Interstage heating or cooling

The next simplest reactor type is a sequence of adiabatic reactors with interstage heating or cooling between reactor stages. We can thus make simple reactors with no provision for heat transfer and do the heat management in heat exchangers *outside* the reactors.

This mode is used industrially for exothermic reactions such as NH_3 oxidation and in CH_3OH synthesis, where exothermic and reversible reactions need to operate at temperatures where the rate is high but not so high that the equilibrium conversion is low. Interstage cooling is frequently accomplished along with separation of reactants from products in units such as water quenchers or distillation columns, where the cooled reactant can be recycled back into the reactor. In these operations the heat of water vaporization and the heat removed from the top of the distillation column provides the energy to cool the reactant back to the proper feed temperature.

Endothermic reactions can also be run with interstage heating. An example we have considered previously is the catalytic reforming of naphtha in petroleum refining, which is strongly endothermic. These reactors are adiabatic packed beds or moving beds (more on these in the next chapter) in which the reactant is preheated before each reactor stage. In the most used current version of catalytic reforming, several moving beds are arranged vertically with the catalyst moving slowly down the reactors as the reactant moves upward. Between stages, the reactant is diverted into a furnace, where it is heated back up to the desired temperature and returned to the reactor. The catalyst moves too slowly for it to carry significant heat as in FCC, but it is circulated slowly (perhaps 1 week residence time) so that it can be withdrawn at the bottom of the reactor (the entrance of the reactants) and the carbon burned off before the catalyst is returned to the reactor. These reactors operate at H_2 pressures of 10–50 atm; so the engineering of catalytic reforming processes with moving catalyst is also fairly complex.

6.8.4   Reactor heating or cooling

The obvious way to heat or cool a chemical reactor is by heat exchange through a wall, either by an external *jacket* or with a *cooling* or *heating coil*. As we discussed previously, the coolant energy balance must be solved along with the reactor energy balance to determine temperatures and heat loads.

The CSTR has obvious virtues in heating and cooling because the fluid can be stirred to promote heat transfer. Thus better heat management can be obtained with minimum heat transfer area. For example, the cooling coil is excellent for heat management in the CSTR, but wall cooling is the only means of heat transfer in the PFTR. Note, however, that a coiled tube in a tank could be regarded as a heat exchanger coil for a CSTR if the reaction occurs in the tank or as a wall heat exchanger for a PFTR if the reaction occurs within the tube. Most of the reactors we deal with are in fact catalytic; so the previous examples would have catalyst either within or outside of the coil.

PROBLEMS

6.1 Show that the reactions $A \rightarrow B \rightarrow C$ can exhibit up to five steady states in a CSTR. How many of these are stable?

6.2 How many steady states are possible for $A \rightarrow B, r = k$ in an adiabatic CSTR if the reaction is exothermic? endothermic?

6.3 In the preceding problem use your graph to estimate the feed temperature at which an operating temperature jump might occur, all other things being left the same. What would be the magnitude of this jump in temperature?

6.4 The aqueous reaction $A \rightarrow B, r = kC_A$ is to be run in an adiabatic reactor to 90% conversion. The rate constant is

$$k \text{ (min}^{-1}) = 10^{10}e^{-9200/T}$$

The heat of reaction is $-20,000$ cal/mole. The solution is pressurized to remain liquid at all temperatures. Assume $\tau = 1$ min and $T_o = 300$ K.

(a) For what C_{Ao} could multiple steady states occur in a CSTR?

(b) If $C_{Ao} = 4$ moles/liter and $\tau = 1$ min, how would you start up the reactor to obtain a high conversion?

6.5 Plot $X(\tau)$ versus T for the aqueous reaction $A \rightarrow$ products, $r = kC_A$, $\Delta H_R = -20$ kcal/mole, $k = 0.05$ min^{-1} at 300 K, $E = 30$ kcal/mole in an adiabatic CSTR for $\tau = 0.5, 1.0,$ and 2 min. Use this graph to determine the range of residence times for multiple steady states in a CSTR for $C_{Ao} = 2$ and $T_o = 300$ K.

6.6 Find the range of feed concentration and residence time in which the reaction $A \rightarrow B$ with $r = 2C_A/(1 + C_A)^2$ will exhibit multiple steady states in an isothermal CSTR. A graphical procedure is suggested.

6.7 Show that the intermediate steady state in a CSTR will always be unstable by considering the slopes of the heat generation and heat removal curves.

6.8 For the exothermic reactions $A \rightarrow B \rightarrow C$ in an adiabatic CSTR the total conversion $(X_1 + X_2)$ versus T curves might look as shown in Figure 6–14.

(a) From the graph determine for what values of T_o this system could exhibit five steady states.

(b) How many of these steady states are stable?

(c) Sketch T versus T_o for this situation, analogous to Figure 6–8.

(d) What startup procedure or procedures could you use to attain the intermediate steady state in order to maximize production of B? to maximize production of C?

6.9 For the adiabatic PFTR of Problem 6.5 with $T_c = 250$ K, find T and C_A versus τ for T_o varying from 300 to 600 K in 50 K steps. What must be T_o if the reactor temperature is not to exceed 500°C?

6.10 Plot $X(\tau)$ versus T for the aqueous reaction $A \rightarrow$ products, $r = kC_A$, $\Delta H_R = -20$ kcal/mole, $k = 0.05$ min^{-1} at 300 K, $E = 30$ kcal/mole in an adiabatic CSTR for $\tau = 0.5, 1.0,$ and 2 min. Use this graph for the following problems.

(a) For an adiabatic CSTR with $C_{Ao} = 2$ moles/liter and $T_o = 300$ K, what are the steady-state conversions and temperatures for each of these residence times?

(b) What is the maximum C_{A_o} that will produce only a single steady state for any T_o for each of these residence times?

(c) For $\tau = 1$ min, plot T versus T_o for an experiment where T_o is varied to pass up and down through the multiple-steady-state region.

6.11 A reaction $A \rightarrow$ products with unknown kinetics gave the following rate data in an adiabatic PFTR with $C_{Ao} = 1$ mole/liter:

Rate moles/liter min	C_A moles/liter
1×10^{-3}	1.0
1.5×10^{-3}	0.8
4×10^{-3}	0.6
1×10^{-2}	0.4
1×10^{-2}	0.2
1×10^{-3}	0.1
0	0.0

(a) From a graph of these data (by hand is OK), plot $X(\tau)$ in the PFTR.

(b) Plot $X(\tau)$ for this reaction in an adiabatic CSTR with the same volume and flow rate.

(c) Estimate the range of τ where multiple steady states can occur in this adiabatic CSTR.

6.12 Analyze the following statement: An adiabatic CSTR usually gives a shorter residence time than a PFTR for the same conversion.

6.13 Consider the reaction $A \rightarrow B \rightarrow C$ in an adiabatic CSTR reactor with the first reaction exothermic and the second reaction endothermic. How many steady states can this system exhibit?

6.14 [Computer] We have a catalyst that causes the oxidative dehydrogenation of ethane to ethylene

$$C_2H_6 + \tfrac{1}{2}O_2 \rightarrow C_2H_4 + H_2O, \qquad \Delta H_R = -25,000 \text{ cal/mole ethane}$$

The process produces only ethylene in a reaction that goes to completion with a feed of 5 mole % C_2H_6 in air in a reactor operated at a pressure of 2 atm. We find that the reaction rate is given approximately by the expression

$$r = k_o \exp(-E/RT)C_{C_2H_6}$$

with $k_o = 1.4 \times 10^8$ mole/liter min and $E = 30,000$ cal/mole. Assume constant density and that the heat capacity is $\tfrac{7}{2}R$ at all temperatures and compositions.

(a) Calculate $X(\tau)$ and $T(\tau)$ in a steady-state adiabatic PFTR with $T_o = 350°C$.

(b) Calculate $X(\tau)$ and $T(\tau)$ in a steady-state adiabatic CSTR with $T_o = 350°C$.

6.15 [Computer] We have a catalyst that causes the oxidative dehydrogenation of ethane to ethylene

$$C_2H_6 + \tfrac{1}{2}O_2 \rightarrow C_2H_4 + H_2O, \qquad \Delta H_R = -25,000 \text{ cal/mole ethane}$$

The process produces only ethylene in a reaction that goes to completion with a feed of 5 mole % C_2H_6 in air in an adiabatic CSTR operated at a pressure of 2 atm with $\tau = 10$ min. We find that the reaction rate is given approximately by the expression

$$r = k_o \exp(-E/RT)C_{C_2H_6}$$

with $k_o = 1.4 \times 10^8$ mole/liter min and $E = 30,000$ cal/mole. Assume constant density and that the heat capacity is $\tfrac{7}{2}R$ at all temperatures and compositions.

(a) Calculate and plot $X(t)$ and $T(t)$ with $T_o = 350°C$ and $\tau = 10$ min, starting initially ($t = 0$) with $X = 0$ or 1 and T_i between 200 and 1000°C in 50°C intervals (34 runs).

(b) Plot these trajectories on the $X(T)$ plane.

6.16 Butadiene and ethylene react to form cyclohexene with pure equimolar feeds at a feed temperature of 520°C in a 100 liter continuous reactor operating at 1 atm. The reaction may be considered irreversible and first order in each reactant with a rate coefficient

$$k(\text{liter/mole sec}) = 6.2 \times 10^{-7} \exp(-14,000/T)$$

The heat of the reaction is $-30,000$ cal/mole, and the heat capacity ρC_p of the fluid is 0.10 cal/liter K, and both may be assumed independent of temperature.

(a) Write out this reaction.

(b) What is the reactor temperature versus conversion?

(c) Calculate the space time necessary to convert 10% of the butadiene to cyclohexane in a CSTR operating at 525°C. You may assume that the density does not change significantly since the conversion is low.

(d) Calculate the space time to convert 10% of the butadiene to cyclohexene in an adiabatic CSTR for $T_o = 520°C$. You may assume that the density does not change significantly since the conversion is low.

(e) Calculate the space time to convert 10% of the butadiene to cyclohexene in a PFTR maintained at 525°C. You may assume that the density does not change significantly since the conversion is low.

(f) [Computer] Calculate the space time to convert 10% of the butadiene to cyclohexene in an adiabatic PFTR with $T_o = 520°C$. You may assume that the density does not change significantly since the conversion is low.

(g) Set up the preceding problems exactly taking into account the density change with conversion.

6.17 A CSTR is used to carry out the aqueous reaction

$$A \rightarrow B, \qquad r = kC_A$$

with $C_{Ao} = 8$ moles/liter, $T_o = 300$ K, $\rho C_p = 1000$ cal/liter K, $k = 10^9 e^{-6000/T}$, $\Delta H_R = 15,000$ cal/mole.

(a) What residence time is required to achieve 50% conversion in a CSTR operating at 300 K?

(b) If the flow rate into the reactor is 100 liter/min, what is the heat removal rate to maintain the reactor isothermal at 300 K for 50% conversion?

(c) In an adiabatic reactor what will be the outlet temperature at 50% conversion?

(d) What residence time is required for 50% conversion in an adiabatic CSTR?

(e) What conversion can we obtain in an adiabatic reactor if the temperature is not to exceed the boiling point of water, 370 K?

6.18 Consider the preceding reaction system to be run in two adiabatic CSTRs operated in series with interstage cooling.

(a) If the temperature in each reactor is 370 K, calculate the residence times and conversions in each reactor.

(b) For a flow rate of 100 liter/min, what is the heat load on the interstage cooler?

6.19 Consider the burning of a spherical carbon particle of radius R in air at temperature T_o with an air concentration C_{Ao}. The temperature of the particle is T and the air concentration at the particle surface is T. The rate of the process is $k_m C_A$ (in appropriate units). [We will consider these processes in more detail in Chapter 9, but even without knowing much about what is going on, you should be able to answer these questions although the problem doesn't look much like an adiabatic CSTR.]

(a) Write the mass and energy balances for an oxidizing particle to show that it can exhibit multiple steady states if mass transfer limited. [Find $C_{Ao} - C_A = \ldots$ and $T_o - T = \ldots$.]

(b) Find an expression for ΔT.

6.20 Figure 6–11 shows the steady states in an adiabatic CSTR for several different values of C_{Ao} for the example worked previously with $T_o = 300$ K.

(a) Sketch X versus C_{Ao} and T versus C_{Ao} for this situation.

(b) For what values of C_{Ao} will multiple steady states exist and what will be the reactor temperatures?

(c) How could you start up this adiabatic CSTR to attain the high conversion steady state for situations where three steady states exist?

Chapter 7

CATALYTIC REACTORS AND MASS TRANSFER

M̲ost real reactors are not homogeneous but use catalysts (1) to make reaction occur at temperatures lower than would be required for homogeneous reaction and (2) to attain a higher selectivity to a particular product than would be attained homogeneously. One may then ask whether any of the previous material on homogeneous reactions has any relevance to these situations. The answer fortunately is yes, because the same equations are used. However, catalytic reaction rate expressions have a quite different meaning than rate expressions for homogeneous reactions.

7.1 CATALYTIC REACTIONS

Catalysts are substances added to a chemical process that do not enter into the stoichiometry of the reaction but that cause the reaction to proceed faster or make one reaction proceed faster than others.

We distinguish between *homogeneous* and *heterogeneous* catalysts. Homogeneous catalysts are molecules in the same phase as the reactants (usually a liquid solution), and heterogeneous catalysts in another phase (usually solids whose surfaces catalyze the desired reaction). Acids, bases, and organometallic complexes are examples of homogeneous catalysts, while solid powders, pellets, and reactor walls are examples of heterogeneous catalysts.

Catalysts provide an alternate path for the reaction to occur. If the catalyst provides a lower energy barrier path, this rate enhancement can be described on a potential energy diagram (Figure 7–1). The energy barrier for reaction (the reaction activation energy) is lowered from E_{homog} to E_{cat} by the use of an appropriate catalyst, and this will enhance the reaction rate. If the selectivity to form a desired product is more important than reactivity (which is usually the case), then a catalyst can provide a lower barrier for the desired reaction, leaving the undesired reaction rate unchanged. The search for this active and selective catalyst in biology or in chemical synthesis is perhaps the central goal in most biological and physical sciences research.

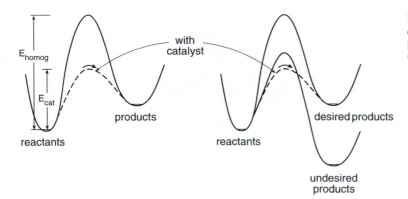

E_{homog}

E_{cat}

with catalyst

products

reactants

desired products

reactants

undesired products

single reaction

multiple reactions

Figure 7–1 Energy diagram showing how a catalyst lowers the barrier for a reaction to improve conversion (left) or lowers the barrier for a desired reaction to improve selectivity (right).

By far the most efficient catalysts are enzymes, which regulate most biological reactions. Biological catalysts are without question the most important catalysts (to us) because without them life would be impossible. Enzymes are proteins that may be either isolated molecules in solution (homogeneous) or molecules bound to large macromolecules or to a cell wall (heterogeneous). We have not yet learned how to create catalysts with nearly the efficiency and selectivity of nature's enzyme catalysts. We will consider biological reactors at the end of this chapter as the example of the most efficient chemical reactor possible.

Catalysts can strongly regulate reactions because they are not consumed as the reaction proceeds. Very small amounts of catalysts can have a profound effect on rates and selectivities. At the same time, catalysts can undergo changes in activity and selectivity as the process proceeds. In biological systems this regulatory function determines which reactions should proceed and in what regions of the body. In industrial reactors the search for active catalysts and the prevention of their deactivation during operation consume much of the engineer's efforts. Promoters and poisons of these catalysts (called cofactors and inhibitors in biology) play a dominant role in all catalytic processes.

We should be clear as to what a catalyst can and cannot do in a reaction. Most important, no catalyst can alter the equilibrium composition in a reactor because that would violate the Second Law of Thermodynamics, which says that equilibrium in a reaction is uniquely defined for any system. However, a catalyst can increase the rate of a reaction or increase the rate of one reaction more than another. One can never use a catalyst to take a reaction from one side of equilibrium to another. The goal in reaction engineering is typically to find a catalyst that will accelerate the rate of a desired reaction so that, for the residence time allowed in the reactor, this reaction approaches equilibrium while other undesired reactions do not. Attempts to violate the laws of thermodynamics always lead to failure, but many engineers still try.

The first complication with catalytic processes is that we need to maintain the catalyst in the reactor. With a homogeneous catalyst (catalyst and reactants in the same phase), the only method to reuse the catalyst is to separate it from the products after the reactor and recycle the catalyst into the feed streams shown in Figure 7–2. Much more preferable and economical is to use a heterogeneous catalyst that is kept inside the reactor such as in a powder that is filtered from the product or as pellets packed in the reactor.

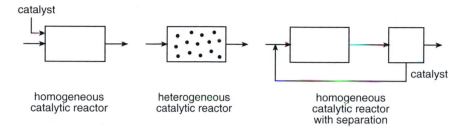

Figure 7–2 Homogeneous catalysts must be separated from the reaction products and returned to the reactor while heterogeneous catalysts can be easily retained in the reactor.

Nature also prefers this method by keeping enzymes within cells, on cell walls, or adsorbed ("bound") on large molecules, which are said to be immobilized. We therefore need to consider carefully the binding of catalysts on supports. The creation of structures and configurations of catalyst, support, and reactants and products to facilitate mass transfer and to provide access of reactants to the catalyst are essential features in designing a reaction system that has appropriate mass transfer characteristics.

7.1.1   Mass transfer

After reactivity and selectivity, the next complication we encounter with all catalytic reactions is that there are essential transport steps of reactants and products to and from the catalyst. Therefore, in practice catalytic reaction rates can be thoroughly disguised by mass transfer rates. In fact, in many industrial reactors the kinetics of individual reactions are quite unknown, and some engineers would regard knowledge of their rates as unimportant compared to the need to prepare active, selective, and stable catalysts. The role of mass transfer in reactions is therefore essential in describing most reaction and reactor systems, and this will be a dominant subject in this chapter.

Mass transfer steps are essential in any multiphase reactor because reactants must be transferred from one phase to another. When we consider other multiphase reactors in later chapters, we will see that mass transfer rates frequently control these processes. In this chapter we consider a simpler example in the catalytic reactor. This is the first example of a multiphase reactor because the reactor contains both a fluid phase and a catalyst phase. However, this reactor is a very simple multiphase reactor because the catalyst does not enter or leave the reactor, and reaction occurs only by the fluid reacting at the catalyst surface.

7.2 CATALYTIC REACTORS

7.2.1   The packed bed reactor

The most used industrial reactor is the catalytic packed bed reactor. This is typically a tank or tube filled with catalyst pellets with reactants entering at one end and products leaving at the other. The fluid flows in the void space around the pellets and reacts on and in the pellets as illustrated in Figure 7–3.

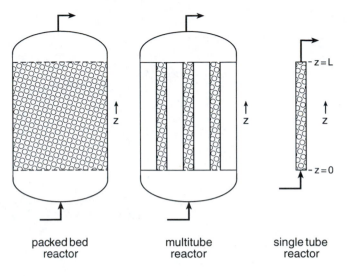

Figure 7–3 Packed bed reactors, which may be a single bed, many tubes, or a single tube filled with catalyst pellets. Multitube reactors allow efficient heat transfer for exothermic or endothermic reactions.

packed bed
reactor

multitube
reactor

single tube
reactor

[Human organs such as the liver are nature's very clever versions of a packed bed reactor, which we will consider at the end of this chapter. Enzymes on surfaces catalyze reactions in blood flowing through this reactor, although the geometry, flow pattern, and reaction selectivity are much more complex and efficient than any reactor that engineers can design.]

The fluid in a packed bed reactor flows from one end of the reactor to the other; so our first approximation will be to assume no mixing, and therefore the mass balance on the fluid will be a PFTR,

$$u \frac{dC_j}{dz} = v_j r \tag{7.1}$$

or for multiple reactions

$$u \frac{dC_j}{dz} = \sum_i v_{ij} r_i \tag{7.2}$$

Here u is an average velocity of the fluid flowing around the catalyst particles. The total reactor volume is V_R, but this is a sum of the volume occupied by the fluid and the volume occupied by the catalyst $V_{catalyst}$,

$$V_R = V_{fluid} + V_{catalyst} \tag{7.3}$$

These can be related by the relations

$$V_{fluid} = \varepsilon V_R \tag{7.4}$$

$$V_{catalyst} = (1 - \varepsilon) V_R \tag{7.5}$$

where V_R is the volume of the reactor, εV_R is the volume occupied by the fluid, and ε is the *void fraction* or the fraction of the reactor volume occupied by the fluid. In a tubular reactor we will call u_o the velocity that would occur in the empty tube. These definitions are formally correct but not always useful, because in many situations the velocity will vary with position in the reactor.

We have to solve the PFTR mass-balance equation to find $C_j(z)$, but the first issue is exactly what form we should use for r or r_i in the mass-balance equations. This is a major topic of this chapter.

7.2.2 The slurry reactor and the fluidized bed reactor

The other major type of catalytic reactor is a situation where the fluid and the catalyst are stirred instead of having the catalyst fixed in a bed. If the fluid is a liquid, we call this a *slurry reactor*, in which catalyst pellets or powder is held in a tank through which catalyst flows. The stirring must obviously be fast enough to mix the fluid and particles. To keep the particles from settling out, catalyst particle sizes in a slurry reactor must be sufficiently small. If the catalyst phase is another liquid that is stirred to maintain high interfacial area for reaction at the interface, we call the reactor an *emulsion reactor*. These are shown in Figure 7–4.

If the fluid is a gas, it is difficult to stir solid particles and gas mechanically, but this can be simply accomplished by using very small particles and flowing the gas such that the particles are lifted and gas and particles swirl around the reactor. We call this reactor a *fluidized bed*.

The moving fluid in these reactors is frequently well mixed within the reactor; so the mass balance on reactant species A is that of a CSTR,

$$C_{Ao} - C_A = \tau r \tag{7.6}$$

The reactor residence time is now

$$\tau = \frac{V_{\text{fluid}}}{v} = \frac{\varepsilon V_R}{v} \tag{7.7}$$

Figure 7–4 Slurry reactor (left) for well-mixed gas–solid reactions and fluidized bed reactor (center) for liquid–solid reactions. At the right is shown a riser reactor in which the catalyst is carried with the reactants and separated and returned to the reactor. The slurry reactor is generally mixed and is described by the CSTR model, while the fluidized bed is described by the PFTR or CSTR models.

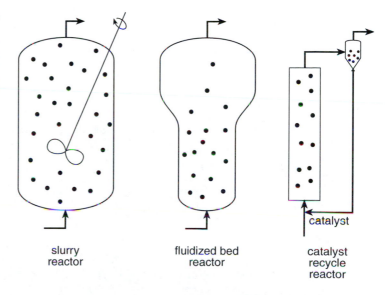

slurry reactor

fluidized bed reactor

catalyst

catalyst recycle reactor

7.2.3 Riser reactor

A variant on the fluidized bed is the *riser reactor*. In this reactor the flow velocity is so high that the solids are entrained in the flowing fluid and move with nearly the same velocity as the fluid. The solids are then separated from the effluent gases at the top of the reactor by a cyclone, and the solids are returned to the reactor as shown in Figure 7–4. The FCC reactor is an example where the catalyst is carried into the regenerator, where carbon is burned off and the catalyst is heated before returning to the reactor.

The flow profile of this reactor is incompletely mixed, and in the limit there is no mixing at all to yield the PFTR,

$$u \frac{dC_A}{dz} = -r \tag{7.8}$$

with $C_A = C_{Ao}$ at $z = 0$. Thus the riser reactor behaves identically to the packed bed reactor (a PFTR) with suitable modification of the definition of τ.

7.3 SURFACE AND ENZYME REACTION RATES

We will develop the rest of this chapter assuming that the catalyst is in a solid phase with the reactants and products in a gas or liquid phase. In Chapter 14 we will consider some of the more complex reactor types, called multiphase reactors, where each phase has a specific residence time. Examples are the riser reactor, the moving bed reactor, and the transport bed reactor.

In catalytic reactors we assume that there is no reaction in the fluid phase, and all reaction occurs on the surface of the catalyst. The surface reaction rate has the units of moles per unit area of catalyst per unit time, which we will call r''. We need a homogeneous rate r to insert in the mass balances, and we can write this as

$$\boxed{r = \left(\frac{\text{area}}{\text{volume}} \right) r''} \tag{7.9}$$

where now r is a "pseudohomogeneous rate" in units of moles/volume time, and (area/volume) is the surface area of catalyst per volume of reactor. We need to find the value of (area/volume) in order to find r in terms of r'' to describe any catalytic reactor, but we will just write out this ratio rather than define separate symbols for it.

Suppose we have a packed bed reactor filled with pellets with surface area $S_g \rho_c$ per unit volume of pellet (units of cm^{-1}, for example). Here we define S_g as the surface area per unit weight of catalyst (area per gram is a common set of units) and ρ_c as the density of a catalyst pellet. We therefore have

$$\left(\frac{\text{area}}{\text{volume}} \right) = S_g \rho_c \frac{V_{\text{catalyst}}}{V_{\text{fluid}}} = S_g \rho_c (1 - \varepsilon) \tag{7.10}$$

because $S_g \rho_c$ is the surface area of the catalyst based on pellet volume.

The pellets leave a fraction ε unoccupied as they pack into the reactor; so the fraction $1 - \varepsilon$ is occupied by the catalyst. The pellet is usually porous, and there is fluid (void space) both between catalyst pellets and within pellets. We measure the rate per unit area of pellet

TABLE 7–1
Homogeneous and Surface Reaction Rates

Rate	Significance	Units
r	homogeneous or pseudohomogeneous rate	moles/volume time
r''	rate per catalyst area	moles/area time
r'	rate per weight of catalyst	moles/weight time

of assumed geometrical volume of pellet so we count only the void fraction external to the pellet.

For the packed bed reactor we now solve the PFTR equation to yield

$$\tau = \frac{\varepsilon V}{v} = \frac{\varepsilon L}{u} = -\int_{C_{Ao}}^{C_A} \frac{dC_A}{r(C_A)} = -\int_{C_{Ao}}^{C_A} \frac{dC_A}{S_g \rho_c (1 - \varepsilon) r''} \tag{7.11}$$

It is more common to specify not the catalyst surface area per unit volume of pellet (we frequently have a problem in determining the pellet volume and ε), but rather its area per unit weight of catalyst. We then measure its rate per weight of catalyst r' (units of moles/weight time) rather than its rate per unit area of catalyst r''. The pseudohomogeneous rate r is now

$$r = \left(\frac{\text{weight}}{\text{volume}}\right) r' = \rho_c r' \tag{7.12}$$

We now can write the PFTR equation in terms of r' to yield

$$\tau = \frac{\varepsilon V_R}{v} = \frac{\varepsilon L}{u_o} = -\int_{C_{Ao}}^{C_A} \frac{dC_A}{r(C_A)} = -\int_{C_{Ao}}^{C_A} \frac{dC_A}{\rho_c r'} \tag{7.13}$$

Note that, if we measure the rate per weight of catalyst, we do not need to worry about the packing of the catalyst in the reactor (determined by ε) but just the density of the catalyst ρ_c in weight/volume.

We summarize the rates we need in a catalytic reactor in Table 7–1. We always need r to insert in the relevant mass-balance equation. We must be given r'' or r' as functions of C_j and T from kinetic data.

7.4 POROUS CATALYSTS

If the rate is proportional to the surface area, we want a catalyst with the highest possible area to attain the highest rate with minimum total reactor volume (high S_g). This is usually achieved by formulating the catalyst as a powder, which is then pressed into porous pellets that are packed into the packed bed reactor. These porous pellets can be powder grains which are compressed into a pellet or a porous network prepared by drying a slurry or a gel as shown in Figure 7–5.

Figure 7–5 Porous catalyst pellets consisting of spheres or grains pressed together into spheres or cylinders. Reactants must diffuse into the pellet in the space between spheres or grains and products must diffuse out of the pellet for reaction to occur. At the right is shown a monolith catalyst in which a ceramic is coated with a wash coat (gray) of porous catalyst.

Our emphasis here is not on catalyst preparation and structure, but we need to describe briefly the preparation and properties of several major catalysts: amorphous silica, γ-alumina, zeolites, activated carbon, and supported metals.

Amorphous silica (SiO_2) can be prepared with very high surface area by precipitating silica from an aqueous silicate solution (a gel) and drying the precipitate. Silica gels have surface areas up to $\sim$500 m^2/g, and they are widely used as supports for catalysts. These particles are generally noncrystalline and are typically monodisperse spheres that are packed into a pellet with gases migrating through the voids between these spheres.

High-area aluminas (Al_2O_3) can be easily prepared by precipitating alumina from aqueous solution and drying and heating as for silica. This material is crystalline, and the major phase is called γ-alumina with a surface area up to $\sim$200 m^2/g. Alumina forms a porous network as the water is removed.

A pellet is typically composed of grains of these porous materials, and gases migrate both between grains and within the porous network inside the grains. Alumina prepared in this way has surface sites that are acidic, and alumina and alumin osilicates are widely used as hydrocarbon cracking catalysts and also as supports for noble metals. These alumina catalysts are generally crystalline, but they form different crystal structures than the most stable α-alumina (corundum or sapphire), which has a very low surface area and little catalytic activity.

Zeolites (aluminosilicates) are crystalline clays that can be either natural minerals or prepared synthetically by crystallizing silica and alumina solutions. Zeolites have regular pore sizes because they are crystalline, and this offers the possibility of "engineering" catalyst geometry by preparing a particular crystal to catalyze reactions with shape selectivity. There are many zeolite crystals, some occurring naturally in clays and some created synthetically in the laboratory. These zeolite materials can be used as catalysts by themselves as acidic catalysts, by ion-exchanging cations that have catalytic activity, or by depositing salts that decompose to form metal particles within the zeolite structure.

Carbon is one of the easiest materials to prepare with high surface area, and this is simply done by partially burning an organic material such as wood or coconut shells to volatilize and pyrolyze the organic components and leave behind a porous carbon. The cellular structures of the wood fiber are left intact to attain very high surface areas, up to 500 m^2/g of charcoal. *Thus ten grams of such a catalyst has an area greater than a football field*! Activated carbons are the easiest and cheapest solids to prepare with high surface areas, and they find wide use as adsorbents and as supports for catalytically active metals and oxides. Since carbon reacts rapidly with oxygen, charcoal catalysts can only be used in the absence of O_2. [Why?]

Supported noble metal catalysts (Pt, Pd, Ag, Rh, Ni, etc.) are an important class of catalysts. Depositing noble metals on high-area oxide supports (alumina, silica, zeolites) disperses the metal over the surface so that nearly every metal atom is on the surface. A critical property of supported catalysts is that they have high *dispersion* (fraction of atoms on the surface), and this is a strong function of support, method of preparation, and treatment conditions. Since noble metals are very expensive, this reduces the cost of catalyst. It is fairly common to have situations where the noble metals in a catalyst cost more than $100,000 in a typical reactor. Fortunately, these metals can usually be recovered and recycled when the catalyst has become deactivated and needs to be replaced.

Powders of these materials are first prepared and then formed into spherical pellets in a ball mill or into cylindrical pellets by extruding a paste. A catalytically active component such as a noble metal or a multivalent oxide is then frequently added to one of these porous supports by depositing ions from salt solution and heating to dry and decompose the salt. We are then interested not in the total surface area of the pellet but the area of the catalytically active component.

Thus most catalysts consist of porous catalyst pellets that may be spherical, cylindrical, or planar and have a characteristic size chosen for desired reactor properties, as we will consider later in this chapter. In most situations reaction occurs throughout the porous pellet, not just on its external surface.

Example 7–1   Consider a silica gel catalyst that has been pelletized into spherical pellets 4 mm diameter. The surface area is measured to be 100 m²/g, and the density of silica is 3 g/cm³. Find the ratio of the area of the catalyst to the external surface area of the pellet.

The volume of a pellet is $V_p = \frac{4}{3}\pi R^3 = \frac{4\pi}{3}0.2^3 = 0.0335$ cm³, and the external area of a pellet is $A_p = 4\pi R^2 = 4\pi 0.2^2 = 0.502$ cm². Each pellet weighs 0.10 g; so the surface area of one pellet is 10^5 cm² $= 10$ m². Thus the total area of the porous pellets is 2×10^5 greater than the external area of the pellet.

Find the surface area of 1 liter of these pellets.

If pellets are packed in a cubic configuration, then there are $2.5^3 \times 10^3$ catalyst pellets per liter. The external area of packed spheres in 1 liter is 8×10^5 m². The internal area of these porous pellets is 1.6×10^{11} m² $= 1.6 \times 10^5$ km² for 1 liter of pellets, an almost unimaginably large surface area!

7.5 TRANSPORT AND REACTION

When a catalytic reaction occurs on the surfaces within a catalyst pellet in a packed bed, there are inevitably concentration gradients around and within the pellet. We have thus far considered only the overall variation in concentrations as a function of distance along the direction of flow z in the reactor. However, in order to account for the variations around and within catalyst pellets, we now need to find $C_j(x, y, z)$ rather than just $C_j(z)$. Stated differently, we need ultimately to reduce $C_j(x, y, z)$ to $C_j(z)$ to solve for the performance of the reactor. This is a major task in describing any catalytic reactor.

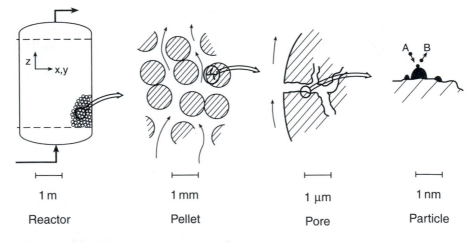

Figure 7–6   Different size scales in a packed bed catalytic reactor. We must consider the position z in the bed, the flow around catalyst pellets, diffusion within pores of pellets, and adsorption and reaction on reaction sites. These span distance scales from meters to angstroms.

7.5.1   Length scales in the reactor

It is important to note that we need to consider several length scales in attempting to describe catalytic reactors. The reactor is on the order of 1 meter diameter and length, the pellet is typically 1 cm diameter, the pores within the pellet are 0.1 mm (10^{-4} m) or smaller in diameter, and catalyst particles might be 100 Å or 10^{-6} m in diameter, and the reactant molecule might be 3 Å or 10^{-10} m in diameter, as shown in Figure 7–6.

It is very important for us to understand the different length scales with which we must be concerned. We need to consider all scales in developing a picture and model of overall reactor performance. We proceed in the reverse order of Figure 7–6

$$\text{catalyst particle} \;\rightarrow\; \text{catalyst pore} \;\rightarrow\; \text{pellet} \;\rightarrow\; \text{reactor} \qquad (7.14)$$

In the reactor we are interested in the position in the bed z or height of the bed L, in the pellet we are interested in the position x in the pellet with radius R, in the pore we are interested in distance x down the pore diameter d_{pore}, and on the walls of the pore we are interested in reactions on the catalyst particle diameter d_{particle}.

7.5.2   Gradients in the reactor

We are also concerned with gradients in composition throughout the reactor. We have thus far been concerned only with the very small gradient dC_j/dz down the reactor from inlet to exit, which we encounter in the species mass balance, which we must ultimately solve. Then there is the gradient in C_j around the catalyst pellet. Finally, there is the gradient within the porous catalyst pellet and around the catalytic reaction site within the pellet. As we consider each of these, we assume that all gradients on larger distance scales are negligible. Then we finally find expressions in which each of these gradients is eliminated in terms of the rate coefficients and mass transfer coefficients and geometry parameters. Figure 7–7 illustrates these reactant concentrations around and within pellets.

Figure 7–7 Reactant concentration profiles in direction x, which is perpendicular to the flow direction z expected for flow over porous catalyst pellets in a packed bed or slurry reactor. External mass transfer and pore diffusion produce the reactant concentration profiles shown.

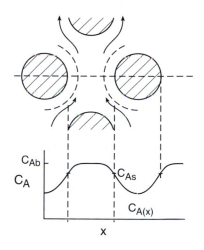

To summarize the goal of this section, we must start with the microscopic description of the catalytic reaction, then consider diffusion in pores, and then examine the reactant composition around and within the pellet, in order finally to describe the reactor mass-balance equations in terms of z alone. The student should understand the logic of this procedure as we go from micrscopic to macroscopic, or the following sections will be unintelligible (or even more unintelligible than usual).

7.5.3 Transport steps

There are essential transport steps for reactants to get to the surface for reaction and for products to escape. These can drastically alter the reaction rate, and the engineering to handle these effects is a major part of catalytic reaction engineering.

Consider first the unimolecular reaction

$$A \rightarrow \text{products}, \qquad r'' = k'' C_{As} \tag{7.15}$$

which occurs on a catalyst surface. We write the surface reaction rate coefficient as k'', using units of k'' (length/time) to satisfy the dimensions of r'' (moles/area time) and C_{As} (moles/volume).

The steps that must be involved in a catalytic reaction on a surface are shown in Figure 7–8. Reactant A_b in the bulk of the flowing fluid must migrate through a boundary

Figure 7–8 Elementary steps that must occur in a catalytic reaction on a surface. All catalytic processes involve transport through a boundary layer, adsorption, surface diffusion, reaction, and desorption.

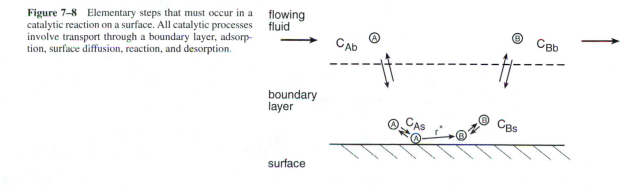

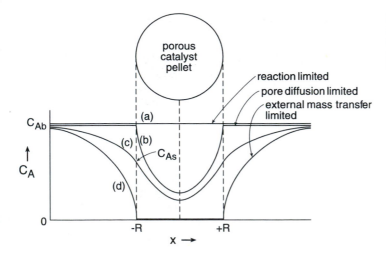

Figure 7–9   Reactant concentration profiles around and within a porous catalyst pellet for the cases of reaction control, external mass transfer control, and pore diffusion control. Each of these situations leads to different reaction rate expressions.

layer over the pellet at the external surface of the pellet. It must then migrate down pores within the pellet to find surface sites where it adsorbs and reacts to form B, which then reverses the process to wind up in the flowing fluid, where it is carried out of the reactor.

The structures and concentration profiles of reactant near and in a catalyst particle might look as illustrated in Figure 7–9.

For the reactant to migrate into the particle, there must be a concentration difference between the flowing bulk fluid C_{Ab}, the concentration at the external surface C_{As}, and the concentration within the pellet $C_A(x)$. We write the concentration within the pellet as $C_A(x)$, indicating that it is a function of position x, which will be different for different geometries, as we will consider later.

To repeat, we don't know and can't measure the gas-phase concentrations near the catalyst surface. We therefore have to eliminate them from any expressions and write the mass balances as functions of bulk concentrations alone.

7.5.4   The overall problem

Our task is to write $r''(C_{Ab})$, that is, to eliminate C_{As} and $C_A(x)$ in terms of reaction rate and mass transfer parameters. The second task is to eliminate all distances except z, the position in the direction of flow the reactor. Thus we will finally arrive at the equations we have solved previously for PFTR or CSTR, which we have considered in previous chapters. The major issue is exactly what these averaged macroscopic quantities mean in terms of microscopic parameters in the catalytic reactor.

7.6   MASS TRANSFER COEFFICIENTS

Here we review some of the correlations of convective mass transfer. We will find that many reactors are controlled by mass transfer processes; so this topic is essential in describing many chemical reactors. This discussion will necessarily be very brief and qualitative, and we will summarize material that most students have encountered in previous courses in mass transfer. Our goal is to write down some of the simple correlations so we can work examples. The assumptions in and validity of particular expressions should of course be checked if one is interested in serious estimations for particular reactor problems. We will only consider here the mass transfer correlations for *gases* because for liquids the correlations are more complicated and cannot be easily generalized.

If there is a difference in concentration of species A between two locations 1 and 2 in a flowing fluid, the mass transfer flux J_A of species A is given by the expression

$$J_A = k_{mA}(C_{A1} - C_{A2}) \tag{7.16}$$

This is simply the definition of the mass transfer coefficient k_m, and the subject of mass transfer courses is to find suitable correlations in order to estimate k_{mA} (units of length/time). The mass transfer coefficient is in turn defined through the *Sherwood number*,

$$Sh_\ell = \frac{[\text{convective mass transfer rate}]}{[\text{diffusive mass transfer rate}]} = \frac{k_{mA}\ell}{D_A} \tag{7.17}$$

where ℓ is a characteristic length in the system and D_A is the diffusion coefficient of molecule A. We therefore find k_{mA} by first computing the Sherwood number and using the equation

$$k_{mA} = \frac{Sh_\ell D_A}{\ell} \tag{7.18}$$

to find k_m.

We next summarize Sherwood numbers for several simple geometries.

7.6.1 Flow over a flat plate

For flow over a flat plate of length L the flow is laminar if $Re_L < 10^5$, and

$$Sh_L = 0.66 \, Re_L^{1/2} Sc^{1/3} \tag{7.19}$$

where the Reynolds number is

$$Re_L = \frac{uL}{\nu} \tag{7.20}$$

and the Schmidt Sc number (no relation) is

$$Sc = \frac{\nu}{D_A} \tag{7.21}$$

with ν the kinematic viscosity and D_A the diffusivity of species A, both of which have units of length^2/time. For gases we usually assume that $Sc = 1$, and we sometimes omit it from expressions for Sh.

If $Re_L > 10^5$, then the flow is turbulent and the Sherwood number is described through the correlation

$$Sh_L = 0.036 \, Re_L^{0.8} Sc^{1/3} \tag{7.22}$$

7.6.2 Flow over a sphere

Flow over a sphere is an important geometry in catalyst spheres, liquid drops, gas bubbles, and small solid particles. In this case the characteristic length is the sphere diameter D, and

$$Sh_D = 2.0 + 0.4\,(Re_D^{1/2} + 0.06\,Re_D^{2/3})Sc^{0.4} \tag{7.23}$$

In the limit of slow flow over a sphere $Sh_D = 2.0$, and this corresponds to diffusion to or from a sphere surrounded by a stagnant fluid. When the sphere diameter is sufficiently small, Re becomes sufficiently small that $Sh_D = 2.0$ in many common situations of flow around spheres.

7.6.3 Flow through a tube

In this case the characteristic length is the tube diameter D. For $Re_D < 2100$, the flow is laminar and

$$Sh_D = \frac{8}{3} \tag{7.24}$$

This expression is appropriate for sufficient distances from the entrance of the tube that the laminar flow profile is fully developed. If $Re_D > 2100$, the flow becomes turbulent, and the most used correlation is

$$Sh_D = 0.023\,Re_D^{0.8}Sc^{1/3} \tag{7.25}$$

7.6.4 Flow over a cylinder

Here the transition from laminar to turbulence depends on position around the cylinder. A simple correlation for $Re_D < 4$ is

$$Sh_D = 0.98Re_D^{1/3} \tag{7.26}$$

Many other expressions have been found to correlate data for different situations.

7.6.5 Tube banks and packed spheres

There are many correlations for these geometries, both of which are important for mass transfer in chemical reactors. Tube banks could be catalyst wires or tubes in a reactor over which reactants flow, and packed spheres are a common geometry of catalyst particles in a packed bed over which reactants flow.

7.6.6 Heat transfer

Convective heat transfer rates $\dot{Q}$ are given by the relation

$$\dot{Q} = hA(T_1 - T_2) \tag{7.27}$$

with A the area and h the heat transfer coefficient defined through the *Nusselt number*

$$Nu_\ell = \frac{h\ell}{k_T} \tag{7.28}$$

where k_T is the thermal conductivity of the fluid. For gases the Sherwood and Nusselt numbers are nearly identical, $Sh = Nu$, because the *Lewis number* is given by $Le = Nu/Sh = k_T/D_A = 1$; so all the above correlations for Sh can also be used for Nu.

In the previous chapters on temperature variations in reactors, we needed heat transfer coefficients h to calculate rates of heat transfer within the reactor. For a more detailed examination of temperature variations, we must use Nusselt number correlations similar to those developed here for mass transfer.

7.6.7 Liquids

For liquids the situation is more complicated for all these flow geometries, and mass and heat transfer coefficients cannot be found from the same relations. One major difference with liquids is that Sc, Pr, and Le are frequently not close to unity so they cannot be ignored in these expressions. Correlations developed for heat transfer in gases are frequently not directly applicable to mass transfer for liquids. One should find suitable sources of correlations for any particular fluid and geometry if accurate estimations of mass and heat transfer rates are needed.

7.6.8 Natural convection

Gravity and natural convection can play an important role in determining flows and mass and heat transfer in many chemical reactors. Whenever there are large temperature gradient, one may expect natural convection. Also, if there is a density change caused by the chemical reaction (whenever there is a change in the number of moles and especially when H_2 is a reactant or product), then the density may change with conversion sufficiently to cause natural convection to overcome the forced convection impressed by pumping the fluids through the reactor. These phenomena are correlated through the dependences of Sh and Nu on the Grasshof number, as described in texts on mass and heat transfer.

In multiphase reactors we frequently exploit the density differences between phases to produce relative motions between phases for better contacting and higher mass transfer rates. As an example, in trickle bed reactors (Chapter 14) liquids flow by gravity down a packed bed filled with catalyst, while gases are pumped up through the reactor in countercurrent flow so that they may react together on the catalyst surface.

7.7 EXTERNAL MASS TRANSFER

We first consider the situation where reaction occurs only on the external surface of a pellet. The simplest example is that of a nonporous pellet so that only the external surface is reactive.

Reactant molecules must get to the surface to find the catalytic sites on which to react. In steady state there cannot be any accumulation on the surface, and therefore we require that

$$[\text{rate of transport to surface}] = [\text{rate of reaction at surface}] \qquad (7.29)$$

For a first-order irreversible reaction on a single spherical pellet of radius R the total rate of reaction on the external surface of the pellet is

$$4\pi R^2 k_{mA}(C_{Ab} - C_{As}) = 4\pi R^2 r'' = 4\pi R^2 k'' C_{As} \qquad (7.30)$$

This says simply that the rate of reaction on the pellet with surface area $4\pi R^2$ is equal to the rate of mass transfer from the bulk fluid to the surface. We call k_{mA} the mass transfer coefficient of reactant species A and note that it has the dimensions of length/time, the same dimensions as k''.

It is a fundamental fact that in steady state nothing is accumulating anywhere around the catalyst and therefore these rates must be exactly equal.

Now from the preceding we can immediately solve for C_{As} to obtain

$$C_{As} = \frac{C_{Ab}}{1 + k''/k_{mA}} \tag{7.31}$$

we can now eliminate C_{As} from the rate expression and solve for the rate to yield

$$r'' = \frac{k'' C_{Ab}}{1 + k''/k_{mA}} = k''_{eff} C_{Ab} \tag{7.32}$$

so that the reaction rate has an effective rate constant

$$k''_{eff} = \frac{k''}{1 + k''/k_{mA}} \tag{7.33}$$

We now examine limiting cases of this expression.
If $k'' \ll k_{mA}$, then

$$r'' \cong k'' C_{Ab} \tag{7.34}$$

so the reaction is proportional to C_{Ab} with a rate coefficient k''. This situation is said to be *reaction limited*.

However, if $k_{mA} \ll k''$, then

$$r'' \cong k_{mA} C_{Ab} \tag{7.35}$$

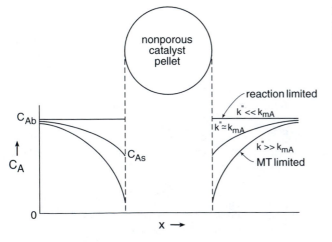

Figure 7–10   Reactant concentration profiles around a catalyst pellet for reaction control ($k_m \gg k''$) and for external mass transfer control ($k_m \ll k''$).

so the reaction is proportional to C_{Ab} with a rate coefficient k_{mA}. This case is said to be mass transfer limited. The quantity k_{mA}/k'' is dimensionless, and the reactor performance depends strongly on whether this ratio is greater or smaller than unity.

Thus we find situations where either k'' or k_{mA} *vanish* from the expression, and the larger quantity disappears. We call these either reaction limiting (k_{mA} large) or mass transfer limited (k'' large), but also note that both rates are always exactly equal in steady state! The key idea is that if a rate coefficient is large, the concentration difference is small. The concentration profiles expected in these cases are shown in Figure 7–10.

7.8 PORE DIFFUSION

In the preceding example we assumed that reaction occurred on the external surface so we did not have to be concerned with diffusion within the catalyst pellet. Now we consider the effect of pore diffusion on the overall rate. We have to do considerable mathematical manipulation to find the proper expressions to handle this, and before we begin, it is worthwhile to consider where we are going. As before, we want the rate as a function of bulk concentration C_{Ab}, and we need to know the rate coefficient for various approximations (Figure 7–11).

The basic problem is that the rate is a function of position x within the pellet, while we need an *average concentration* to insert into our reactor mass balances. We first have to solve for the concentration profile $C_A(x)$ and then we eliminate it in terms of the concentration at the surface of the pellet C_{As} and geometry of the pellet. We will find that we can represent the reaction as

$$r'' = k''C_{As}\eta$$ (7.36)

where diffusion effects are handled through an *effectiveness factor* η, which is a function of rate coefficients and geometry of the catalyst.

7.8.1 Single pore

First we consider a single pore of length ℓ and diameter d_p. The walls are uniformly reactive with $r'' = k''C_A(x)$. The concentration at the pore mouth ($x = \ell$) is C_{As} and the end of the

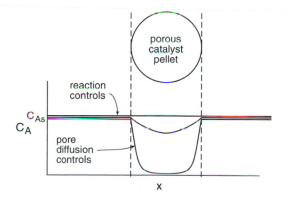

Figure 7–11 Reactant concentration profiles within a porous catalyst pellet for situations where surface reaction controls and where pore diffusion controls the reactions.

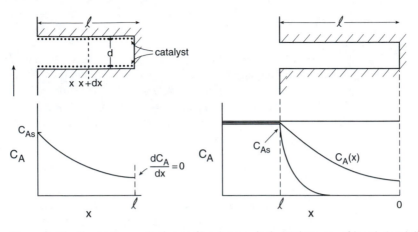

Figure 7–12 Reactant concentration profiles down a single catalyst pore of length ℓ and diameter d_p with a catalytic reaction occurring on the walls of the pore. The concentration is C_{As} at the pore mouth, $x = \ell$, and the gradient is zero at the end of the pore because the end is assumed to be unreactive and there is therefore no flux of reactant through the end.

pore is assumed to be unreactive so that $dC_A/dx = 0$ at $x = 0$, as sketched in Figure 7–12. A shell balance between x and $x + dx$ yields

$$[\text{net flux in at } x] - [\text{net flux out at } x + dx]$$

$$= [\text{rate of reaction on wall between } x \text{ and } x + dx] \tag{7.37}$$

or for a first-order reaction

$$\frac{\pi d_p^2}{4} D_A \left(\frac{dC_A}{dx} \right)_x - \frac{\pi d_p^2}{4} D_A \left(\frac{dC_A}{dx} \right)_{x+dx} = -\pi d_p \, dx \, k'' C_A \tag{7.38}$$

where the terms on the left represent the fluxes of A at x and at $x + dx$ and $\pi d_p \, dx$ is the wall area of the cylinder between x and $x + dx$. Now letting $dx \to 0$ and dividing by dx yields

$$\frac{d^2 C_A}{dx^2} = \frac{4k''}{D_A d_p} C_A = \lambda^2 C_A \tag{7.39}$$

where we have defined

$$\lambda = \left(\frac{4k''}{d_p D_A} \right)^{1/2} \tag{7.40}$$

We next need to solve this second-order differential equation subject to the boundary conditions

$$C_A = C_{As}, \qquad \text{at } x = \ell \tag{7.41}$$

$$\frac{dC_A}{dx} = 0, \qquad \text{at } x = 0 \tag{7.42}$$

The equation

$$\frac{d^2 C_A}{dx^2} = \lambda^2 C_A \tag{7.43}$$

has the general solution

$$C_A = C_1 e^{\lambda x} + C_2 e^{-\lambda x} \tag{7.44}$$

where C_1 and C_2 are constants to be determined from boundary conditions. At $x = \ell$ the solution becomes

$$C_{As} = C_1 e^{\lambda \ell} + C_2^{-\lambda \ell} \tag{7.45}$$

while at $x = 0$

$$\frac{dC_A}{dx} = \lambda(C_1 - C_2) = 0 \tag{7.46}$$

Therefore, $C_1 = C_2$, and the solution becomes

$$C_A(x) = C_{As} \frac{e^{\lambda x} + e^{-\lambda x}}{e^{\lambda \ell} + e^{-\lambda \ell}} = C_{As} \frac{\cosh \lambda x}{\cosh \lambda \ell} \tag{7.47}$$

We now have found $C_A(x)$, but this isn't useful yet because C_A is a function of position x within the pore. To find the average rate within the pore we must integrate the rate along the length of the pore

$$\text{actual rate} = \pi d_p \int_{x=0}^{\ell} k'' C_A(x)\, dx \tag{7.48}$$

We want to compare this to the rate in the pore if the concentration remained at C_{As},

$$\text{ideal rate} = [\text{area}] \times r'' = \pi d_p \ell k'' C_{As} \tag{7.49}$$

because the area of the wall of the pore is $\pi d_p \ell$

The ratio of these rates (actual rate/ideal rate) is the fraction by which the rate is reduced by pore diffusion limitations, which we call η, and this is the definition of the effectiveness factor,

$$\eta = \frac{\text{actual rate}}{\text{ideal rate}} \tag{7.50}$$

for the single pore η is found from the relation

$$\eta = \frac{\pi d_p \int_{x=0}^{\ell} k'' C_A(x)\, dx}{\pi d_p \ell k'' C_{As}} \tag{7.51}$$

$$\eta = \int_{x=0}^{\ell} \frac{\cosh \lambda x\, dx}{\ell \cosh \lambda \ell} \tag{7.52}$$

$$\eta = \frac{\tanh \lambda \ell}{\lambda \ell} \tag{7.53}$$

and finally

$$\eta = \frac{1}{\phi} \frac{e^{\phi} - e^{-\phi}}{e^{\phi} + e^{-\phi}} = \frac{\tanh \phi}{\phi} \tag{7.54}$$

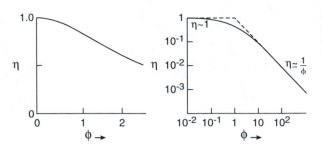

Figure 7–13 Plots of effectiveness factor η versus Thiele modulus ϕ for diffusion in a single catalyst pore or in a porous catalyst pellet. On a log–log plot the effectiveness factor is seen to give $\eta = 1$ if $\phi \ll 1$ and $\eta = 1/\phi$ if $\phi \gg 1$.

where

$$\phi = \lambda \ell = \left(\frac{4k''}{d_p D_A} \right)^{1/2} \ell \tag{7.55}$$

There is a lot of mathematical manipulation in these steps which the student may (or may not) want to verify.

Thus the rate of reaction in a porous catalyst is equal to the rate if the concentration had remained constant at C_{As} multiplied by η, $r'' = k'' C_{As} \eta$. We have defined the dimensionless quantity ϕ which is called the *Thiele modulus*. This function is plotted on linear and log–log plots in Figure 7–13.

The limits of η are instructive:

$$\phi \ll 1, \qquad \eta = 1, \qquad \text{no pore diffusion limitation}$$
$$\phi = 1, \qquad \eta = 0.762, \qquad \text{some limitation}$$
$$\phi \gg 1, \qquad \eta = 1/\phi, \qquad \text{strong pore diffusion limitation}$$

Thus we can say that if the Thiele modulus is much less than unity, there is no pore diffusion effect on the reaction rate because the reactant concentration remains at C_{As} down the pore, but if the Thiele modulus is much greater than unity, then the rate is proportional to $1/\phi$ because C_A decreases strongly within the pore so that the overall rate is much lower than if the concentration remained as C_{As}.

7.8.2   Honeycomb catalyst

The single cylindrical pore is of course not the geometry we are interested in for porous catalysts, which may be spheres, cylinders, slabs, or flakes. Let us consider first a honeycomb catalyst of thickness 2ℓ with equal-sized pores of diameter d_p, as shown in Figure 7–14. The centers of the pores may be either open or closed because by symmetry there is no net flux across the center of the slab. (If the end of the pore were catalytically active, the rate would of course be slightly different, but we will ignore this case.) Thus the porous slab is just a collection of many cylindrical pores so the solution is exactly the same as we have just worked out for a single pore.

7.8.3   Porous catalyst slab

Consider next a slab of thickness 2ℓ but containing irregular pores, such as would be obtained with a foam or a pressed pellet of spheres with the voids between the spheres the

Figure 7–14 Structure and concentration profiles for a planar porous catalyst, which could have straight (honeycomb) or tortuous pores.

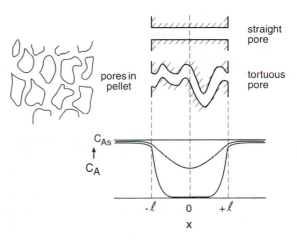

channels down which the reactant must travel. If these pores had an average diameter d_p and length ℓ, we would have exactly the problem we have just solved. If these are irregular, we replace length and diameter by averages containing the total surface area in defining the Thiele modulus, but the solution is otherwise the same,

$$\eta = \frac{\tanh \phi}{\phi} \qquad (7.56)$$

where

$$\phi = \left(\frac{S_g \rho_c k''}{D_A} \right)^{1/2} \ell \qquad (7.57)$$

For the single straight cylindrical pore the surface-to-volume ratio was $4/d_p$, but for a porous catalyst slab, this ratio is $S_g \rho_b$, where $S_g \rho_b$ is the catalyst surface area per unit volume of catalyst, as defined previously.

7.8.4 Porous spheres and cylinders

For a spherical pellet we must find $C_A(R)$, where R is a radius within the pellet, which has total radius R_o. Then we proceed as before to find η. For a sphere the shell balance is set up in spherical coordinates to yield

$$\frac{1}{R^2} \frac{d}{dR} \left(R^2 D_A \frac{dC_A}{dR} \right) = S_g \rho_b k'' C_A \qquad (7.58)$$

This equation has the solution

$$C_A(R) = C_{As} \frac{R_o}{R} \frac{\sinh \lambda R}{\sinh \lambda R_o} \qquad (7.59)$$

where sinh is the hyperbolic sine. This expression for concentration gives

$$\eta = \frac{3}{\phi} \frac{\phi \coth \phi - 1}{\phi} \qquad (7.60)$$

as the effectiveness factor for a porous catalyst sphere, and the Thiele modulus ϕ is now defined as

$$\phi = \left(\frac{S_g \rho_c k''}{D_A}\right)^{1/2} R_o = \lambda R_o \tag{7.61}$$

For a long cylindrical pellet, the differential equation describing C_A versus radius R is

$$\frac{1}{R}\frac{d}{dR}\left(RD_A\frac{dC_A}{dR}\right) = k''C_A \tag{7.62}$$

This equation can be integrated, but it gives a fairly complicated expression for $C_A(R)$. The effectiveness factor for a cylinder is given by the expression

$$\eta = \frac{2}{\phi}\frac{I_1(\phi)}{I_o(\phi)} \tag{7.63}$$

where I_o and I_1 are modified Bessel functions.

Thus we have expressions for the effectiveness factor for different catalyst geometries. The Thiele modulus can be computed from catalyst geometry and surface area parameters. The characteristic size is 2ℓ for a porous slab and $2R_o$ for a cylinder or sphere. While the expressions for $\eta(\phi)$ appear quite different, they are in fact very similar when scaled appropriately, and they have the same asymptotic behavior,

$$\eta = 1, \qquad \phi \ll 1 \tag{7.64}$$

$$\eta = 0.8, \qquad \phi = 1 \tag{7.65}$$

$$\eta = 1/\phi, \qquad \phi \gg 1 \tag{7.66}$$

Thus we can in general write for the pseudohomogeneous rate of a catalytic reaction in a reactor with porous catalyst pellets

$$r = r_{\text{ideal}}\eta(\phi) \tag{7.67}$$

where r_{ideal} is the rate per unit volume if there is no pore diffusion influence on the catalytic reaction rate.

7.9   TEMPERATURE DEPENDENCE OF CATALYTIC REACTION RATES

We have considered three limiting rate expressions for catalytic reaction rates, depending on which rate coefficients control the overall process,

$$r \approx (\text{area/volume})k''C_{Ab}, \qquad \text{reaction limited}$$

$$r \approx (\text{area/volume})k_{mA}C_{Ab}, \qquad \text{external mass transfer limited}$$

$$r \approx (\text{area/volume})k''C_{Ab}\eta, \qquad \text{pore diffusion limited}$$

It is interesting to consider the temperature dependence of the reaction rates predicted by these limiting expressions, which are contained in the effective rate coefficients. The true surface reaction rate coefficient has the temperature dependence

$$k''(T) = k_o''e^{-E/RT} \tag{7.68}$$

Figure 7–15   Plots of r versus T and log r versus $1/T$. We expect the rate to exhibit breaks on the $1/T$ plot as the reaction process goes from reaction limited at low temperature, pore diffusion limited at intermediate temperature, and external mass transfer limited at high temperature.

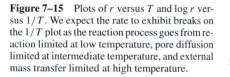

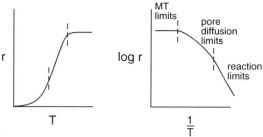

with k_o'' the usual pre-exponential factor and E the reaction activation energy. The mass transfer coefficient has only a weak temperature dependence, typically varying as $T^{1/2}$. For the pore-diffusion-limited case the effective rate coefficient is

$$k''\eta \approx k''/\phi \sim k''/k''^{1/2} = k''^{1/2} \sim e^{-E/2RT} \tag{7.69}$$

Thus the reaction-limited case has a strong temperature dependence with activation energy E, the mass transfer limited case is nearly temperature independent, and the pore diffusion limited case has an activation energy $E/2$:

Rate-limiting step	Temperature dependence
reaction	activation energy E
mass transfer	nearly constant
pore diffusion	activation energy $E/2$

As the temperature is varied in a reactor, we should expect to see the rate-controlling step vary. At sufficiently low temperature the reaction rate coefficient is small and the overall rate is reaction limited. As the temperature increases, pore diffusion next becomes controlling (D_A is nearly independent of temperature), and at sufficiently high temperature external mass transfer might limit the overall process. Thus a plot of log rate versus $1/T$ might look as shown in Figure 7–15.

7.10   THE AUTOMOTIVE CATALYTIC CONVERTER

The largest market for chemical reactors by far is the automotive catalytic converter (ACC), both in number of reactors in existence (many million sold per year) and in amount of reactants processed (millions of tons per year). There are >50 million automotive catalytic converters operating (or at least existing) throughout the world, and everyone owns one if he or she has a car less than 10 years old.

We begin by considering the engine and exhaust system of an automobile (Figure 7–16). The engine is basically a batch reactor [what is the reaction time?] in which the combustion reaction

$$C_8H_{18} + \tfrac{25}{2}O_2 \rightarrow 8CO_2 + 9H_2O \tag{7.70}$$

occurs. This is a strongly exothermic reaction that attains a high temperature and pressure and is converted into mechanical energy (with ~10% efficiency). As we will consider in

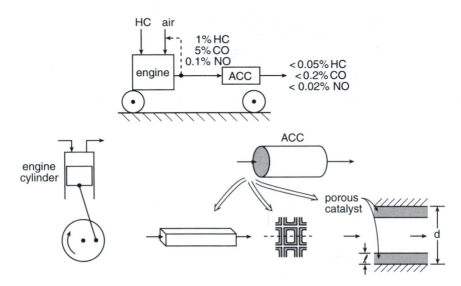

Figure 7–16   A highly simplified sketch of an automobile engine and catalytic converter with typical gas compositions indicated before and after the automotive catalytic converter. The catalytic converter is a tube wall reactor in which a noble-metal-impregnated wash coat on an extruded ceramic monolith creates surface on which reactions occur.

Chapter 10, this is a complex chain reaction that forms an explosion within the cylinder. In addition to CO_2 and H_2O, combustion processes produce small amounts of CO and partially burned hydrocarbons such as smaller alkanes, olefins, and oxygen-containing hydrocarbon molecules such as aldehydes. Finally, these combustion processes produce NO at high temperatures because the free radicals react with N_2 from air to produce equilibrium in the reaction

$$N_2 + O_2 \rightleftarrows 2NO \qquad (7.71)$$

which was one of the first reaction examples we considered in this book. The NO remains at nearly thermodynamic equilibrium in this reaction until the temperature falls to $\sim 1300°C$, where it remains "frozen" as the combustion gases cool.

Thus the engine of an automobile produces approximately 5% of various hydrocarbons, 1% of CO, and 0.1% of NO. These gases are major pollutants in urban areas, and clean air regulations require that they be removed to approximately 0.2, 0.05, and 0.03%, respectively.

The purpose of the converter is to reduce unburned hydrocarbon emissions to 4 g/mile, CO to 40 g/mile, and NO to 0.4 g/mile. (The mixed units are those in several Federal Clean Air Acts, which mandate these maximum emissions.) In California these standards are even more stringent, and lower levels are mandated to be required in several years throughout the country.

The ACC is a stainless steel cylinder 3 in. in diameter and 1 ft long through which exhaust gases pass before they enter the muffler and tailpipe. The converter contains an extruded ceramic honeycomb structure with square pores 0.5 mm in diameter through which gases pass. The walls of these pores are coated with porous γ-Al_2O_3 on which Pt,

Pd, and Rh are deposited, along with several catalyst promoters such as cerium oxide. The residence time in the ACC is ~0.05 sec, because the volume of gas that must be processed and the volume of the converter that will fit under an automobile will not permit use of longer times. Consequently, engineers must devise a chemical reactor that will achieve ~95% conversion of these pollutants within these very short residence times.

Modern converters contain approximately $30 of these precious metals, and the retail cost of the entire converter is $200, most of which is the stainless steel reactor.

The desired reactions are *oxidation* of hydrocarbons and CO,

$$C_nH_{2n+2} + \frac{3n+1}{2}O_2 \rightarrow nCO_2 + (n+1)H_2O \qquad (7.72)$$

$$CO + \tfrac{1}{2}O_2 \rightarrow CO_2 \qquad (7.73)$$

Removal of NO is a *reduction* process, which occurs mostly through the reaction with CO

$$NO + CO \rightarrow \tfrac{1}{2}N_2 + CO_2 \qquad (7.74)$$

It would be simpler if NO could be decomposed through the reaction

$$NO \rightarrow \tfrac{1}{2}N_2 + \tfrac{1}{2}O_2 \qquad (7.75)$$

but this reaction is far too slow under conditions of the ACC.

We have discussed NO_x pollution from combustion previously, and we will consider pollution and environmental modeling later in Chapter 13. These are situations where we need chemical reactors to eliminate the pollutants we have made in other reactors.

The major problems in the engineering the ACC are the facts that (1) they must operate under transient conditions and (2) customers treat different cars very differently. The feed and catalyst temperatures vary widely, from ambient temperature upon startup (which can be −30°C in Minnesota) to >500°C during high-speed desert or mountain driving.

7.10.1   Three-way catalysts

Since the oxidation reactions require O_2 while NO reduction requires CO, it is essential that the gases entering the ACC be at nearly the stoichiometric composition. This has led to the "three-way catalyst" in which oxidation and reduction processes are accomplished simultaneously by maintaining the air–fuel mixture near stoichiometric.

The first requirement in developing the ACC was to control air–fuel mixtures, and in contemporary cars this is maintained at precisely the 19:1 stoichiometric ratio for all engine loads, using an exhaust O_2 sensor that controls the fuel injection system. The composition and location of the catalyst within the reactor are also crucial to proper design.

The problem of a "window" of operation where oxidation and reduction reactions occur simultaneously requires that the air–fuel ratio be carefully controlled so that the conversions of all species remains high. This is shown in Figure 7–17.

Thus hydrocarbons and CO are products of incomplete combustion, which occurs primarily when the engine (and the catalyst) are cold. NO_x is produced primarily by free-radical reactions with atmospheric N_2 in the high temperatures (2500°C) within the engine cylinders. The overall reaction can be written as

$$\tfrac{1}{2}N_2 + \tfrac{1}{2}O_2 \rightarrow NO \qquad (7.76)$$

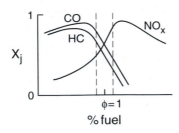

Figure 7–17 Conversions of hydrocarbons, CO (oxidation reactions), and NO (reduction reactions) versus air–fuel ratio. The engine should operate near the stoichiometric ratio to obtain maximum conversions in all reactions, and cars are tuned to operate within this window of composition (dashed lines).

which we discussed as our first example in Chapter 2. The equilibrium concentration of NO in air is negligible at room temperature, but the composition becomes "frozen" at about 1300°C, where the equilibrium composition of NO in air is ~100 ppm. $\phi = 1$ is the stoichiometric composition for complete combustion.

7.10.2 Startup

Most of the hydrocarbons and CO are emitted during startup while the engine is cold and operating inefficiently, while most of the NO_x is emitted during high-speed driving, where free radicals decompose N_2.

The catalyst has very low activity during warmup until the catalyst temperature attains ~200°C. Since the oxidation reactions are exothermic, the catalyst provides heat, which further heats the catalyst and increases the rate. Therefore, the ACC can exhibit *multiple steady states*, where the catalyst has essentially no activity until a certain temperature where the catalyst ignites. This is described as *lightoff*, and a good catalyst system has a low lightoff temperature. Lightoff characteristics are sketched in Figure 7–18.

7.10.3 Aging

Probably the major problem in designing the ACC is the maintenance of activity as the catalyst ages (law requires performance within tight specifications to at least 50,000 miles). Rapid activation of the cold catalyst in startup is related to aging because the inactive catalyst generates less heat.

Aging results in sintering of the metal clusters, their dissolution into the support, and their burial by accumulating solids such as Pb (from leaded gasoline) and phosphates (from lubricating oil). Rapid activation requires rapid heating by the exhaust gases. The converter

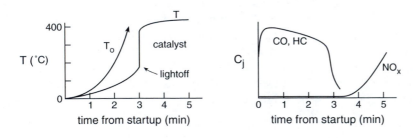

Figure 7–18 Since the reactions in the ACC are strongly exothermic, the catalyst can exhibit multiple steady states as the catalyst suddenly lights off as the car is started and the temperature rises to the ignition temperature, which then causes a rapid increase in temperature.

actually exhibits a lightoff ignition (multiple steady states driven by the exothermic oxidation reactions) as it heats up, and active catalysts rapidly lightoff to $>200°C$, when the feed gases reach $100°C$.

There is great interest in new ACC designs that will heat up more rapidly upon startup. Small mass catalysts, close coupling to the hot exhaust temperature, and electrical heating of the catalyst are all being developed.

The gases move down the monolith ceramic cells of the ACC in plug flow with the noble metal catalyst deposited on the walls. This reactor is therefore a PFTR with catalyst on the wall, which is a catalytic wall reactor whose equations we will develop in the next section.

7.11 THE CATALYTIC WALL REACTOR

Next we consider a tubular reactor with no homogeneous reaction but a wall that is catalytic with a rate r'' (Figure 7–19).

This problem may look at first to be the same geometry as for diffusion in a single pore, but this situation is quite different. In the single pore we had reaction controlled by diffusion down the pore, while in the tube wall reactor we have convection of reactants down the tube.

The mass-balance equation on C_A is simply

$$u \frac{dC_A}{dz} = -(A/V)r'' \tag{7.77}$$

Here (A/V) is the wall area per unit volume of the tube. For a cylindrical tube the area per unit volume is $4/d$. The mass balance for the reaction $A \rightarrow B$, $r'' = k''C_{As}$, in a cylindrical tube wall reactor is therefore

$$u \frac{dC_A}{dz} = -\frac{4k''}{d} C_A \tag{7.78}$$

The solution with $C_A = C_{Ao}$ at $z = 0$ is

$$C_A(z) = C_{Ao} \exp\left(-\frac{4k''}{d}\frac{z}{u}\right) \tag{7.79}$$

Figure 7–19 Sketch of a catalytic tube wall reactor with gases flowing down the tube and reaction occurring on the walls of the tube, which are coated with a wash coat of catalyst (dots).

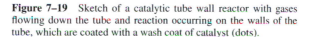
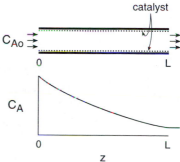

which predicts the usual exponential decrease of reactant with distance down the tube, just as for homogeneous reaction. For a tube of length L this expression becomes

$$C_A(L) = C_{Ao} \exp\left(-\frac{4k''}{d}\frac{L}{u}\right) = C_{Ao} \exp\left(-\frac{4k''}{d}\tau\right) \tag{7.80}$$

Note the different ds in these problems. While d always signifies diameter, D is reactor diameter (or diameter of a cylinder or sphere because that notation is used in mass transfer rather than the radius R), d_p is the diameter of a single pore, and d is the diameter of a single tube in a tube wall reactor.

There are a number of examples of tube wall reactors, the most important being the automotive catalytic converter (ACC), which was described in the previous section. These reactors are made by coating an extruded ceramic monolith with noble metals supported on a thin wash coat of γ-alumina. This reactor is used to oxidize hydrocarbons and CO to CO_2 and H_2O and also reduce NO to N_2. The rates of these reactions are very fast after warmup, and the effectiveness factor within the porous wash coat is therefore very small. The reactions are also external mass transfer limited within the monolith after warmup. We will consider three limiting cases of this reactor, surface reaction limiting, external mass transfer limiting, and wash coat diffusion limiting. In each case we will assume a first-order irreversible reaction.

7.11.1 Reaction limiting

If the process is reaction limiting, the composition is given directly by the preceding expression,

$$C_A(z) = C_{Ao} \exp\left(-\frac{4k''}{ud}z\right) \tag{7.81}$$

7.11.2 Mass transfer limiting

In this case the amount of catalyst within the washcoat is irrelevant (as long as there is enough to make the process mass transfer limited), and the rate is controlled by the rate of mass transfer radially across the tube to the external surface of the wash coat where reaction occurs rapidly. In this limit the reactant concentration falls to near zero at the wall and the rate is controlled by the mass transfer coefficient in the tube. If species A is limiting, we replace k'' by k_{mA} to obtain

$$C_A(z) = C_{Ao} \exp\left(-\frac{4k_{mA}}{ud}z\right) \tag{7.82}$$

If the flow is laminar ($\mathrm{Re}_d < 2100$ in a straight tube), then the Sherwood number is $\frac{8}{3}$ and the mass transfer coefficient is given by the expression

$$k_{mA} = \frac{\mathrm{Sh}_d D_A}{d} = \frac{8}{3}\frac{D_A}{d} \tag{7.83}$$

The effective rate coefficient should be given by the expression

$$k_{\mathrm{eff}} = \frac{k''}{1 + k''/k_m} \tag{7.84}$$

In the mass-transfer-limited situation the conversion versus z should be given by given by the expression

$$C_A(z) = C_{Ao}\, \exp\!\left(-\frac{4\mathrm{Sh}_d D_A}{ud^2}z\right) = C_{Ao}\, \exp\!\left(-\frac{32 D_A}{3ud^2}z\right) \qquad (7.85)$$

Note that, since k_m depends on the tube diameter d, the dependence of the conversion on d is different for reaction- and mass-transfer-limited cases. If the flow is turbulent, the Sherwood number is given by the correlation $\mathrm{Sh}_d = 0.023\mathrm{Re}_d^{0.8}\,\mathrm{Pr}^{1/3}$, which predicts a different dependence on the tube diameter.

7.11.3   Porous catalyst on tube wall reactor

In the previous section we considered the situation where there was a film of catalyst on the wall of the tube, so that the surface area was the geometric area of the tube wall. Another common situation is where a porous catalyst film is coated on the walls of the tube as shown in Figure 7–20. Finally we consider the case where the gas-phase concentration is uniform across the cross section of the tube and mass transfer rates are large, but C_A falls within the layer of porous catalyst on the wall. The thickness of the film is ℓ, and the surface area of the catalyst per unit volume of catalyst is $S_g\rho_c$. If we assume that ℓ is much less than the tube diameter, then

$$\frac{\text{catalyst area}}{\text{reactor volume}} = \frac{4\ell S_g\rho_c}{d}$$

This is exactly the situation we considered previously for diffusion in porous catalysts; so we multiply r'' in the porous wash coat by the effectiveness factor η to obtain

$$C_A(z) = C_{Ao}\, \exp\!\left(-\frac{S_g\rho_c \ell k'' \eta}{d}\frac{z}{u}\right) \qquad (7.86)$$

because we simply need to multiply the rate for the reaction-limited case by the effectiveness factor η. If the thickness of the porous catalyst layer is small compared to the diameter of the pore, then we have a planar porous catalyst geometry so that $\eta = (\tanh\phi)/\phi$, where now ϕ is based on the thickness of the catalyst layer on the walls of the tube.

Figure 7–20   Sketch of tube wall reactor with a porous catalyst film of thickness ℓ on walls. Expected reactant concentration profiles with reaction-limited, mass-transfer-limited, and pore-diffusion-limited reaction.

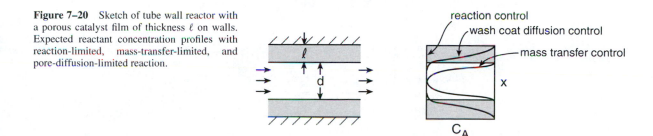

7.12 LANGMUIR–HINSHELWOOD KINETICS

We have thus far written unimolecular surface reaction rates as $r'' = kC_{As}$, assuming that rates are simply first order in the reactant concentration. This is the simplest form, and we used it to introduce the complexities of external mass transfer and pore diffusion on surface reactions. In fact there are many situations where surface reactions do not obey simple rate expressions, and they frequently give rate expressions that do not obey simple power-law dependences on concentrations or simple Arrhenius temperatures dependences.

Surface reactions are in principle multistep processes, and previously we saw how diffusion to the surface of a porous pellet or diffusion within the pores of a pellet can lead to rate expressions with rate coefficients much different than k''. However, so far all were first order in the reactant partial pressure.

Another reason for describing surface reaction kinetics in more detail is that we need to examine the processes on a *microscopic* scale. While we are interested primarily in the *macroscopic* description of catalytic reactor behavior, we cannot do this intelligently until we understand these processes at a molecular level.

7.12.1 Densities of molecules on surfaces

We will need to describe concentrations of species in the gas or liquid above the catalyst surface as well as adsorbed on the surface.

Densities of molecules adsorbed on surfaces are in moles per unit area, which we will give the symbol n_j for species j. Typical units of n_j are moles per cm². There is a maximum density n_{jo} of molecules packed in a two-dimensional layer on a flat surface when all molecules "touch" or reach liquidlike or solidlike densities. We call this density a *monolayer*, whose density is approximately the inverse of the square of the molecular diameter, which is less than 1×10^{15} molecules/cm² for all molecules (Figure 7–21). We will find it convenient also to define a *coverage* of adsorbed molecules as the fraction of the monolayer density θ_j,

$$\theta_j = \frac{n_j}{n_{jo}}$$

a quantity that is zero for a "clean" surface and unity for a "saturated" surface.

Another convention in catalytic reactions at surfaces is to use the partial pressure P_j rather than the concentration C_j in describing densities in the bulk phase above catalyst surfaces, although there are of course many situations where the reactants and products are in the liquid phase. These are simply related for ideal gases by the expression $C_j = P_j/RT$.

Therefore, we confess that in this section we violate a principle of this book in that we promised that we would only use concentration as a variable throughout.

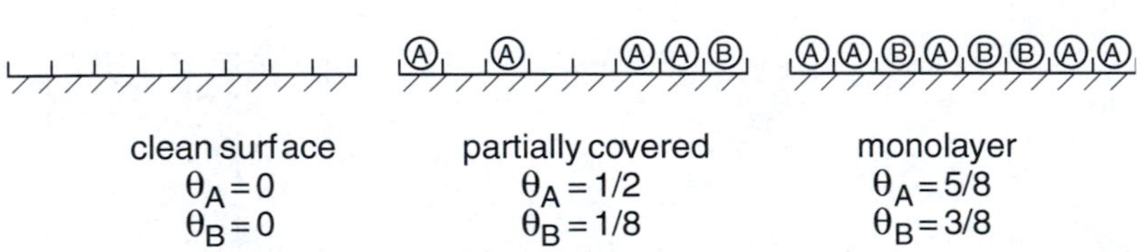

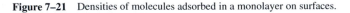

clean surface
$\theta_A = 0$
$\theta_B = 0$

partially covered
$\theta_A = 1/2$
$\theta_B = 1/8$

monolayer
$\theta_A = 5/8$
$\theta_B = 3/8$

Figure 7–21 Densities of molecules adsorbed in a monolayer on surfaces.

7.12.2 Reactions of NO on noble metals

Some of the important reactions in contemporary technology involve NO_x, which is a designation of N_2O, NO, and NO_2, and was one of the first examples in this book. The formation of these molecules in combustion processes is a major source of air pollution, and the catalytic oxidation of NH_3 to NO on Pt surfaces is used to produce nitric acid, a major industrial chemical. The decomposition of NO_x to N_2 is a major process in the automotive catalytic converter.

We will use catalytic reactions of NO to illustrate surface reaction rates.

7.12.3 Unimolecular reactions

NO will decompose on noble metal surfaces at sufficiently high temperatures to form N_2 and O_2 in the overall reaction

$$NO \rightarrow \tfrac{1}{2}N_2 + \tfrac{1}{2}O_2 \tag{7.87}$$

The elementary steps in this reaction are thought to be

$$NO + S \rightleftarrows NO\text{–}S \tag{7.88}$$

$$NO\text{–}S + S \rightleftarrows N\text{–}S + O\text{–}S \tag{7.89}$$

$$2N\text{–}S \rightleftarrows N_2 + 2S \tag{7.90}$$

$$2O\text{–}S \rightleftarrows O_2 + 2S \tag{7.91}$$

The species in this sequence are gas molecules NO, N_2, and O_2 and adsorbed NO, O atoms, and N atoms, which we designate as NO–S, N–S, and O–S, respectively. In fact, this reaction sequence is not effective in decomposing NO, even though the equilibrium is favorable in the presence of O_2 (air), and it is necessary to use the bimolecular reaction of NO with CO to remove NO from the automobile exhaust.

Consider a surface-catalyzed unimolecular reaction $A_g \rightarrow B_g$ with A and B gases with partial pressures P_{As} and P_{Bs} above the catalytic surface and coverages θ_A and θ_B on the surface. The elementary steps and their rates for a simple unimolecular reaction might be

$$A_g \rightleftarrows A_s, \qquad r'' = r''_{aA} - r''_{dA} \tag{7.92}$$

$$A_s \rightarrow B_s, \qquad r'' = r''_R \tag{7.93}$$

$$\underline{B_s \rightarrow B_g, \qquad r'' = r''_{dB} - r''_{aB}} \tag{7.94}$$

$$A_g \rightarrow B_g$$

where the subscripts g and s refer to gas and adsorbed molecules. Another way of writing these steps is

$$A + S \rightleftarrows AS \tag{7.95}$$

$$AS \rightarrow BS \tag{7.96}$$

$$\underline{BS \rightarrow B + S} \tag{7.97}$$

$$A \rightarrow B$$

where in this notation A and B are gas molecules, S is a surface site, and AS and BS are adsorbed species.

7.12.4 Catalytic reaction cycles

These elementary steps add up to give the overall reaction $A \rightarrow B$. It is instructive to consider these steps as a cyclic process in which the surface site is S, AS, BS, and back to S. This cycle is illustrated in Figure 7–22. We will encounter similar cycles in chain reactions in combustion and polymerization processes in Chapters 10 and 11 where a chemical species acts as a catalyst to promote the reaction around a closed cycle.

In these expressions r''_{aj} is the rate of adsorption of species j, which for A may be written as $A + S \rightarrow AS$, where A is the gas-phase molecule, S is an empty site on the surface, and AS is the adsorbed molecule. We can consider adsorption as a bimolecular chemical reaction that is proportional to the densities of the two "reactants" A and S to give

$$r''_{aA} = k_{aA} P_A \theta_S \tag{7.98}$$

with k_{aA} the adsorption rate coefficient of species A, and the two densities are as defined above.

The surface coverage of empty sites $[S]$ is the fraction of total sites (density $[S_o]$) not occupied by adsorbed molecules. Since the surface is assumed to saturate when $\theta_A = 1$, then $\theta_S = 1 - \theta_A$, or

$$r''_{aA} = k_{aA} P_A (1 - \theta_A) \tag{7.99}$$

If species A and B compete for the same site, then $\theta_S = 1 - \theta_A - \theta_B$ to give

$$r''_{aA} = k_{aA} P_A (1 - \theta_A - \theta_B) \tag{7.100}$$

The rate expression r''_d is the rate of desorption, which can be regarded as the elementary reaction $AS \rightarrow A + S$, which should have the expression

$$r''_d = k_{dA} \theta_A$$

Note that the units of ks in these expressions are somewhat unusual units in that they must be consistent with r'' (molecules/area time), P_A (torr), and θ_A (dimensionless).

The surface reaction rate r''_R is assumed to be a unimolecular chemical reaction $AS \rightarrow BS$, which has the rate expression

$$r''_R = k_R \theta_A \tag{7.101}$$

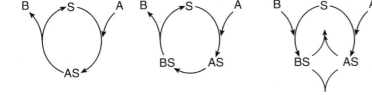

unimolecular surface reaction   bimolecular reaction

Figure 7–22 Catalytic cycle in which reactant A adsorbs on a vacant surface site S to form the adsorbate AS, which reacts to form BS. The product B then desorbs to form the original vacant site S, which allows the catalytic process to continue indefinitely in a cycle.

with k_R the surface reaction rate coefficient. We have arbitrarily written the first and third steps as reversible (adsorption and desorption) and the surface reaction step as irreversible.

We now proceed to write rate expressions for all species and solve them. Mass balances on adsorbed A and B yield

$$\frac{d\theta_A}{dt} = k_{aA} P_A (1 - \theta_A - \theta_B) - k_{dA} \theta_A - k_R \theta_A \tag{7.102}$$

$$\frac{d\theta_B}{dt} = +k_{aB} P_B (1 - \theta_A - \theta_B) - k_{dB} \theta_B + k_R \theta_A \tag{7.103}$$

The terms and their signs come simply by applying the formula $dC_j/dt = \sum v_{ij} r_i$ for θ_A and θ_B.

We assume that no species are accumulating and therefore the derivatives of coverages with respect to time are zero.

These are two algebraic equations in P_A, P_B, θ_A, and θ_B. They can be solved exactly to eliminate θ_A and θ_B for this simple case to give

$$r_R'' = \frac{k_R K_A' P_A}{1 + K_A' P_A + K_B' P_B} \tag{7.104}$$

where $K_A' = k_{aA}/(k_{dA} + k_R)$ and $K_B' = k_{aB}/(k_{dB} + k_R)$. We put primes on the Ks because they are similar to quantities K_j, which will be the adsorption–desorption equilibrium constant. We already see that our rate expressions are not simple power-law expressions.

For any more complicated rate expressions the equations become polynomials in coverages and pressures, and the general solution is uninstructive. As we saw for any multistep reaction with intermediates whose concentrations we do not know and that may be small (now we have surface densities rather than free radical concentrations), we want to find approximations to eliminate these concentrations from our final expression. The preceding solution was for steady state (a very good approximation for a steady-state reactor), but the expression becomes even simpler assuming thermodynamic equilibrium.

7.12.5 Adsorption–desorption equilibrium

If the reaction step is slower than the rates of adsorption and desorption of A and B, then we can assume thermodynamic equilibrium in adsorption and desorption of A and B. This gives

$$r_{aA}'' - r_{dA}'' = r_{dB}'' - r_{aB}'' = 0$$
$$= k_{aA} P_A (1 - \theta_A - \theta_B) - k_{dA} \theta_A = k_{dB} \theta_B - k_{aB} P_B (1 - \theta_A - \theta_B) \tag{7.105}$$

In fact the differences between adsorption and desorption are *exactly* equal to the reaction rate; the pss approximation requires that r_a and r_d be individually much greater than r_R. We now have two equations from which we can solve explicitly for θ_A. This gives

$$\theta_A = \frac{K_A P_A}{1 + K_A P_A + K_B P_B} \tag{7.106}$$

where

$$K_A = k_{aA}/k_{dA} \tag{7.107}$$

$$K_B = k_{aB}/k_{dB} \tag{7.108}$$

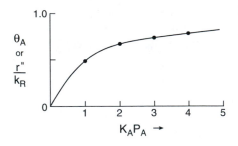

Figure 7–23 Langmuir isotherm of coverage θ_A versus $K_A P_A$ and rate of reaction r'' for a unimolecular surface-catalyzed reaction $A \rightarrow$ products.

If K_B or P_B is zero, we can neglect the adsorption of B, and the adsorption isotherm becomes

$$\theta_A = \frac{K_A P_A}{1 + K_A P_A} \qquad (7.109)$$

This is called a *Langmuir adsorption isotherm* for a species A, and the function $\theta_A(P_A)$ is shown in Figure 7–23. The K_js are the *adsorption–desorption equilibrium constants* for species A and B. By historical convention we call these the adsorption isotherms. Before we proceed let us note that this is a true thermodynamic equilibrium relation so that

$$K_A = K_{oA} e^{-E_A/RT} \qquad (7.110)$$

where K_{oA} is the pre-exponential factor and E_A is the heat of adsorption of A. Again by historical convention we define the "heat of adsorption" as a positive quantity even though the adsorption process is exothermic.

7.12.6   Measurement of adsorption–desorption equilibrium

The relation $\theta_A(P_A)$ is called the *adsorption isotherm*. It is used to determine surface areas of solids and catalysts as well as to determine the adsorption–desorption equilibrium constant K_A. This is measured by determining the amount of a gas that can be adsorbed by a known weight of solid, as shown in Figure 7–24.

The results of N_A versus P_A are analyzed by an equation first derived by Brunauer, Emmet, and Teller, and the resultant isotherm is called the *BET isotherm*. Typically one measures the amount of N_2 adsorbed for a particular pressure at 78 K (the boiling point of N_2 at a pressure of 1 atm) as sketched in Figure 7–24. There are several regimes of an adsorption isotherm. At low densities the density increases linearly with pressure. When the density approaches one monolayer, the surface saturates. As the pressure approaches the saturation pressure of the gas, bulk condensation of liquid occurs. This condensation can occur preferentially in pores of the solid due to capillary condensation, and the amount of gas and pressure where this occurs can be used to determine the *pore volume* of the catalyst.

Figure 7–24 Adsorption equilibrium apparatus to determine adsorption isotherms and surface areas of catalysts. From the saturation of a sample of known weight, the surface area can be determined if the area occupied by a molecule is known. (Adsorbed molecules are dots.)

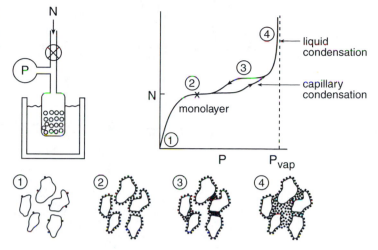

7.12.7 Langmuir–Hinshelwood kinetics

We can now write the Langmuir–Hinshelwood rate expression for this unimolecular reaction $A \rightarrow B$,

$$r_R = k_R \theta_A = \frac{k_R K_A P_A}{1 + K_A P_A + K_B P_B} \qquad (7.111)$$

which is the prototype expression for the simplest surface-catalyzed reaction. We sometimes further simplify this rate expression by assuming that the product B desorbs immediately (equivalent to saying that $K_B P_B = 0$) to give

$$r_R = \frac{k_R K_A P_A}{1 + K_A P_A} \qquad (7.112)$$

We see that in this simplest possible rate expression we could have for a unimolecular catalytic reaction there is no simple power-law dependence on partial pressures (Figure 7–25).

Figure 7–25 Langmuir–Hinshelwood kinetics for a unimolecular surface-catalyzed reaction $A \rightarrow$ products. The rate is first order in P_A at low coverages and zeroth order at high coverage.

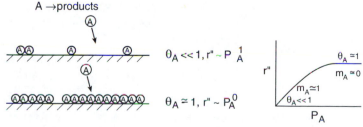

7.12.8 Approximate rate expressions

We are interested in obtaining approximate rate expressions that do have power-law dependences on partial pressures. We are interested in writing these rates as

$$r''_{\text{eff}} = k_{\text{eff}} \prod_j P_j^{m_j} = k_{\text{o,eff}} \exp(E_{\text{eff}}/RT) \prod_j P_j^{m_j} \tag{7.113}$$

where k_{eff}, $k_{\text{o,eff}}$, E_{eff} are effective parameters and m_j is the order of the reaction with respect to species j.

Since there are three terms in the denominator, we recover a power-law expression if one of these terms is so large to be able to neglect the others.

If $1 \gg K_A P_A$, $K_B P_B$, then this expression becomes

$$r_R = k_R K_A P_A \tag{7.114}$$

so we recover first-order kinetics in P_A. However, the reaction rate coefficient $k'' = k_R K_A$ or the rate coefficient is the actual k_R multiplied by the adsorption–desorption equilibrium constant!

If $K_A P_A \gg 1$, $K_B P_B$, then the expression becomes

$$r_R = k_R \tag{7.115}$$

The reaction rate coefficient is now k_R, but the reaction is *zeroth order*!

Finally if $K_B P_B \gg 1$, $K_A P_A$, then the expression becomes

$$r_R = k_R K_A K_B^{-1} P_A / P_B \tag{7.116}$$

so the reaction is first order in P_A but is *negative order* in P_B even though we assumed an irreversible reaction! The reaction rate coefficient is now $k''_{\text{eff}} = k_R K_A K_B^{-1}$.

7.12.9 Mechanism

Why do we obtain such complex rate expressions for such a simple process? The answer is found simply in the $1 - \theta$ terms we added in the adsorption rate expression. We said that the density of adsorbed molecules could never exceed one monolayer because that was the maximum density with which molecules could be "packed." The rate is assumed to be always proportional to the coverage of species A,

$$r_R = k_R \theta_A \tag{7.117}$$

When the pressure of A is sufficiently low, its coverage is much less than one monolayer and $\theta_A = K_A P_A$ to give first-order kinetics. As P_A increases and θ_A approaches unity, the coverage stops increasing $[\theta_A = K_A P_A/(1 + K_A P_A)]$; so the reaction becomes zeroth order in P_A as $\theta_A \to 1$, where $r_R = k_R$. Finally if the product B is sufficiently strongly adsorbed to build up a sufficient coverage, then adsorbed B blocks the adsorption of A ($\theta_A = K_A P_A/K_B P_B$), and the rate becomes first order in P_A but negative order in the product B.

7.12.10 Adsorption of inerts

Consider next the situation where an additional molecule is in the system that does not participate in the reaction but strongly adsorbs on the same sites occupied by A and B. One can easily show that the coverage of A is affected by this species as

$$\theta_A = \frac{K_A P_A}{1 + K_A P_A + K_B P_B + K_C P_C} \tag{7.118}$$

where K_C is the adsorption equilibrium constant of C. Now if $K_C P_C \gg 1$, then K_A, K_B, then $\theta_A = K_A P_A / K_C P_C$), and the rate becomes

$$r_R = \frac{k_R K_A}{K_C} \frac{P_A}{P_C} \tag{7.119}$$

The rate therefore becomes reduced by species C (as P_C^{-1}), and the effective rate coefficient is $k'' = k_R K_A / K_C$.

This shows that the "inert" C can strongly inhibit the unimolecular reaction of A, both by creating a rate inversely proportional to its partial pressure and by reducing the rate coefficient. Analysis of the coverages shows the cause of this. Species C blocks the sites on which A must adsorb to react, and lowers its coverage to a small value. This species is therefore a "poison" for the reaction.

It is a common phenomenon to find that apparently innocuous species that do not participate in catalytic reactions can strongly poison or inhibit catalytic reactions. As examples, sulfur compounds are severe poisons for most catalysts, and the lead formerly added to gasoline as an antiknock agent completely "kills" the catalyst by covering its surface sites.

7.12.11 Temperature dependence

We next consider how surface reaction rates should vary with temperature. Since the rate coefficients are groups of reaction and equilibrium constants, one expects that they are different for different limiting cases.

The reaction rate coefficient varies as

$$k_R = k_{oR} \exp(-E_R / RT) \tag{7.120}$$

and the adsorption equilibrium constants are

$$K_j = K_{jo} \exp(+E_j / RT) \tag{7.121}$$

where E_R is the activation energy for the surface reaction and E_j is the heat of adsorption of species j. We use E rather than ΔH_j because, as noted previously, it is a convention to define the heat of adsorption as a positive quantity. In the above expression E_j is a positive quantity and K_j *decreases* with increasing temperature (coverages decrease as temperature increases).

We can summarize the cases and parameters expected in the four cases in Table 7–2, which indicates all the orders and activation energies that may be expected for the Langmuir–Hinshelwood model. Note particularly how activation energies vary by subtracting or adding heats of adsorption to E_R (a high activation energy implies a low rate). Subtracting a heat of adsorption will increase the rate considerably, while subtracting a heat of adsorption by the presence of a poison will decrease a rate, perhaps enough to make it negligible by adding a "poison."

Returning to our example of NO decomposition, we need to make only slight modification of these equations to fit experimental results. The product molecules N_2 and O_2 adsorb dissociatively in the step

$$O_2 + 2S \rightleftarrows 2O\text{–}S \tag{7.122}$$

TABLE 7–2
Effective Rate Coefficients for Catalytic Reaction

Approximation	k''_{eff}	E_{eff}	Pressure dependence	Coverages
$1 \gg K_A P_A, K_B P_B$	$k_R K_A$	$E_R - E_A$	P_A^{+1}	$\theta_A, \theta_B \ll 1$
$K_A P_A \gg 1, K_B P_B$	k_R	E_R	P_A^0	$\theta_A = 1$
$K_B P_B \gg 1, K_A P_A$	$k_R K_A / K_B$	$E_R - E_A + E_B$	$P_A^{+1} P_B^{-1}$	$\theta_B = 1$
$K_C P_C$ large	$k_R K_A / K_C$	$E_R - E_A + E_B$	$P_A^{+1} P_C^{-1}$	$\theta_C = 1$

which has a rate

$$r = r_a - r_d = k_a P_{O_2}(1 - \theta_O - \theta_N - \theta_{NO})^2 - k_d \theta_O^2 \qquad (7.123)$$

With adsorption–desorption equilibrium we set $r = 0$ and solve for θ_o to obtain the adsorption isotherm of oxygen assume $\theta_N = \theta_{NO} = O$ and

$$\theta_O = \frac{K_{O_2}^{1/2} P_{O_2}^{1/2}}{1 + K_{O_2}^{1/2} P_{O_2}^{1/2}} \qquad (7.124)$$

so the normal KP of a nondissociated species is replaced by $(KP)^{1/2}$ for a dissociated species. If nondissociated NO competes for sites with O_2 and with dissociated N_2, then the isotherm for NO becomes

$$\theta_{NO} = \frac{K_{NO} P_{NO}}{1 + K_{NO} P_{NO} + K_O^{1/2} P_{O_2}^{1/2} + K_N^{1/2} P_{N_2}^{1/2}} \qquad (7.125)$$

The rate of NO decomposition should therefore be

$$r = k_R \theta_{NO}(1 - \theta_{NO} - \theta_N - \theta_O) = \frac{k_R K_{NO} P_{NO}}{(1 + K_{NO} P_{NO} + K_O^{1/2} P_{O_2}^{1/2} + K_N^{1/2} P_{N_2}^{1/2})^2} \qquad (7.126)$$

At sufficiently high temperature the denominator is unity because all coverages are low,

$$r = k_R K_{NO} P_{NO} \qquad (7.127)$$

while at low temperature the $K_O^{1/2} P_{O_2}^{1/2}$ term in the denominator dominates to give a rate

$$r = \frac{k_R K_{NO} P_{NO}}{K_O P_{O_2}} \qquad (7.128)$$

Here we see that the rate is strongly inhibited by adsorbed oxygen, which blocks sites for NO to adsorb into. This is the reason that NO cannot be decomposed in a unimolecular process at moderate temperatures in the presence of O_2.

7.12.12 Bimolecular reactions

NO must be decomposed by bimolecular reactions to keep the product oxygen from blocking the surface sites on which NO must adsorb. In the automotive catalytic converter the NO in fact reacts with CO, which is another pollutant reduced in combustion. The reactions may be written as

$$NO + S \rightleftarrows NO\text{--}S \tag{7.129}$$

$$CO + S \rightleftarrows CO\text{--}S \tag{7.130}$$

$$NO\text{--}S + CO\text{--}S \rightleftarrows N - S + CO_2\text{--}S \tag{7.131}$$

$$2N\text{--}S \rightarrow N_2 + 2S \tag{7.132}$$

$$CO_2\text{--}S \rightleftarrows CO_2 + S \tag{7.133}$$

This is a simple example of a bimolecular surface reaction, which we will use to illustrate bimolecular reactions.

Consider an irreversible bimolecular reaction

$$A + B \rightarrow C \tag{7.134}$$

of gas molecules A and B, which occurs on a surface to form a gas C. The rate of the surface reaction we will assume to be

$$r'' = k_R \theta_A \theta_B \tag{7.135}$$

and our task is to write this rate as a function of partial pressures.

We assume that the elementary steps are

$$A_g \rightleftarrows A_s, \qquad r'' = r_{aA} - r_{dA} \tag{7.136}$$

$$B_g \rightleftarrows B_s, \qquad r'' = r_{aB} - r_{dB} \tag{7.137}$$

$$A_s + B_s \rightarrow C_s, \qquad r'' = r_R \tag{7.138}$$

$$\underline{C_s \rightleftarrows C_g, \qquad r'' = r_{dC} - r_{aC}} \tag{7.139}$$

$$A_g + B_g \rightarrow C_g$$

Alternatively, writing these steps as "reactions" between A and B molecules and surface sites S, the elementary steps become

$$A + S \rightleftarrows AS \tag{7.140}$$

$$B + S \rightleftarrows BS \tag{7.141}$$

$$AS + BS \rightarrow CS \tag{7.142}$$

$$\underline{CS \rightleftarrows C + S} \tag{7.143}$$

$$A + B \rightarrow C$$

Now, as for the unimolecular reaction, we could also write the mass-balance equations for θ_A, θ_B, and θ_C and solve the algebraic equations for the rate in terms of partial pressures. However, for these expressions we obtain a quadratic equation whose solution is not very instructive.

Instead we assume adsorption–desorption equilibrium ($r_a = r_d \gg r_R$), and write Langmuir isotherms. These are exactly as for equilibrium among three adsorbed species in the unimolecular reaction

$$\theta_A = \frac{K_A P_A}{1 + K_A P_A + K_B P_B + K_C P_C} \tag{7.144}$$

with similar expressions for θ_B and θ_C.

The reaction rate can now be written as a function of partial pressures to yield

$$r'' = k_R \theta_A \theta_B = \frac{k_R K_A K_B P_A P_B}{(1 + K_A P_A + K_B P_B + K_C P_C)^2} \qquad (7.145)$$

Note that the denominator contains the square of the denominators in the θ denominator for both A and B. We therefore have a more complex function of partial pressures than for the unimolecular reaction, but as before we want to find out what these expressions become if each of the three terms in the denominator dominate.

We summarize these in the following table, but let us first consider several of these cases.

If $1 \gg K_j P_j$, then $r'' = k_R K_A K_B P_A P_B$, then the orders of the reaction with respect to P_A and P_B are "normal," but the rate coefficient has both of the adsorption equilibrium constants in it. If K_A is large, then $r'' = k_R K_B K_A^{-1} P_B P_A^{-1}$, and the rate is in fact *negative order* with respect to a reactant.

We can summarize the effective rates and parameters expected in the four cases for the bimolecular reaction in the following table:

Approximation	k''_{eff}	E_{eff}	Pressure dependence	Coverages
$1 \gg K_A P_A, K_B P_B$	$k_R K_A K_B$	$E_R - E_A - E_B$	$P_A^{+1} P_B^{+1}$	$\theta_A, \theta_B \ll 1$
$K_A P_A \gg 1, K_B P_B$	$k_R K_A^{-1} K_B$	$E_R + E_A - E_B$	$P_A^{-1} P_B^{+1}$	$\theta_A = 1$
$K_B P_B \gg 1, K_A P_A$	$k_R K_A K_B^{-1}$	$E_R - E_A + E_B$	$P_A^{+1} P_B^{-1}$	$\theta_B = 1$
$K_C P_C$ large	$k_R K_A K_B K_C^{-2}$	$E_R - E_A - E_B + 2E_C$	$P_A^{+1} P_B^{+1} P_C^{-2}$	$\theta_C = 1$

Physically, these limits correspond to saturation of the surface with different species, as sketched in Figure 7–26.

Note especially the strong inhibition of the "inert" or product C in this irreversible process. The reaction could be proportional to P_C^{-2}, and the activation energy is increased by *twice* the heat of adsorption of C. These features come from the *squared* denominator in the rate expression for a bimolecular reaction.

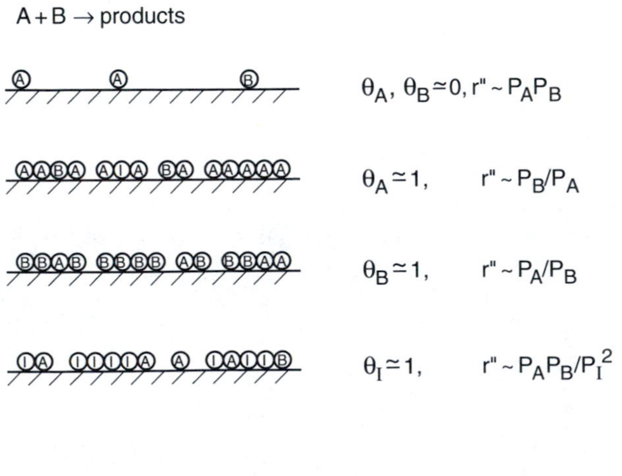

A + B → products

$\theta_A, \theta_B \approx 0, r'' \sim P_A P_B$

$\theta_A \approx 1, \quad r'' \sim P_B / P_A$

$\theta_B \approx 1, \quad r'' \sim P_A / P_B$

$\theta_I \approx 1, \quad r'' \sim P_A P_B / P_I^2$

Figure 7–26 Coverages of species A, B, C, and I corresponding to the limiting rate expressions. Saturation of a species can lead to negative-order kinetics because it blocks other species from adsorbing.

Returning to the NO example, of the reaction of NO with CO and assuming that the adsorbed species are CO, NO, O, and N, the rate of reaction of NO with CO is

$$r = k_R \theta_{NO} \theta_{CO} = \frac{k_R K_{NO} K_{CO} P_{NO} P_{CO}}{(1 + K_{NO} P_{NO} + K_{CO} P_{CO} + K_O^{1/2} P_{O_2}^{1/2} + K_N^{1/2} P_{N_2}^{1/2})^2} \tag{7.146}$$

This reaction is sufficiently fast that it proceeds at low enough temperature to be useful to eliminate NO in the automotive catalytic converter. We assumed that the surface reaction step is

$$NO\text{–}S + CO\text{–}S \rightarrow N\text{–}S + CO_2\text{–}S \tag{7.147}$$

More likely it involves several steps such as

$$NO\text{–}S + S \rightarrow N\text{–}S + O\text{–}S \tag{7.148}$$

followed by

$$2N\text{–}S \rightleftarrows N_2 + S \tag{7.149}$$

$$CO\text{–}S + O\text{–}S \rightleftarrows CO_2\text{–}S \tag{7.150}$$

Thus the reactions are similar to those listed for the unimolecular reaction except that now the presence of adsorbed CO oxidizes off the oxygen, which would otherwise poison the surface.

These rate expressions are for Langmuir–Hinshelwood kinetics, which are the simplest forms of surface reaction rates one could possibly find. We know of no reactions that are this simple. LH kinetics requires several assumptions:

1. Adsorbed species are competitive for sites but do not otherwise interact.
2. There is only one type of adsorption site on the surface for each species.
3. Adsorbed molecules form a maximum coverage of one monolayer.
4. All properties of all adsorbates are independent of the coverage of all species.

None of these is found to be true experimentally for any adsorption system of interest. Therefore, we do not expect LH rate expressions to be quantitative for any catalytic reaction system. The real situation will always be much more complicated, and in most cases we do not yet know what rate expressions might describe catalytic reactions with more than empirical accuracy.

7.13   SUMMARY OF SURFACE REACTION KINETICS

The previous derivations describe what must be expected in the simplest catalytic reactions: Orders of reaction and rate coefficients change strongly with conditions. For a unimolecular reaction $A \rightarrow B$ on a surface we expect the rate expression of the form

$$\boxed{r_R = \frac{k_R K_A P_A}{1 + K_A P_A + K_B P_B}} \tag{7.151}$$

while for a bimolecular reaction $A + B \rightarrow C$ we expect

$$r'' = \frac{k_R K_A K_B P_A P_B}{(1 + K_A P_A + K_B P_B + K_C P_C)^2} \qquad (7.152)$$

Thus, for all reactions that are catalyzed by surfaces we should expect

1. Nonintegral orders and even negative orders with respect to a reactant species in catalytic reactions.

2. Rate expressions obtained in one set of pressures and temperatures might not describe a reactor in another situation because rate expressions of catalytic reactions do not obey the power law.

3. Traces of impurities should drastically alter catalytic reaction rates even when they do not enter into the reaction.

For example, in problems we will consider an interesting rate expression

$$r'' = \frac{k' P_A}{(1 + K' P_A)^2} \qquad (7.153)$$

for a unimolecular reaction $A \rightarrow C$. This form of a rate expression could be obtained for a bimolecular competitive reaction $A + B \rightarrow C$ if B is in excess so that its partial pressure does not change. The standard form of a bimolecular reaction can then be transformed into this expression if $k' = k_R K_A K_B P_B$ and $K' = K_A$.

7.14 DESIGNING CATALYTIC REACTORS

We now can begin to see how we choose catalyst parameters in a catalytic reactor if we want the pseudohomogeneous rate to be as high as possible. We write the "general" expression for a catalytic reaction rate as

$$r = (\text{area/volume})r'' = (\text{area/volume})\frac{k''}{1 + k''/k_m}C_{Ab}\eta(\phi) \qquad (7.154)$$

where we can see the reaction rate represented by k'', external diffusion represented by k_m, and pore diffusion represented by $\eta(\phi)$. We write these as

$$k'' = k''_o \exp(-E/RT) \qquad (7.155)$$

$$k_m = \frac{\text{Sh}_D D_A}{D} \qquad (7.156)$$

and $\eta(\phi)$ through the Thiele modulus ϕ as

$$\phi = \left(\frac{S_g \rho_c k''}{D_A}\right)^{1/2} D \qquad (7.157)$$

where all of these quantities have been defined previously.

We want the pseudohomogeneous rate as a function of the parameters in the system,

$$r(u, k'_o, E, D, d_p, S_g\rho_c, D_A, \text{Sh}_D) \tag{7.158}$$

The optimum performance of a reactor will occur when the rate is as high as possible. We can always increase k'' by increasing the temperature, assuming there is only one reaction so that selectivity is not important. However, we will find that at some temperature the mass transfer and/or pore diffusion processes will begin to limit the rate.

The ideal reactor will usually be obtained when the reactor just begins to be mass transfer limited,

$$k_m = k'' \tag{7.159}$$

and when pore diffusion just begins to affect the rate

$$\phi = 1 \tag{7.160}$$

The catalyst loading in the reactor should be as small as required to attain the desired conversion to save cost of the catalyst, and the catalyst pellet size should be as large as possible to minimize pressure drop in the reactor.

One can increase the external mass transfer rate by *increasing the flow velocity u* over the catalyst pellet. In a slurry this can be accomplished by increasing the stirring, and in a packed bed or fluidized bed it can be increased somewhat by increasing the flow velocity (although with the same catalyst this decreases τ, which lowers the conversion).

Pore diffusion can be increased by choosing a catalyst with the proper *geometry*, in particular the pellet size and pore structure. Catalyst size is obvious ($r \sim R^{1/2}$ if pore diffusion limited for the same total surface area). The diameter of pores can have a marked influence on η because the diffusion coefficient of the reactant D_A will be a function of d_p if molecular flow in the pore dominates. Porous catalysts are frequently designed to have different distributions of pore diameters, sometimes with macropores to promote diffusion into the core of the catalyst and micropores to provide a high total area.

One can begin to diagnose the performance of a catalytic reactor very quickly with very limited knowledge of its properties by considering the effects of (1) temperature, (2) stirring, and (3) pellet size. Reactor engineers should always be aware of these effects in any catalytic process.

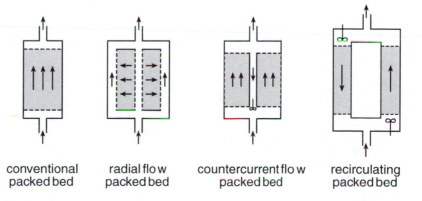

conventional
packed bed

radial flow
packed bed

countercurrent flow
packed bed

recirculating
packed bed

Figure 7–27 Fluid flow patterns in fixed bed reactors, which allow the variation of velocity profiles within the catalyst bed.

7.14.1 Flow patterns in catalytic reactors

In addition to using different catalyst flow patterns in packed and slurry reactors, the flow can be varied to attain different catalyst contacting patterns. As shown in Figure 7–27, many flow patterns such as radial flow and fluid recirculation can be used. These allow variation of the flow velocity u for a given reactor size and residence time τ. These recirculation flow patterns approach the flow of recycle reactors; so the reactor performance approaches that of a CSTR at high recirculation.

7.15 ELECTROCHEMICAL REACTORS

Electrochemical reactions are catalytic reactions in which strong electric fields near surfaces are used to alter chemical equilibrium. By applying an external voltage V between two electrodes and creating an electric field $\mathcal{E}$ in an ionically conducting solution, we can cause current to flow and reaction to occur at electrodes. The Gibbs free energy is altered from its value in the absence of an electric field by the Faraday relation

$$\Delta G = G^{\circ} + n_e F \mathcal{E} \qquad (7.161)$$

where ΔG° is the standard state Gibbs free energy, n_e is the number of electrons transferred, and F the Faraday, $F = 96{,}500$ coulombs/mole and 1 coulomb=1 amp sec. This allows us to "cheat" normal equilibrium constraints in a chemical reaction. For example, if we apply a voltage $V > \mathcal{E}_O$ to an aqueous solution containing electrolyte, we can actually decompose water molecules

$$H_2O \rightarrow H_2 + \tfrac{1}{2}O_2 \qquad (7.162)$$

and generate H_2 gas bubbles at the negative electrode and O_2 bubbles at the positive electrode (Figure 7–28).

The electrochemical reactor has two "magic" properties: (1) It causes chemical reaction where equilibrium does not permit reaction in the absence of an electric field; and (2) it causes a separation of products because they are generated at different electrodes.

The electrode surfaces are catalysts to promote certain reactions. In H_2O electrolysis the reaction

$$2H^+ + 2e^- \rightarrow H_2 \qquad (7.163)$$

occurs on a Pt electrode at negative potential (the cathode), while the reaction

$$2OH^- \rightarrow \tfrac{1}{2}O_2 + H_2O + 2e^- \qquad (7.164)$$

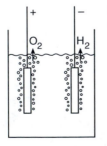

Figure 7–28 Electrochemical reactor in which water is electrolyzed into H_2 and O_2 by application of a voltage sufficient to dissociate H_2O in a batch reactor configuration.

Figure 7–29 Salt electrolysis to produce Cl_2 gas and NaOH solution in a chloralkali process. This is the process by which most alkali and chlorine are produced. The reactor operates continuously with brine solution flowing into the cell and NaOH and Cl_2 gas flowing out.

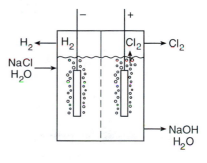

occurs at the positive electrode (the anode). At each electrode a catalytic reaction occurs, the first with addition of two electrons from the electrode into the solution, and the second by transfer of two electrons from the solution into the electrode. When the above electrode reactions are added, we obtain H_2O decomposition into its elements.

Thus electrochemical reactions are two coupled surface catalytic reactions, with the rates at each electrode identical

$$r''_{anode} = r''_{cathode} = \frac{n_e F I}{A} \tag{7.165}$$

because in steady state the electrons must flow through the complete circuit. Here A is the electrode area and I is the current that flows through the circuit.

By far the most important application of electrochemical reactors is the chloralkali industry, which produces most industrial Cl_2 and NaOH. (Recall our discussion of the origins of Dow Chemical Company in Chapter 3.) In this process graphite electrodes are used, and concentrated brine (NaCl) is electrolyzed by the electrode reactions

$$2Cl^- \rightarrow Cl_2 + 2e^- \tag{7.166}$$

$$2H_2O + 2e^- \rightarrow 2OH^- + H_2 \tag{7.167}$$

to give an overall reaction

$$NaCl + H_2O \rightarrow NaOH + \tfrac{1}{2}Cl_2 \tag{7.168}$$

A chloralkali reactor might look as shown in Figure 7–29.

Another important industrial electrochemical reaction is the production of aluminum metal from its ore bauxite, $Al(OH)_3$. The molten bauxite is electrolyzed with carbon electrodes, Al metal bars are plated onto one electrode, and O_2 gas is created at the other electrode.

Batteries are an old example of using an electrochemical reaction to produce electrical energy from solids, and fuel cells are a new example of using an electrochemical reaction to produce energy directly from chemicals.

7.16 REAL CATALYTIC REACTIONS

We have finished our discussions of the fundamentals of catalytic reactions, catalytic reactors, and mass transfer effects. While we noted that most catalytic reactions can be made to exhibit complicated kinetics, we have confined our considerations to the almost trivial reaction $r'' = k''C_A$. We did this because the algebra was messy enough with the simplest kinetics.

If we begin to complicate the kinetics, the problems seldom have analytical solutions. While we can seek numerical solutions, it is very easy to get really lost in the complexities of the equations and the pages of computer printouts. Remember, we want first to figure out what is going on and what steps are controlling the process. Then we later may want to find accurate numbers to predict reactor performance.

Multiple reaction systems further complicate our lives. Recall that selectivity issues can dominate reactor performance, and minimally we have to solve one equation for each chemical reaction. Now when we add the different kinetics of each reaction and recall that diffusivities and mass transfer coefficients are different for each chemical species, we see that we have an even more difficult problem.

Recall finally that chemical engineers are the only people who are capable of understanding what is going on in chemical reactors, and that suitable design of catalytic reactors frequently determines whether we have a money-making or money-losing process. Chemical engineers earn their salaries by being the only ones who can provide answers, and even then we frequently have to guess.

7.17 BIOREACTORS

This chapter has focused on inorganic and heterogeneous catalysts, because historically these are the major systems with which chemical engineers have been concerned. There are number of important homogeneous catalytic processes such as the Wacker process to make vinyl acetate from ethylene and acetic acid, and there are many acid and base homogeneous catalyst systems.

However, there are also many important biological catalytic systems. Fermentation of sugars into ethanol has been practiced since the Stone Age, and we still have not improved these catalysts much from the naturally occurring enzymes that exist in the yeasts that are found naturally growing in natural sugar sources.

Fermentation is an anaerobic process, but in the presence of O_2 the oxidation of ethanol to acetic acid (vinegar) accompanies the biological reactions and produces less valuable products. All biological decay processes are enzyme catalyzed, as single-celled bacteria produce enzymes to digest dead plants and animals. The metabolic products of these reactions are frequently numerous, and some of the organic products contain sulfur and nitrogen, which have unpleasant odors. There are therefore few examples of aerobic biological reactions except sewage treatment. This reaction is one of the most important chemical reaction processes in our civilization, but we have generally relegated this chemical process to civil engineers.

Production of high-fructose corn syrup is an important enzyme process. Because fructose tastes much sweeter than glucose, sucrose is converted into fructose by enzymes, and most soft drinks contain HFCS prepared by enzyme catalysts. Many small-volume enzyme processes are practiced (penicillin is a very important one), and much research and development is underway to find biological processes to make fine chemicals and pharmaceuticals.

One of the major issues in biological processes is that physiological systems create and use *chiral molecules* in very specific ways. Only one of the six isomers of sugars tastes sweet and can be metabolized. Thalidomide is a harmless tranquilizer, but the other of its isomers causes serious birth defects. Many common drugs are sold as both chiral isomers, with only one having any beneficial value. Scientists have been only modestly successful in creating synthetic catalysts to prepare chiral molecules, and we will continue to rely mostly on natural enzymes to produce these products. A major growth industry for chemical engineers is in the discovery and development of new chiral molecules because the potential value of these new products can be billions of dollars.

7.18   THE HUMAN REACTOR

Living organisms are incredible examples of well-designed chemical reactors. All animals have a food consumption–processing system coupled to a circulating reacting system with blood as the fluid. A simplified (one of the greatest possible understatements) schematic diagram of us as chemical reactors is illustrated in Figure 7–30.

7.18.1   The food processing system

The food and water processing reactor system has a semibatch feed, continuous chemical reactors, and a semibatch elimination system. Food is chewed in the mouth to small pieces of solid and enters the stomach every 8 hours or so. In the stomach food is further mixed with water, acidified, and mixed with enzymes, which begin reacting it into small particles and molecules. The food mixes in the stomach (volume 0.5 liter), but its feed is semibatch; so we describe it as a transient CSTR.

Next the acidified food passes into the small intestine (a reactor 3/4 in. in diameter and 20 ft long), where it is neutralized and mixed with more enzymes from the pancreas.

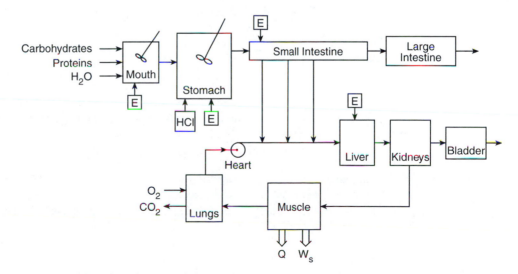

Figure 7–30   A simplified flow sheet of the digestive and circulatory systems of animals such as humans (not to scale). HCl and enzymes (E) catalyze most reactions.

This is the primary chemical reactor of the body, operating with secreted enzymes and with *E. coli* bacteria catalysts.

Reactions in the stomach and small intestine are primarily the breakdown of carbohydrates and dissaccharide sugars into monosaccharides

$$C_{12}H_{22}O_{11} \rightarrow 2C_6H_{12}O_6 \qquad (7.169)$$

and the breakdown of proteins into amino acids, which is the hydrolysis of amide linkages into amine and carboxylic acid. [Proteins are Nature's nylon]. All these reactions are catalyzed by acids, bases, and enzymes. Bacteria break large proteins and complex carbohydrates into sufficiently small molecules that these hydrolysis reactions can occur.

Peristalsis is the intestinal pump that maintains a suitable velocity and enhances mixing. We describe this reactor as nearly plug flow, although the backmixing of bacteria is essential in keeping them in the reactor.

The wall of the small intestine is permeable to water and to small molecules such as the amino acids produced by protein breakdown and sugars produced by carbohydrate breakdown; so this system is a reactor–separator combination, a membrane reactor. Finally the undigested food passes into the large intestine, where more water is removed through the permeable wall before exiting the reactor.

7.18.2   The circulatory system

Amino acids, sugars, and minerals pass through the small intestine into the circulatory system, where they are mixed with blood. The primary reactor organs in processing blood are muscle and the kidneys. The fluid flows in nearly total recycle through arteries and veins, which are basically the pipes in the system, and capillaries, where most of the transfer to and from the reactors and separators occurs.

7.18.3   The combustion and cell formation system

Muscles and other cells use sugars as their energy source by oxidizing them into CO_2 and water

$$sugar + O_2 \rightarrow CO_2 + H_2O \qquad (7.170)$$

or, if the body does not need energy, by converting it to fat, which is stored for future use. The other reaction is the building and decomposing of muscle and other cells from amino acids,

$$amino\ acids \rightarrow proteins \qquad (7.171)$$

This reaction is of course just the reverse of the hydrolysis reactions that occur in the digestive system, but the proteins formed are different from those in the food.

Combustion occurs by adding O_2 to the bloodstream adsorbed in hemoglobin,

$$HGB + O_2 \rightarrow HGB-O_2 \qquad (7.172)$$

transporting both fuel and oxidant to the cells where they diffuse through cell walls, and are reacted by ATP enzymes to produce energy and CO_2, which diffuses and is carried back out through the lungs.

The lungs are the gas–liquid separation units that transport O_2 into the body and remove the CO_2 oxidation product (definitely no partial oxidation to CO).

Cells are built by combining amino acids back into proteins, mostly within cells. This process uses RNA template catalysts, whose formation from DNA and "expression" are the major marvels of living organisms. The enzyme catalysts are also proteins that have structures to catalyze specific reactions; these molecules are being continuously created and decomposed as needed.

Waste products (cells only live for hours to days) pass back to the blood, where they are carried through the liver, which is the other major chemical reactor in the circulatory system. Here enzymes bound to surfaces in the liver further react the decomposition product molecules into even smaller molecules that can pass through the body's major filter, the kidneys. Liquid wastes and excess water then are stored in the bladder until needed.

The circulatory system is nearly continuous, with many short and long time constant transients, depending on age, activity, time of day, etc.

Each of these reactor systems can be described by a residence time or time constants; for example, $\tau_{digestion}$ is the time for food to pass through the digestive system, and $\tau_{circulation}$ is the time for blood to circulate. Consumption and metabolism of medicines and alcohol are described by the transient response of the concentration in the bloodstream or in tissues.

7.18.4   Human factory

These chemical reactor systems are of course under high levels of process control through the brain and the genetic code. The body is a very inefficient factory to produce materials (although chicken factories produce a pound of chicken meat for approximately 3 pounds of grain consumed).

However, the primary intent of physiological systems is not to increase individual size but to produce a functioning organism that can *reproduce* itself. The reproductive capacity of living systems involves another complicated set of chemical reactors that we will not consider here. However, we will examine the kinetics of populations of living systems in Chapter 8 in connection with environmental modeling.

Another way of looking at physiological systems is not just as chemical reactors but as a *complete chemical plant*, which contains a control room (with its own supercomputer), a lunchroom, recreational facilities, and of course facilities for constructing a new chemical plant when the present one corrodes away or become obsolete.

We suggest that consideration of Nature's ways of running and controlling chemical reactors provides lessons of the "ideal reactor" of which we can create only primitive copies from our education as chemical engineers. As the Bible says, we are "fearfully and wonderfully made."

7.19   REFERENCES

Aris, Rutherford, *Introduction to the Analysis of Chemical Reactors*, Prentice-Hall, 1965.
Aris, Rutherford, *Elementary Chemical Reactor Analysis*, Prentice-Hall, 1969.
Boudart, Michel, *The Kinetics of Chemical Processes*, Prentice-Hall, 1968.
Butt, John B., *Reaction Kinetics and Reactor Design*, Prentice-Hall, 1980.

Carberry, James J., *Chemical and Catalytic Reaction Engineering*, McGraw-Hill, 1976.

Clark, Alfred, *The Theory of Adsorption and Catalysis*, Academic Press, 1970.

Doraiswamy, L. K., and Sharma, M. M., *Heterogeneous Reactions, Volume I: Gas–Solid and Solid–Solid Reactions*, Wiley, 1984.

Doraiswamy, L. K., and Sharma, M. M., *Heterogeneous Reactions, Volume II: Fluid–Fluid–Solid Reactions*, Wiley, 1984.

Farrauto, R. J., and Bartholomew, C. H., *Fundamentals of Catalytic Processes*, Chapman and Hall, 1997.

Frank-Kamenetskii, D. A., *Diffusion and Heat Exchange in Chemical Kinetics*, Princeton University Press, 1955.

Froment, Gilbert F., and Bischoff, Kenneth B., *Chemical Reactor Analysis and Design*, 2nd ed., Wiley, 1990.

Gardiner, William C., Jr., *Combustion Chemistry*, Springer-Verlag, 1984.

Gates, Bruce C., *Catalytic Chemistry*, Wiley, 1992.

Gates, Bruce C., Katzer, James R., and Schuit, G. C. A., *Chemistry of Catalytic Processes*, McGraw-Hill, 1979.

Hougen, O. A., and Watson, K. M., *Chemical Process Principles, Volume III: Kinetics and Catalysis*, Wiley, 1947.

Lee, Hong H., *Heterogeneous Reactor Design*, Butterworths, 1985.

Ponec, Vladimir, and Bond, Geoffrey C., *Catalysis by Metals and Alloys*, Elsevier, 1995.

Satterfield, Charles N., *Heterogeneous Catalysis in Practice*, McGraw-Hill, 1980.

Smith, J. M., *Chemical Engineering Kinetics*, 3rd ed., McGraw-Hill, 1981.

Spitz, Peter H., *Petrochemicals: The Rise of an Industry,* Wiley, 1988.

Thomas, J. M., and Thomas, W. J., *Introduction to the Principles of Heterogeneous Catalysis*, Academic Press, 1967.

Twigg, Martyn V., *Catalyst Handbook*, Wolfe Publishing, 1989.

PROBLEMS

7.1 Design a continuous coffee maker to produce 100 gallons per hour of decaffeinated liquid coffee with a quality as close as possible to conventional batch processes starting with fresh coffee beans and water using 0.1 lb of coffee per gallon of water. Coffee beans are first dried and fermented by heating for 3 days at 40°C. Then the beans are roasted for approximately 15 min in an oven at 400°C before grinding. Caffeine is removed with supercritical CO_2 extraction. Use your own recipe for coffee brewing. Describe raw material needs, reactor sizes, and flow patterns in reactors. The solution need be no more than one page in length.

7.2 A slurry reactor contains 100-μm-diameter spherical particles that catalyze the reaction $A \rightarrow B$ with a rate $r'' = 10^{-3} C_A$ in moles/cm^2 sec with C_A in moles/liter. There are 10^5 particles/cm^3, and only the external surface of the particles is catalytic. What flow rate of $C_{Ao} = 2$ moles/liter can be processed in a 10 liter reactor if the conversion is 90%?

7.3 The reaction $r'' = kC_{As}$ occurs on the external surface of a sphere of diameter D suspended in a stagnant fluid in which the diffusion coefficient of the reactant is D_A. Find the total rate of the reaction in terms of these quantities. How does the rate depend on particle diameter? How would this influence the design of a slurry reactor with this catalyst?

7.4 If the reaction in the previous problem is exothermic and mass transfer limited, show that multiple steady states could occur.

7.5 A 1-m^3 packed bed reactor is filled with 5-mm-diameter catalyst spheres that occupy 0.7 of the total reactor volume. The feed concentration is 2 moles/liter and the flow rate is 1 liter/sec.

(a) In separate experiments using a small reactor a pseudohomogeneous rate coefficient of 4×10^{-3} sec^{-1} based on the reactor volume was obtained. Find the conversion in the packed bed.

(b) Find the conversion if the reaction occurs only on the external surface of the catalyst spheres with a rate r'' of 2×10^{-6} moles cm^{-2} sec^{-1}.

7.6 A reactant of bulk concentration C_{Ao} reacts on the external surface of catalyst spheres of radius R in a slurry reactor. The first-order surface reaction rate coefficient is k'', and the diffusivity of A in the solution is D_A. Find the effective rate coefficient k''_{eff} in terms of these quantities, assuming that stirring is sufficiently slow that the fluid around particles is stagnant.

7.7 Calculate the effectiveness factor of a single catalyst pore and of a catalyst slab for zeroth-order kinetics.

7.8 Find the effectiveness factor for a spherical catalyst particle with first-order kinetics.

7.9 Find η in the limits of $\phi \to 0$ and $\phi \to \infty$ for a spherical catalyst pellet. How should η depend on catalyst temperature? Show that the effective activation energy of a catalyst reaction is one-half of the time activation energy when $\phi \to \infty$.

7.10 Derive an expression for the effectiveness factor of a porous catalyst slab of thickness $2L$ that has an effective pore diameter $d_e = 4\varepsilon/\rho_c S$ with a first-order reaction $r'' = k''C_A$. Can you justify $d_e = 4\varepsilon\rho_c S$?

7.11 What equation would have to be solved for the catalyst effectiveness factor for a second-order reaction? What is the apparent activation energy of a second-order reaction in the limit of $\phi \to \infty$?

7.12 A first-order catalytic reaction occurs on the walls of a tube 0.5 cm in diameter with a flow rate of 0.1 liter/sec. The rate of the reaction is r'' (moles/cm^2 sec) $= 10^{-2}C_A$ with C_A in moles/liter.

(a) How long must the tube be to obtain 90% conversion if the concentration is uniform across the tube?

(b) Would this be a good approximation in water ($\nu = 1 \times 10^{-6}$ m^2/sec)? In air at atmospheric pressure ($\nu = 2 \times 10^{-5}$ m^2/sec)?

(c) Find an expression for the pseudohomogeneous reaction rate coefficient in terms of k'', D_A, and D if flow is laminar.

(d) Find an expression for the pseudohomogeneous rate coefficient if the flow is turbulent and the boundary layer is small compared to the tube diameter.

7.13 A first-order irreversible catalytic reaction ($r'' = k''C_A$) occurs in a slurry reactor (or fluidized bed reactor). Spherical catalyst particles have diameter R, density ρ_b, surface area/mass S_g, and a fraction ε of the reactor is occupied by catalyst. The average pore diameter is d. Find an expression for $r(C_A)$ in terms of these quantities.

7.14 A reactant of bulk concentration C_{Ao} reacts on the external surface of catalyst spheres of radius R in a slurry reactor. The first-order surface reaction rate coefficient is k'' and the diffusivity of A in the solution is D_A. Find the effective rate coefficient k''_{eff} in terms of these quantities, assuming that stirring is sufficiently slow that the fluid around particles is stagnant.

7.15 A particular catalyst has a surface area $S = 100 \text{ m}^2/\text{g}$, a density $\rho_c = 3 \text{ g/cm}^3$, and a void fraction $\varepsilon = 0.3$.

(a) The pseudohomogeneous rate of the irreversible reaction $A \rightarrow B$ in a test of this catalyst was 1×10^{-2} moles/liter sec when the reactant concentration was 1 mole/liter and 2.5×10^{-3} when the reactant concentration was 0.5 moles/liter. What is the surface reaction rate expression in moles/cm^2 sec?

(b) What reactor volume will be necessary to process 100 liters/sec of 2 molar A to 90% conversion in a packed-bed reactor?

7.16 We have a catalyst 1/8 in. in diameter in a slurry reactor that gives 90% conversion in a slurry reactor. Diagnose the following observations:

(a) The conversion increases when we double the stirring rate but doesn't change further when we double it again.

(b) The conversion is unchanged when we replace the catalyst by pellets 1/4 in. in diameter, holding the total weight of catalyst the same.

(c) The conversion decreases when we replace the catalyst by pellets 1/4 in. in diameter, holding the total weight of catalyst the same.

(d) The conversion increases when we replace the catalyst by pellets 1/4 in. in diameter, holding the total weight of catalyst the same.

(e) The reaction rate increases by a factor of 10 when the temperature is increased by 10 K.

(f) The reaction rate is unchanged when the temperature is increased by 10 K.

(g) In (e) the Thiele modulus was 0.1. How would the rate change if the Thiele modulus were 10? How does Thiele modulus vary with temperature?

7.17 A catalytic reaction $A \rightarrow B$ occurs on the walls of a tube 1 cm in diameter with a flow rate of 0.1 liters/sec. The rate of the reaction r'' (moles/cm^2 sec) is $5 \times 10^{-4} C_A$ with C_A in moles/liter. Assume $D_A = 0.1$ cm^2/sec.

(a) How long must the tube be to obtain 95% conversion if the concentration is uniform across the tube?

(b) How long must the tube be if the flow is laminar with $Sh_D = 8/3$?

7.18 A certain spherical porous catalyst with a pellet diameter of 1/8 in. has a Thiele modulus of 0.5 for a first-order reaction and gives 90% conversion in a packed bed reactor. It is proposed to replace this catalyst by the exact same catalyst but with pellets of 1/4 or 1/2 in. to reduce the pressure drop. How will the conversion change with these catalysts?

7.19 A catalytic reaction has an activation energy of 20 kcal/mole for a given catalyst pellet size and gives a 90% conversion in a slurry reactor at 50°C.

(a) What will be the conversion if the temperature is increased by 10 K if the Thiele modulus is small?

(b) What will be the conversion if the temperature is increased by 10 K if the Thiele modulus is large?

(c) What might you infer if the conversion does not change when the temperature is increased by 10 K? How might you test this hypothesis?

7.20 The catalyst in the previous problem had a Thiele modulus of 0.1. A new catalyst was found to have a Thiele modulus of 5 with all other properties unchanged. What will be the conversion in the above reactor with this new catalyst? What will be the appropriate kinetics?

7.21 What are the maximum and minimum orders of reaction predicted for an enzyme reaction $A + B \rightarrow C$ if both A and B require two competitive enzyme sites?

7.22 A packed bed has spherical pellets that pack in a square close-packed configuration.

(a) What is the void fraction in the reactor?

(b) Calculate the external surface area per liter of fluid volume for pellet diameters of 1/4, 1/8, 1/16, 1/32, and 1/64 in.

(c) If a first-order irreversible reaction $A \rightarrow$ products occurs on these pellets with a conversion of 90% using 1/16 in. pellets, what conversions would be obtained for each of these pellet diameters, assuming that the reaction is always reaction limited?

7.23 In the spherical pellet reactor of the previous problem assume that the reactants are completely mixed in the void spaces between particles, so that each interpellet space behaves as a CSTR. For a reactor height of 1 in., compare the conversions for each of the above pellet sizes with the 90% conversion in the PFTR approximation. You can find the formula for conversion in a service of CSTRs in Chapter 8.

7.24 Electrolysis can be used to produce very high-purity H_2 and O_2.

(a) If electricity costs $0.02/kWh, what is the cost/ton of producing these gases assuming no ohmic losses?

(b) Look up the current prices of H_2 and O_2 and compare with this price.

(c) Electrolytic cells can draw a maximum of ~ 0.5 amps/cm^2 in steady state without an unacceptable overvoltage. Devise a reactor system to produce 1 ton/h of these separated gases. Make assumptions about cell sizes and spacing. Include heat removal provisions. What is its size?

(d) Under what conditions might electrolytic H_2 and O_2 be economical?

7.25 The amount of oxygen adsorbed in a monolayer on 1 g of a sample of silica gel at low temperature, measured as the volume of oxygen adsorbed, was 105 cm^3 at 300 K and 1 atm. If the cross-sectional area for oxygen is taken as 14 Å^2, determine the surface area per gram for this silica sample.

7.26 H_2 adsorbs on metals as atoms. Show that the Langmuir isotherm for H_2 with dissociation should be

$$\theta = \frac{K_{H_2}^{1/2} P_{H_2}^{1/2}}{1 + K_{H_2}^{1/2} P_{H_2}^{1/2}}$$

7.27 The reaction $H_2 + \frac{1}{2}O_2 \rightarrow H_2O$ has a rate-limiting step $H-S + O-S \rightarrow HO-S$, and the rate of the reaction is $r = k_R \theta_H \theta_O$. Derive a reaction rate expression $r(P_{H_2}, P_{O_2})$ for this reaction assuming Langmuir–Hinshelwood kinetics with competitive adsorption. What are the maximum and minimum orders for this reaction predicted by this expression?

7.28 Consider the reaction engineering of a four-cycle spark ignition automobile. Assume a 1.5-liter, 4-cylinder engine with a compression ratio of 10 driving at 60 mph with a tachometer reading of 2400 rpm and burning octane in 20% excess air.

(a) Estimate v_0, ignoring the vaporized fuel, assuming intake at 20°C.

(b) Estimate the gas mileage and heat generation rate assuming complete combustion if the engine is 20% efficient in producing power from thermal energy.

(c) The actual combustion occurs in a batch reactor. Calculate the residence time of this reactor.

(d) If intake is at 25°C, calculate the temperature at the top of the compression stroke before combustion occurs.

(e) Calculate the maximum temperature assuming that upon spark ignition the reaction goes to completion before expansion in the power stroke occurs and that no heat losses occur.

(f) If the reaction front propagates from the spark plug at the velocity of sound, estimate the reaction time.

(g) Calculate the temperature at the end of the power stroke before the exhaust valve opens, assuming no heat losses in the cylinder.

7.29 A 10-cm-diameter 25-cm-long extruded monolith catalytic converter consisting of 2-mm-square channels is used to remove pollutants from an automobile. Assuming that there is 1% CO in the exhaust and that the surface reaction rate is infinite with excess O_2, calculate the conversion of CO to be expected for the automobile engine described above.

7.30 On platinum H_2 adsorbs dissociatively and CO adsorbs without dissociation.

(a) Write reasonable Langmuir isotherms for these gases assuming competitive adsorption.

(b) What would be a reasonable rate expression for the surface reaction

$$H_2 + CO \rightarrow HCHO$$

on Pt?

7.31 The pre-exponential of a first-order adsorption process is 10^{17} molecules cm^{-2} sec^{-1} torr^{-1} and the desorption pre-exponential is 10^{13} sec^{-1}. The heat of adsorption of CO on Pt is 30 kcal/mole.

(a) Write a reasonable numerical expression for $\theta(P_{CO}, T)$.

(b) At what surface temperature will the CO coverage be 0.1 at a CO pressure of 1 atm?

(c) At what pressure will the CO coverage be 0.5 at 800 K? At what pressure will the CO coverage be 0.01?

(d) [Graduate students only] Show that these numerical values of pre-exponentials are reasonable.

7.32 A unimolecular catalytic reaction is frequently assumed to depend on temperature and partial pressure of reactant P_A according to a Langmuir–Hinshelwood rate expression

$$r_R = \frac{k_R(T)K_A(T)P_A}{1 + K(T)P_A}$$

$$k_R(T) = k_{Ro}\, \exp(-E_R/RT)$$

$$K_A(T) = K_{Ao}\, \exp(+E_A/RT)$$

(a) What must be the relation between the reaction activation energy E_R and the heat of adsorption of the reactant E_A for this expression to predict a maximum in the rate as a function of temperature?

(b) Show that the effective activation energy E^* for this reaction is given by the expression

$$E^* = E_R - E_A(1 - \theta_A)$$

where θ_A the coverage of A is given by a Langmuir isotherm and E^* is defined by the expression

$$E^* = -R \frac{\partial \ln r_R}{\partial (1/T)}$$

(c) At what coverage can E^* be zero?

(d) Show that the temperature of the rate maximum T_m is given by the expression

$$T_m = E_A/R \ \ln \left[(E_A - E_R)/E_R P_A\right]$$

7.33 In catalytic distillation a catalyst is placed on the trays of a distillation column.

(a) What are the principles and possible advantages of catalytic distillation compared to a conventional packed bed catalytic reactor? Consider temperature variations, recycle in a distillation column, and the fact that species have different boiling points. Don't look up anything.

(b) CD is the major industrial process to produce MTBE. The boiling points of methanol, isobutane, and MTBE are 65, -12, and 56°C, respectively, and the reaction is reversible. Describe how this process should give a better yield of MTBE than a conventional packed bed.

7.34 Moving catalyst beds are used in catalytic reactors continuously to remove and replace spent catalyst. As examples, FCC operates with a catalyst residence time in the riser of ~1 sec, while in catalytic reforming the catalyst is recycled with a residence time of ~1 week. The spent catalyst has a lower activity. Should the catalyst flow be cocurrent or countercurrent with the feed?

(a) Consider an exothermic reaction with wall cooling where it is important to minimize and spread out the hot spot to avoid thermal deactivation of the catalyst.

(b) Consider an endothermic reaction with wall heating where the temperature should be high throughout the reactor to maintain a high rate throughout the bed.

(c) Consider the situation where the feed contains catalyst poisons that deactivate the catalyst by adsorbing on it and where the strategy is to maintain activity throughout the bed.

7.35 Considerable research and engineering are being devoted to improving the performance of the automotive catalytic converter to reduce emissions that occur during startup.

(a) One solution to emissions in startup involves adding a small catalytic converter placed nearer the engine than the conventional catalytic converter. Why should this reduce emissions?

(b) Another solution to emissions in startup involves adding a zeolite adsorber section in the exhaust manifold just before the catalytic converter. Why should this reduce emissions?

7.36 The water gas shift reaction

$$CO + H_2O \rightarrow CO_2 + H_2$$

is thought to proceed on a certain catalyst by the following sequence of elementary steps:

$$H_2O + S \rightleftarrows H_2 + O\text{–}S, \qquad r_1 = k_{1f}[H_2O][S] - k_{1b}[H_2][O\text{–}S]$$

$$CO + O\text{–}S \rightleftarrows CO_2 + S, \qquad r_2 = k_{2f}[CO][O\text{–}S] - k_{2b}[CO_2][S]$$

(a) Find an expression for θ_O, the coverage of adsorbed O.

(b) If step 2 is rate limiting but reversible, find an expression for the rate in terms of the gas-phase concentrations and the ks.

(c) Find an expression for θ_O, the coverage of adsorbed O with only the steady-state assumption.

(d) Derive an expression for the surface reaction rate making only the steady-state assumption in the adsorbed species.

7.37 The catalytic dehydrogenation of ethane on supported Pt catalysts proceeds through the reaction steps

(1) $C_2H_{6g} \rightleftarrows C_2H_{6s}$ $r_1'' = k_{a1}P_{C_2H_6}(1 - \Sigma\theta_j) - k_{d1}\theta_{C_2H_6}$

(2) $C_2H_{6s} \rightarrow C_2H_{4s} + 2H_s$ $r_2'' = k_R\theta_{C_2H_6}$

(3) $C_2H_{4s} \rightleftarrows C_2H_{4g}$ $r_3'' = k_{d3}\theta_{C_2H_6} - k_{a3}P_{C_2H_4}(1 - \Sigma\theta_j)$

(4) $2H_s \rightleftarrows H_{2g}$ $r_4'' = k_{d4}\theta_H^2 - k_{a4}P_{H_2}(1 - \Sigma\theta_j)^2$

(a) Formulate a reasonable rate expression assuming that step 2 is irreversible and slow compared to the adsorption–desorption steps.

(b) The heats of adsorption of ethane and ethylene are 5 and 8 kcal/mole, respectively, and the reaction activation energy is 3 kcal/mole What should be the orders of the reaction with respect to these partial pressures, and what is the effective activation energy of the process at high temperatures where all coverages are low? Assume that pre-exponential factors for adsorption equilibrium are identical, and assume that the hydrogen coverage is small.

(c) What should be the orders of reaction and effective activation energy if the temperature is low enough that coverages are high?

(d) Repeat parts (b) and (c) assuming a heat of adsorption of hydrogen of 15 kcal/mole. Assume all pre-exponential factors identical.

7.38 The catalytic decomposition of NO discussed in the text could have a rate-limiting reaction step

$$NO\text{-}S + S \rightarrow N\text{-}S + O\text{-}S, \qquad r'' = k_R'' \,\theta_{NO}\theta_S$$

in which a vacant site S on the surface is required for the dissociation of adsorbed NO. Assuming adsorption-desorption equilibrium of NO, N, and O, what reaction rate expression, $r''(P_{NO},\ P_{N2},\ P_{O_2})$ might be expected? [Answer: Possibly negative order in P_{NO}.]

7.39 The catalytic reaction between NO and CO discussed in the text may have the surface reaction rate steps

$$NO\text{-}S + S \rightarrow N\text{-}S + O\text{-}S$$
$$CO\text{-}S + O\text{-}S \rightarrow CO_2 + 2S$$
$$2N\text{-}S \rightleftarrows N_2 + 2S$$

If the first step is rate-limiting and the following steps are fast, what rate expression $r''(P_{NO}P_{CO}, P_{O_2})$ should be expected? What possible orders of reaction are expected with respect to these species?

7.40 Suppose that the surface reaction $A \rightarrow B$ proceeds by the steps

$$A + S \rightleftarrows A\text{-}S$$
$$A\text{-}S + xS \rightarrow B\text{-}S + xS$$
$$B\text{-}S \rightleftarrows B + S$$

in which x vacant sites are required for the reaction step to occur. What is a reasonable rate expression, $r''(P_A, P_B)$, with this mechanism?

7.41 Suppose that the catalytic decomposition of NH_3

$$NH_3 \rightarrow \tfrac{1}{2}N_2 + \tfrac{3}{2}H_2$$

actually proceeds by the steps

(1) $NH_3 + S \rightleftarrows NH_3\text{-}S$

(2) $NH_3\text{-}S + 3S \rightarrow N\text{-}S + 3H\text{-}S$

(3) $2H\text{-}S \rightleftarrows H_2 + 2S$

(4) $2N\text{-}S \rightleftarrows N_2 + 2S$

(a) Sketch the process assumed by step (2).

(b) Formulate a reasonable expression for $r''(P_{NH_3}, P_{H_2}, P_{N_2})$ for this mechanism. What is the range of reaction orders with respect to P_{NH_3}, P_{H_2}, and P_{N_2}?

(c) Step (2) actually involves the sequential dissociation of H atoms from NH_3,

$$NH_3 \rightarrow NH_2 \rightarrow NH \rightarrow N$$

which involves four steps. If one of these steps is slow and all before and after it are fast, the rate-limiting step can be written as

$$NH_3\text{-}S + xS \rightarrow NH_{3\text{-}x} + xH\text{-}S$$

What rate expressions should one expect for different values of x?

(d) What kinetic experiment could you do to test what is the rate-limiting step in part (c)? If one could measure the coverages of adsorbed NH_x-S species, what would one conclude is the slow step if one found NH-S to have the largest coverage?

7.42 An automobile with a 2.2-liter, 4-cylinder, 4-cycle gasoline engine is traveling at 55 mph. The tachometer reads 2500 rpm. The fuel is octane. The automobile has a catalytic converter, which is a cylinder 20 cm long with 8-cm diameter.

(a) Sketch a flow diagram of the air and fuel as they pass through the engine and catalytic converter.

(b) What models would you use to describe this chemical reactor? What is the reaction time for the combustion process? The flow is pulsed. What is the frequency of the pulsing in the exhaust manifold?

(c) What is the molar ratio of air to fuel if the engine operates at the stoichiometric ratio? What is the volumetric ratio after vaporization and mixing?

(d) Calculate the volumetric flow rate of air into the engine.

(e) Calculate the volumetric flow rate through the exhaust manifold if the temperature is at 800°C.

(f) Calculate the space velocity in the catalytic converter at 800°C.

(g) Calculate the residence time in the catalytic converter if the temperature of the exhaust gases is 450°C.

7.43 Why will a catalyst manufacturing company go broke if it makes perfect catalysts? How might one build a desired lifetime into a catalyst? What are the appropriate strategies for developing a successful catalyst?

7.44 One characteristic of catalysts is that they are intrinsically sensitive to poisons and promoters.

(a) Discuss the mechanisms of operation of "physical" poisons and "chemical" poisons.

(b) How might promoters be added to a catalyst to counteract the effects of poisons?

(c) In enzyme catalysts the typical goal is to create extremely high selectivities for a specific reaction rather than to cause a high rate. Why is this so and why do most enzyme systems use chiral molecules?

7.45 A surface reaction $A \rightarrow B$ proceeds by the reaction steps

$$A + S \rightleftarrows AS$$

$$AS + 2S \rightarrow B + 3S$$

What Langmuir–Hinshelwood rate expression $r''(P_A)$ does this mechanism predict?

7.46 Consider a catalyst pellet with a fast exothermic reaction occurring on the external surface.

(a) Show that the mass balance equation should be

$$C_{Ao} - C_{As} = \frac{k''}{k_m} C_{As}$$

(b) Noting that this looks very much like the CSTR equation $C_{Ao} - C_A = \tau k C_A$, set up the relevant energy balance:

$$T_s - T_o =$$

(c) Sketch C_A and T profiles to be expected around the particle.

(d) Indicate how multiple steady states can occur in exothermic reactions in mass transfer-limited catalyst particle.

PART II

APPLICATIONS

DESIGNING A CHEMICAL REACTOR AND INTRODUCTION TO APPLICATIONS

We have thus far considered all of the principles necessary to design a chemical reactor. In this brief section we summarize these principles and then discuss where we are going in the rest of the book.

Design Principles

Chemistry The first topic to examine is the chemical reactions one wants to run and all the reactions that can occur. One immediately looks up the ΔH_{Ri} and ΔG_{Ri} to determine the heat release or absorption and the equilibrium composition. Equilibrium considerations also govern the temperature and pressure necessary for an acceptable equilibrium yield. This was the subject of Chapter 2.

Flow pattern Next one decides whether a batch or continuous reactor is suitable and, if flow, whether a mixed of unmixed reactor is preferred. Initially one may do calculations for PFTR and CSTR to bracket all flow patterns. This is the subject of Chapters 3 and 4. The choice of catalyst and heat removal method will be very important in deciding the best flow pattern.

Heat management The method of heat removal or addition must be considered early in reactor design. The heat load is easily calculated from the estimated selectivity and production rates. In designing for heat management the worst-case scenario of maximum heat release must be considered. Jackets or cooling coils will be important in preliminary design, and attainable heat transfer rates will determine the needed heat exchanger area and configuration. This was the subject of Chapters 5 and 6.

Reaction rates   Next we need kinetics. These are usually of the Arrhenius form

$$r_i = k_i C_A^{m_{iA}} C_B^{m_{iB}} \cdots$$

where ms are orders with respect to reactant and product concentrations. One must of course consider reversibility (especially if the equilibrium constant is not very large), and reaction rates may not obey the simple power-law expressions. Thus far, we have implicitly assumed that accurate rate expressions were available, a situation seldom encountered in industrial practice.

Catalysts   The major problem with obtaining rate expressions is that most interesting processes employ catalysts to attain high rates and selectivities, and catalytic kinetics depend sensitively on the details of the catalyst chemistry. Aspects such as promoters, poisons, activation, and deactivation play crucial roles in determining catalyst performance. With catalytic processes we expect complex rate expressions and fractional orders of reaction. This was the subject of Chapter 7.

Separations   Separation processes and staged processes must be considered along with the performance of the individual reactor. The performance of the *integrated system* must be considered in all stages of design. Very important in the overall process are the yield of the desired product and the separation and disposal of byproducts, and separation capabilities can dominate these issues.

You have learned or will learn about the separation components of a chemical process in mass transfer and separations courses. The energy requirements of separation processes, the purities of different streams from separation equipment, and possible integration of heat flows between units are frequently important in design.

Feedstocks and products   Next (or perhaps first) we need to consider the markets of products and the availability of feedstocks. The prices of these depend sensitively on purity levels that can be tolerated. For the reactants these are usually determined by the effects of impurities as catalyst poisons and on product distributions. For products different markets demand specific impurities. All byproducts and unused reactants must be disposed of, either sold, recycled into the reactor, or incinerated.

Economics   The costs of equipment, operations, and feedstocks obviously are crucial in any process, and these must be considered in all stages of design, from preliminary to final.

Transients   All processes must be started and shut down, and the transients in these periods must be considered. Feedstocks can vary in quality and availability, and demand can vary in quantity and in required purities; so the reactor system must be "tunable" to meet these variations. These aspects are frequently considered after the steady-state operation has been decided.

Safety   Safety must be considered in all stages of process design. Potential hazards in equipment components and in operations must be considered carefully. The toxicity of reactants, products, and intermediates must be thought through before anything is constructed. All possible temperature and pressure runaway events must be considered, and responses thought through to minimize the chances of unforseen events. These events include (in order of increasing undesirability): performance outside specifications, unscheduled shutdowns, chemical releases, fires, and explosions.

Auxiliary equipment   The reactor and associated separation equipment are frequently a small part of the cost of designing and operating a process. Pumps and storage vessels can be very expensive, and overhead consumes an incredibly large portion of any budget. Offsite costs can be large. The salary of the engineers (you) who design the process must be paid.

STAGES OF DESIGN

When considering the design of a chemical process, one first does *preliminary process design*, and then, if a processes looks promising, one begins *detailed process design*. We in fact teach these subjects in the reverse order, and you will only really begin to consider preliminary process design in your process design projects course. In early courses (with the words "Principles of . . ." in their title) you worked problems related to the detailed design of chemical processes: problems like the pressure drop in a pipe or the vapor pressure of a mixture or the mass flux across a membrane.

Detailed design is in fact much easier than preliminary design because in detailed design you are given all variables in a process except the one you are asked to compute. In preliminary design you first have to *guess* all of the units, their configurations, and sizes of equipment. Then you examine each one of these and try to determine if they are feasible and, if so, what combination will be the most profitable.

In this book we have begun to introduce you to the preliminary design of a chemical reactor by starting with the simple material (a single reaction in an ideal isothermal reactor with no catalyst). Then we have begun to add complexity with multiple reactions, variable temperature, and catalysts. This material has taken us through Chapter 7. In the following chapters we will introduce the additional complexities needed in obtaining realistic design of real reactors.

Detailed design takes more time than preliminary because it requires many stages of calculations, but the difficult part is in laying out a feasible and clever preliminary design from which to do the detailed calculations. This part also requires the most *imagination* and *ingenuity* of the engineer, and only with experience in chemical processes can you begin design with confidence and skill. We cannot teach you these skills in courses.

The additional complexity of real chemical processes comes from the fact that they are invariably *multistage processes* with many reactors and separation units linked with complex configurations. One must then design each unit individually and then see how its design must be modified to be integrated into the entire process.

It is also interesting to note that the texts in chemical process design describe mostly separations and economics, with little or no coverage of the chemical reactors. This is because the chemical reactor issues (selectivity, yield, energy management, catalysts, multiple phases) are so complex that the authors have decided to omit them. Yet the chemical reactors dominate any chemical process design situation. For chemical reactors the designer is required to guess configurations and performance or to assume a process similar to existing ones described in the literature. In fact, in any new process the reactor and the reactions are frequently designed only by constructing and operating a pilot plant to test and refine these assumptions.

Optimization

With these aspects in mind, the design begins. One must play out many scenarios and determine their relative economics and acceptabilities. This involves an almost endless series of iterations and reiterations to try to find one or a small number of possible designs for more detailed design considerations. This requires all the skills of the engineer, and considerable experience and ingenuity are essential in deciding among alternatives.

The basic idea is to examine operating parameters to find the optimum combination of them for optimum performance. A short list of the most important might include the following: F_j, C_{jo}, C_j, v, V, T, T_o, u, P, and, of course $. For catalytic processes additional variables include D, d_p, S_g, ε, shape, and catalyst chemical properties such as chemical composition, activity, and selectivity. Most catalytic reactors operate with significant mass transfer limitations because one usually wants to raise the temperature until mass transfer becomes noticeable in order to attain the highest rate possible. In all cases one determines the effects of these variables on reactor performance.

Many aspects of a process can be evaluated with a chemical process flowsheet program such as Aspen. These programs handle mass balances and heat loads on each component with great accuracy. Separation components can also be handled accurately as long as they are rather straightforward. Cost databases also exist on these programs that allow rapid costing of many components.

However, the chemical reactor still requires considerable thinking and hand calculations because there are few situations where databases exist on reaction kinetics. Especially for catalytic reactions, data either do not exist or are unreliable far from the conditions under which the data were obtained. The number of possibilities of catalyst chemical and geometrical properties are so large that no flowsheet programs can handle anything beyond extremely simple situations.

What's next?

Chapters 1 through 7 have dealt with what we regard as the "fundamental" aspects of chemical reactors. Without consideration of all of this material, you simply cannot design any chemical process. (Or at least you would be dangerous if you were allowed to.)

However, the processes with which you will be called on to work are invariably very complex. We have mentioned some of these in the "Industrial Processes" we have interspersed throughout the previous chapters.

In the last half of this text we will consider some of the additional aspects that are encountered in many actual chemical processes. These include

Nonideal reactors (Chapter 8),

Solids processing (Chapter 9),

Combustion processes (Chapter 10),

Polymerization (Chapter 11),

Biological reactions (Chapter 12),

Environmental reactions (Chapter 13), and

Multiphase reactors (Chapter 14).

Our organization and intent in this part of the text is quite different from that in the previous sections. Previously we have tried to be systematic in developing the ideas so that they flow logically as increasing degrees of complexity are introduced. In the following sections we try to show how the principles developed in the previous sections can be applied to these types of processes.

These chapters should be considered the *optional* part of this text. This part is written so that your instructor can choose particular sections for consideration as she or he has time and expertise, and sees interests of the students in these applications. These chapters will also be very brief and superficial. For one reason, we promised earlier that we would use only a single notation throughout the book, and we will try to introduce very few new quantities in the rest of the text. In many applications (bioprocesses and foods are the most notorious) quite different nomenclatures have developed than those that have become common in the chemical and petroleum process industries. We can therefore only delve into these applications as far as our limited notation will allow.

Our assumption is that your instructor will choose specific chapters and sections of chapters and will omit or skim large sections. If she or he is an expert in some of these topics, you may see much additional or substituted material and notation that extends and corrects the naive or wrong material in these sections. The material and notation in these chapters are sufficiently simple that you should be able to work many of homework problems without having previous lectures on that chapter. Half of the challenge in chemical reaction engineering is recognizing that the previous concepts can be directly applied to these problems.

We close by apologizing that you may find the following chapters described by some combinations of superficial, incomplete, random, and wrong. However, we need make no apology for this in an introductory course if you find the material challenging and fun. There are many complete texts and monographs on each of these subjects, and any student wanting to know more about them should locate these books or take relevant courses.

However, we urge all students at least to skim all the following chapters to get a flavor of some of the subjects to which *The Engineering of Chemical Reactions* can be applied. The subject is both extremely challenging and a great deal of fun.

Chapter **8**

NONIDEAL CHEMICAL REACTORS

Thus far we have considered only two flow patterns: the completely mixed reactor and the completely unmixed reactor. This is because only for these flow patterns can we completely ignore the fluid flow configurations in the reactor. In this chapter we will begin to see how reactors that have more complex flow patterns should be treated. We will not attempt to describe the fluid mechanics completely. Rather, we will hint at how one would go about solving more realistic chemical reactor problems and examine the errors we have been making by using the completely mixed and unmixed approximations.

We will also see that residence time distribution notions lead to some interesting examples of pollution reactions, biological populations, and environmental modeling.

This chapter begins the "optional" portion of this book. The previous material was "essential" to describe chemical reactors. With this chapter we begin to cover some topics that illustrate some of the applications of chemical reaction engineering. Some of this material your instructor may want to skip or skim. However, we suggest that you should still look over this material because it contains some interesting applications that you may need to be aware of for the jobs you will be required to do later in your career.

Finally, we remark that this material varies between boring and fun. In this chapter we begin by examining the basic equations and residence time distributions (pretty boring), and in later chapters we discuss flames, explosions, and environmental and ecological modeling (topics like chemical spills, global warming, and fish populations, which tend to get one's attention). We cannot make significant contributions to important applications unless we know how to describe complex chemical systems quantitatively.

8.1 THE "COMPLETE" EQUATIONS

Any fluid flow situation is described completely by momentum, mass, and energy balances. We have thus far looked at only simplified forms of the relevant balance equations for our simple models, as is done implicitly in all engineering courses. It is interesting to go back to the basic equations and see how these simple approximations arise. We need to examine the full equations to determine the errors we are making in describing real reactors with these models and to see how we would have to proceed if we wanted to solve a chemical reactor problem "correctly."

8.1.1   Momentum balance

The momentum balance is a version of Newton's Second Law of Mechanics, which students first encounter in introductory physics as $F = ma$, with F the force, m the mass, and a the acceleration. For an element of fluid the force becomes the stress tensor acting on the fluid, and the resulting equations are called the Navier–Stokes equations,

$$\rho \frac{\partial \mathbf{u}}{\partial t} + \rho \, \mathbf{u} \cdot \nabla \mathbf{u} = -\nabla P + \nabla \cdot (\mu \nabla \mathbf{u}) + \rho \mathbf{g} \tag{8.1}$$

where ρ is the mass density, P the total pressure, μ the viscosity, and $\mathbf{g}$ the acceleration of gravity. Each term in bold is a vector with x, y, and z coordinates; so this equation must be rewritten as three scalar equations. As an example, the velocity $\mathbf{u}$ has three components, and in our PFTR species balance equation we have been concerned only with u_z, the velocity component in the direction of flow down the tube.

We also have to solve this equation if we need to calculate the pressure drop in a reactor.

8.1.2   Total mass balance

We next have to consider the continuity equation, which students first encounter seriously in introductory chemistry and physics as the principle of mass conservation. For any fluid we require that the total mass flow into some element of volume minus the flow out is equal to the accumulation of mass, and we either write these as integral balances (stoichiometry) or as differential balances on a differential element of volume,

$$\frac{\partial \rho}{\partial t} + \nabla \cdot (\rho \mathbf{u}) = 0 \tag{8.2}$$

When we assume a steady-state constant-density reactor, we obtain the simple form of this equation ($\nabla \cdot \rho \mathbf{u} = 0$ or ρu constant), which is just total mass conservation as required from stoichiometry.

These equations apply to the total mass or mass density of the system, while we use moles when describing chemical reaction. Therefore, whenever we need to solve these equations simultaneously, we must transform our species mass balances into weight fraction when including momentum and total mass-balance equations.

8.1.3   Species mass balances

For a multicomponent fluid (the only situation of interest with chemical reactions) we next have to solve mass balances for the individual chemical species. This has been implicitly the subject of this book until now. The species balance is written as flow in minus flow out plus change due to chemical reaction equals the accumulation of that species. We have developed the integral version of these equations (the CSTR) and the differential version (PFTR) in previous chapters because our interest centered on the effects due to the changes in species caused by chemical reaction.

There is of course an equation for each species, subject to conservation of total mass given by the continuity equation, and we implicitly used conservation of total mass as "stoichiometric constraints," $(N_{jo} - N_j)/\nu_j$ are equal for all j in a batch (closed) system or $(F_{jo} - F_j)/\nu_j$ are equal for all j in a steady-state continuous reactor.

The "complete" (but still containing many approximations) species balance equations are

$$\frac{\partial C_j}{\partial t} + \mathbf{u} \cdot \nabla C_j = D_j \, \nabla^2 C_j + \sum_{i=1}^{R} \nu_{ij} r_i \qquad (8.3)$$

In the derivation of this equation we assume that the density is constant (we use molar density N_j/V rather than mass density ρ_j), and that the diffusivity D_j is independent of composition, and we simplify the convection term somewhat. The gradient (∇) and Laplacian (∇^2) operators should be familiar to students from fluid mechanics and heat and mass transfer courses.

8.1.4 Energy balance

Finally we have the energy balance, which was encountered in physics as Newton's First Law of Motion and in thermodynamics as the First Law of Thermodynamics. We divide energy into potential (gravity), kinetic ($\frac{1}{2}\rho m u^2$), and thermal (temperature) energies. In most systems of interest here thermal energy considerations are most important, leading to temperature balances. We have to consider kinetic energy for very fast fluid motion, and the equations become especially complicated near sonic velocities. Energy dissipation from flow in pipes and the energy input of stirring a reactor can be added to the energy balance, although the continuity and momentum equations must be solved simultaneously to describe these situations. We have thus far ignored these additional terms in Chapters 5 and 6 when we considered the energy balance in nonisothermal reactors.

The energy-balance equation (more properly an enthalpy balance) is

$$\rho C_p \left(\frac{\partial T}{\partial t} + \mathbf{u} \cdot \nabla T \right) = k_T \, \nabla^2 T + \sum_{i=1}^{R} (\Delta H_{Ri}) r_i + \tfrac{1}{2}\rho u^2 + \rho \mathbf{g} \cdot \mathbf{h} \qquad (8.4)$$

where again several implicit simplifications have been made in its derivation such as that the thermal conductivity of the fluid k_T is independent of T. The Q' term in the one-dimensional energy-balance equation of the PFTR must be obtained by solving for the temperature profile in the radial direction to obtain our one-dimensional shell balance for the tubular reactor with uniform properties across the cross section.

Thus to describe a chemical reactor (or almost any process unit in chemical engineering) we have to solve these partial differential equations with appropriate boundary and initial conditions in order to describe the composition, temperature, and velocity within the reactor versus position and time, $C_j(x, y, z, t)$, $T(x, y, z, t)$, $\rho(x, y, z, t)$, and $\mathbf{u}(x, y, z, t)$.

8.1.5   Economic balances

Finally, since engineers must be concerned with making money in a chemical process, the economic balance equations must also be solved along with the above equations. Managers frequently believe that the preceding equations are not important at all, and that one should only worry about profit and loss. The problem with this idea is that these equations underlie any economic considerations.

There should be some sort of a money continuity equation for a chemical process, which might look like

$$\frac{d\$}{dt} = \$_o - \$ + \text{ generation } - \text{ losses} \tag{8.5}$$

where $\$_o$ and $\$$ represent money flow into and out of the reactor or process, which can be written as $\$ = \sum \$_j$ with $\$_j$ the price of component j. The money generation and loss terms are the subjects of process design courses. They involve topics such as labor (your salary), overhead, capital costs, depreciation, and taxes. Also, in the long term no one wants to operate a chemical plant in steady state where economics is involved, because growth and improved profitability are the major goals of most companies.

One fundamental problem with economic balances is that there is no underlying conservation law for any economic system larger than your check book. Equations such as the preceding are written as integral balances or difference equations. In macroeconomics no one seems to understand what the governing equations are (or even if they exist), and in all economic systems the definitions and accounting practices vary widely, with the goal being to make more money than the competition and stay out of jail.

8.2   REACTOR MASS AND ENERGY BALANCES

8.2.1   The PFTR

We can easily recover the mass- and energy-balance equations used in previous chapters from these equations. First we assume variation in only one spatial dimension, the direction of flow z, to obtain

$$\frac{\partial C_j}{\partial t} + u \frac{\partial C_j}{\partial z} = D_j \frac{\partial^2 C_j}{\partial z^2} + \sum_{i=1}^{R} v_{ij} r_i \tag{8.6}$$

$$\rho C_p \left(\frac{\partial T}{\partial t} + u \frac{\partial T}{\partial z} \right) = k_T \frac{\partial^2 T}{\partial z^2} + \sum_{i=1}^{R} \Delta H_{Ri} r_i - Q' \tag{8.7}$$

where now u is the velocity in the direction of flow (the z direction), and all gradients are assumed to be in the axial, not radial, direction.

The heat removal term Q' has to be inserted in the one-dimensional energy balance by implicitly solving the energy balance in the radial direction. The heat flux in the radial direction is $q = -k_T \, \partial T / \partial R$, and we replace the radial flux to the wall by assuming that the fluid at any position z is at a constant temperature T while the wall is at temperature T_w to obtain

$$Q' = p_w U (T - T_w) \tag{8.8}$$

with p_w the wall perimeter in a differential element of volume. We developed this term in the one-dimensional energy balance in Chapter 5 from a one-dimensional energy balance between z and $z + dz$.

Now if we assume no diffusion, $D_j = 0$, these equations become

$$\frac{\partial C_j}{\partial t} + u \frac{\partial C_j}{\partial z} = \sum_{i=1}^{R} v_{ij} r_i \tag{8.9}$$

$$\rho C_p \left(\frac{\partial T}{\partial t} + u \frac{\partial T}{\partial z} \right) = \sum_{i=1}^{R} \Delta H_{Ri} r_i - Q' \tag{8.10}$$

which are the transient mass- and energy-balance equations for the PFTR. In steady state (no time dependence) these are

$$u \frac{dC_j}{dz} = \sum_{i=1}^{R} v_{ij} r_i \tag{8.11}$$

$$u \frac{dT}{dz} = \sum_{i=1}^{R} \frac{\Delta H_{Ri}}{\rho C_p} r_i - \frac{Q'}{\rho C_p} \tag{8.12}$$

where we replace the partial derivatives by total derivatives because the only gradient is in the z direction. We recognize these equations as our steady-state PFTR mass- and energy-balance equations, which we have derived many times in previous chapters.

8.2.2 The batch reactor

Note also that these equations can be simplified to obtain the batch reactor mass- and energy-balance equations by setting $D_j = 0$ and $u = 0$ to give

$$\frac{dC_j}{dt} = \sum_{i=1}^{R} v_{ij} r_i \tag{8.13}$$

$$\rho C_p \frac{dT}{dt} = \sum_{i=1}^{R} \Delta H_{Ri} r_i - Q' \tag{8.14}$$

where we now have only the total derivative with respect to time because we assume no variations in concentrations or temperature with position.

8.2.3   CSTR

These "general" equations cannot be transformed in a straightforward way into the CSTR mass- or energy-balance equations. We obtained the CSTR equations from integral balances, assuming that stirring was sufficiently rapid that there were no gradients in composition or temperature. This can be thought of as the limit of $u \to \infty$ in the preceding equations so that the terms $\mathbf{u} \cdot \nabla C_j$ and $\mathbf{u} \cdot \nabla T$ become indeterminate to give $\nabla C_j = 0$ and $\nabla T = 0$ for spatially uniform reactors.

In the previous chapters we have been using simplified forms of these general equations to obtain the equations we have derived and solved previously. We now examine in a bit more detail the solutions to them in more realistic situations.

8.3   RESIDENCE TIME DISTRIBUTION

We will not attempt to solve the preceding equations except in a few simple cases. Instead, we consider nonideal reactors using several simple models that have analytical solutions. For this it is convenient to consider the residence time distribution (RTD), or the probability of a molecule residing in the reactor for a time t.

For any quantity that is a function of time we can describe its properties in terms of its distribution function and the moments of this function. We first define the probability distribution function $p(t)$ as the probability that a molecule entering the reactor will reside there for a time t. This function must be normalized

$$\int_0^{\infty} p(t)\, dt = 1 \tag{8.15}$$

because the molecule must leave the reactor sometime. The fraction of molecules that have left the reactor after time t is

$$\int_0^{t} p(t)\, dt \tag{8.16}$$

The average value of the time $\bar{t}$ spent in the reactor is our definition of the residence time τ, which is given by the expression

$$\bar{t} = \int_0^{\infty} t p(t)\, dt \tag{8.17}$$

The average value of any function $f(t)$ is given by the integral

$$\bar{f} = \int_0^{\infty} f(t) p(t)\, dt \tag{8.18}$$

The standard deviation σ of a function $f(t)$ is given by the expression

$$\sigma^2 = \int_0^{\infty} [\bar{f} - f(t)]^2 p(t)\, dt \tag{8.19}$$

All these statistical quantities are introduced in courses in statistics and analysis of data, but we will only need them to describe the effects of RTD, and will just use them in these forms without further explanation.

8.3.1   Response to pulse or step tracer injection

The RDT or $p(t)$ is the probability of a molecule residing in the reactor for a time t. This can be found from a tracer experiment in which we inject a tracer of a nonreacting species (perhaps ink) with concentration $C_o(t)$ into a reactor with fluid flowing at a steady flow rate, and we measure its concentration $C(t)$ (perhaps from the absorbance of the ink) as it flows out of the reactor. The flow could be pure solvent, or a reaction could be taking place as long as the tracer does not participate in the reaction. The concentration of the tracer leaving the reactor versus time gives $p(t)$ with suitable normalization.

In this thought experiment we inject a perfectly sharp pulse $\delta(0)$ whose total amount is unity. This pulse is a Dirac delta function defined as a rectangle centered about time zero with a height $1/\varepsilon$ and a width ε in the limit $\varepsilon \to \infty$ so that the area of the function remains unity as the height goes to infinity.

Another common tracer injection is a step that is zero for $t < 0$ and a constant C_o for $t > 0$. This is an easier tracer to inject,

Graphically the pulse and step tracer injections and their responses in a reactor are indicated in Figure 8–1.

8.3.2   RTD in a CSTR

For a CSTR we solve the transient CSTR mass-balance equation because we want the time dependence of a pulse injection without reaction. The transient CSTR equation on a species of concentration C is

$$\tau \frac{dC}{dt} = C_o(t) - C \tag{8.20}$$

Note that this is simply the transient CSTR equation with the reaction term omitted, which implies either no reaction ($r = 0$) or that the species in question is an inert in a reacting system so that its stoichiometric coefficient is zero.

Since the tank was previously flowing pure solvent at $t < 0$, the initial conditions on C within the CSTR in this equation are $C = 0$ at $t = 0$. The feed is our δ function, which

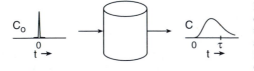

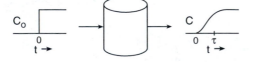

Figure 8–1   Sketch of response to a pulse injection $\delta(t)$ of a tracer (upper) and response to a step injection of a tracer (lower) in a steady-state chemical reactor with an arbitrary flow pattern. The response to a pulse injection is the residence time distribution $p(t)$, and the derivative of the step response is $p(t)$.

is C_o in the above equation, $C_o = \delta(0)$. The concentration in the feed is zero for all $t > 0$; so the transient equation becomes

$$\tau \frac{dC}{dt} = -C, \qquad C = C_i, t = 0 \tag{8.21}$$

or, upon integration,

$$\int_{C_i}^{C} \frac{dC}{C} = -\frac{t}{\tau} = \ln \frac{C}{C_i} \tag{8.22}$$

where C_i is the concentration of the tracer inside the tank at time 0 but just after the pulse has been injected (and mixed instantly because we assume a perfect CSTR). The concentration in the tank following injection of a pulse is therefore

$$C(t) = C_i e^{-t/\tau} \tag{8.23}$$

We now must normalize $C(t)$ to obtain $p(t)$, and we do this by the relation

$$\int_0^\infty C(t) \, dt = 1 = C_i \int_0^\infty e^{-t/\tau} \, dt = C_i \tau \tag{8.24}$$

so that $C_i = 1/\tau$. This normalized concentration gives

$$\boxed{p(t) = \frac{1}{\tau} e^{-t/\tau}} \tag{8.25}$$

which is the residence time distribution in a CSTR. The fraction of the fluid that has left the reactor by time t is $\int_0^t p(t) \, dt$, and the fraction remaining in the reactor after time t is $\int_t^\infty p(t) \, dt$. These integrals are shown in Figure 8–2.

Next we compute the average value of t, which is the standard deviation of $p(t)$ in the CSTR. From the preceding formulas we obtain

$$\bar{t} = \int_0^\infty t p(t) \, dt = \int_0^\infty t \frac{1}{\tau} e^{-t/\tau} \, dt = \tau \tag{8.26}$$

This shows that the average residence time $\bar{t}$ in a CSTR is the residence time τ, as we have assumed all along.

Figure 8–2   Fractions of fluid that have left a reactor and remain in the reactor in terms of $p(t)$.

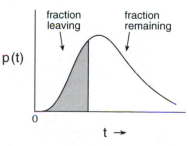

The standard deviation σ is given by the formula

$$\sigma^2 = \int_0^\infty (t - \tau)^2 \frac{1}{\tau} e^{-t/\tau}\, dt = \tau^2 \qquad (8.27)$$

which shows that σ is also equal to τ. This shows that the standard deviation of the residence time distribution in a CSTR is equal to the residence time. No distribution could be broader than this.

8.3.3 RTD in a PFTR

Next consider the response of a PFTR with steady flow to a pulse injected at $t = 0$. We could obtain this by solving the transient PFTR equation written earlier in this chapter, but we can see the solution simply by following the pulse down the reactor. (This is identical to the transformation we made in transforming the batch reactor equations to the PFTR equations.) The $\delta(0)$ pulse moves without broadening because we assumed perfect plug flow; so at position z the pulse passes at time z/u and the pulse exits the reactor at time $\tau = L/u$. Thus for a perfect PFTR the RTD is given by

$$\boxed{p(t) = \delta(\tau)} \qquad (8.28)$$

which shows that $p(t) = 0$ except at $t = \tau$. The standard deviation of $p(t)$ for a PFTR from the above definition is

$$\sigma = 0 \qquad (8.29)$$

since the pulse remains perfectly sharp.

These RTDs for the perfect PFTR and CSTR are shown in Figure 8–3.

8.3.4 Calculation of conversion from the RTD

From our average value formula we can determine the average conversion in a reactor if we know how it varies with time because we can let $f(t) = X(t)$ to obtain

$$\overline{X} = \int_0^\infty X(t) p(t)\, dt \qquad (8.30)$$

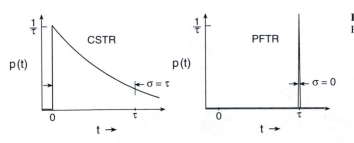

Figure 8–3 Residence time distributions $p(t)$ in an ideal PFTR and CSTR.

For a first-order irreversible reaction in an isothermal batch reactor $X(t) = 1 - e^{-kt}$ (Chapter 2); so the average value of X is

$$\overline{X} = \int_0^\infty (1 - e^{-kt}) p(t)\, dt \qquad (8.31)$$

For the PFTR $p(t)$ is zero for all times except $t = \tau$; so this integral yields $\overline{X} = 1 - e^{-k\tau}$.

8.3.5 Segregated flow in CSTR

For the CSTR we can find $\overline{X}$ by insertion of $p(t)$ for the CSTR into the average value formula to obtain the expression

$$\overline{X} = \int_0^\infty (1 - e^{-kt}) \frac{1}{\tau} e^{-t/\tau}\, dt = \frac{k\tau}{1 + k\tau} \qquad (8.32)$$

These are exactly the expressions for conversions for a first-order irreversible reaction in PFTR and CSTR reactors that we derived much earlier (and with much less mathematical manipulation).

However, the solution for the CSTR obtained by the RTD equation is correct only for first-order kinetics. For other rate expressions the conversion predicted by the RTD is incorrect for a mixed reactor because molecules do not simply react for time t, after which they leave the reactor. Rather, the fluid is continuously mixed so that the "history" of the fluid is not describable in these terms. This expression for conversion in the CSTR is applicable for segregated flow, in which drops of fluid enter the reactor, swirl in the reactor, and exit after time t because then each drop behaves as a batch reactor with the RTD describing the probability distribution of the drops in the CSTR.

This situation describes an *emulsion reactor* in which reacting drops (such as oil drops in water or water drops in oil) flow through the CSTR with stirring to make the residence time of each drop obey the CSTR equation. A *spray tower* (liquid drops in vapor) or *bubble column* or *sparger* (vapor bubbles in a continuous liquid phase) are also segregated-flow situations, but these are not always mixed. We will consider these and other multiphase reactors in Chapter 14.

The composition in each drop now changes with the time that the drop has been in the CSTR in the same way as if the drop were in a batch reactor. However, since the drops are in a CSTR, the time each drop stays in the reactor is given by the previous expression for $p(t)$, and the overall conversion for any reaction is given by solving the preceding equation.

8.4 LAMINAR FLOW TUBULAR REACTORS

Consider the injection of a pulse into an arbitrary reactor with a partially mixed flow pattern. Now $p(t)$ will be somewhere between the limits of unmixed and mixed, and $0 < \sigma < \tau$. We will now find the RDT for several models of tubular reactors. We noted previously that the perfect PFTR cannot in fact exist because, if flow in a tube is sufficiently fast for turbulence ($\mathrm{Re}_D > 2100$), then turbulent eddies cause considerable axial dispersion, while if flow is slow enough for laminar flow, then the parabolic flow profile causes considerable

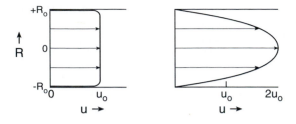

Figure 8–4   Velocity profiles in a tube in plug-flow and in laminar flow reactors.

deviation from plug flow. We stated previously that we would ignore this contradiction, but now we will see how these effects alter the conversion from the plug-flow approximation.

If flow is laminar with no radial or axial mixing (diffusion coefficient zero), we can write $p(t)$ by finding p at each annulus that flows along the streamline in plug flow, as shown in Figure 8–4. From fluid mechanics we know that the velocity profile $u(R)$ [really $u_z(R)$] in laminar flow is given by the expression

$$u(R) = 2u_o \left[1 - \left(\frac{R}{R_o} \right)^2 \right] \tag{8.33}$$

Therefore the residence time at radius R is

$$\tau(R) = \frac{L}{u(R)} \tag{8.34}$$

The amount of fluid flowing between R and $R + dR$ is $2\pi R \, dR$, and $p(t)$ is found by integrating from $R = 0$ at the center of the tube to $R = R_o$ at the pipe wall. We will leave the derivation of this equation as a homework problem. This gives

$$p(t) = \frac{\tau^2}{2t^3}, \qquad t > \tau/2 \tag{8.35}$$

$$p(t) = 0, \qquad t < \tau/2 \tag{8.36}$$

which is plotted in Figure 8–5. Note that no fluid leaves the LFTR until time $\tau/2$, which is the residence time for fluid flowing along the axis, $R = 0$, where the velocity is $2u_o$ or twice the average velocity. Fluid molecules near the wall have an infinite residence time in a LFTR, and thus $p(t)$ has a significant "tail."

We can use this expression to obtain the conversion in an LFTR and compare it with the PFTR. For first-order kinetics again $X(t) - (1 - e^{-kt})$ so the average X is given by

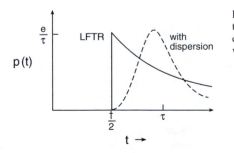

Figure 8–5   Residence time distribution in a laminar flow tubular reactor. The dashed curve indicates the $p(t)$ curve expected in laminar flow after allowing for radial diffusion, which makes $p(t)$ closer to the plug flow.

$$\overline{X}_{\mathrm{LF}} = \int_{\tau/2}^{\infty} (1 - e^{-kt}) \frac{\tau^2}{2t^3} \, dt \tag{8.37}$$

which can be integrated only numerically. For second-order kinetics the conversion in a batch reactor is

$$X(t) = \frac{C_{Ao}}{1 + C_{Ao}kt} \tag{8.38}$$

Inserting this conversion into the residence distribution formulas

$$\overline{X}_{\mathrm{LF}} = \int_{0}^{\infty} X(t) p(t) \, dt = \int_{\tau/2}^{\infty} \frac{C_{Ao}}{1 + C_{Ao}kt} \frac{\tau^2}{2t^3} \, dt \tag{8.39}$$

which can be integrated analytically as the exponential integral, which is a tabulated function.

The difference between the LFTR and the PFTR can be determined by examining the ratio of X obtained, which for first-order kinetics is

$$\frac{X_{\mathrm{LFTR}}}{X_{\mathrm{PFTR}}} = \frac{\int_{\tau/2}^{\infty} (1 - e^{-kt})(\tau^2/2t^3) \, dt}{1 - e^{-k\tau}} \tag{8.40}$$

This ratio must be < 1 for any positive-order kinetics because plug flow must give better conversion than any other reactor with partial backmixing. This ratio is plotted versus conversion in Figure 8–6. When one solves these equations numerically, one finds that this ratio is nearly unity. In fact the *smallest* this ratio becomes is 0.88. This shows that the loss in conversion with laminar flow compared to perfect plug flow is only 12% at most. This is very reassuring, because it says that the simple PFTR equation we have been solving (and will continue to solve) does not lead to large errors, even though plug flow is a flow profile that cannot ever be realized exactly.

In fact, the plug-flow approximation is even better than this calculation indicates because of *radial mixing*, which will occur in a laminar flow reactor. A fluid molecule near the wall will flow with nearly zero velocity and have an infinite residence time, while a molecule near the center will flow with velocity $2\overline{u}$. However, the molecule near the wall

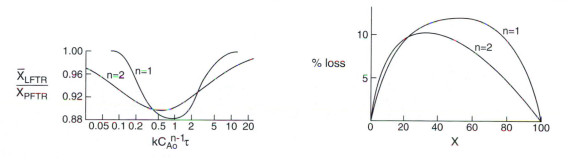

Figure 8–6   Plots of the ratio of conversions in a tubular reactor with laminar flow to that in a perfect PFTR for first- and second-order kinetics. The right panel shows the percent loss in conversion from laminar flow compared to plug flow.

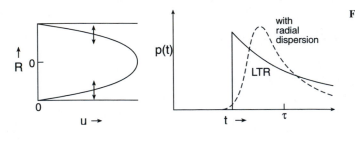

Figure 8–7   Effect of radial mixing on flow profile and $p(t)$.

will diffuse toward the center of the tube, and the molecule near the center will diffuse toward the wall, as shown in Figure 8–7. Thus the "tail" on the RTD will be smaller, and the "spike" at $\tau/2$ will be broadened. We will consider diffusion effects in the axial direction in the next section.

8.5 DISPERSION IN A TUBULAR REACTOR

We have thus far considered the tubular reactor only for perfect plug flow and for laminar flow. We have completely ignored diffusion of the fluid. Diffusion will cause the fluid to disperse axially. It will also sharpen the laminar-flow profile by radial diffusion, as discussed in the previous section. In this section we consider the effect of axial diffusion on the residence time distribution and thus its effect on conversion in a tubular reactor. As shown in Figure 8–8, $p(t)$ is broadened by dispersion as the fluid travels down the tube.

A shell balance on species j between z and $z + dz$ yields the expression

$$\frac{\partial C_j}{\partial t} + u\,\frac{\partial C_j}{\partial z} = D_j\,\frac{\partial^2 C_j}{\partial z^2} + \sum_{i=1}^{R} v_{ij}r_i \tag{8.41}$$

where the terms are now familiar. The term $D_j\,\partial^2 C_j/\partial z^2$ describes *axial dispersion* in the tubular reactor. In steady state with a single first-order irreversible reaction $r = kC_A$ we write this equation as

$$D_A\,\frac{d^2 C_A}{dz^2} - u\,\frac{dC_A}{dz} - kC_A = 0 \tag{8.42}$$

(Again note that we recover the PFTR equation if $D_A = 0$.)

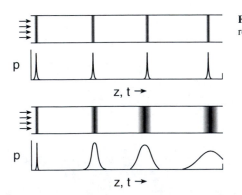

Figure 8–8   Dispersion broadening of $p(t)$ in a tubular reactor.

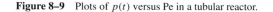

Figure 8–9 Plots of $p(t)$ versus Pe in a tubular reactor.

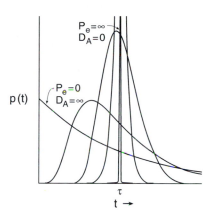

It will be convenient to make this equation dimensionless. We first replace z by a dimensionless length $x = z/L$ and $dx = dz/L$, with L the length of the tube. We next divide each term by D_A to obtain

$$\frac{d^2 C_A}{dx^2} - \frac{uL}{D_A}\frac{dC_A}{dx} - \frac{kL^2}{D_A}C_A = 0 \tag{8.43}$$

or

$$\frac{d^2 C_A}{dx^2} - \text{Pe}_L\frac{dC_A}{dx} - \text{Da}_L C_A = 0 \tag{8.44}$$

where we define

$$\text{Pe}_L = \frac{uL}{D_A} \tag{8.45}$$

as the dimensionless *Peclet number* and

$$\text{Da}_L = kL^2/D_A \tag{8.46}$$

as a dimensionless *Damkohler number*.

We now have to solve this equation for different values of Pe_L. Solutions are plotted in Figure 8–9 for $p(t)$ with $\text{Da}_L = 0$ and the conversion versus position with a first-order reaction.

With laminar flow, there is an effective "axial dispersion coefficient" E_{aA}, called Aris Taylor diffusion, which is driven by radial mixing,

$$E_{aA} = D_A + \frac{u^2 R^2}{48 D_A} \tag{8.47}$$

and this produces much more dispersion in the turbulent tubular reactor than does the molecular diffusivity of a species D_j. Nevertheless, the Peclet and Damkohler numbers can still be defined as $\text{Pe}_L = kL^2/E_a$ and $\text{Da}_L = uL^2/E_a$.

8.6 RECYCLE REACTORS

Next we consider the PFTR with recycle in which a portion of the product stream is fed back into the feed stream as shown in Figure 8–10. A pump is needed to recycle the fluid back to the entrance and control R, the amount of recycle.

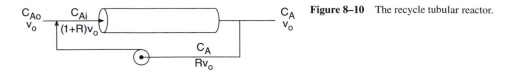

Figure 8–10   The recycle tubular reactor.

There is of course no point in adding recycle to a CSTR because reactants and products are assumed to be already mixed instantly, and recycle would not change the performance at all except for adding unnecessary pipes and pumps.

The volumetric flow rate into the reactor system (tubular reactor and associated pipes for recycle) is v_o and the feed concentration is C_{Ao} as before. However, a portion of the exit stream from the reactor is fed back and mixed with the feed stream. We call this portion R the *recycle ratio*. (Note that in this section we use R as recycle ratio while everywhere else R means tube radius.) The volumetric flow rate into the reactor is now

$$v = (1 + R)v_o \tag{8.48}$$

This indicates that the feed into the PFTR is therefore diluted with the product stream. A balance on the flow of A at the mixing point yields

$$vC_{Ai} = (1 + R)v_oC_{Ai} = v_oC_{Ao} + Rv_oC_A \tag{8.49}$$

which can be solved for the inlet concentration C_{Ai} into the PFTR section of the reactor to give

$$C_{Ai} = \frac{C_{Ao} + RC_A}{1 + R} \tag{8.50}$$

as the feed concentration of A into the plug-flow tubular reactor.

The residence time in the plug-flow reactor is *shortened* by the addition of recycle and is given by

$$\tau_{\text{PFTR}} = \frac{V}{(1 + R)v_o} \tag{8.51}$$

Thus in the RTR the PFTR section "sees" (1) a diluted feed, (2) a higher flow rate, and (3) a shorter residence time if the recycle ratio R is greater than zero. However, the mass balance on the PFTR itself is unchanged,

$$\tau_{\text{PFTR}} = -\int_{C_{Ai}}^{C_A} \frac{dC_A}{r(C_A)} = \frac{V}{(1 + R)v_o} \tag{8.52}$$

with the only difference being the C_{Ai} rather than C_{Ao} as the lower limit in the integration.

We now write the mass balance for the recycle tubular reactor (RTR). This is

$$\tau_{\text{RTR}} = \frac{V}{v_o} = -(1 + R)\int_{C_{Ai}}^{C_A} \frac{dC_A}{r(C_A)} \tag{8.53}$$

We have to write C_{Ai} in terms of C_{Ao} in order to express the conversion in terms of the feed conditions and reactor parameters, and the preceding expression for C_{Ai} in terms of C_{Ao}, C_A, and R allows us to do this.

We do this for the first-order irreversible reaction, $r = kC_A$, for which the integral becomes

$$\tau_{RTR} = \frac{V}{v_o} = -(1+R) \int_{C_{Ai}}^{C_A} \frac{dC_A}{kC_A} = \frac{1+R}{k} \ln \frac{C_{Ai}}{C_A} = \frac{1+R}{k} \ln \frac{C_{Ao} + RC_A}{C_A(1+R)} \quad (8.54)$$

Recycle will always give a *lower conversion* in a PFTR for positive-order kinetics because it produces *backmixing*, in which product mixes with reactant at the entrance to the plug-flow section.

Let us examine the forms of this expression for $R = 0$ and for $R \to \infty$. For $R = 0$ we recover the PFTR equation $\tau = (1/k) \ln(C_{Ao}/C_A)$ as expected. For $R \to \infty$ we examine the expression

$$\tau_{RTR} = \lim_{R \to \infty} \left(\frac{1+R}{k} \ln \frac{C_{Ao} + RC_A}{C_A(1+R)} \right) \quad (8.55)$$

We can write this as

$$\tau_{RTR} = \frac{1+R}{k} \ln \left(\frac{C_{Ao}/C_A R}{1+R} \right) = \frac{1+R}{k} \ln \left(\frac{C_{Ao}/C_A - 1 + 1 + R}{1+R} \right)$$

$$= \frac{1+R}{k} \ln \left(\frac{C_{Ao}/C_A - 1}{1+R} + 1 \right) \quad (8.56)$$

Now recalling that $\ln(1 + x) = x$ for small x, we have

$$\tau_{RTR} = \lim_{R \to \infty} \left[\frac{1+R}{k} \ln \left(\frac{C_{Ao}/C_A - 1}{1+R} + 1 \right) \right] = \frac{1+R}{k} \frac{C_{Ao}/C_A - 1}{1+R} = \frac{C_{Ao} - C_A}{kC_A} \quad (8.57)$$

which is exactly the expression we obtained long ago for the first-order reaction in the CSTR. Thus recycle in a PFTR increases fluid mixing such that for sufficiently large recycle the reactor becomes completely backmixed and approaches the performance of a CSTR. Of course this is a very inefficient way to produce a CSTR because infinite recycle needs an infinite flow rate in the recycle stream, which would require infinite pumping costs.

However, the RTR is an efficient way to introduce a small amount of backmix into a reactor. This is particularly useful for an autocatalytic reactor where the rate is proportional to the concentration of a product. In the enzyme-catalyzed reactions discussed previously and modeled as $A \to B, r = kC_A C_B$, we need to seed the reactor with enzyme, and this is easily accomplished using recycle of a small amount of product back into the feed.

Example 8–1   Plot the RTD for the RTR by sketching the pulses we should obtain at the exit following a single pulse injected at $t = 0$ for $R = 0, 1, 2, 9$, and 100.

For $R = 0$ we have the pulse at τ at the same height as the pulse injected, for the perfect PFTR. For $R = 1$ the first pulse containing one-half of the tracer will leave at time $\tau/2$, and the rest will recycle; one-half of this will exit at time τ, one-fourth will recycle, etc. Thus we will observe pulses at intervals $\tau/2$ with heights $\frac{1}{2}, \frac{1}{4}, \frac{1}{8}, \frac{1}{16}$, etc. The sum of all of the infinite number of pulses must be the height of the initial pulse.

We can proceed to sketch the pulses for any R. They will occur at time $\tau/(1 + R)$, and the corresponding expression for pulse height of each pulse can easily be derived. These are sketched in Figure 8–11. For $r \to \infty$, $p(t)$ will be a series of closely spaced pulses, and the envelope of these pulses is an exponential. This is $p(t)$ for the CSTR, because for infinite recycle the reactor is completely backmixed.

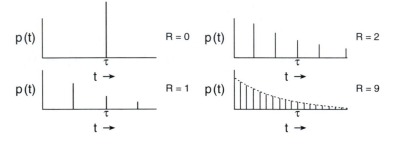

Figure 8–11 $p(t)$ for a tubular reactor with increasing recycle.

8.7   CSTRs IN SERIES

The RTD of equal-volume CSTRs in series we considered previously in Chapter 3 can easily be obtained by finding the response of a pulse tracer into the first, allowing that tracer to pass into the second, etc. We inject tracer at concentration C_1 into the first reactor, it travels as C_2 into the second, etc. We normalize these concentrations to p_1, p_2, etc., rather than writing expressions for concentrations. We also assume that each CSTR has the same residence time τ.

For the first reactor we have

$$\tau \frac{dp_1}{dt} = p_o(t) - p_1 \tag{8.58}$$

which can be integrated to give

$$p_1(t) = \frac{1}{\tau} e^{-t/\tau} \tag{8.59}$$

For the second reactor we have feed $p_1(t)$, so that $p_2(t)$ is given by the differential equation

$$\tau \frac{dp_2}{dt} = p_1(t) - p_2 = \frac{1}{\tau} e^{-t/\tau} - p_2 \tag{8.60}$$

This equation is *linear* in p_2 so it can be integrated to yield

$$p_2(t) = \frac{t}{\tau^2} e^{-t/\tau} \tag{8.61}$$

This is the feed to the third reactor,

$$\tau \frac{dp_3}{dt} = p_2(t) - p_3 = \frac{t}{\tau^2} e^{-t/\tau} - p_3 \tag{8.62}$$

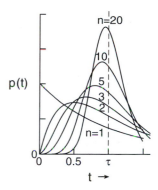

Figure 8–12   Residence time distribution of n CSTRs of equal residence times in series. The distribution sharpens as n increases and eventually approaches the perfect PFTR.

which gives

$$p_3(t) = \frac{t^2}{2\tau^3} e^{-t/\tau} \tag{8.63}$$

We can continue this for n reactors in series

$$p_n(t) = \frac{t^{n-1}}{(n-1)!\tau^n} e^{-t/\tau} \tag{8.64}$$

This function is plotted in Figure 8–12. The residence time indicated is the total residence time $\tau = n\tau$. It is seen that $p(t)$ sharpens as n increases such that, for large n, τ approaches that for the PFTR.

8.8   DIAGNOSING REACTORS

Until this chapter all reactors were assumed to be either totally unmixed or totally mixed. These are clearly limits of a partially mixed reactor, and thus these simple calculations establish bounds on any chemical reactor. We have developed models of several simple partially mixed reactors, and we showed that $p(t)$ is somewhere between the δ function of the PFTR and the exponential of the CSTR. We show these models in Figure 8–13 for increasing backmixing.

8.9   SUMMARY

To summarize, whenever one needs to model a chemical process, the first step is to predict the performance of the PFTR or CSTR for assumed kinetics, the subjects of Chapters 2 through 4. If the reactors are not isothermal, we need to add the complexities of Chapters 5 and 6.

The next level of sophistication is to use the nonideal reactor models developed in this chapter. These are fairly simple to calculate, and the results tell us how serious these nonidealities might be.

Finally, when we need accurate predictions of a chemical reactor, we must do the "complete" simulations by solving the differential equations written at the start of this

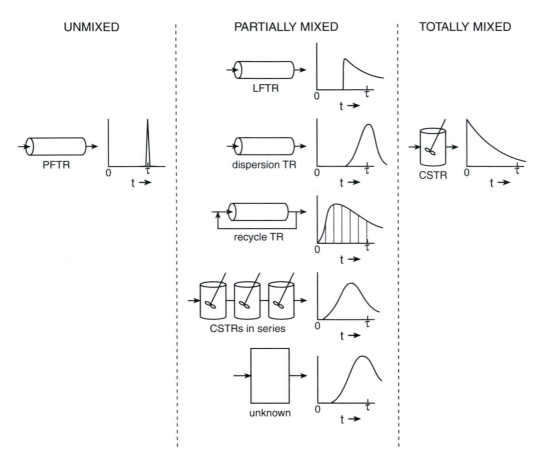

Figure 8–13   Residence time distributions for the nonideal reactors considered in this chapter.

chapter. This almost always requires numerical solutions. However, whenever we are dealing with complex systems (which environmental and biological population problems certainly are), we must always resort to approximations and intuition. A wrong or inadequate set of approximations can lead to inaccurate or nonsensical solutions, and we must always be creative in our models and skeptical of our calculations.

PROBLEMS

8.1 Sketch the probability $p(t)$ of the following objects leaving or failing at time t and describe what failure modes cause different features of this distribution function.

(a) a ball point pen (same as PFTR)

(b) a wine glass (same as CSTR)

(c) a university engineering student

(d) a university liberal arts student

(e) an American car

(f) a Japanese car

(g) the wall of a chemical reactor

(h) a pump

(i) a process control computer

8.2 Calculate the conversion of $A \rightarrow B, r = kC_A$ in two CSTRs using the residence time distribution and compare the result with that obtained by integrating the CSTR mass balances. Repeat this problem for zeroth-order kinetics.

8.3 Real Madeira wine is made only on the island of Madeira, and it is made by fermentation in ten 50-gallon casks with fresh juice continuously poured into the first, and the effluent passing successively into each cask and the product from the tenth bottled.

(a) What is the production of 1 liter bottles if the fermentation time is 1 month?

(b) We suddenly find that we fed pure vinegar into the first cask for a 10-min period before switching back to juice. Describe the concentration of vinegar versus time if flow is unchanged. If the quality of the wine is affected noticeably by a concentration of 1% of the original vinegar in the tank, how many bottles must be given to the employees?

8.4 What equations would have to be solved to determine transient behavior in a tubular reactor? How would you guess that temperatures might vary with time for step changes in T_f, T_{co}, or flow rates?

8.5 We have a ten-stirred-tank system (20 liters each) that is contaminated so we need to flush it with fresh water (1 liter/min). How long must we flush with fresh water so that the concentration is 0.1% of its present value:

(a) if each tank is fed fresh water and discharges its own waste?

(b) if the tanks are connected in series, the first tank is fed fresh water, and only the last tank discharges waste?

8.6 For laminar flow in an isothermal tubular reactor of length L and velocity v, calculate the average extent $\overline{X}$ for the reaction $A \rightarrow B, r = kC_A$, and compare this result with the extent obtained for plug flow. Can you satisfy yourself that $\overline{X}$ will always be tangent with plug flow?

8.7 A sewage treatment plant consists of several tanks in series with inlet and outlet pipes at opposite sides. The liquid and sludge are stirred by a weir that rotates slowly to keep solids suspended and to incorporate oxygen.

(a) Sketch $p(t)$ if rotation is too slow for complete CSTR mixing. Indicate the time interval between passage of the weir blades and the residence time.

(b) Suppose there are blades that divide the tank into ten nearly isolated compartments and that the rotation time is much shorter than the residence time in each compartment. Sketch $p(t)$ for this reactor.

8.8 Derive the expression for $p(t)$ for the laminar-flow tubular reactor.

8.9 A 20-gallon fish tank contain 3.5% NaCl. If 1 cup of water is replaced each day by fresh water, how many days are required for the salt concentration to be reduced to 0.1%?

8.10 Figure 8–14 shows several reactor configurations and some $p(t)$ curves. Sketch $p(t)$ for those not given or a reactor configuration that will give the $p(t)$ shown.

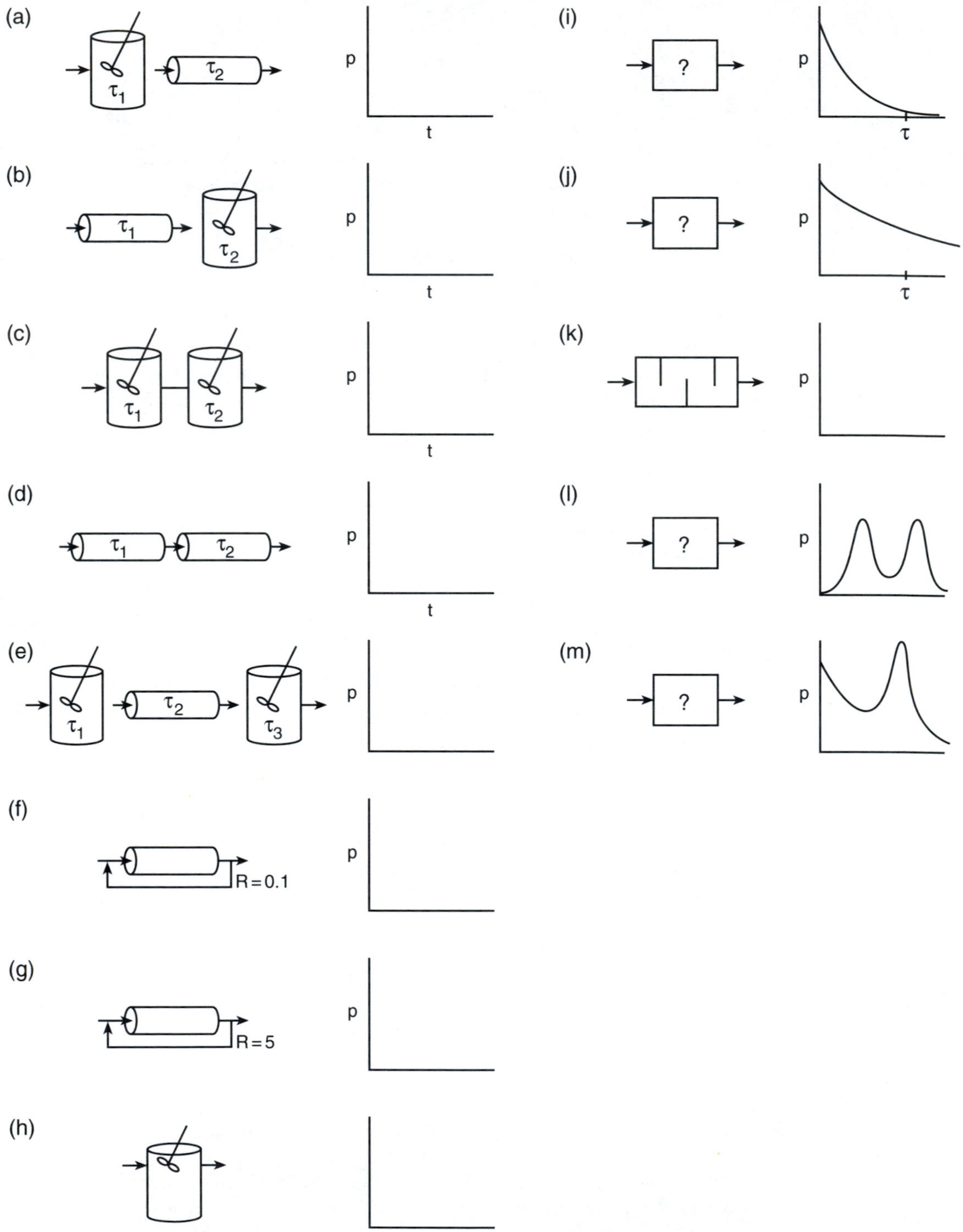

(a)
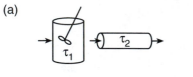 

(b)
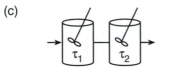

(c)
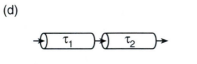

(d)

(e)
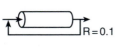

(f)
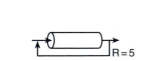

(g)
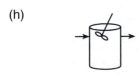

(h)

(i)
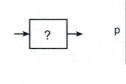 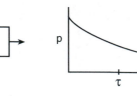

(j)
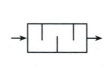 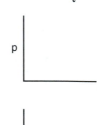

(k)

(l)
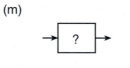

(m)

Figure 8–14 Residence time distribution and reactor configurations for several nonideal chemical reactors.

Chapter 9

REACTIONS OF SOLIDS

9.1 REACTIONS INVOLVING SOLIDS

In Chapter 7 we considered catalytic reactions on solid surfaces and found that transport steps are essential in describing these reactions because mass transfer of reactants and products between phases must accompany reaction. In this chapter we consider the reactions of solids in which the solid enters the stoichiometry of the reaction as a reactant or product or both. We remark that the texts of Levenspiel give excellent and thorough descriptions of the reactions of solids, and we will only summarize some of the features of reactions involving solids here.

The reaction of solids is not regarded as a traditional aspect of chemical reaction engineering, which usually concerns mostly fluids (gases and liquids). However, chemical processing of solids represents an increasingly important part of reaction engineering. The material in this chapter also introduces gas–liquid and liquid–liquid processes because similar geometries are involved.

In Figure 9–1 we sketch several examples of the type of process we will consider in this chapter.

The examples in Figure 9–1 all involve roughly spherical solid or liquid particles in a fluid containing reactive solute. If the solid promotes a reaction of fluid molecules but reactants and products can only form a monolayer on the solid surface so that the solid doesn't grow, we have a *catalyst particle*, the subject of Chapter 7. If the "reactant" is a supersaturated solute that can deposit on or dissolve from the solid, we have a *crystallization* or *dissolution* problem, the subject of mass-transfer courses. If in the previous case we replace the solid by a liquid drop, we have a growing or evaporating drop. If the reactant from the fluid phase deposits a product on the solid with possible formation of another volatile product, we have chemical vapor deposition and solid reactions. Particular examples of these processes are those in the rising and toasting of food. A final example is the burning of carbon particle where the reaction generates so much heat that the particle glows, a situation we will consider in more detail in Chapter 10.

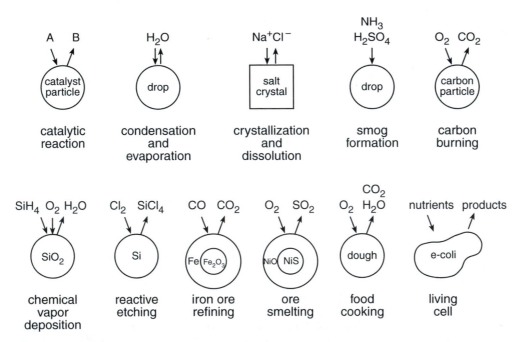

Figure 9–1   Examples of fluid–solid and fluid–liquid reactions of approximately spherical solid or liquid particles.

A major point we emphasize is that, from a reaction engineering viewpoint, these are all essentially identical problems. The fluid may be a gas or a liquid, and the particle may be a liquid, gas, or a solid, but the geometries and the reactions are very similar. The interfaces may be gas–liquid, liquid–liquid, gas–solid, or liquid–solid. For gas–liquid problems we have drops and bubbles rather than particles, but the geometries are identical. These systems are the subject of Chapter 14. The applications are also quite different, but, once one realizes the similarities, the same ideas and equations unify all these problems.

Reactions involving solids are very important in many technologies such as microelectronics processing, ceramics, ore refining, electrochemical deposition and etching, chemical vapor deposition and etching, and food processing. We will consider some of these applications in problems, but we first note several important examples.

The previous examples were all for drops, bubbles, or particles, all objects small in three directions. The other examples in this chapter are for solids that are small only in one direction. These are *solid films*, a class of solids with many important applications.

9.2   CHEMICAL VAPOR DEPOSITION AND REACTIVE ETCHING

The prototype processes we consider in this chapter are chemical vapor deposition and reactive etching, and the other examples listed such as crystallization and catalytic reactions are basically simplifications of these processes. In *chemical vapor deposition* (CVD), gases react to form solid films in microelectronic chips and in wear protective coatings.

Examples are the deposition of polycrystalline silicon coatings in microelectronic circuit fabrication by reactions such as the decomposition of silane,

$$SiH_{4g} \rightarrow Si_s + 2H_{2g} \tag{9.1}$$

The deposition of protective silica films on these circuits occurs by reactions such as

$$SiH_{4g} + 2O_{2g} \rightarrow SiO_{2s} + 2H_2O_g \tag{9.2}$$

and the deposition of hard TiC films on machine tool surfaces involves the decomposition of titanium tetrachloride,

$$TiCl_{4g} + CH_{4g} \rightarrow TiC_s + 4HCl_g \tag{9.3}$$

In these applications one needs to form a uniform film of specified grain size that grows and adheres strongly to its substrate.

Gallium arsenide for solid-state lasers and fast memory chips can be formed by molecular beam epitaxy through the reaction

$$(CH_3)_3Ga_g + AsH_{3g} \rightarrow GaAs_s + 3CH_{4g} \tag{9.4}$$

where the precursors trimethylgallium and arsine are highly toxic gases, GaAs is a single crystal solid film, and methane is an unreactive gas. Now, however, one desires a single crystal film of very high purity with very few defects.

These are reactions of the type

$$A_g \rightarrow B_s + C_g \tag{9.5}$$

or

$$A_g + B_g \rightarrow C_s + D_g \tag{9.6}$$

These processes can also involve reactants in the liquid phase, and we call this *liquid-phase epitaxy*, although the equations will be the same as for gases.

In *reactive etching* of silicon a patterned film is selectively etched by reacting with a gas such as chlorine

$$Si_s + 2Cl_2 \rightarrow SiCl_{4g} \tag{9.7}$$

In the preparation of silicone polymers, an important monomer is SiH_2Cl_2, which is formed by attacking high-purity silicon powder with HCl in the presence of copper catalyst

$$Si_s + 2HCl_g \rightarrow SiH_2Cl_{2g} \tag{9.8}$$

The rates and selectivities of these processes are frequently enhanced by the presence of *plasmas*, in which a high electric field in a gas causes ionization of molecules, and the reactions of these ions and the increased transport alters reaction rates. We will not consider these processes in this chapter.

These reactions are of the type

$$A_g + B_s \rightarrow C_g \tag{9.9}$$

In *plating* of metals a film of metal such as chromium is deposited from a salt solution onto a reactive metal electrochemically,

$$Cr_{soln}^{+3} + 3e^- \rightarrow Cr_s \tag{9.10}$$

In iron ore *smelting* spherical pellets of iron oxide (taconite) are reacted with CO (from coke) in a blast furnace

$$Fe_2O_{3s} + \tfrac{3}{2}CO \rightarrow 2Fe_s + \tfrac{3}{2}CO_2 \tag{9.11}$$

In nickel smelting particles of nickel sulfide ore are first oxidized,

$$NiS_s + \tfrac{3}{2}O_{2g} \rightarrow NiO_s + SO_{2g} \tag{9.12}$$

and then the oxide is reduced to metal

$$NiO_s + H_{2g} \rightarrow Ni_s + H_2O_g \tag{9.13}$$

in a second-stage process.

In these processes one solid is converted into another in reactions that can be represented as

$$A_g + B_s \rightarrow C_s + D_g \tag{9.14}$$

In the food processing industry flour doughs are *baked*. This involves many processes such as (1) *rising of dough* by evolution of CO_2 by yeast-catalyzed oxidation of sugar to vapor in pockets or by decomposition of baking powder,

$$NaHCO_{3s} \rightarrow NaOH_s + CO_{2g} \tag{9.15}$$

(one adds an acid such as vinegar or lemon juice in this process to promote the reaction and because the NaOH product would taste bitter), (2) *drying* by evolution of water vapor, which may be physically or chemically bound in the dough, and (3) *toasting* or oxidation of a surface layer of carbohydrate with O_2

$$\text{carbohydrate} + O_{2g} \rightarrow \text{oxidized carbohydrate} \tag{9.16}$$

In *crystallization* a solid is formed from a supersaturated solution, frequently with chemical reactions such as formation of complex crystals from ions in the solution. The dissolution of solids is the reverse of crystallization. We can write this process as

$$A_s \rightleftarrows A_{\text{solution}} \tag{9.17}$$

where we could consider this either as a chemical reaction or as a physical phase change of species A.

In *combustion* of solids such as coal, wood, and charcoal the reaction of O_2 occurs with solid carbon, sometimes leaving a solid ash residue. Fuel oil drops react with O_2 in a similar process in boilers and in diesel engines. Since the exothermicity of these processes creates very large temperature differences, we will describe them in the next chapter.

In this chapter we write the "generic" reaction as

$$\boxed{A_g + B_s \rightarrow C_s + D_g} \tag{9.18}$$

so that A and D are the fluid reactant and product while B and C are the solid reactant and product. Not all these species exist in the examples listed previously, so reactive etching is $A_g + B_s \rightarrow D_g$, crystal formation is $A_g \rightarrow C_s$, and sublimation is $B_s \rightarrow D_g$. Thus, throughout this chapter C_A is the concentration of the fluid species A_g that is reacting with a solid reactant species B_s in a film or particles to make different products. If the solid B_s reacts to form another solid, we will call this product solid C_{Cs}. The subscript s in this chapter always applies to the species in the dispersed phase (except that C_{As} means the reactant species concentration at the surface of the solid, just as for a catalyst surface).

We also make the simplifying assumption that stoichiometric coefficients are $+1$ for all products and -1 for all reactants. Other stoichiometries are easy to include, but we want to make the equations as simple as possible.

We also want these designations of A, B, C, and D to be more general than gases and solids. The ideas developed in this chapter apply to any continuous fluid reacting with any dispersed phase. Thus the fluid and rigid phases could be gas, liquid, or solid, for example, gas bubbles (dispersed) reacting with a liquid solution (continuous) or a solid film. Examples such as these are important in most multiphase reactors, the subject of Chapter 14.

9.3 SOLIDS REACTORS

Each of these processes can be modeled with chemical reactions such as those shown, and in this chapter we will consider their engineering in sufficient detail that we can describe their rates of formation or dissolution so that we can control the size and quality of the products we are forming.

Reactions such as these usually take place in continuous reactors such as chip (memory or potato) processing lines, retorts, kilns, and conveyer belts in which the solids pass through an oven with reactive gases such as air or perhaps in a fluidized solid bed reactor as in the "shot from guns" puffed rice and other grain cereals or in a slurry reactor such as a vegetable oil or fat fryer. Some typical solids reactors are sketched in Figure 9–2.

These processes can obviously be modeled as PFTR, CSTR, batch, or semibatch reactors. However, we now must consider the flows of both fluid and solid phases; so we have a *multiphase reactor* because there are distinct residence times of solid and fluid phases,

$$\tau_s = \frac{V_s}{v_s} \tag{9.19}$$

$$\tau_f = \frac{V_f}{v_f} \tag{9.20}$$

where V and v are the volumes and volumetric flow rates on fluid f and solid s phases. These reactors are examples of multiphase reactors, which we will consider formally in Chapter 14. In this chapter we will be concerned mostly with developing suitable rate expressions for the reaction rates of solids; in Chapter 14 we will consider the reactor mass and energy equations.

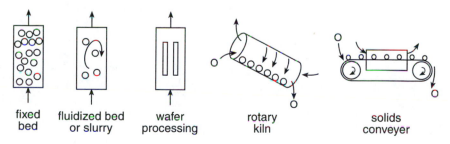

| fixed bed | fluidized bed or slurry | wafer processing | rotary kiln | solids conveyer |

Figure 9–2 Typical continuous reactors for processing solids. Solids handling and particular solid products frequently require configurations not common in processing fluids.

Batch and semibatch processes are particularly common for processing solids because the conditions of the process can be varied as the reaction proceeds to control the quality of the solid formed. Burned or raw food and defective computer chips are intolerable.

9.4 REACTION RATES OF SOLIDS

As with catalytic reactions, our task is to develop pseudohomogeneous rate expressions to insert into the relevant mass-balance equations. For any multiphase reactor where reaction occurs at the interface between phases, the reactions are primarily surface reactions (rate r''), and we have to find these expressions as functions of concentrations and rate and transport coefficients and then convert them into pseudohomogeneous expressions,

$$r = (\text{area}/\text{volume})r''(ks, C_j s) \tag{9.21}$$

which we considered in Chapter 7 in connection with heterogeneous catalytic reactions where the solid phase remained unchanged as the reaction proceeds in the fluid phase.

However, now the solid is shrinking, growing, or transforming as the process proceeds, rather than being unchanged as for solid catalysts. Thus there is a time dependence within the solid in the reactor.

The reaction of solids occurs in the monolayer of molecules adsorbed on the surface of the solid B, as sketched in Figure 9–3. Therefore, this process can be regarded as a surface reaction with rate r'', just like the catalytic reactions of Chapter 7 except that now the molecules of the solid react with the gaseous molecules to form a gaseous product and remove solid molecules. The rate of the reaction should therefore be

$$r'' = k''_{\cdot} C_{As} C_{Bs} \tag{9.22}$$

where C_{As} and C_{Bs} are the concentrations of A and B at the reacting surface of B. Since C_{As} is independent of the conversion of the solid, C_{As} is constant as long as any solid is present. Further, the concentration of solid B at the surface is *constant* because new surface is exposed continuously so that the concentration of exposed solid B, C_{Bs}, is always one monolayer. We therefore write the reaction rate as

$$
\begin{aligned}
r'' &= k'' C_{As}, & \text{reacting solid present} \\
&= 0, & \text{reaction complete}
\end{aligned}
\tag{9.23}
$$

In this rate expression we have lumped C_{Bs} into the effective surface rate coefficient by defining $k'' = k''_{\cdot} C_{Bs}$. All solid reactions have reaction steps similar to those in catalytic reactions, and the rate expressions we need to consider are basically Langmuir–Hinshelwood

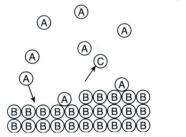

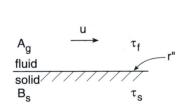

Figure 9–3 Sketch of molecules of A_g reacting with solid atoms or molecules B_s to form gaseous and solid products in the reaction $A_g \rightarrow B_s + C_g$. Reaction occurs by species A first adsorbing on the surface of B; so the process is a surface reaction. At the right are shown the quantities of interest in this process.

kinetics, which were considered in Chapter 7. Our use of a first-order irreversible rate expression is obviously a simplification of the more complex rate expressions that can arise from these situations.

9.4.1 Drops and bubbles

We title this chapter the reactions of solids and we deal mostly with gaseous and solid reactants and products, but the same ideas and equations apply to gas–liquid and liquid–liquid systems. For example, in frying of foods, the fluid is obviously a liquid that is transferring heat to the solid and carrying off products. The same equations also apply to many gas–liquid and liquid–liquid systems. The drops and bubbles change in size as reactions proceed; so the same equations we derive here for transformation of solids will also apply to those situations.

In Chapter 14 we will consider multiphase reactors in which drops or bubbles carry one phase to another continuous fluid phase. In fact, these reactors frequently have a solid also present as catalyst or reactant or product to create a three-phase reactor. We need the ideas developed in this chapter to discuss these even more complicated reactors.

For the student who is thinking that catalytic reactors and nonideal reactors were very complicated and that this topic is becoming so horribly complicated that they can never understand it, we offer two comments. First, we will only consider simple problems where the interpretation and the mathematics are relatively simple. Second, the chemical engineer gets paid (rather well) to deal with real problems, and no problem dealing with the engineering of chemical reactions should be too messy or complex for us to handle. The willingness to deal with complex issues (intelligently) is what makes a good engineer, and if you aren't willing to try to solve a particular problem, then someone else will (and will take your job or be your boss).

9.5 FILMS, SPHERES, AND CYLINDERS

Just as for catalyst pellets we can divide the shapes of solids into films, spheres, and cylinders, each with a particular one-, three-, or two-dimensional geometry (Figure 9–4). (We won't do much with cylinders on this chapter, although fiber spinning is an important chemical process.) The notation we will use is the same as with catalyst pellets or slabs (now called

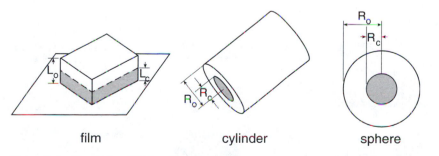

film   cylinder   sphere

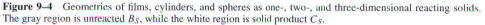

Figure 9–4 Geometries of films, cylinders, and spheres as one-, two-, and three-dimensional reacting solids. The gray region is unreacted B_S, while the white region is solid product C_S.

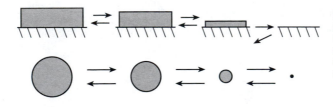

Figure 9–5 Sketches of shrinking or growing films and particles, with arrows going either forward or backward as the process progresses. Reaction is assumed to occur only at the interface of the solid with the fluid.

films). We need to describe the evolution of structure in the reacting particle. For dissolving or growing objects, we assume that everything occurs on the external surface of thickness ℓ or radius R. Note, however, that these can actually be rough or porous, depending on growth conditions. We of course used ℓ and R to designate these quantities before, but now we need to see how these vary with time as the reaction proceeds, that is, to find $\ell(t)$ and $R(t)$.

9.5.1 Shrinking and growing solids

The first category is that of solid films or particles that shrink or grow during the process, as shown in Figure 9–5. Here the thickness of the film or the radius of the particle are the primary variables.

9.5.2 Transforming solids

The second class of solid reactions involves situations where the solid does not disappear or appear but rather transforms from one solid phase into another as the reaction proceeds, as shown in Figure 9–6. For transformations of solids there are several models that may be appropriate, depending on the microstructure of the reacting solid. Limiting cases of concentration profiles within the solid are (1) uniform reaction and (2) film formation. Concentration profiles within the solid for these situations are shown in Figure 9–7.

The first two profiles are limiting cases of solid concentration profiles, and most situations may be somewhere between those limits, as shown in the third panel.

The growing or dissolving of a porous solid is a variant on the porous catalyst, but this situation combines transport through a porous solid with changing solid phases and is too complex to be considered here.

As sketched for spheres and films, the processes involve either a totally reacted external region and a totally unreacted core or a spatially uniform solid in which the

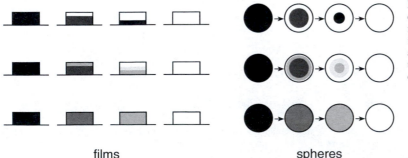

films   spheres

Figure 9–6 Sketches of transforming films or particles, where solid B_s (black) is transformed into C_s (colorless). Sketches shown involve a sharp boundary between reacted and unreacted layers, with reaction assumed to occur only at the interface between the core B and the film C.

Figure 9–7 Sketch of possible concentration profiles (indicated by shading) as solids react.

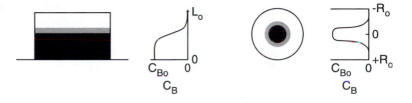

transformation occurs microscopically (perhaps in each single crystal grain) such that macroscopically the solid appears to be spatially uniform at any degree of conversion.

9.5.3 Solid fractional conversion

We need to describe the conversion of the solid from reactant to product. We can use concentrations in the solid to describe the solid transformations. However, the concentrations are varying *within* each particle, while we want the *overall* conversion to describe the transformation. Therefore, we will use a variable describing the solid conversion X_s to describe the conversion of a solid, in analogy with the conversion X of homogeneous fluids used previously, so that X_s goes from 0 to 1 as the reaction proceeds.

In the shrinking core model a film of initial thickness ℓ_o transforms with an unreacted core of thickness ℓ_c. The initial volume of a planar solid film is

$$V_o = A_B \ell_o \tag{9.24}$$

while the volume of the unreacted core is

$$V_c = A_B \ell c \tag{9.25}$$

where A_B is the external area of the film of B. If the solid is being reacted from an initial thickness ℓ_o to a thickness ℓ_c, the conversion of the solid X_s is

$$X_s = 1 - \frac{V_c}{V_o} = 1 - \frac{\ell_c}{\ell_o} \tag{9.26}$$

For a sphere of initial radius R_o transforming with a core of radius R_c, the total volume is

$$V_o = \tfrac{4}{3}\pi R_o^3 \tag{9.27}$$

while the unreacted core volume is

$$V_c = \tfrac{4}{3}\pi R_c^3 \tag{9.28}$$

so that the conversion is given by

$$X_s = 1 - \frac{V_c}{V_s} = 1 - \left(\frac{R_c}{R_o}\right)^3 \tag{9.29}$$

For a cylinder with an original radius R_o reacting with a core radius R_c, the conversion is

$$X_s = 1 - \left(\frac{R_c}{R_o}\right)^2 \tag{9.30}$$

Thus we will have to solve the solid and gas mass-balance equations to find expressions for $R(t)$ and $\ell(t)$ to insert them into these equations to find $X_s(t)$. This is the major subject of this chapter.

We will consider later the situation where the solid does not shrink but rather transforms to another solid. In that situation the outer dimension ℓ_o or R_o remains constant but the core shrinks. For this, we will need positions ℓ_c and R_c versus time.

We will next develop expressions for $X_s(t)$ and the residence time τ_s for complete reaction of the solid in terms of parameters of systems for different approximations of rate parameters. In a continuous reactor solid particles are fed into reactor at a constant rate, and each transforms as a function of the time it has been in the reactor. Therefore, we would have to use the probability distribution function to compute the average conversion,

$$\overline{X}_s = \int_0^\infty X_s(t)p(t)\,dt \tag{9.31}$$

a subject we considered in Chapter 8, where we dealt with the distribution of residence times and conversions. Further, since most batches of solids will not all have the same size but rather a distribution of sizes, we must frequently consider the particle size distribution and the resultant residence time distribution $p(X_s, t)$ and the average particle size distribution in calculating average conversion. These complexities are beyond the scope of topics considered in this chapter.

9.5.4 Steady-state solid composition profiles

Another simplifying assumption we will make in solid–gas reactions is that the gas composition profile around and through the solid remains in steady state as the solid conversion increases. This is obviously not precisely true because the composition of the solid is varying with time.

However, the solid density is approximately 1000 times the density of a gas at atmospheric pressure, and molecules in gases and liquids have much higher diffusivities than in solids. Therefore, the reacting boundary (ℓ or R) moves very slowly compared to the motion of gas molecules to and from the boundary, and we can assume that concentration profiles near this boundary remain in steady state while we calculate the steady-state concentration profiles in the reaction.

After we have calculated these profiles assuming a stationary boundary, we will then proceed to calculate the movement of the reacting boundary.

9.6 MACROSCOPIC AND MICROSCOPIC SOLIDS

Recall that for a heterogeneous catalytic reactor we had to consider the microscopic concentration profiles around and within a catalyst particle and then eliminate them in terms of the macroscopic position variable z in the reactor design equations (mass transfer limits and effectiveness factors).

For solids reactors we must add the complication that there is frequently a variation within the solid particle or film. This occurs because solids usually consist of small single crystals that have grain boundaries between them. Also, solids are frequently formed as *composites* of single crystal or polycrystalline grains that grow or are compacted into a macroscopic film or pellet with void spaces between them, as shown in Figure 9–8. Thus there can be several distance scales that we must be concerned with in a solid film or

Figure 9–8   Microstructures of solid films and pellets. Variations can occur across a film or sphere as well as within individual particles in the film or sphere.

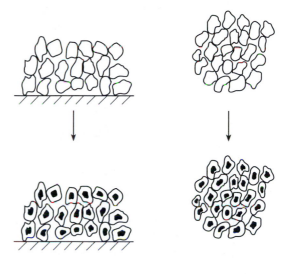

pellet, and these must all be considered sequentially starting from the smallest distance and winding up with the average rate $\bar{r}$ at an average position $\bar{z}$ in the overall reactor, just as we noted for the catalytic reactor.

These microstructures also can frequently change as a solid is processed. As examples, when an oxide ore is reduced to metal, there is usually a decrease in density, and this can produce changes in a porous metal as the oxide is reduced. In the baking of foods, an early process is the rising of dough by formation of CO_2 and H_2O bubbles that may or may not break to allow air to penetrate and oxidize the internal surfaces of the solid. As a final example, in etching of alloys one metal is usually removed preferentially, leaving a rough or porous surface.

Possible solid composition profiles of a partially reacted solid film or pellet are shown in Figure 9–9. It is evident that there are multiple possible fluid concentration profiles within the film or pellet such as within pores (where diffusivities are high), down grain boundaries where some solutes have high diffusivities, and within single crystal grains where diffusion can be very low.

In all these examples the external and internal surface areas of the solid increases (sometimes decreases) as the process proceeds, and different processing conditions (such as oven temperature in baking) can cause different microstructures to develop. The consideration of these effects goes far beyond the scope of this text, and information from courses in ceramics, metallurgy, corrosion, electrochemistry, food engineering, and biology must be included to consider microstructures and their transformations during reactions of solids.

Figure 9–9   Solid concentration profiles expected for partially reacted films or pellets that have both crystal grains and pores.

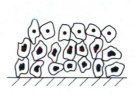

However, all of these involve reactions between gases and solids with rate coefficients k and diffusivities D; so the general principles described here allow us to consider them if we know the relevant microstructures.

9.7 DISSOLVING AND GROWING FILMS

We begin with the etching or dissolution of a uniform solid film described by initial thickness ℓ_0. We assume a reaction of the type

$$A_g + B_s \rightarrow C_g \tag{9.32}$$

for the solid film of B_s being etched by reaction with gas A_g to form gaseous product C_g (for example, the reaction of Cl_2 with a silicon film to form $SiCl_4$). A related process would be the thinning of a solid Si film by evaporation, which we would write as

$$B_s \rightarrow B_g \tag{9.33}$$

which is not a true chemical reaction but is described in the same way. We assume that the reaction occurs by molecule A_g impinging on the surface of the film B and reacting chemically to form a product molecule C_g, which immediately evaporates.

We now construct a mass balance on the total amount of B_s in the film. The volume of solid is $A_B \ell$ (area × length) with A_B the area of the film. If the density of the solid is ρ_B (mass of B_s per unit volume) and the molecular weight of B is M_B, then the number of moles of solid is given by

$$N_B = \frac{\rho_B}{M_B} A_B \ell \tag{9.34}$$

The rate of change in the number of moles of solid is

$$\frac{dN_B}{dt} = -[\text{area}]\, r'' \tag{9.35}$$

$$\frac{dN_B}{dt} = A_B \frac{\rho_B}{M_B} \frac{d\ell}{dt} = -A_B r'' \tag{9.36}$$

with the $-$ sign arising because $\nu_B = -1$ if species B is a reactant. In this mass-balance equation the area of the film A_B is assumed to be constant so it cancels to yield

$$\frac{\rho_B}{M_B} \frac{d\ell}{dt} = -r'' = -k'' C_{A_s} \tag{9.37}$$

or

$$\boxed{\frac{d\ell}{dt} = -\frac{M_B k'' C_{A_s}}{\rho_B}}$$
$$= \text{constant} \tag{9.38}$$

We assume that the right-hand side of this equation is constant because C_{A_s} usually does not change much as the reaction proceeds. The rate expression could be more complicated, but we will assume a first-order irreversible reaction of the reactant gas at the surface, $r'' = k'' C_{A_s}$, throughout this chapter.

Since the right-hand side of this equation is independent of time, we can immediately integrate for an initial film thickness ℓ_o

$$\int_{\ell_o}^{\ell} d\ell = \frac{-M_B k'' C_{A_s}}{\rho_B} \int_0^t dt \qquad (9.39)$$

to find $\ell(t)$,

$$
\begin{aligned}
\ell(t) &= \ell_o - \frac{M_B k'' C_{A_s}}{\rho_B} t, \qquad t < \ell_o \rho_B / M_B k'' C_{As} \\[2mm]
\ell &= 0, \qquad t > \ell_o \rho_B / M_B k'' C_{As}
\end{aligned}
\qquad (9.40)
$$

In terms of fraction dissolved X_s, starting at $\ell = \ell_o$ at $t = 0$, we can write

$$
\begin{aligned}
X_s(t) &= 1 - \frac{\ell}{\ell_o} = \frac{M_B k'' C_{A_s}}{\ell_o \rho_B} t, \qquad t < \ell_o \rho_B / M_B k'' C_{As} \\[2mm]
&= 1, \qquad t > \ell_o \rho_B / M_B k'' C_{As}
\end{aligned}
\qquad (9.41)
$$

Finally, the time for complete reaction of the film of B, τ_B, is

$$\tau_B = \frac{\rho_B}{M_B k'' C_{As}} \ell_o \qquad (9.42)$$

9.7.1 Film growth

For film growth we assume that the chemical reaction is

$$A_g \rightarrow B_s + C_g, \qquad r = k'' C_{As} \qquad (9.43)$$

for example, the decomposition of gaseous $SiCl_4$ on a surface to form solid Si and gaseous Cl_2. We could have simply written this reaction as the reverse of the dissolution process, but we prefer always to call species A the gaseous reactant. The mass balance on solid B is

$$\frac{d\ell}{dt} = +\frac{M_B k'' C_{As}}{\rho_B} = \text{constant} \qquad (9.44)$$

with the only difference being the change of sign because now B is a product; so $\nu_B = +1$.

The solution is also very similar to the preceding except that $\ell_o = 0$ at $t = 0$. Therefore, $\ell(t)$ is given by

$$\ell(t) = +\frac{M_B k'' C_{As}}{\rho_B} t \qquad (9.45)$$

These problems are deceptively simple, both conceptually and mathematically, and the student should verify the logic of the derivation and the solution.

Example 9–1   A silicon film is to be deposited on a planar substrate by chemical vapor deposition of silane with a rate $r'' = k'' C_{SiH_4}$. If $T = 200°C$, $P_{SiH_4} = 0.1$ atm, and $k'' = 10$ cm/min, calculate the time necessary to deposit a 10-μm-thick Si film. The density of Si is 2.33 g/cm³, and its molecular weight is 32.

The concentration of silane is

$$C_{SiH_4} = \frac{P_{SiH_4}}{RT} = \frac{0.1}{82 \times 473} = 2.6 \times 10^{-6} \text{ moles/cm}^3 \tag{9.46}$$

Using the formula for the film thickness versus time, we obtain

$$\tau_B = \frac{\rho_B \ell}{M_B k'' C_{Ab}} = \frac{2.33 \times 10^{-3}}{32 \times 10 \times 2.6 \times 10^{-6}} = 2.80 \text{ min} \tag{9.47}$$

Note that we want concentrations in moles/cm³ because we use lengths in centimeters throughout.

9.7.2 Mass transfer

Next we assume that the mass transfer rate of A to or from the surface limits the rate of etching or growth, or that the concentration of A in the flowing or bulk fluid C_{Ab} is not identical to the concentration in the fluid near the solid surface C_{As}. (Recall that we used the same notation in the catalytic reactions of Chapter 7. Note also that we have a problem with subscript s designating either the gas concentration of A at the surface or the fact that we have a solid.)

As with external mass transfer near catalyst pellets, we obtain a steady-state mass balance on species A,

$$k'' C_{As} = k_{mA}(C_{Ab} - C_{As}) \tag{9.48}$$

to give the rate expression

$$\ell(t) = \ell_o \pm \frac{k''}{1 + k''/k_{mA}} \frac{M_B C_{Ab}}{\rho_B} t \tag{9.49}$$

with the signs depending on whether we have growth or etching. The effective rate coefficient is

$$k_{eff} = \frac{k''}{1 + k''/k_{mA}} \tag{9.50}$$

in the reaction-limited case $k_{eff} \simeq k''$, while in the mass-transfer-limited case is $k'' \simeq k_{mA}$, so that

$$\ell(t) = \ell_o \pm \frac{k_m C_{Ab} M_B}{\rho_B} t \tag{9.51}$$

9.7.3 Mass transfer and film uniformity

With a reaction-limited deposition process, the film should have uniform thickness as long as the partial pressures of reactants do not vary with position. In a tubular reactor, the conversion of reactants must be kept small or the film thickness will depend on the location of the solid in the reactor, with upstream regions having a greater deposition rate. It is therefore common to use a gas recirculation reactor (a recycle PFTR) so that the composition of the reactants is independent of position in the reactor to assure uniform film thickness.

Mass transfer can produce films of nonuniform thickness because the deposition rate can depend on the velocity field u over the solid. Regions with high k_m will have the highest deposition rates in a mass-transfer-limited process. For flow over a flat plate of length L the average Sherwood number for laminar flow is given by the expression

$$Sh_L = 0.66 Re_L^{1/2} Sc^{1/3} \tag{9.52}$$

as discussed in Chapter 7. The local Sh is given by the expression

$$Sh_x = 0.33 Re_x^{1/2} Sc^{1/3} \tag{9.53}$$

where x is the distance from the leading edge of the plate of length L. Since $Re_x = ux/\nu$ with ν the kinematic viscosity, we see that

$$k_m = \frac{Sh_x D_A}{x} \sim x^{-1/2} \tag{9.54}$$

This expression predicts that k_m is infinite at the leading edge of a plate and decreases as $x^{-1/2}$ down the plate. Therefore, for mass-transfer-limited film deposition of reactants flowing over a flat plate, the film will be thickest near the leading edge. This is illustrated in Figure 9–10. If k'' is comparable to k_m, the film may be uniform near the leading edge, where $k_m > k''$ and the deposition is reaction limited, but since k_m decreases with x, the reaction can become mass transfer limited and the film will be thinner downstream as sketched in Figure 9–10(c).

Chemical etching processes have similar mass-transfer-limited possibilities, with etching rates highest in regions of high flow velocities. In electrochemical etching of solids, natural convection of the electrolyte caused by rising gas bubbles causes more rapid etching near the bottom of a vertical plate.

Obviously, if one wants to produce very uniform films, one should assure that $k_m \gg k''$, and this is best accomplished by (1) heating the system to increase k'' and (2) by maintaining high relative fluid velocities over the solid on which deposition or etching is occurring. Uniform deposition on irregularly shaped solid objects can be especially difficult because of unavoidable variations in the velocity field.

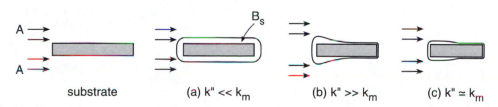

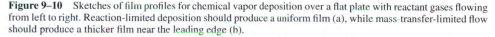

Figure 9–10 Sketches of film profiles for chemical vapor deposition over a flat plate with reactant gases flowing from left to right. Reaction-limited deposition should produce a uniform film (a), while mass-transfer-limited flow should produce a thicker film near the leading edge (b).

9.8 DISSOLVING AND GROWING SPHERES

Next we consider similar problems as the growth or etching of solid films, but now the solid is a sphere instead of a planar film. For growth or dissolution of a sphere with initial radius R_o, the volume is $\frac{4}{3}\pi R^3$ and $dV = 4\pi R^2\, dR$, so that the number of moles of the solid B is

$$N_B = \frac{4\pi R^3 \rho_B}{3M_B} \tag{9.55}$$

and the rate of change of the total number of moles of B in a single sphere is

$$\frac{dN_B}{dt} = \frac{\rho_B}{M_B}\frac{dV_{Bs}}{dt} = \pm 4\pi R^2 \frac{\rho_B}{M_B}\frac{dR}{dt}$$

$$= [\text{area}]\, r''$$

$$= \pm 4\pi R^2 k'' C_{As} \tag{9.56}$$

with the $+$ sign representing growth and the $-$ sign representing dissolution.

The difference between the growing or dissolving sphere and the growing or dissolving film is that for the sphere the surface area where the reaction occurs ($4\pi R^2$) changes as growth occurs. Therefore, we need to consider the dependence of properties on the radius R of the sphere. This gives interesting functional dependences on size that are not found with planar films.

Concentration profiles around the sphere might look as shown in Figure 9–11.

9.8.1 Reaction-limited dissolution

If $C_{As} = C_{Ab}$, then all quantities on the right side in the preceding equation are independent of time and size so the equation can be rearranged

$$dR = -\frac{k'' C_{Ab} M_B}{\rho_B}\, dt \tag{9.57}$$

and integrated from R_o at $t = 0$ to R at t,

$$
\boxed{
\begin{aligned}
R(t) &= R_o - \frac{k'' C_{Ab} M_B}{\rho_B}t, \qquad t < \rho_B R_o / M_B k'' C_{Ab} \\[2mm]
R(t) &= 0, \qquad t > \rho_B R_o / M_B k'' C_{Ab}
\end{aligned}
}
\tag{9.58}
$$

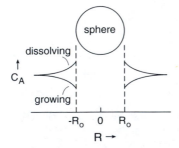

Figure 9–11 Concentration profiles around growing or dissolving sphere. The dissolving case implies a product species.

The conversion versus time for the dissolving sphere is

$$X_s(t) = 1 - \left(\frac{R}{R_o}\right)^3 = 1 - \left(1 - \frac{k'' C_{Ab} M_B}{R_o \rho_B} t\right)^3, \qquad t < \rho_B R_o / M_B k'' C_{Ab}$$

$$\text{(9.59)}$$

$$= 1, \qquad t > \rho_B R_o / M_B k'' C_{Ab}$$

The time τ_B required for complete dissolution of B is

$$\tau_B = \frac{\rho_B}{M_B k'' C_{Ab}} R_o \qquad \text{(9.60)}$$

9.8.2 Sphere growth

Here the equation for the radius versus time is

$$dR = +\frac{k'' C_{Ab} M_B}{\rho_B} dt \qquad \text{(9.61)}$$

with a $+$ sign because the film is now growing. This can be integrated to yield

$$R(t) = R_o + \frac{k'' C_{Ab} M_B}{\rho_B} t \qquad \text{(9.62)}$$

if mass transfer of reactant is sufficiently fast that $C_{As} = C_{Ab}$.

Example 9–2   Silicon particles are to be grown on Si seed crystals by chemical deposition of silane with a rate $r'' = k'' C_{\text{SiH}_4}$. If $T = 200°C$, $P_{\text{SiH}_4} = 0.1$ atm, and $k'' = 10$ cm/min, calculate the time necessary to grow Si particles from 1 to 10 μm, assuming particles are spherical and that the reaction is reaction limited. The density of Si is 2.33 g/cm^3 and its molecular weight is 32.

For reaction-limited growth of a sphere the preceding equation gives

$$t = \frac{\rho_B}{M_B k'' C_{Ab}} (R - R_o)$$

$$= \frac{2.33}{32 \times 10 \times 2.6 \times 10^{-6}} (10^{-3} - 10^{-4})$$

$$= 2.77 \text{ min} \qquad \text{(9.63)}$$

9.8.3 Mass-transfer-limited dissolution and growth

If the surface reaction rate coefficient is large compared to the mass transfer coefficient, $k'' \gg k_{mA}$, then the process is mass transfer limited. For growth this requires that $C_{As} \ll C_{Ab}$, and for dissolution $C_{As} \gg C_{Ab}$.

For mass transfer around a sphere we have from Chapter 7

$$\text{Sh}_D = 2 + 0.4\text{Re}_{D^{1/2}} \tag{9.64}$$

and for slow fluid flow around the sphere (Re $\to$ 0) we have $\text{Sh}_D = 2$ to give

$$k_{mA} = \frac{2D_A}{D} = \frac{D_A}{R} \tag{9.65}$$

with R the radius of the sphere. Thus around a reacting sphere in a stagnant fluid there is a boundary layer of thickness R across which the concentration gradient is formed. As before, we set up the problem to solve for $R(t)$ by writing the differential equation

$$4\pi R^2 \frac{\rho_B}{M_B} \frac{dR}{dt} = \pm 4\pi R^2 r'' = +4\pi R^2 k_{mA} C_{Ab} \tag{9.66}$$

for growth and

$$4\pi R^2 \frac{\rho_B}{M_B} \frac{dR}{dt} = -4\pi R^2 k_{mA} C_{Ab} \tag{9.67}$$

for dissolution. Note that solid species B is either dissolving or forming, but gaseous species A is causing the reaction by migrating between the solid and the fluid.

For growth in a stagnant gas we have

$$4\pi R^2 \frac{\rho_B}{M_B} \frac{dR}{dt} = \frac{4\pi R^2 D_A C_{Ab}}{R} \tag{9.68}$$

or

$$\frac{dR}{dt} = \frac{D_A M_B C_{Ab}}{\rho_B} \frac{1}{R} \tag{9.69}$$

This can be rearranged and integrated to yield

$$\int_{R_0}^{R} R\,dR = \frac{D_A M_B C_{Ab}}{\rho_B} \int_0^t dt \tag{9.70}$$

or

$$\boxed{R^2 = R_o^2 - \frac{2D_A M_B C_{Ab}}{\rho_B} t} \tag{9.71}$$

Thus we see that the square of the radius of a sphere in a stagnant fluid is proportional to time. The time to react or dissolve a sphere of initial radius R_o completely is

$$\boxed{\tau_B = \frac{\rho_B}{2D_A M_B C_{Ab}} R_o^2} \tag{9.72}$$

For mass-transfer-limited growth of a sphere in a stagnant fluid the radius varies with time as

$$\boxed{R^2 = R_o^2 + \frac{2D_A M_B C_{Ab}}{\rho_B} t} \tag{9.73}$$

Example 9–3 In the previous example we calculated the time necessary to grow Si particles from 1 to 10 μm, assuming that the process is reaction limited with $k'' = 10$ cm/min. Calculate the time to form these particles assuming that the process is mass transfer limited with $D_A = 0.1$ cm²/sec $= 6$ cm²/min.

From the text with mass-transfer-limited growth we have

$$R^2 = R_o^2 + \frac{2D_A M_B C_{Ab}}{\rho_B} t \tag{9.74}$$

or

$$t = \frac{\rho_B}{2D_A M_B C_{Ab}}(R^2 - R_o^2) \tag{9.75}$$

The reactant concentration is

$$C_{Ab} = \frac{P_A}{RT} = \frac{0.1}{82 \times 473} = 2.6 \times 10^{-3} \text{ moles/cm}^3 \tag{9.76}$$

and therefore

$$t = \frac{2.33}{2 \times 6 \times 32 \times 2.6 \times 10^{-6}}[(10^{-3})^2 - (10^{-4})^2]$$

$$= 0.0023 \text{ min } = 0.14 \text{ sec} \tag{9.77}$$

The mass-transfer-limited growth time is much shorter than the reaction-limited growth time, and the reaction should therefore be nearly reaction controlled with a reaction time of 2.77 min. If we heated the reactor to increase k'', then the limiting time if $k'' \to \infty$ may be the mass-transfer-limited rate of <1 sec.

9.8.4 Condensation and evaporation

For crystallization or for dissolution of a crystal in a solution C_{As} is usually the saturation concentration of the solute in the solution and C_{Ab} is the concentration in the bulk solution. For growth the solution must be supersaturated ($C_{As} < C_{Ab}$), and for dissolution the solution must be subsaturated ($C_{As} > C_{Ab}$). We call the corresponding gas–solid processes *sublimation* and liquid–solid processes *dissolution*, but the equations we obtain are of course identical for gas or liquid phases.

For formation or evaporation of a liquid from vapor, the equilibrium shape of the liquid is a spherical drop; so the sphere approximation is more appropriate than for a solid particle, where we approximate it as a sphere to make the mathematics simple.

Of course, the growth or dissolution of bubbles from liquid solution has the same geometry and equations. In both drops and bubbles the situation could involve chemical reaction or a simple phase-change process. The equations are the same, but the chemistry is quite different.

9.9 DIFFUSION THROUGH SOLID FILMS

Next we consider the situation where the reaction of the solid is controlled by the rate of a surface reaction at a solid interface rather than at the external surface of the solid. In this situation the rate of transformation of the solid can be limited by diffusion of the reactant *through* the growing or shrinking solid film.

In this section we consider the general reaction

$$A_g + B_s \rightarrow C_s \tag{9.78}$$

with A_g the reactant that must diffuse through a film of C_s to react with another solid B_s to form more C_s. The situation could also be

$$A_g + B_s \rightarrow C_s + D_g \tag{9.79}$$

as long as the gaseous product D_g is rapidly removed from the film.

We listed many examples early in this chapter, and they can be quite different in practice, from the rusting of iron to the toasting of dough to the roasting of ore. The reactant A could be a gas or a liquid, and the film could be a solid or liquid. The migration of A through the reacted film could be diffusion of A dissolved in C or permeation of A through a porous film of C. We describe this by a diffusion coefficient D_{As}, but the value of D_{As} and the mechanism by which transport occurs will not be discussed here.

We emphasize that we are interested in the *fluid-phase* diffusion coefficient of the reactant A, which we call D_A, and also the *solid-phase* diffusion of this species $\mathbf{D}_{As}$ (bold and subscript s). The diffusion of reactant A in either situation can limit the reaction process.

We describe this situation of a reacting film in Figure 9–12. The unreacted solid B_s forms a planar film of thickness ℓ_c on an impermeable substrate, and the reacted layer C_s forms another film over B_s. We shall assume that, as the reaction occurs, the total film thickness ℓ_o remains unchanged, so that, as the reaction proceeds, the position ℓ_c moves from the substrate at $x = 0$ initially to the position $x = \ell_o$ when the reaction has gone to completion.

The reactant concentration is C_{Ab} in the bulk fluid outside the films, and its concentration is C_{As} at the external surface of the film. The reactant concentration falls further within the growing film, and at the interface between B_s and C_s where the concentration is C_{Ac} (the subscript c means core). The reaction occurs at the boundary between solid reactant B and solid product C with a rate

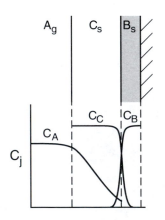

Figure 9–12   Concentration profiles of films where one solid B_s reacts with a gas A_s to form another solid C_s, which forms a film that covers the shrinking film of B_s. In the sketch we assume that C_A goes to zero at the C–B boundary and that this boundary is fairly sharp.

$$r'' = k'' C_{Ac} \tag{9.80}$$

This is a surface reaction, just as for reaction at the external surface of the solid, but now the product is another solid C_s.

If transport of A_g through the film of C_s is very fast compared to r'', then the rate of the process is exactly the same as if no solid film were formed, and this would be the growing (or dissolving) film problem we considered previously. However, now we have the possibility that the rate of the reaction slows down as the film of product thickens.

The subject of this section is to find how the rate should depend on the thickness of the growing (or shrinking) product film, that is, to find expressions for $\ell_c(t)$ as the reaction proceeds. Recall that for planar single films the thickness always changed *linearly* with conversion so that the rate was constant.

We divide the problem into three parts:

1. Find $C_A(x)$ in the reacted film of C_s.
2. Find $r''(\ell)$.
3. Find $\ell_c(t)$.

Once we have done this, we can obtain the conversion

$$X_B = 1 - \frac{\ell_c}{\ell_o} \tag{9.81}$$

to determine the progress of the reaction.

9.9.1 Concentration of A in reacted film

For a planar film of C with no reaction of A within the film, the steady-state flux of A is constant at all values of x; so we solve the equation

$$D_{As} \frac{d^2 C_A}{dx^2} = 0 \tag{9.82}$$

subject to the boundary conditions $C_A = C_{As}$ at the external surface of the solid at $x = 0$ and $C_A = C_{Ac} = 0$ at the reacting interface between B_s and C_s at $x = \ell$. We use the symbol D_{As} to designate the diffusion coefficient of species A within the solid film of product C; this is to distinguish it from the diffusion coefficient D_A in the gas phase outside the solid.

This second-order differential equation can be integrated twice to yield

$$C_A(x) = C_1 x + C_2 \tag{9.83}$$

where C_1 and C_2 are constants to be determined by the boundary conditions. Applying the boundary conditions, this equation becomes

$$C_A(x) = C_{As} \frac{\ell - x}{\ell} \tag{9.84}$$

where $\ell = \ell_o - \ell_c$ as the thickness of the reacted film of C.

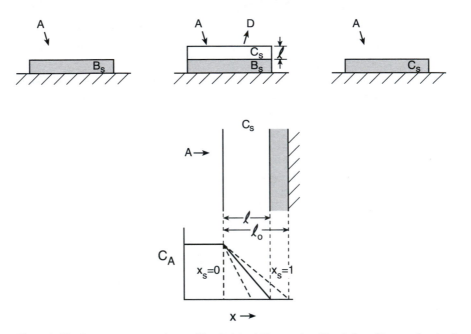

Figure 9–13 Reactant concentration profiles $C_A(x)$ within a product film C_s for a film transforming from B_s to C_s. The profile of the diffusion-limited migration of A through the film remains a straight line; so the growth rate (proportional to the gradient) decreases as the film thickens, leading to the parabolic law of film growth.

9.9.2   Reaction rate at the interface

The rate of the reaction at the B–C interface is equal to the flux to the interface or

$$r'' = -D_{As} \frac{dC_A}{dx} = \frac{D_{As} C_{As}}{\ell} \tag{9.85}$$

which gives the reaction rate in terms of the concentration of reactant at the external surface C_{As} and D_{As} and ℓ.

    We plot the concentration profiles of the reactant in Figure 9–13. It is seen that the gradient of the concentration within the film decreases as the film thickens. This means that the rate of film growth decreases as the film thickens.

9.9.3   Film thickness versus time

Finally we can calculate the variation of the thickness of the growing film with the time of reaction. We assume that at $t = 0$ the film is pure B_s of thickness ℓ_o and that at completion the film has transformed to pure C_s of thickness ℓ_o.

    We make a mass balance on the rate of disappearance of B_s

$$-\frac{dN_{Bs}}{dt} = A_B \frac{\rho_B}{M_B} \frac{d\ell}{dt} = A_B r'' \tag{9.86}$$

where we see that this is identical to the dissolving or growing film except that the surface of B_s (area A_B) is now at the interface with C_s.

Substitution for the rate of the reaction r'' yields

$$\frac{\rho_B}{M_B}\frac{d\ell}{dt} = r'' = -D_{As}\frac{dC_A}{dx} = \frac{D_{As}C_{As}}{\ell} \tag{9.87}$$

or

$$\frac{d\ell}{dt} = \frac{D_{As}M_BC_{Ab}}{\rho_B\ell} \tag{9.88}$$

In this equation the only variables are assumed to be ℓ and t, so the equation can be separated

$$\ell \, d\ell = \frac{D_{As}M_BC_{As}}{\rho_B} \, dt \tag{9.89}$$

and integrated from $\ell = 0$ at $t = 0$ to ℓ at $t = t$,

$$\int_0^\ell \ell \, d\ell = \frac{D_{As}M_BC_{As}}{\rho_B}\int_0^t dt \tag{9.90}$$

This gives

$$\boxed{\ell^2 = \frac{2D_{As}M_BC_{As}}{\rho_B}t} \tag{9.91}$$

The time τ required for a film of B_s to completely transform into C_s is

$$\boxed{\tau_B = \frac{\rho_B}{2D_{As}M_BC_{As}}\ell_o^2} \tag{9.92}$$

which predicts that the time to completely react a film is proportional to the square of the film thickness.

Thus we see that the thickness of the film of C_s increases as

$$\ell \sim t^{1/2} \tag{9.93}$$

according to these calculations. This is an important result for many thin film growth problems, and it is frequently called the parabolic law of film growth.

9.10 TRANSFORMATION OF SPHERES

Another important example of a solid transformation where growth requires diffusion through the product film is the transformation of solid spheres. The principles of this process are similar to those for planar films, but now the concentration profile is not linear, and the expression one obtains for the transformation and the solid conversion is more complex.

9.10.1 Solid loss or formation

We write this mass balance as [rate of transformation of sphere] = [rate of reaction of sphere] = [total flux of A to the surface], or

$$-\frac{dN_{Bs}}{dt} = -4\pi R^2 r''$$ (9.94)

Further, since the system has spherical symmetry, the total flux into the sphere is identical at all radii,

$$\text{total flux of } A = 4\pi R_o^2 \boldsymbol{D}_{As} \left(\frac{dC_A}{dR}\right)_{R_o} = 4\pi R_c^2 \boldsymbol{D}_{As} \left(\frac{dC_A}{dR}\right)_{R_c}$$ (9.95)

$$= 4\pi R^2 \boldsymbol{D}_{As} \left(\frac{dC_A}{dR}\right)_R$$

at the outer radius of the sphere R_o, at the core radius R_c, and at any radius R.

We now combine the above two equations to obtain

$$\frac{dN_{Bs}}{dt} = -4\pi R^2 \boldsymbol{D}_{As} \left(\frac{dC_A}{dR}\right)_R$$ (9.96)

Next we assume that the concentration profile of the reactant A within the reacted coating on the sphere remains in steady state, even though we know that the interface at R_c is moving slowly inward. In this approximation dN_{Bs}/dt is constant; so the above equations can be combined to give

$$\left(\frac{-dN_{Bs}}{dt}\right) = 4\pi R^2 \boldsymbol{D}_{As} \left(\frac{dC_A}{dR}\right)_R$$ (9.97)

This equation can be rearranged,

$$\left(-\frac{dN_{Bs}}{dt}\right) \frac{dR}{R^2} = 4\pi \boldsymbol{D}_{As} dC_A$$ (9.98)

and integrated between R_o, where $C_A = C_{As}$ to R_c, where $C_A = 0$,

$$\left(-\frac{dN_{Bs}}{dt}\right) \int_{R_o}^{R_c} \frac{dR}{R^2} = 4\pi \boldsymbol{D}_{As} \int_{C_{As}}^{0} dC_A$$ (9.99)

which gives

$$\left(-\frac{dN_{Bs}}{dt}\right) \left(\frac{1}{R_o} - \frac{1}{R_c}\right) = -4\pi \boldsymbol{D}_{As} C_{As}$$ (9.100)

so that the loss of B_s in the sphere becomes

$$-\frac{dN_{Bs}}{dt} = -4\pi \boldsymbol{D}_{As} C_{As} / \left(\frac{1}{R_o} - \frac{1}{R_c}\right)$$ (9.101)

which is the steady state rate of loss of B_s in terms of the position of the core R_c.

9.10.2 Boundary motion

Finally we want to know how the number of moles of solid N_{Bs} and the position of the boundary R_c vary with time. The number of moles of solid in a sphere is given by the expression

$$N_{Bs} = \frac{4\pi}{3} R_c^3 \frac{\rho_B}{M_B}$$ (9.102)

Therefore, we can write dN_B/dt in terms of R_c

$$-\frac{dN_{Bs}}{dt} = -\frac{\rho_B}{M_B}4\pi R_c^2 \frac{dR_c}{dt} \tag{9.103}$$

If we equate dN_{Bs}/dt from the previous expression, we obtain

$$-\frac{dN_{Bs}}{dt} = -\frac{\rho_B}{M_B}4\pi R_c^2 \frac{dR_c}{dt} = -4\pi D_{As}C_{As}/\left(\frac{1}{R_o} - \frac{1}{R_c}\right) \tag{9.104}$$

This can be rearranged and solved for the time dependence of R_c,

$$R_c^2\left(\frac{1}{R_o} - \frac{1}{R_c}\right)dR_c = \frac{M_B D_{As}C_{As}}{\rho_B}dt \tag{9.105}$$

or

$$\int_{R_c=R_o}^{R_c} R_c^2\left(\frac{1}{R_o} - \frac{1}{R_c}\right)dR_c = \frac{M_B D_{As}C_{As}}{\rho_B}\int_0^t dt \tag{9.106}$$

This expression can be integrated to yield

$$t = \frac{\rho_B}{M_B D_{As}C_{As}}\int_{R_c=R_o}^{R_c} R_c^2\left(\frac{1}{R_o} - \frac{1}{R_c}\right)dR_c \tag{9.107}$$

$$= \frac{\rho_B R_o^2}{6M_B D_{As}C_{As}}\left[1 - 3\left(\frac{R_c}{R_o}\right)^2 + 2\left(\frac{R_c}{R_o}\right)^3\right] \tag{9.108}$$

Since the conversion of solid $X_s = 1 - (R_c/R_o)^3$, this expression becomes

$$t = \frac{\rho_B R_o^2}{6M_B D_{As}C_{As}}[1 - 3(1 - X_s)^{2/3} + 2(1 - X_s)] \tag{9.109}$$

and the time for complete conversion of a sphere is

$$\boxed{\tau_B = \frac{\rho_B R_o^2}{6M_B D_{As}C_{As}}} \tag{9.110}$$

which shows that the time for complete transformation of a sphere depends on the *square* of the original radius R_o.

9.11   MASS BALANCES IN SOLID AND CONTINUOUS PHASES

In this chapter we have described mass balances of solids (species B_s) in reactions with gases (species A_g). Implicitly we have assumed that we do not have to deal with mass balances in the continuous gas phase because the concentration C_A in the continuous phase was constant. Now we briefly consider how the more general situation must be handled.

Consider the etching reaction

$$A_g + B_s \rightarrow \text{products}, \qquad r'' = k''C_{As} \tag{9.111}$$

in a batch reactor. We assume that B is in the form of spheres of initial radius R_o and that A in the continuous phase has initial concentration C_{Ao}. We assume that there are n_B solid spheres per unit volume.

The mass balance on a single sphere of B is

$$\frac{dN_B}{dt} = 4\pi R^2 \frac{M_B k'' C_{As}}{\rho_B} = -4\pi R^2 r'' = -4\pi R^2 k'' C_{As} \tag{9.112}$$

where the reaction rate is assumed to be proportional to the concentration of A at the surfece C_{As}. If C_{As} is constant, this leads to

$$r(t) = R_o - \frac{M_B k'' C_{As}}{\rho_B} t \tag{9.113}$$

However, if C_A in the gas phase is changing because it is a reactant, we must also solve the mass balance in this phase. The changes required by considering both phases are indicated in Figure 9–14. For a batch reactor this is simply

$$\frac{dC_A}{dt} = -kC_A = -\frac{area}{volume} k'' C_{As} \tag{9.114}$$

This comes from the batch reactor mass balance of Chapter 2 for a process that is first order in C_A and from the modification for a surface reaction from Chapter 7. For n_B particles per unit volume, each with area $4\pi R^2$, the area of the solid per unit volume is

$$\frac{area}{volume} = 4\pi R^2 n_B \tag{9.115}$$

so the mass balance on the continuous phase becomes

$$\frac{dC_A}{dt} = -4\pi R^2 n_B k'' C_{As} \tag{9.116}$$

Thus we have to solve these two coupled equations simultaneously to find $R(t)$ and $C_A(t)$.

If the process is reaction rate-limited ($k'' \ll k_m$), then $C_{As} = C_A$, so the equations become

$$\frac{dR}{dt} = -\frac{M_B}{\rho_B} k'' C_A \tag{9.117}$$

and

$$\frac{dC_A}{dt} = -4\pi R^2 n_B k'' C_A \tag{9.118}$$

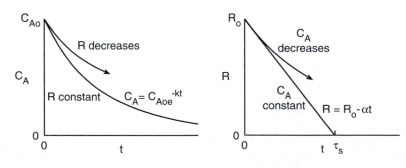

Figure 9–14   Plot of gas-phase reactant concentration C_A (left) and sphere radius R (right) versus time t for a multiphase reaction $A_g + B \rightarrow$ products in a batch reactor. The lower curves are behaviors expected when the other variable is constant while the upper curves represent the coupled situation where both C_A and R are changing.

These equations are of the form

$$\frac{dR}{dt} = \alpha C_A \qquad (9.119)$$

and

$$\frac{dC_A}{dt} = \beta R^2 C_A \qquad (9.120)$$

where α and β are constants.

In the other hand, if the process is mass transfer controlled with $Sh_D = 2.0$, then

$$k'' C_{As} = k_m C_A = \frac{D_A}{R} C_A \qquad (9.121)$$

and the equations have the form

$$\frac{dR}{dt} = \alpha \frac{C_A}{R} \qquad (9.122)$$

and

$$\frac{dC_A}{dt} = \beta R^2 C_A \qquad (9.123)$$

where α and β are constants.

In this situation there are two independent conversions

$$X_s = 1 - \left(\frac{R}{R_o}\right)^3 \qquad (9.124)$$

and

$$X_A = \frac{C_{Ao} - C_A}{C_{Ao}} \qquad (9.125)$$

While the reaction of solids can go to completion, $X_s = 1$, if C_A is constant, the situation becomes more complex if both mass balances must be included. In fact, either A_g or B_s may be the limiting reactant, determined by the values of C_{Ao} and $n_B R_o^2$. Either conversion may approach unity while the other conversion remains less than 1.

In Chapter 14 we will consider multiphase reactors in more detail. The above example is a gas–solid process, but the same analysis applies to any multiphase process such as liquid–solid, liquid–liquid, or gas–liquid. The major complexity introduced in multiphase reactors is that mass balances must be solved simultaneously in both phases. The above example was for reaction of solids in a batch reactor, but in a continuous process there can be different residence times in each phase and each phase may be mixed or unmixed, so the analysis can become much more complicated.

For multiphase reaction systems there is one mass balance equation for each chemical reaction and one for each phase. For nonisothermal reactors, one must add one or more energy balance equations, depending on whether different phases are at different temperatures. Thus an actual reactor situation can become very complex to solve exactly (for C_j, R, T, and conversions) because many equations must be solved simultaneously. However, the material in this chapter should give you the tools to find approximate solutions and to examine qualitative features of designing multiphase reactors.

9.12 ELECTRICAL ANALOGY

The ideas of reaction steps in series and rate-limiting steps can be best understood by the idea of resistors in series. If a voltage V is applied across a resistor with resistance R, the current flow I is given by Ohm's Law,

$$I = \frac{V}{R} \tag{9.126}$$

For three resistors in series, the current through each resistor is identical so that

$$I = \frac{V_3 - V_2}{R_3} = \frac{V_2 - V_1}{R_2} = \frac{V_1 - 0}{R_1} \tag{9.127}$$

It is a simple exercise to eliminate the intermediate voltages to write an expression in terms of V_4 alone:

$$I = \frac{V_3 - 0}{R_3 + R_2 + R_1} \tag{9.128}$$

If one of these resistances is much larger than the others, we say that it is the *limiting resistance* because most of the voltage drop is across it.

Now consider mass transfer and film diffusion in series with reaction. The rate r'' is a current flow (of molecules of reactant), the concentration C_A is a potential, and the rate coefficients k'' are conductivities (the reciprocal of a resistance). Therefore, in analogy with the expression for current flow through resistors in series, we write

$$r'' = k_{\mathrm{m}}(C_{Ab} - C_{As}) = \frac{D_A}{\ell}(C_{As} - C_{Ac}) = k''(C_{Ac} - 0) \tag{9.129}$$

Now eliminating intermediate concentrations (voltages), we obtain

$$r'' = \frac{C_{Ab}}{1/k_{\mathrm{m}} + \ell/D_A + 1/k''} \tag{9.130}$$

Consider in turn the situation where one of the three rate coefficients (conductances) is much smaller than the others. The rate then can be written as

$$r'' = k_{\mathrm{m}}C_{Ab} \tag{9.131}$$

or

$$r'' = \frac{D_A}{\ell}C_{Ab} \tag{9.132}$$

or

$$r'' = k''C_{Ab} \tag{9.133}$$

just as we derived previously. Analogies such as this allow one quickly to write expressions for the effective rate coefficient k_{eff} for any rate processes in series,

$$k_{\mathrm{eff}} = 1/\sum_n \frac{1}{k_n} \tag{9.134}$$

where k_n is the rate coefficient of the nth resistance to flow of A. For diffusion steps we write this rate coefficient as D_A/ℓ. This analogy is only valid if the rate is proportional to the concentration, and this requires first-order reaction kinetics (Figure 9–15).

Figure 9–15 Analogy of voltage drop and concentration changes for processes in series. As long as chemical processes are linear in concentration, these can be written simply.

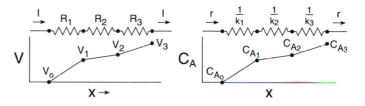

Figure 9–16 Analogy of series and parallel resistors for transformations in composite films.

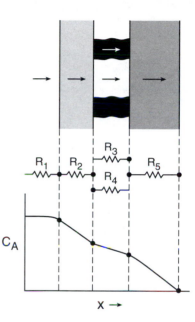

9.12.1 Processes in parallel

If a film is heterogeneous, there may be parallel paths of solid transformations (Figure 9–16). A transformation process can be "short circuited" by the presence of a parallel process, and this can greatly accelerate solid-state transformations. These can also be simplified by considering them as series and parallel processes, and, as long as they are linear (first order in concentrations), the electrical analogs can be used (Figure 9–16).

For resistors in parallel, the current (flux) branches between paths while the voltage (concentration at each end) is constant. The electrical analogy of resistors in series and parallel allows one to solve these problems quite simply.

9.13 SUMMARY

In this chapter we have developed rate expressions $r'' = k_{\text{eff}} C_{Ag}$ for reactions in which gases react with solids to form other gases and solids. We have written this as a fractional conversion X_s and the time τ to react the solid completely, $X_s = 1$.

An important feature of solids reaction is that there are essential mass transport steps of reactants and products either in convective boundary layers or within the reacting solids. These can cause solids reaction processes to be mass transfer limited where the surface reaction rate coefficient does not appear in the reaction rate expression.

Solids reactors are multiphase reactors; so we must consider both the conversions of solid X_s and of fluid phases X to specify these reactors. In this chapter we have not considered the reactor equations that must be solved to predict the overall conversion of reactants in the reactor. As might be expected, we assume PFTR and CSTR reactors, and solve for conversion X in the reactor.

We will consider these multiphase reactors in Chapter 14. We note also that the equations we developed in this chapter for the thickness of a film $\ell(t)$ and radius of a sphere $R(t)$ are also applicable to liquid films and liquid or vapor drops in gas–liquid and liquid–liquid multiphase reactors.

9.14 REFERENCES

Campbell, Stephen A., *Science and Engineering of Microelectronic Fabrication*, Oxford University Press, 2001.

Hess, Dennis, and Jensen, Klavs F., *Microelectronics Processing: Chemical Engineering Aspects*, American Chemical Society, 1989.

King, Alan G., *Ceramics Processing and Technology*, William Andrew Publishing, 2001.

May, Gary S., and Sze, Simon M., *Fundamentals of Semiconductor Fabrication*, Wiley, 2003.

PROBLEMS

9.1 When a powder of spheres of uniform size is stirred in water, 10% dissolves in 10 min. If the dissolution rate at the surface is rate controlling, after what time will the particles completely dissolve? [Answer: ∼5 h.]

9.2 When a powder of spheres of uniform size is stirred in water that is supersaturated with a solute that grows on the surface, the mass of the crystals grows by 10% in 10 min. If the condensation rate at the surface is rate controlling, after what time will the particles double in size?

9.3 You find that the rust layer from a spot on your car fender is 0.3 mm thick 6 months after the paint was chipped off. If the sheet metal is 1 mm thick, when do you have to sell it to be able to tell the buyer it has no rust holes?

(a) Assume dry oxidation where oxidation is parabolic.

(b) Assume that salt (as in a Minnesota winter) makes the film conductive so that the growth rate is linear in time.

9.4 Describe in less than one page each the chemical reactions and processes by which the following solids processes are carried out. Each involves a multistep process with physical separation steps between chemical steps.

(a) Manufacture of aluminum cans from bauxite;

(b) Manufacture of white paint pigment from titanium ore;

(c) Manufacture of silicon wafers from sand;

(d) Manufacture of silicone adhesive from sand;

(e) Manufacture of steel from hematite;

(f) Manufacture of fissionable ^{235}U from pitchblende;

(g) Manufacture of toasted breakfast cereal from wheat flour.

9.5 Develop a model of chemical transformation of a heterogeneous solid film (B_s) assuming a regular perfect packing of solid cubes in a cubic array, each of dimension L packed with a gap space d between cubes.

(a) Describe the film thickness and void fraction in terms of these quantities.

(b) Assume that reaction with a gas on the external surface (concentration C_{Ab}) of the film is very fast so that mass transfer of the gas down the gaps between the cubes limits the reaction of A_g with B_s. Find an expression for the rate of reaction in terms of L, d, and the diffusivity of A in the gas phase.

(c) Assume that reaction has occurred to an extent where the reacted film has a thickness that is a significant fraction of D. How does conversion depend on D and on the diffusivity of A through the solid film?

(d) Assume that each cube is in fact granular (picture sugar cubes) with the migration of A much faster down these micrograins. Develop a model to describe the transformation of this microgranular film.

9.6 Corn flakes of uniform thickness are to be toasted (oxidized) in a conveyer oven. In separate batch experiments it is found that in air at 1 atm 50% of the dough flake is oxidized in 10 min. By cutting the flakes, it is noted that the process proceeds by the shrinking core model.

(a) How long is required completely to oxidize the flakes under these conditions if the process is controlled by the rate of oxidation at the dough surface? By diffusion through the toasted layer? By external mass transfer?

(b) When the reaction is carried out in the oven at particular air flow rate with the conveyer stationary (a batch process), it is found that initially the oxygen content of the air leaving the tube is 2%. How would you predict the time for complete reaction?

(c) In a continuous process how should the roasting depend on the speed of the conveyer and the air flow rate?

(d) How would you model this process including dough heating in the oven and the heat of reaction? How would you modify this to include the temperature profile in the oven and radiation?

(e) How would you modify the situation for a drying process where you want a uniform product?

9.7 One-fourth-in. diameter pellets of a sulfide ore are to be reacted with oxygen to produce the oxide. It is found that 0.1 of the ore is reacted in 10 min. Calculate the time for complete reaction if the process is controlled by:

(a) Uptake of oxygen on the external surface of the pellet;

(b) Diffusion of oxygen through a shrinking core of oxide;

(c) Reaction of oxygen at the boundary of the shrinking core;

(d) Mass transfer at the external surface.

9.8 A 2-cm-diameter tube is coated with a 0.1-mm-thick film of carbon that is to be reacted off by heating in air. The reaction is mass transfer limited with $D_{O_2} = 0.1$ cm²/sec.

(a) At a velocity of 1 cm/sec, how long will be required to remove completely the carbon?

(b) If the velocity is increased to 5 cm/sec, how long will be required to burn off the carbon?

(c) At what flow rate will the flow become turbulent? In turbulent flow how does the time to burn off completely the carbon depend on flow rate?

9.9 In CVD a surface is heated in a gas that decomposes to form a film. At low pressures the reactor behaves as a CSTR because gaseous diffusion is fast enough to maintain a uniform composition throughout the reactor. For silicon formation from silane the reaction

$$SiH_4 \rightarrow Si + 2H_2$$

forms a film 1 μm thick after 10 min. If this reaction is limited by the flux of silane to the surface, which is given by the expression

$$r = \frac{P_j}{(2\pi M_j RT)^{1/2}}$$

where M_j is the molecular weight of the gas. What will be the deposition rates of tetrachlorosilane and tetrabromosilane at the same pressure if reactions are flux limited?

9.10 Crystals of uniform size are prepared by seeding a supersaturated solution with nuclei.

(a) If 1 μm crystals grow to 50 μm in 10 min at constant supersaturation, how long will be required to form 100 μm crystals assuming spherical crystals with $Sh_D = 2$?

(b) How should the growth rate change if the supersaturation of the solution is doubled?

(c) How might you estimate when the growth rate becomes a function of the stirring rate?

9.11 Describe the reaction engineering of a kitchen toaster. The reactions are thermal drying (which alone makes stale bread) and thermal decomposition and oxidation of starch, which requires temperatures of ~300°C. [Recall that the bread appears white until it is nearly done, and then it browns quickly and blackens if left in too long. Recall also that good toast has a brown layer <1 mm thick while the interior is still white and soft. The control of temperature profiles and heat transfer aspects are essential in producing good toast.]

9.12 Design a PFTR toaster. [This is a major type of reactor in the food industry.]

9.13 Design a PFTR solids reactor in which heat from the walls requires a finite time to heat the solid particles to the reaction temperature and compare its operation with the isothermal situation.

9.14 How is the preceding problem changed if the solids are continually mixed? How is the product quality changed from plug flow?

9.15 An important application of formation of a solid from a sphere is the spinning of solid fibers from liquids, as sketched in Figure 9–17. These are usually formed by forcing the liquid through a circular orifice, beyond which it solidifies. Describe the processes and reactions for forming the following fibers.

(a) spaghetti

(b) molten liquid injected into a fluid below the melting point

(c) a liquid solution injected into a fluid in which the solvent is soluble but the fiber is insoluble

(d) a liquid that will react chemically with the fluid beyond the orifice to form a solid

For each of these cases sketch the process, list the reactions, and sketch concentration and temperature profiles parallel and perpendicular to the direction of fiber flow. Consider profiles both around and within the fibers.

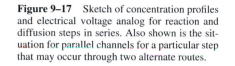

Figure 9–17 Sketch of concentration profiles and electrical voltage analog for reaction and diffusion steps in series. Also shown is the situation for parallel channels for a particular step that may occur through two alternate routes.

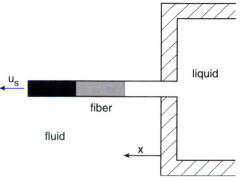

9.16 Consider formation of a solid cylindrical fiber by solvent rejection in which the solvent initially in the liquid being injected into another solvent will dissolve, leaving behind the solute, which solidifies. Consider a process in which the radius of the fiber R_o does not change, but the core radius R_c shrinks as the solvent at concentration $C_A(R)$ diffuses out of the fiber. Sketch solvent concentration profiles expected in steady-state fiber spinning versus distance from the spinneret.

9.17 An important application of formation of solids is *blow molding* to form thin sheets of polymer. This is done by forcing molten polymer such as polyethylene through an annulus to form a tube. The top of the tube is closed and air is forced into the tube as the top is stretched to form a continuous cylindrical balloon typically many feet in diameter. This balloon is then flattened and slit to form a solid film. Sketch this reactor and the temperature profile.

9.18 We need to convert solid Si particles to $SiCl_4$ by reacting them with Cl_2 gas at 300°C in a fluidized bed reactor. Assuming that the process is mass transfer limited with $Sh_D = 2.0$, how long will be required to react completely 10-μm-diameter particles if $P_{Cl_2} = 0.02$ atm? The density of Si is 2.33 g/cm^3, and its atomic weight is 32. Because of the costs of grinding Si into particles, 5 μm particles cost 2 times as much as 10 μm particles, and 20 μm particles cost half as much. Assume $D_{SiCl4} = 0.1$ cm^2/sec. What size particles will be most economical if the reactor conditions remain constant?

9.19 We need to etch a planar Si film by reacting it with Cl_2 gas at 300°C to form $SiCl_4$. Assuming that the process is reaction limited with $k'' = 10$ cm/sec, how long will be required to react completely a 10 μm film if $P_{Cl_2} = 0.02$ atm? The density of Si is 2.33 g/cm^3 and its atomic weight is 32.

9.20 We are depositing a wear resistant titanium carbide coating on steel by the reaction

$$TiCl_4 + C_3H_8 \rightarrow TiC + \text{ other products}$$

The process occurs in excess propane; so the rate is given by the expression $r'' = k'' C_{TiCl_4}$. Calculate the deposition rate if $k'' = 0.1$ cm/sec, $P_{TiCl_4} = 10$ torr, and T=600°C. The density of TiC is 4.94 g/cm^3.

9.21 We need to deposit a silica film on 200 μm wires by reacting tetraethyl siloxane. In the reactor configuration proposed the silane in N_2 at a partial pressure of 10 torr flows over the wires at 1 cm/sec at 200°C. Assuming that the reaction is mass transfer limited and that the wires are widely separated, estimate the rate of the surface reaction and the film growth rate. The density of SiO_2 is 2.66 g/cm^3.

9.22 We need to deposit a silica film on a flat plate by reacting with tetraethyl siloxane. In the reactor configuration proposed the silane in N_2 at a partial pressure of 10 torr flows over the film at 1 cm/sec at 200°C. Assuming that the reaction is mass transfer limited, calculate the average surface reaction rate r'' and the average rate of film growth for a plate 3 cm wide. The density of SiO_2 is 2.66 g/cm^3.

9.23 In electrochemical etching of metals the plate and a counterelectrode are placed in a liquid electrolyte and a voltage is applied to create an electric field between the plates to dissolve the metal as soluble ions. Discuss the following issues and sketch configurations and profiles.

(a) How should the etching rate vary across a plate because of the electric field variations between the plate and a counterelectrode that is the same shape and is parallel to it?

(b) In electrochemical etching gas bubbles are frequently produced. How should the etching rate vary with position for a vertical plate because of natural convection driven by the bubbles?

(c) Discuss the analogous processes with electrochemical deposition of metal films from electrolyte solution.

9.24 Summarize the reasons for the following statements regarding design of solids processing reactors. Indicate situations where these statements may not be appropriate.

(a) Solids should be in plug flow.

(b) The fluid should be in mixed flow.

(c) The reactant fluid conversion should be low.

(d) The fluid velocity should be high.

(e) The solids should be stirred.

(f) There is an optimum residence time for the solid.

(g) Mass transfer limitation should be avoided.

9.25 A carbon particle is growing by decomposition of methane,

$$CH_4 \rightarrow C_s + 2H_2.$$

If the methane partial pressure is 0.01 atm in an inert gas at atmospheric pressure, what time is required for the particle to grow from 0.0001 cm in diameter to 0.01 cm in diameter? Assume that the reaction is mass transfer-limited with $D_{CH_4} = 0.1$ cm^{2}sec^{-1} and assume the concentration of the gas to be 1/22,400 moles/cm^3. Assume $\rho_c = 2.25$ g/cm^3.

9.26 Acetylene is being decomposed on a surface to form a carbon film on the inside walls of a tube of diameter $D = 2R$. The reaction is mass transfer-limited with $Sh_D = 8/3$, and $D_A = 0.2$ cm^2/sec.

(a) Find an expression for the concentration of acetylene C_A versus position in the tube in terms of the above quantities starting with $C_A = C_{Ao}$ at $z = 0$ assuming plug flow with a velocity u.

(b) Find an expression for the film thickness versus time and position in terms of these quantities and ρ_C and M_C.

9.27 Fe particles 100 μm in diameter are to be oxidized to Fe_2O_3 in air,

$$Fe_s + \tfrac{3}{2}O_2 \rightarrow Fe_2O_3s, \qquad r' = k'C_{O_{2s}}$$

Assume that the process occurs by the shrinking core model and that the particles are spheres whose outside radius R_o does not change as reaction proceeds.

(a) Sketch C_{O_2} versus position for the following situations.

    1. Reaction at the metal–oxide interface is limiting.

    2. Reaction at the external surface of the oxide film is limiting.

    3. Mass transfer around the particle is limiting.

    4. Diffusion within the oxide film is limiting.

(b) Find an expression in terms of ρ_{Fe}, M_{Fe}, D_{O_2}, R_o, and C_{O_2} for the time for complete reaction, assuming that the reaction is limited by external mass transfer to the particle and that the air around the particle is stagnant.

9.28 A silicon film is being grown on a planar substrate by reacting dichlorosilane gas in the reaction

$$SiH_2Cl_2 \rightarrow Si + 2HCl, \qquad r = k''C_{silane}$$

The density of silicon is 3 g/cm^3, the molecular weight of Si is 32, and the concentration of dichlorosilane is 0.02 moles/cm^3.

(a) Set up the expression for the rate of film growth in terms of ρ_{Si}, M_{Si}, k'', and C_{silane}.

(b) Calculate the time to grow a film 0.01 cm thick if mass transfer is fast and $k'' = 2 \times 10^2$ cm/sec.

(c) How long will be required if $k_m = 1 \times 10^2$ cm/sec?

9.29 A sphere is being chemically etched, and we find that one-half is reacted away in 10 minutes.

(a) How long will be required for the sphere to completely disappear if the process is reaction-limited?

(b) How long will be required for the sphere to completely disappear if the process

(c) Assuming $k_{mA} \ll k''$, find $R(t)$ in terms of these quantities.

Chapter **10**

CHAIN REACTIONS, COMBUSTION REACTORS, AND SAFETY

C ombustion processes are fast and exothermic reactions that proceed by free-radical chain reactions. Combustion processes release large amounts of energy, and they have many applications in the production of power and heat and in incineration. These processes combine many of the complexities of the previous chapters: complex kinetics, mass transfer control, and large temperature variations. They also frequently involve multiple phases because the oxidant is usually air while fuels are frequently liquids or solids such as coal, wood, and oil drops.

In addition to the importance of combustion reactors in chemical processes, uncontrolled combustion reactions create the greatest potential safety hazard in the chemical industry. Therefore, all chemical engineers need to understand the basic principles of combustion reactors to recognize the need for their proper management and to see how improper management of combustion can cause unacceptable disasters.

Polymerization also involves chain reactions; so this chapter also introduces some of the concepts we will need in the following chapter on polymerization processes.

10.1 CHAIN REACTIONS

We begin with a classic example from kinetics, the decomposition of acetaldehyde to methane and carbon monoxide

$$CH_3CHO \rightarrow CH_4 + CO \qquad (10.1)$$

We are repeating a discussion from Chapter 4, but here we emphasize that this is a chain reaction. This reaction is found experimentally to be irreversible and proportional to the $\frac{3}{2}$ power of the acetaldehyde concentration,

$$r = kC_A^{3/2} \qquad (10.2)$$

rather than being first order as might be expected from stoichiometry. [We will use the symbol A for acetaldehyde.] This is a simple chain reaction, and it is a prototype of many chain reactions. The mechanism by which this reaction proceeds is thought to involve four major steps

$$A \rightarrow CH_3\cdot + CHO\cdot, \qquad r_i = k_i[A] \qquad\qquad \text{(i)} \qquad (10.3)$$

$$A + CH_3\cdot \rightarrow CH_4 + CH_3CO\cdot, \qquad r_{p1} = k_{p1}[A][CH_3\cdot] \qquad \text{(p1)} \qquad (10.4)$$

$$CH_3CO\cdot \rightarrow CO + CH_3\cdot, \qquad r_{p2} = k_{p2}[CH_3CO\cdot] \qquad\qquad \text{(p2)} \qquad (10.5)$$

$$2CH_3\cdot \rightarrow C_2H_6, \qquad r_t = k_t[CH_3\cdot]^2 \qquad\qquad\qquad \text{(t)} \qquad (10.6)$$

Note first that if we consider only the reaction steps named p1 and p2, we obtain the overall reaction by simply adding them,

$$A + CH_3\cdot \rightarrow CH_4 + CH_3CO\cdot \qquad\qquad\qquad (10.7)$$

$$\underline{CH_3CO\cdot \rightarrow CO + CH_3\cdot} \qquad\qquad\qquad (10.8)$$

$$A \rightarrow CH_4 + CO$$

Thus the overall reaction to make these products is obtained by *omitting* the first and fourth reactions. However, these reactions would not proceed without the first reaction to generate $CH_3\cdot$, which is required for the other steps, and the propagation reactions would have zero rate if $CH_3\cdot$ and $CH_3CO\cdot$ were not present. We have anticipated this by naming the first step initiation (i), the second and third steps propagation (p1) and (p2), and the fourth step termination (t). Physically this means that the propagation steps occur much faster than initiation and termination steps. Therefore, the side products CH_4 and CO of the propagation steps are formed in much larger amounts than the products $CHO\cdot$ and C_2H_6 from initiation and termination steps, and the concentrations of these species may be immeasurably small.

We also note that some of the products are stable molecules (CH_4, CO, and C_2H_6), while others are free-radical molecules ($CH_3\cdot$, $CH_3CO\cdot$, and $CHO\cdot$), which are highly unstable and will quickly react with most stable molecules in the system or with other radical species whenever they collide. We will write dots in the symbols for these molecules to indicate that they have an unpaired electron.

In this reaction system traces of ethane formed by the termination and probably traces of formaldehyde (formed by a hydrogen abstraction by formyl) are also formed, but their concentrations are always much less than those of CH_4 and CO.

We can of course write the six mass balances on each species and solve them in batch or continuous reactors to find the species concentrations as a function of residence time: $[A](\tau)$, $[CH_3\cdot](\tau)$, $[CH_4](\tau)$, etc. The mass-balance equations in a PFTR (or in a batch reactor by replacing τ by t) are

$$\frac{d[A]}{d\tau} = -k_i[A] - k_{p1}[A][CH_3\cdot] \qquad\qquad (10.9)$$

$$\frac{d[CH_3\cdot]}{d\tau} = +k_i[A] - k_{p1}[A][CH_3\cdot] + k_{p2}[CH_3CO\cdot] - 2k_t[CH_3\cdot]^2 \qquad (10.10)$$

$$\frac{d[CH_3CO\cdot]}{d\tau} = +k_{p1}[A][CH_3\cdot] - k_{p2}[CH_3CO\cdot] \qquad\qquad (10.11)$$

$$\frac{d[CO]}{d\tau} = k_{p2}[CH_3CO\cdot] \qquad\qquad (10.12)$$

$$\frac{d[CH_4]}{d\tau} = k_{p1}[A][CH_3\cdot] \qquad\qquad (10.13)$$

In a CSTR these mass-balance equations become

$$\frac{[A] - [A_o]}{\tau} = -k_i[A] - k_{p1}[A][CH_3\cdot] \tag{10.14}$$

$$\frac{[CH_3\cdot] - [CH_{3o}\cdot]}{\tau} = +k_i[A] - k_{p1}[A][CH_3\cdot] + k_{p2}[CH_3CO\cdot] - 2k_t[CH_3\cdot]^2 \tag{10.15}$$

$$\frac{[CH_3CO\cdot] - [CH_3CO\cdot_o]}{\tau} = +k_{p1}[A][CH_3\cdot] - k_{p2}[CH_3CO\cdot] \tag{10.16}$$

$$\frac{[CO] - [CO_o]}{\tau} = k_{p2}[CH_3CO\cdot] \tag{10.17}$$

$$\frac{[CH_4] - [CH_{4o}]}{\tau} = k_{p1}[A][CH_3\cdot] \tag{10.18}$$

Note that the right-hand sides of these equations are identical in any reactor.

We can immediately see several features of this reaction system. Since $[CH_3\cdot]$ and $[CH_3CO\cdot]$ are very reactive, and their concentrations are always very low. Therefore, we eliminate terms in the mass-balance expressions for these species,

$$\frac{[CH_3\cdot] - [CH_3\cdot_o]}{\tau} = 0 = +k_i[A] - k_{p1}[A][CH_3\cdot] + k_{p2}[CH_3CO\cdot] - 2k_t[CH_3\cdot]^2 \tag{10.19}$$

$$\frac{[CH_3CO\cdot] - [CH_3CO\cdot_o]}{\tau} = 0 = +k_{p1}[A][CH_3\cdot] - k_{p2}[CH_3CO\cdot] \tag{10.20}$$

to obtain

$$k_{p1}[A][CH_3\cdot] = k_{p2}[CH_3CO\cdot]$$
$$k_i[A] = 2k_t[CH_3\cdot]^2 \tag{10.21}$$

Thus in this chain reaction system we find that it is a good approximation to assume that (1) the rates of the two propagation steps are exactly equal and (2) the initiation and termination steps are exactly equal.

It is instructive to sketch this reaction as in Figure 10–1. The chain is "fed" by initiation reactions that create $CH_3\cdot$ and "terminated" by reactions that destroy $CH_3\cdot$ to form C_2H_6. However, the major processes by which the reaction proceeds are the two propagation steps that alternately create and destroy the two chain-propagating radicals

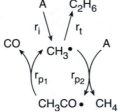

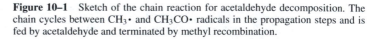

Figure 10–1 Sketch of the chain reaction for acetaldehyde decomposition. The chain cycles between $CH_3\cdot$ and $CH_3CO\cdot$ radicals in the propagation steps and is fed by acetaldehyde and terminated by methyl recombination.

$CH_3 \cdot$ and $CH_3CO \cdot$. Thus we have the notion of a "chain" of reactions, which is a kinetic chain of propagation reaction steps that feed reactants into the chain and spit out stable products. This is a *kinetic chain* reaction, which is different from the *polymerization chain* of reactions, which we will introduce in the next chapter (although those reactions form both kinetic and polymer chains).

10.1.1 Autocatalytic reactions

In another view of this reaction system, we can simplify the two propagation steps by adding the p1 and p2 steps to yield a single reaction

$$A + CH_3 \cdot \rightarrow CO + CH_4 + CH_3 \cdot \tag{10.22}$$

in which $CH_3 \cdot$ is seen as the molecule that catalyzes the overall reaction (the rate is proportional to the $CH_3 \cdot$ concentration), but $CH_3 \cdot$ is neither consumed or generated in the chain reaction steps; so this molecule is a proper *catalyst*.

We call this type of reaction autocatalytic because the initiation step generates the catalyst $[CH_3 \cdot]$, which is necessary to make the overall process proceed, and the concentration of the catalyst molecule is not changed by the propagation reaction steps. It is called *autocatalytic* because the reaction itself generates the catalyst that promotes the reaction. Note also that the overall rate is enhanced by large k_i and inhibited by large k_t even though k_i and k_t are much smaller than k_p.

Another characteristic of chain reactions is that they generate *minor products* (C_2H_6 and $CHO \cdot$) as well as the major products (CO and CH_4). The major products are made by the propagation steps and the minor products by the initiation and termination steps. In fact, the ratio of CH_4 to C_2H_6 gives the ratio of r_p to r_t,

$$\frac{[CH_4]}{[C_2H_6]} = \frac{r_p}{r_t} \tag{10.23}$$

This ratio is sometimes called the *kinetic chain length*.

10.1.2 Pseudo-steady-state approximation

Students may have seen the acetaldehyde decomposition reaction system described as an example of the application of the pseudo steady state (PSS), which is usually covered in courses in chemical kinetics. We dealt with this assumption in Chapter 4 (along with the equilibrium step assumption) in the section on approximate methods for handling multiple reaction systems. In this approximation one tries to approximate a set of reactions by a simpler single reaction by invoking the pseudo steady state on suitable intermediate species.

We can do this in either batch or PFTR by setting $d[CH_3 \cdot]/d\tau$ and $d[CH_3CO \cdot]/d\tau = 0$ or in CSTR by setting $([CH_3 \cdot] - [CH_3 \cdot])/\tau = ([CH_3CO \cdot] - [CH_3CO \cdot])/\tau = 0$. Solving the resulting algebraic equations for $[CH_3 \cdot]$ yields

$$[CH_3 \cdot] = \left(\frac{k_i}{2k_t} \right)^{1/2} [A]^{1/2} \tag{10.24}$$

Substitution into the rate of production of CH_4 formed by this reaction yields

$$r = \frac{d[CH_4]}{dt} = k_{p1}[A][CH_3\cdot] = \left(\frac{k_i}{2k_t}\right)^{1/2} k_{p1}[A]^{3/2} \tag{10.25}$$

which is the experimentally observed rate expression.

Textbooks state that the pseudo-steady-state approximation will be valid if the concentration of a species is small. However, one then proceeds by setting its *time derivative* equal to zero ($d[CH_3CO\cdot]/dt = 0$) in the batch reactor equation, not by setting the concentration ($CH_3CO\cdot$) equal to zero. This logic is not obvious from the batch reactor equations because setting the derivative of a concentration equal to zero is not the same as setting its concentration equal to zero.

However, when we examine the CSTR mass balance, we see that the pseudo-steady-state approximation is indeed that the concentration be small or that $[CH_3CO\cdot]/\tau = 0$. Thus by examining the CSTR version of the mass-balance equations, we are led to the pseudo-steady-state approximation naturally. This is expressly because the CSTR mass-balance equations are developed assuming steady state so that the pseudo-steady-state approximation in fact implies simply that an intermediate species is in steady state and its concentration is small.

10.1.3 The generic chain reaction

Consider the preceding reaction as an example of a general reaction process

$$A \rightarrow B + C \tag{10.26}$$

which is propagated by a single radical $R\cdot$. We write the mechanism as

$$n_i A \rightarrow R\cdot, \qquad r_i = k_i[C_A]^{n_i} \qquad \text{(i)} \tag{10.27}$$

$$A + R\cdot \rightarrow B + C + R\cdot, \qquad r_p = k_p[C_A][C_R] \qquad \text{(p)} \tag{10.28}$$

$$n_t R\cdot \rightarrow X, \qquad r_t = k_t[C_R]^{n_t} \qquad \text{(t)} \tag{10.29}$$

where n_i and n_t are the number of molecules that react in the initiation and termination steps, respectively. Since these steps are assumed to be elementary, the orders of the reactions are as indicated.

Now if we make the PSS approximation on $[C_R]$, we obtain

$$\boxed{r = \left(\frac{k_i}{n_t k_t}\right)^{1/n_t} k_p[C_A]^{1+n_i/n_t}} \tag{10.30}$$

as the overall rate expression for this reaction in the pseudo-steady-state approximation.

This reaction is of the form

$$r = k_{eff}[C_A]^{n_{eff}} \tag{10.31}$$

and we see that the effective order of the reaction with respect to the reactant A is

$$n_{eff} = 1 + \frac{n_i}{n_t} \tag{10.32}$$

which can easily have values of 1, $\frac{3}{2}$, 2, $3, \ldots$ for fairly reasonable assumptions on the initiation and termination steps. The effective rate coefficient is

$$k_{\text{eff}} = \left(\frac{k_i}{n_t k_t}\right)^{1/n_t} k_p \tag{10.33}$$

which is a product of individual rate coefficients raised to appropriate exponents. By writing each of these ks for initiation, propagation, and termination as

$$k = k_o \exp\left(\frac{-E}{RT}\right) \tag{10.34}$$

we see that the effective activation energy becomes

$$E_{\text{eff}} = \frac{E_i}{n_t} - \frac{E_t}{n_t} + E_p \tag{10.35}$$

These examples illustrate several features of all chain reactions. For example, if $n_i = n_t = 1$, we obtain a rate of the overall reaction

$$r = \frac{k_i}{k_t} k_p [A]^2 \tag{10.36}$$

which gives an effective rate coefficient

$$k_{\text{eff}} = \frac{k_i}{k_t} k_p = k_{o,\text{eff}} e^{-E_{\text{eff}}/RT} \tag{10.37}$$

and an activation energy

$$E_{\text{eff}} = E_i - E_t + E_p \tag{10.38}$$

Thus, even if the propagation steps have very low activation energies and are very fast, the effective activation energy still contains the term E_i, which can be very large, perhaps as large as 100 kcal/mole.

This generic chain reaction can be sketched similarly to the acetaldehyde decomposition reaction as shown in Figure 10–2. The circular chain propagates itself indefinitely with a rate r_p once initiated by rate r_i, but it is terminated by rate r_t, and in steady state r_i and r_t control how "fast" the cycle runs. The overall reaction rate is controlled by the concentration or the chain-propagating radical C_R because this controls how many molecules are participating in the chain. This is why r_i and r_t are so important in determining the overall rate.

Figure 10–2 Sketch of a simple generic chain reaction $A \rightarrow B + C$ propagated by a radical species $R\cdot$.

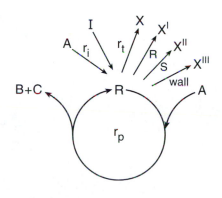

Example 10–1   Consider the chain reaction

$$A \rightarrow R, \qquad r_i = k_i C_A \tag{10.39}$$

$$A + R \rightarrow B + C + R, \qquad r_p = k_p C_A C_R \tag{10.40}$$

$$R \rightarrow X, \qquad r_t = k_t C_R \tag{10.41}$$

in a CSTR assuming constant density.

(a) Write the mass-balance equations for A, B, R, and X in a CSTR assuming constant density.

The CSTR mass-balance equations for these species are

$$\frac{C_{Ao} - C_A}{\tau} = +k_i C_A + k_p C_A C_R \tag{10.42}$$

$$\frac{C_R}{\tau} = k_i C_A - k_t C_R \tag{10.43}$$

$$\frac{C_B}{\tau} = k_p C_A C_R \tag{10.44}$$

$$\frac{C_X}{\tau} = k_t C_R \tag{10.45}$$

(b) What is the overall reaction rate with respect to C_A?

First we need to find C_R in terms of C_A,

$$\frac{C_R}{\tau} = k_i C_A - k_t C_R = 0 \tag{10.46}$$

so that

$$C_R = \frac{k_i}{k_t} C_A \tag{10.47}$$

Therefore the rate becomes

$$r = k_p C_A C_R \tag{10.48}$$

$$r = \frac{k_i k_p}{k_t} C_A^2 \tag{10.49}$$

and exhibits second-order kinetics.

(c) Find τ for 90% conversion of A in a CSTR assuming pseudo steady state if $C_{Ao} = 2$ moles/liter, $k_i = 0.001$ sec^{-1}, $k_p = 20$ liter/mole sec, and $k_t = 0.1$ sec^{-1}.

Solving the CSTR equation, we obtain a residence time of

$$\tau = \frac{k_t}{k_i k_p} \frac{C_{Ao} - C_A}{C_A^2} = \frac{0.1 \times (2.0 - 0.2)}{0.001 \times 20 \times 0.2^2} = 225 \text{ sec} \tag{10.50}$$

(d) What are C_R and C_X for this conversion?

From the above, we obtain the concentrations

$$C_R = \frac{k_i}{k_t}C_A = \frac{0.001 \times 0.2}{0.1} = 0.002 \text{ mole/liter} \tag{10.51}$$

$$C_X = \tau k_t C_R = 225 \times 0.1 \times 0.002 = 0.045 \text{ mole/liter} \tag{10.52}$$

Since $C_B = C_C = 1.8$ moles/liter from stoichiometry, we see that C_R and C_X are indeed small compared to the major product, as expected.

10.2   CHARACTERISTICS OF CHAIN REACTIONS

From these simple examples we see that chain-radical-propagated combustion processes frequently have very complicated kinetics.

Chain reactions can have extremely large temperature dependences. This produces reactions that are negligible at low temperatures but can be extremely fast at high temperatures. Chain reactions are also very sensitive to trace impurities that can alter the initiation and termination rates and control the overall processes even though they do not participate in the propagation steps by which most of the products are formed. *Scavengers* and *initiators* (also called *poisons* and *promoters*) can have large influences on the characteristics of chain reactions.

10.2.1   Initiators

We wrote the initiation reaction as caused by decomposition of the reactant

$$n_i A \rightarrow R, \qquad r_i = k_i C_A^{n_i} \qquad \text{(i)} \tag{10.53}$$

Chain reactions can also be initiated by intentionally adding a species I that easily forms radicals and acts to initiate the reaction faster than the reactant. We write this reaction as

$$I \rightarrow R, \qquad r_{iI} = k_{iI} C_I \qquad \text{(iI)} \tag{10.54}$$

and the rate coefficient k_{iI} can be much larger than k_i initiated by the reactant. We will leave for a homework problem the formulation of kinetics with added initiators.

10.2.2   Scavengers

We also wrote the termination step as if the radical species decomposed or reacted with other radical species to form an inactive species X,

$$n_t R \rightarrow X \qquad r_t = k_t C_R^{n_t} \qquad \text{(t)} \tag{10.55}$$

We can also terminate a chain reaction by intentionally adding a scavenger species S that readily scavenges the chain propagator R. We write this reaction step as

$$R + S \rightarrow X, \qquad r_{tS} = k_{tS} C_R C_S \qquad \text{(tS)} \tag{10.56}$$

and again, this reaction rate coefficient may be very large so that this termination rate may be much larger than the "natural" termination rate without intentional addition of an additional species. We will also save for a homework problem the calculation of rate expressions with terminators.

It is also interesting to ask "what is X?", the species formed by the termination reaction. Basically, we seldom care or know what this species is because its concentration is extremely small and we seldom measure it. These species are the products of termination reactions, which are very important in determining the overall reaction rate but are unimportant in determining the major products formed.

10.2.3   Wall termination reactions

Surface reaction steps are frequently very important in controlling chain reactions. [Catalytic reactions again!] For example, the termination steps by which free-radical intermediates are removed will frequently occur readily on surfaces simply by *adsorption*. We can write this reaction as

$$R + S \rightarrow X, \quad r_{tS} = k_{tS} C_R C_S \tag{10.57}$$

but now S signifies a surface reaction site rather than a scavenger molecule in the same phase as the other species. Just as with homogeneous scavengers, this rate may be many orders of magnitude smaller than the propagation rates, and a small surface area in a reactor can have a large effect on the overall rate.

Wall termination reactions immediately introduce a complexity to all chain reactions, namely, that the overall reaction rate can be a strong function of the size of the reactor. In a small reactor where the surface-to-volume ratio is large, termination reactions on surfaces can keep the radical intermediate pool small and thus strongly inhibit chain reactions (nothing appears to happen), while in a large reactor the surface-to-volume ratio is smaller so that the termination rate is smaller and the effective rate increases by a large factor (and the process takes off).

The rate of termination of the chain reaction by surface scavenging of radicals can be estimated because r_t is frequently a mass transfer rate if the surface reaction rate is very large. Just as in Chapter 7, we write this rate as

$$r_t = \left(\frac{\text{area}}{\text{volume}} \right) r_{t''} \tag{10.58}$$

with

$$r_{t''} = k_{t''} C_{Rs} = k_m (C_{Rb} - C_{Rs}) \tag{10.59}$$

but now the surface reaction is only a part of the overall homogeneous chain reaction, because it affects only the termination step. Therefore, we can estimate k_t by calculating the Sherwood number for mass transfer since

$$k_{t''} = k_m = \frac{\text{Sh}_D D_A}{D} \tag{10.60}$$

and therefore the termination rate coefficient for a process controlled by surface termination of a chain reaction is given by

$$k_t = \frac{[\text{area}]}{[\text{volume}]} k'' = \frac{[\text{area}]}{[\text{volume}]} k_m \tag{10.61}$$

where D is a characteristic length such as the vessel diameter and Sh must be calculated for the reactor flow pattern. [Mass transfer again!]

In the simplest version of a surface reaction, the rate of termination of the chain reaction by reaction of R on a surface should be proportional to D^{-2}, the square of the vessel diameter! We will also leave this calculation for a homework problem, but the implications are profound for a chain reaction. When the vessel size increases, the rate of a chain reaction can increase drastically, from a slow surface-quenched process in small vessels to very fast process whose only quenching steps are homogeneous reactions.

Chain reactions can also be very sensitive to the materials used for constructing the walls of the reactor because the adsorption and reaction properties of the walls in promoting or retarding termination and initiation steps depend on the nature of these surfaces. These effects have caused much difficulty in establishing the "true" rates of chain reactions and, more important, have led to accidents when chain reactions propagated unexpectedly because of unpredicted kinetics, which can be quite different from those obtained in a small lab reactor. One common difference is that lab data are frequently obtained in glass-walled vessels, which are quite unreactive, while large-scale reactors are frequently made of steel, whose walls have different reactivities.

10.2.4   Scaleup of chain reactions

Chemical engineers need to be very aware of these size effects in scaling up any process that might involve chain reaction processes. The lab data, even if carefully obtained, may not explain large-reactor behavior. The rate of a chain reaction may be much higher in a large reactor. This might be good news unless the rate is too high and results in reactor runaway, where the temperature cannot be controlled. Similarly, one may be planning to run a process that does not occur by chain reactions, but in a large reactor a "side reaction" that occurs by a chain reaction could take over and dominate the process. Combining complicated kinetics with the possibility of thermal runaway can make engineers old before their time.

10.3   AUTOOXIDATION AND LAB SAFETY

Every student who works in a chemical laboratory is warned that organic chemicals should never be stored for long times because they can explode without warning or apparent provocation. Bottles of organic chemicals should always be disposed of within a few years, and, if a chemical is found that might be very old, a "bomb squad" of safety experts should be called to remove it and properly dispose of it.

Why do all organic chemicals just sitting around the laboratory at room temperature have the potential to explode spontaneously? The answer is that they can react with O_2 in the air at room temperature in chain reactions to form organic peroxides, which can spontaneously react explosively upon shaking or opening the cap. Organic peroxides are examples of compounds that have fuel (C and H atoms) in the same compound with the oxidant, and these are solid and liquid explosives similar to dynamite and TNT, which we will discuss later in this chapter.

In this section we will consider the mechanism by which a harmless organic molecule can react with O_2 at room temperature to form a potentially lethal system. Then we will discuss positive applications of this process, which are the primary processes by which such common chemicals such as acetone, phenol, isobutylene, styrene, and propylene glycol are made commercially.

10.3.1   A simple model of autooxidation

Consider any organic molecule, which we will call R–H, where $R\cdot$ could be an alkyl or any fragment containing C, H, and O atoms. In the presence of O_2 this molecule can undergo the autooxidation reaction

$$R\text{–}H + O_2 \rightarrow ROOH \qquad (10.62)$$

This reaction never happens significantly as a single step at any temperature because the R–H bond strength is >80 kcal/mole for any C–H bond. However, consider the chain reaction sequence

$$R\text{–}H \rightarrow R \cdot + H\cdot, \qquad \text{initiation} \qquad (10.63)$$

$$R \cdot + O_2 \rightarrow ROO\cdot, \qquad \text{oxidation} \qquad (10.64)$$

$$ROO \cdot + R - H \rightarrow ROOH + R \cdot \qquad \text{radical transfer} \qquad (10.65)$$

If we add the second and third steps, we obtain the chain reaction

$$R\text{–}H + O_2 \rightarrow ROOH \qquad (10.66)$$

by which a hydrocarbon is converted into a peroxide.

The first step is a very slow dissociation step, with an activation energy >80 kcal/mole. However, once the radical $R\cdot$ is made, it will rapidly react with O_2 to form the alkylperoxy radical, which is also very reactive and can abstract an H atom from any organic molecule in the solution to form the relatively stable (until someone shakes the bottle) alkyl peroxide.

We call this type of reaction *autooxidation* because it is a an autocatalytic process (the reaction generates radical intermediates that propagate chain reactions) and it is an oxidation that converts alkanes into alkyl peroxides.

This chain reaction is just like those discussed previously, except now the reactants are an organic molecule and O_2, and the product is a potentially explosive hydroperoxide. It can be sketched as a chain in Figure 10–3. The features of this chain reaction that make it so explosive are

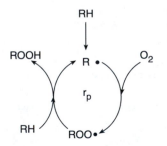

Figure 10–3   Chain reaction for alkane oxidation by autooxidation in which the alkyl and alkylperoxy radicals propagate the chain, which is fed by the alkane RH.

1. Organic peroxides have very exothermic heats of decomposition because they carry both fuel and oxidant in one molecule. We will discuss this part of the process (the explosion) later in this chapter.

2. The reactions can only occur if O_2 is present. A bottle of an organic chemical is perfectly safe for centuries if the cap is impermeable to O_2 and is kept tightly fastened. (A cork is not a suitable cap.) However, one never knows whether someone may have left the cap off or loose for some time to allow the peroxide to form significantly.

3. This reaction depends crucially on the initiation step, and different molecules have vastly different capabilities of dissociation. The major problem here is that the initiation step can be initiated by a trace amount of an *impurity* that no one knows is in the bottle. As examples, all tertiary C–H bonds are fairly weak, as are the C–H bonds in aldehydes and ethers. The reaction can be initiated by traces of these compounds, and the chain propagated throughout all molecules in the otherwise quite unreactive liquid.

4. The initiation step can be *photoinduced*. If a bottle is sitting in sunlight, UV photons (fluorescent lights are also more dangerous than incandescent lights) can cause photodissociation to initiate the chain reaction much faster than in the dark. [Guess why many chemicals are sold in brown bottles?]

Since one cannot be sure about the history of any old bottle of a chemical, one should always dispose of old chemicals very carefully.

10.3.2   Spoilage of food

Spoilage of food is a very important engineering issue in the food industry. This has two main forms: bacterial and yeast growth and oxidation.

In bacterial and yeast growth, improper sterilization during processing leaves organisms that reproduce and cause problems by creating large concentrations of organisms that taste bad or, even worse, make the consumer sick. Living organisms grow autocatalytically (modeled as $A \rightarrow 2A$, $r = kC_A$); so the concentrations in foods are inherently unpredictable, as we discussed in Chapter 12.

These organisms have two varieties, aerobic and anaerobic, signifying with or without oxygen. Thus one type of microorganism growth is basically an oxidation process, but now the organism can be regarded as a catalyst for oxidation. Anaerobic organisms are in fact more dangerous than aerobic organisms because they tend to produce more dangerous toxins, such as botulism and salmonella.

Another significant spoilage mechanism of food is *autooxidation*, the subject of this section. Here O_2 from air at room temperature causes oxidation of ingredients in food to produce molecules such as aldehydes and organic acids, both of which generally taste bad, although they are not usually toxic. The mechanism of food turning rancid is the same as we have discussed in this section: peroxide formation, which is autocatalytic, leading to rapid growth. As in all these processes, the rate is controlled by traces of particular compounds that initiate peroxide formation, which then propagates throughout the food sample.

No amount of sterilization will prevent or even slow autooxidation, and there are only two defenses: *removal of O_2 and addition of inhibitors*. Oxygen barriers in food packaging are a major topic in the engineering of polymer films. The barrier properties of various polymers are very important in food applications, and many of these are multilayer polymers that have a thin layer of an impermeable polymer (such as polyacrylonitrile and ionic polymers) on a cheaper but O_2-permeable polymer such as a polyolefin, which gives mechanical strength to the film.

The other method of oxidation prevention in foods is the addition of chemicals that act as antioxidants. Nitrite solutions on meat or lettuce preserve the red or green color and make the package look fresher. The long lists of "additional ingredients" you see on packages of foods are either antioxidants or chemicals that slow bacterial growth. The phrase "all natural ingredients" on a package means that you need to be especially careful regarding spoilage.

One example of autooxidation you have experienced is the room-temperature oxidation of alcohol in wine. Within days and sometimes within minutes of opening a bottle of wine, the taste begins to deteriorate because of autooxidation. This converts the ethanol into acetaldehyde and to acetic acid, both of which taste bad. Wine lovers talk about letting the wine "breathe" after opening; so apparently some oxidation actually helps the taste. Distilled vinegar is made by the intentional oxidation of the alcohol in fermented apple juice into acetic acid, which can then be distilled from the juice and pulp.

Both biological reactions and autooxidation reactions in food are complex problems because both are highly autocatalytic. Their occurrence is therefore very variable between nominally identical samples of food. When a reaction is both unpredictable and intolerable, the chemical engineer must be especially alert.

10.3.3 Antioxidants

There is considerable discussion in the popular press about the role of antioxidants in promoting human health, from preventing cancer to slowing aging. These molecules presumably act to prevent autooxidation within our bodies. The processes are very complex and not well understood because they are different in different regions of our bodies and because the foods and additives we consume may be rapidly transformed into other molecules that may or may not be antioxidants. These molecules act as free-radical scavengers within our bodies, which prevent the chain reactions of autooxidation. Because these processes are so complex, no one knows how a particular chemical will act when ingested, and therefore additives can be sold that have no beneficial effect (as long as they are not the "synthetic chemicals" to which we are told we should never be exposed).

Room-temperature oxidation processes supply the energy by which we exist, but it is important that only the proper oxidation reactions occur in our bodies and not the ones that cause our skin to wrinkle and our brains to perform badly on exams. If you could come up with a chemical to assist in either of these, you could become very wealthy or get in trouble with the FDA.

10.4 CHEMICAL SYNTHESIS BY AUTOOXIDATION

Now we consider some very positive examples of this type of reaction sequence. Some organic molecules have weak C–H bonds that are easily broken. This fact has been exploited for some key industrial reactions. Some of these weak chemical bonds are listed

TABLE 10–1
Bond Energies

Bond	E (kcal/mole)
H–H	104
CH_3–H	103
C_2H_5–H	98
$(CH_3)_3C$–H	91
HO–H	119
HO–OH	51
CH_3O–H	104
CH_3–OH	90
C_2H_5O–H	105
C_2H_5O–OH	43

in Table 10–1. We will refer to these molecules and bonds throughout this chapter because weak bonds cause fast initiation steps in chain reactions.

As indicated in Table 10–1, the tertiary C–H bond in isobutane is fairly weak (91 kcal/mole) compared to other bonds in the molecule, and this can be exploited to make two very important chemicals.

10.4.1   Propylene oxide and isobutylene

Isobutane and O_2 in water in the presence of a suitable catalyst (which promotes the initiation step) at $\sim100°C$ is rapidly converted to isobutyl hydroperoxide $(CH_3)_3COOH$. This molecule is purified by distillation (very carefully) and is then reacted with propylene to form propylene oxide and tertiary butanol, which is then dehydrated to isobutylene. All these steps take place at $\sim100°C$ in the liquid phase. The steps in the first reactor are

$$(CH_3)_3C–H \rightarrow (CH_3)_3C \cdot + H \cdot \tag{10.67}$$

$$(CH_3)_3C \cdot + O_2 \rightarrow (CH_3)_3COO \cdot \tag{10.68}$$

$$(CH_3)_3COO \cdot + (CH_3)_3CH \rightarrow (CH_3)_3COOH + (CH_3)_3C \cdot \tag{10.69}$$

In the second reactor the overall reaction is

$$(CH_3)_3COOH + C_3H_6 \rightarrow C_3H_6O + (CH_3)_3COH \tag{10.70}$$

After separation, the tertiary butyl alcohol is dehydrated,

$$(CH_3)_3COH \rightarrow \text{isobutylene} + H_2O \tag{10.71}$$

Isobutylene now finds considerable use in producing methyl t-butyl ether (MTBE) by reacting it with methanol,

$$iso\text{-}C_4H_8 + CH_3OH \rightarrow MTBE \tag{10.72}$$

which is an important gasoline additive because it enhances the octane rating of gasoline and (maybe) reduces pollution from automobiles. These reaction steps are shown in Figure 10–4.

Propylene oxide is widely used in polyurethanes and in other products such as a nontoxic antifreeze. This route of propylene oxide synthesis is replacing the old chlorohydrin process.

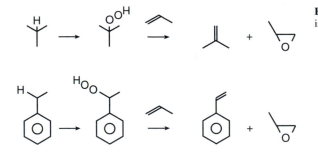

Figure 10–4   Reaction steps to make propylene oxide and either isobutylene or styrene by autooxidation.

$$C_3H_6 + HOCl \rightarrow HOC_3H_6Cl \rightarrow \text{ propylene oxide } + HCl \qquad (10.73)$$

because it eliminates the need for Cl_2 and the disposal of byproduct HCl.

10.4.2   Propylene oxide and styrene

Another related process to prepare propylene oxide uses ethyl benzene as the peroxidizable species, and it produces a coproduct that is even more valuable than isobutylene. This process will be left for a homework problem.

10.4.3   Acetone and phenol

These molecules are made by the autooxidation of cumene (what's that?), which is made by alkylation (what's that?) of propylene and benzene.

One starts by reacting refinery petrochemicals over an acid catalyst to form isopropyl benzene, which is also called cumene. Cumene has a weak tertiary C–H bond that can be easily reacted with O_2 (industrially a catalyst is used to speed up the process and secure suitable patents) to make the hydroperoxide. This is then distilled to separate the hydroperoxide.

In another reactor with a different catalyst this molecule decomposes with water to form acetone and phenol (Figure 10–5). Most industrial acetone and phenol are now made by this process, in which the key step is autooxidation.

10.4.4   Adipic acid

Adipic acid is a key monomer in Nylon 6,6. The molecule [HOOC(CH₂)₄COOH] must be prepared with high purity (exactly 6 carbon atoms with carboxylic acid groups on

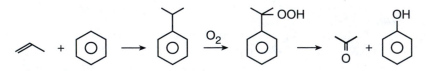

Figure 10–5   Reaction steps to make acetone and phenol by autooxidation of isopropyl benzene (also called cumene).

Figure 10–6 Reaction steps to make adipic acid by autooxidation of cyclohexane. Adipic acid is a key ingredient in Nylon.

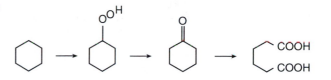

both ends and no alkyl branching). It is made by careful oxidation of cyclohexane. By opening the ring, one has a six-carbon atom chain with functional groups on each end. This process proceeds by air oxidation of liquid cyclohexane at ~140°C in a sparger reactor to make a mixture of cyclohexylperoxide, cyclohexanol, and cyclohexanone. This mixture is then further oxidized to adipic acid by reaction with HNO_3 in a two-phase liquid–liquid reactor. The first stage of this process is autooxidation by O_2 to form the alkylperoxy and alkylperoxide (Figure 10–6).

Both air oxidation of liquid cyclohexane in a sparger reactor and HNO_3 oxidation in a liquid–liquid reactor are two-phase reactors, which will be discussed in Chapter 14.

10.5 COMBUSTION

In the previous examples we considered chain reactions that can occur in liquids at ambient temperature or at least where we assumed that the process could be maintained isothermal. This produces a carefully controlled oxidation process (if the peroxides do not explode).

We will next consider the additional complexities of combustion, which are caused by the fact that combustion reactions are chain reactions, which are extremely fast and exothermic, and therefore, once the reaction is ignited, the process proceeds very quickly and becomes very nonisothermal.

In the next sections we will consider two exothermic chain reactions involving gases, the oxidation of H_2 and alkanes. Then we will consider combustion of liquid and solid fuels, a gas–liquid process. These processes all occur very rapidly and at very high temperatures.

10.6 HYDROGEN OXIDATION

A "simple" prototype of combustion reactions is hydrogen oxidation,

$$2H_2 + O_2 \rightarrow 2H_2O \tag{10.74}$$

However, this reaction is in fact extremely complex, and the standard model describing it consists of 38 reaction steps among 8 species: H_2, O_2, H_2O, H, O, OH, HO_2, and H_2O_2. These reactions are listed in Table 10–2. Listed in Table 10–2 are the rate coefficients of the forward reactions shown. Rate constants are given in the form

$$k = AT^\beta e^{-E/RT} \tag{10.75}$$

All reactions are bimolecular, so the units of k are $cm^3\,mole^{-1}\,sec^{-1}$. The 19 reverse reaction rate coefficients k' are found from the equilibrium constants $k' = k/K_{eq}$.

A simplified reaction scheme captures many features of this reaction. First, the reaction is initiated by generation of free radicals, with the lowest activation energy process being H_2 dissociation,

TABLE 10–2
Reaction Rates of the Elementary Steps in the $H_2 + O_2$ Reaction

Reaction	A (cm³/mole⁻¹ sec⁻¹)	β	E (kJ mole⁻¹)
1. $O_2 + H \rightleftarrows OH + O$	5.10E+16	-0.82	69.1
2. $H_2 + O \rightleftarrows OH + H$	1.80E+10	1.00	37.0
3. $H_2 + OH \rightleftarrows H_2O + H$	1.20E+09	1.30	15.2
4. $OH + OH \rightleftarrows H_2O + O$	6.00E+08	1.30	0.0
5. $H_2 + O_2 \rightleftarrows OH + OH$	1.70E+13	0.00	200.0
6. $H + OH + M \rightleftarrows H_2O + M$	7.50E+23	-2.60	0.0
7. $O_2 + M \rightleftarrows O + O + M$	1.90E+11	0.50	400.1
8. $H_2 + M \rightleftarrows H + H + M$	2.20E+12	0.50	387.7
9. $H + O_2 + M \rightleftarrows HO_2 + M$	2.10E+18	-1.00	0.0
10. $H + O_2 + O_2 \rightleftarrows HO_2 + O_2$	6.70E+19	-1.42	0.0
11. $H + O_2 + N_2 \rightleftarrows HO_2 + N_2$	6.70E+19	-1.42	0.0
12. $HO_2 + H \rightleftarrows H_2 + O_2$	2.50E+13	0.00	2.9
13. $HO_2 + H \rightleftarrows HO + OH$	2.50E+14	0.00	7.9
14. $HO_2 + O \rightleftarrows OH + O_2$	4.80E+13	0.00	4.2
15. $HO_2 + OH \rightleftarrows H_2O + O_2$	5.00E+13	0.00	4.2
16. $HO_2 + HO_2 \rightleftarrows H_2O_2 + O_2$	2.00E+12	0.00	0.0
17. $H_2O_2 + M \rightleftarrows OH + OH + M$	1.20E+17	0.00	190.5
18. $H_2O_2 + H \rightleftarrows H_2 + HO_2$	1.70E+12	0.00	15.7
19. $H_2O_2 + OH \rightleftarrows H_2O + HO_2$	1.00E+13	0.00	7.5

$$H_2 \rightarrow 2H\cdot, \qquad E = 104 \text{ kcal/mole} \qquad \text{(i)} \qquad (10.76)$$

which is reaction 8 in Table 10–2. Next H atoms attack O_2 to produce O· and ·OH, both of which attack the parent molecules in the propagation steps

$$H\cdot + O_2 \rightarrow \cdot OH + O\cdot \qquad \text{(p1)} \qquad (10.77)$$

$$O\cdot + H_2 \rightarrow \cdot OH + H\cdot \qquad \text{(p2)} \qquad (10.78)$$

$$\cdot OH + H_2 \rightarrow H_2O + H\cdot \qquad \text{(p3)} \qquad (10.79)$$

These three reactions are chain reactions, involving a radical attacking a parent to form another radical. Note that only the last reaction produces the stable product H_2O.

If we add another reaction

$$H\cdot + \cdot OH \rightarrow H_2O \qquad \text{(t)} \qquad (10.80)$$

and add all four of the propagation reactions, we see that all the radical species cancel, leaving only the "simple" reaction

$$2H_2 + O_2 \rightarrow 2H_2O \qquad (10.81)$$

which is the stoichiometric reaction producing water from hydrogen and oxygen at high temperatures, although hydrogen peroxide H_2O_2 is also produced in measurable amounts in low-temperature processes. The reactions in this chain reaction are shown in Figure 10–7.

This reaction sequence is similar to that described for acetaldehyde decomposition to methane and carbon monoxide: Reactions that produce stable products actually occur only

Figure 10–7   Sketch of chain reaction in $H_2 + O_2 \rightarrow H_2O$. This is a highly simplified set of the reactions shown in the table, which indicate the dominant chain reaction steps in forming water from three chain-propagating radical species: H, O, and OH.

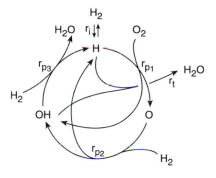

through radicals attacking the parent and eventually producing the stable products (CH_4, CO, and H_2O in these examples).

However, the complete reaction mechanism of the hydrogen oxidation reaction is much more complex, both in its number of reaction steps, number of intermediates (OOH and H_2O_2), and observed behavior. A mixture of H_2 and O_2 can sit in a dry bulb for many years with absolutely no H_2O detected. However, if water is initially present, the reaction will begin, and if a spark is ignited or a grain of platinum is added to the mixture at room temperature, the reaction will occur instantly and explosively.

The reaction cannot begin without initiation of free radicals, and H_2 dissociation is very slow. Therefore, the few H atoms produced by the initiation steps may diffuse to the walls of the vessel and recombine before they can begin chain reactions. However, the presence of H_2O accelerates the rate of the reaction because it causes formation of traces of H_2O_2, which easily dissociates and forms more H· and ·OH, which initiate the reaction by attack of H_2 and O_2, and these reactions rapidly produce more radicals, which strongly accelerate the process.

A spark generates many radical species in the spark gap, and these can propagate radicals throughout the vessel. A platinum surface catalyzes the reaction (H_2 and O_2 readily dissociate on a Pt catalyst surface to begin the reaction), and the surface reaction then heats the Pt to a high temperature, where homogeneous reaction begins rapidly.

Thus we see that premixed H_2 and O_2 mixtures behave very unpredictably. They are usually quite unreactive, but moisture or a trace chemical such as H_2O_2, a spark, a trace of a catalyst (perhaps undetectable), or a hot surface can cause rapid and uncontrolled reaction.

The explosion limits of the $H_2 + O_2$ reaction have been studied thoroughly, and these limits exhibit an interesting variation with pressure. At low pressures the reaction is stable because radicals diffuse to the walls of the chamber and prevent chain branching, which leads to an explosion. At very high pressure radical quenching by diffusion to the walls is slowed down, and the heat released by the reaction leads to a thermal explosion (Figure 10–8).

The boundaries between stable flames and explosions are very "fuzzy" because they depend on traces of chemicals in the vessel, which may act as initiators or scavengers, and they depend strongly on the size and composition of the walls of the vessel. These diagrams are shown to illustrate the complexities in predicting the behavior of any chain reaction, even a "simple" one such as $H_2 + O_2$.

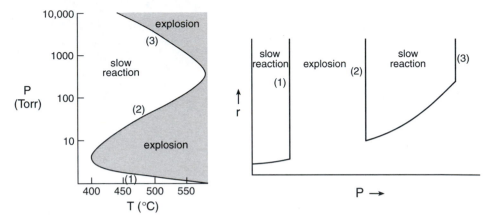

Figure 10–8 Explosion limits of the $H_2 + O_2$ reaction. At low pressures explosion is determined by the quenching of chain branching of radicals at the walls of the vessel, while at high pressures the larger rate of heat generation leads to a thermal explosion.

10.7 CHAIN BRANCHING REACTIONS

In the reaction steps for H_2 oxidation the first two reactions are chain branching reactions, with one radical species producing two other radicals every time a reaction event occurs. The first of these probably proceeds by the steps

$$\cdot H + O_2 \rightarrow \cdot OOH \rightarrow \cdot OH + O\cdot \tag{10.82}$$

and the hydroperoxy radical $\cdot OOH$ reacts immediately at high temperatures but can be observed in finite concentrations at low temperatures.

Note that this reaction produces two reactive free radicals from one. We call these chain branching reactions because the propagation steps produce more radical species than they start with. Thus the propagation steps can also increase the concentration of free-radical species, and further destabilize the kinetics. We call a propagation step in which one radical species is produced for each radical reacting a *linear chain*, while a process that produces more radicals is called a *branched chain*. (In the next chapter on polymers and polymerization we will encounter linear and branched chain reactions, but these will have quite different implications for polymers.) These are sketched in Figure 10–9.

Most combustion reactions involve chain branching reaction steps. Under conditions where these steps are less significant than linear chain reactions, the reaction appears to be

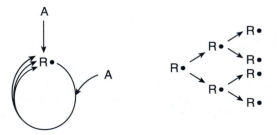

Figure 10–9 Sketch of chain branching reactions where the propagation steps produce more than one radical so that the process tends to grow exponentially, leading to a rapid acceleration of the reaction and possibly a chain-branching explosion.

stable, but when the chain branching steps dominate, the overall reaction rate can accelerate uncontrollably.

Recall that in the previous section on autooxidation, hydroperoxide species were sufficiently stable at low temperatures that only linear chains formed. When the bottle of peroxides explodes, chain branching steps such as

$$ROOH \rightarrow R\cdot + O\cdot + OH\cdot \qquad (10.83)$$

begin to become significant. This is a chain branching reaction because one radical produces three radicals. The rapid rise in the concentration of radical species can accelerate the decomposition reaction of a bottle containing peroxide rapidly and catastrophically.

10.7.1 Model of chain branching reaction

We can construct a simple model of a chain branching reaction that is a simple variation of the linear chain reaction we used previously. Consider the reaction system $A \rightarrow B + C$, which proceeds by the steps

$$A \rightarrow R, \qquad r_i = k_i C_A \qquad (10.84)$$

$$A + R \rightarrow B + C + \alpha R, \qquad r_p = k_p C_A C_R \qquad (10.85)$$

$$R \rightarrow X, \qquad r_t = k_t C_R \qquad (10.86)$$

This looks exactly like the generic chain reaction we wrote down previously, except now the propagation step produces α radical species rather than one.

We will leave the derivation of this rate expression for a homework problem. A steady-state balance on C_R yields

$$\frac{dC_R}{dt} = k_i C_A + (\alpha - 1)k_p C_A C_R - k_t C_R = 0 \qquad (10.87)$$

which gives

$$C_R = \frac{k_i C_A}{k_t - (\alpha - 1)k_p C_A} \qquad (10.88)$$

For a linear chain we obtained $C_R = k_i C_A / k_t$ because $\alpha = 1$, but now we have the additional term $(\alpha - 1)k_p C_A$ in the denominator. The major difference between this expression and many similar rate expressions we have derived throughout this book is that there is a minus sign in the denominator. Whenever this occurs, we need to be suspicious, because the two terms can become equal

$$k_t = (\alpha - 1)k_p C_A \qquad (10.89)$$

There is no reason why this is not possible if $\alpha > 1$ and C_A is large enough, but the consequences are of this situation are enormous. This condition gives $C_R \rightarrow \infty$, and this requires that there is no steady state!

If the propagation term is large compared to the others, we can write this expression as

$$\frac{dC_R}{dt} = +(\alpha - 1)k_p C_A C_R \qquad (10.90)$$

with a $+$ sign on the derivative. This condition in fact predicts a chain branching explosion, where the radical concentration grows exponentially

$$C_R(t) = C_{Ro}e^{kt} \qquad (10.91)$$

which is clearly an unstable situation, both mathematically and physically.

10.8 ALKANE OXIDATION

One of the most important combustion reactions in industry and in home heating is natural gas combustion,

$$CH_4 + 2O_2 \rightarrow CO_2 + 2H_2O \qquad (10.92)$$

The transportability of natural gas in pipelines and the clean combustion in its oxidation make it the fuel of choice in many applications.

As one might guess from our previous example of H_2 oxidation, this reaction is extremely complex. The standard model involves over 300 reaction steps among approximately 30 chemical species. Here even the stable product species can be complex, with CO, HCHO (formaldehyde), and soot (carbon) being among the highly undesired pollutants in CH_4 combustion. Since H_2, O_2, and H_2O are involved as intermediates and products, the 38 reaction steps in the $H_2 + O_2$ reaction listed previously set of the reactions that be involved in CH_4 oxidation.

We will not attempt to describe the mechanism of this process except to note that an important initiation step is CH_4 dissociation,

$$CH_4 \rightarrow CH_3 \cdot + H \cdot, \quad E = 103 \text{ kcal/mole} \qquad (10.93)$$

with the $H \cdot$ and the $CH_3 \cdot$ active to attack O_2 and begin propagation steps.

Mixtures of natural gas and air are similarly unpredictable, being quite stable and unreactive until a spark or a hot surface ignites reaction, and then thermal and chain branching autocatalysis takes off and so does the building.

Similar mechanisms (but even more complex) operate in the combustion of higher hydrocarbons. Initiation is easier because the C–H bonds can be weaker. The $CH_4 + O_2$ reactions are a subset of all hydrocarbons since CH_4 is one of the possible decomposition products of higher hydrocarbons.

One much employed example is combustion of gasoline in the engine cylinders of an automobile. Here the fuel is injected into the combustion chamber as a vapor with air at nearly the stoichiometric composition for complete combustion, and a spark ignites the mixture to create pressure, which forces the piston down. Details of ignition reactions determine issues such as engine knock or flooding, with branched aliphatics and small fuel molecules being advantageous for both of these problems. High-octane alkanes produce fewer chain branching radicals during the compression of the air–fuel mixture, and this avoids engine knock, which will occur if ignition occurs spontaneously before the spark plug fires to ignite the alkane–air mixture at the start of the power stroke.

10.8.1 Liquid alkane oxidation

In the diesel engine, the jet engine, and kerosene combustion, liquid fuel is sprayed into the combustion chamber with air, and the liquid is not totally vaporized before ignition. Therefore, in drop combustion there are flames around the liquid drops that shrink as the reaction proceeds. This and wood and coal combustion are examples of the shrinking sphere problem of Chapter 9, in which we derived expressions for the drop radius versus time $R_o(t)$. These are also multiphase reactions, which will be considered in more detail later in this chapter and in Chapter 14.

We note finally that for most of these processes no one knows their kinetics. Rates have been measured with great precision in the laboratory under specific conditions, but in engineering applications the processes are so complex that these kinetics are not useful. In almost all situations the rates of reaction are controlled by mass and heat transfer rates, and the fluid flow is turbulent. When one remembers that the actual reaction steps depend on mass transfer (ignition by initiators and termination reactions at surfaces), it is evident that one has a very complicated problem.

The complexity and importance of combustion reactions have resulted in active research in *computational chemistry*. It is now possible to determine reaction rate coefficients from quantum mechanics and statistical mechanics using the ideas of reaction mechanisms as discussed in Chapter 4. These rate coefficient data are then used in large computer programs that calculate reactor performance in complex chain reaction systems. These computations can sometimes be done more economically than to carry out the relevant experiments. This is especially important for reactions that may be dangerous to carry out experimentally, because no one is hurt if a computer program blows up. On the other hand, errors in calculations can lead to inaccurate predictions, which can also be dangerous.

10.9   THERMAL IGNITION

Most chemical accidents have occurred because some of these effects were not expected or accounted for, and most of the research on combustion processes originated in an effort to explain the unexpected and potentially hazardous. Thermal ignition contributes another safety hazard in combustion processes.

Our first example of a chain reaction, the decomposition of acetaldehyde to methane and CO, is endothermic; so the reactor tends to *cool* as reaction proceeds. However, the oxidation of H_2 is exothermic by 57 kcal/mole of H_2, and the oxidation of CH_4 to CO_2 and H_2O is exothermic by 192 kcal/mole of CH_4. Thus, as these reactions proceed, heat is released and the temperature tends to increase (strongly!). Thus thermal ignition is very important in most combustion processes.

In addition to ignition of gases, liquids and solids can also exhibit ignition phenomena. A classic example is the "spontaneous combustion" of oily rags containing paint solvents or automobile oil. One does not want to store large quantities of these rags together, because they can suddenly heat up and catch fire. Another example is the "spontaneous combustion" of coal piles and of underground coal fields. A large pile of coal will almost inevitably catch fire and burn, and barges of coal (we ship considerable amounts of U.S. coal to Asia) must be transported and unloaded quickly or the barge will begin to burn, a quite unpleasant experience for the sailors, especially if far from port.

In this section we consider a simple model that explains many features of thermal ignition processes.

We have already dealt with nonisothermal reactors and their effects in Chapters 5 and 6. The major and potentially most dangerous application of these effects is in combustion reactors. Consider our standard CSTR with parameters as indicated in Figure 10–10.

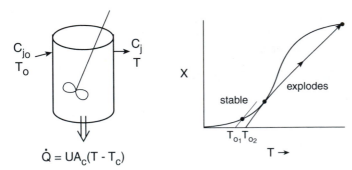

$$\dot{Q} = UA_c(T - T_c)$$

Figure 10–10 Heat generation and removal for combustion reaction in a CSTR. When the feed temperature increases slightly, the rate can go from nearly zero to nearly completion when the heat removal curve becomes tangent to the heat generation curve, and the steady-state temperature jumps to a high value. This simple model shows why these processes can be extremely sensitive to conditions such as size, heat transfer, and the temperature of surroundings.

Recall from Chapter 6 that the steady-state energy-balance equations in a CSTR can be reduced to a single equation, which we wrote by considering the rates of heat generation and removal. We wrote these as

$$T - T_o + \frac{U A_c}{v \rho C_p}(T - T_c) = \tau \frac{-\Delta H_R}{\rho C_p} r(C_A, T) \qquad (10.94)$$

We then rearranged this equation to become

$$\frac{T - T_o + \kappa(T - T_c)}{J} = X(T) \qquad (10.95)$$

which we wrote as

$$\mathbf{R}(T) = \mathbf{U}(T) \qquad (10.96)$$

and we constructed a graphical solution by plotting the left-hand side $\mathbf{R}(T)$ and the right-hand side $\mathbf{G}(T)$ versus T, as repeated in Figure 10–10.

Consider an oxidation reaction in this CSTR. Suppose we begin at the lower steady state, where the reaction rate and the rate of heat generation are very low. If the feed temperature T_o and the wall temperature T_c (c for coolant) are low, there is little reaction, and we would detect negligible temperature change. However, if we increase T_o or T_c slightly, the system can jump to the upper steady state, the situation we discussed in Chapter 6.

If this is a combustion process in which the rate has a very large effective activation energy, ΔH_R is very large and negative, and ρC_p is small (undiluted gases), then these effects become very large. One therefore observes a situation where the rate is effectively zero for one temperature because the effective activation energy is very large, but a slight change in any parameter can cause the system to jump to the upper steady state, where the reaction goes to completion and the temperature goes to the adiabatic flame temperature.

For H_2 or CH_4 in air near stoichiometric compositions (28% H_2 in air and 9.5% CH_4 in air, respectively), the temperature jumps from ambient to >2000° (it doesn't make much difference whether one uses °C or K units for this calculation), and the reactor almost certainly explodes.

For a coal pile or a barrel of oily rags, there is a slow flow of air through the pile or reactor a large residence time τ), and processes are slow enough (an autooxidation process) that the composition is nearly uniform (creating a CSTR geometry). A slight increase in temperature (perhaps a warm day) can cause the rate of oxidation of the alkane in the oil or the carbon in the coal to oxidize, and at some T_c the "reactor" will suddenly ignite and burn. Once this happens, natural convection forces air rapidly through the barrel or coal pile and "fans" the flame. Underground coal fields are especially difficult to extinguish because there are many holes in the ground which act as inlets and outlets for the "reactor."

While the CSTR model is extremely primitive, the ideas of size dependence and differences in τ for air flow before and after ignition explain many of the features of these troublesome chemical reactors.

10.10 THERMAL AND CHEMICAL AUTOCATALYSIS

The preceding situation is an example of *thermal autocatalysis*, and its consequence is frequently a thermal explosion.

We have previously encountered examples of *chemical autocatalysis*, where the reaction accelerates chemically such as in enzyme-promoted fermentation reactions, which we modeled as $A + B \rightarrow 2B$ because the reaction generates the enzyme after we added yeast to initiate the process. The other example was the chain branching reaction such as $H \cdot + O_2 \rightarrow OH \cdot + O \cdot$ just described in hydrogen oxidation. The enzyme reaction example was nearly isothermal, but combustion processes are *both* chain branching and autothermal, and therefore they combine chemical and thermal autocatalysis, a tricky combination to maintain under control and of which chemical engineers should always be wary.

This suggests immediately the definition of *autocatalysis*. A reactive intermediate or heat can act as catalysts to promote the reaction. However, in contrast to conventional catalysis, we do not add the catalyst from outside the system, but the catalyst is generated by the reaction (autocatalysis). We may add promoters or heat to initiate the process, which then accelerates by autocatalysis. Conversely, we may add inhibitors or cool the reactor to prevent both types of autocatalysis.

The best inhibitor of autocatalysis is to have a small system. Surfaces act as both sinks of free radicals to increase termination and as sinks of heat to prevent thermal runaway, and small systems have large surface-to-volume ratios to prevent these runaway possibilities.

10.11 PREMIXED FLAMES

In this section we consider the combustion of premixed gaseous fuel and air mixtures. Consider first the laboratory Bunsen burner, shown in Figure 10–11. Natural gas from the gas supply system enters the bottom of the burner, where it is mixed with air, with flow rates adjusted by the gas valve and holes in the bottom of the burner, where air is sucked

Figure 10–11 Bunsen burner, which is a typical premixed flame in which the flame velocity u_f equals the flow velocity u in the expansion at the end of the tube, creating a stable flame.

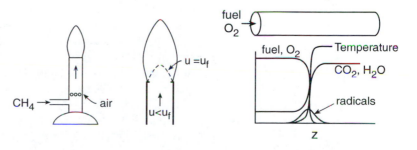

in by natural convection. The premixed gases travel up the barrel of the burner (a *tubular reactor*), and, if flows are suitably adjusted and a match has been used to ignite the mixture, a stable flame forms at the top of the tube.

If the flow rate is too low, the flame will *flash back* and propagate back up the tube to the mixing point and extinguish. If the flow rate is too high, the flame will lift off the top of the tube and either extinguish or burn as a turbulent *lifted flame*.

10.11.1   Stability of a tube flame

Students have probably noticed that is it nearly impossible to maintain a stable flame within the tube of a Bunsen burner; it will either burn at the top or flash back to the mixing point. There is usually no stable steady state for a flame reaction in a straight tube.

This points out several major characteristics of premixed flames: (1) They cannot be controlled stably within a constant-diameter tubular reactor configuration; and (2) the chain branching reactions and the heat release tend to make them either blow out or flash back into the tube and extinguish.

Flames are characterized by a *flame velocity* u_f, which is the velocity with which a flame front will move down a tube containing stagnant premixed gases. The upper and lower compositions where the flame velocity goes to zero are called the *flammability limits*, and the flame velocity is usually its maximum near the stoichiometric composition for CO_2 and H_2O formation.

The reason why it is difficult to sustain a flame within a tube is that the flame will only be stable if

$$u = u_f \qquad (10.97)$$

where u is the flow velocity used throughout this book for the PFTR. It is virtually impossible to adjust the flow velocity to be exactly u_f. Even if this occurs, the flame is only neutrally stable; it will drift up or down the tube.

Flow in the tube will be parabolic if the flow is laminar (Re<2100), and thus gases near the center of the tube may propagate downstream, while gases near the wall propagate upstream toward the mixing point. This further destabilizes a premixed flame within a tube.

10.11.2   Premixed burner flames

Premixed flames can only be stabilized by providing an *expansion section* in the tube where the cross section increases so that u slows, creating a zone where $u = u_f$ and a flame is stable. In a Bunsen burner this is achieved quite simply by having the flame height vary as u changes within the tube so that a burning surface (roughly a cone) is maintained stable over a large range of flow rates. At the exit of the tube the flow velocity decreases over a wide range so that the flame is quite stable.

The other stabilizer of flames is cold surfaces. These act to lower u_f in the region of the surface and thus create a stable edge to the flame at the wall of the burner. This occurs both by thermally quenching the reaction near the surface and by quenching free radicals by adsorption on the surfaces, which stops the combustion reactions.

One very important device that uses this principle is the *flame arrestor*, which is inserted in pipes to prevent flames from propagating between pieces of gas-handling equipment. These are usually metal shavings (steel wool) inserted in tubes at strategic distances to stop any flames that may travel down the tube.

10.12 DIFFUSION FLAMES

When the air inlet is shut off in a Bunsen burner, the flame can still be maintained, but it switches from the blue color of a normal CH_4 flame to the yellow of a diffusion flame. Now pure CH_4 flows up the tube, and the O_2 from outside the tube is mixed with CH_4 just above the tube, with flow of air driven by natural convection below the hot product gas mixture.

This burner, a butane lighter, a candle, a burning log, and a match (after ignition) are all examples of diffusion flames where one generally provides the fuel and relies on natural convection of air to provide the oxidant.

Diffusion flames are much more stable than premixcd flames because the rate of the reactions is limited by the supply of air and its diffusion into the fuel region to form the flame front. Several examples of diffusion flames are given in Figure 10–12. These all typically have *cylindrical symmetry*, with air surrounding the fuel in the center. The cylindrical laminar flame is simplest to sketch because it is two dimensional in composition (direction of flow z and radial position from the center R), with composition and temperature profiles as shown in Figure 10–13. Gaseous fuel and air are admitted through separate slots, and the reaction proceeds along the line of contact. The gases are convected in the vertical flow z direction, and horizontal diffusion (x) between the fuel and air create the flammable mixture.

One can model this system in principle using the mass- and energy-balance equations written with the equations of Chapter 8 with flow in the z direction and diffusion in the x direction to obtain profiles of $C_j(z, x)$ and $T(z, x)$. However, the student can see immediately that this will be a very complex mathematical problem to solve because there

Figure 10–12 Some typical diffusion flames in which gas or liquid reacts with O_2 from air. The reaction occurring at the region where mixing occurs.

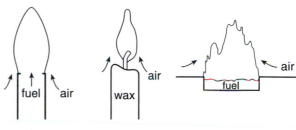

tube burner   candle   pool flame

Figure 10–13 Concentration and temperature profiles in a diffusion flame such as a candle. Radical intermediates that propagate the flame have high concentrations in the flame zone between fuel and O_2.

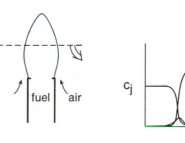

are many species (at least 30 for natural gas flames), but the problem will be made even more complex because of natural convection. Since the temperature in the flame varies from 25 to >2000°C in short distances, the density in and above the flame decreases dramatically (by nearly a factor of 10), and this causes large *buoyancy* effects, which accelerate the gases in the vertical direction. This provides the air flow in many diffusion flames, but it can also create *turbulence*, which causes greatly enhanced mixing above that which would occur by true diffusion. These phenomena are driven by the gravity term in the continuity equation.

10.12.1   Laminar and turbulent flames

This is another distinction between flames—laminar versus turbulent—with small flames generally steady and laminar but large ones exhibiting visible turbulent eddies. Candles and fireplaces are well-known examples where in still air the flame is steady, but, when there is wind or one blows on the flame, the flame flickers.

One can describe these phenomena through the Reynolds number (forced convection) and Rayleigh number (natural convection), but the reader can see immediately that the situations are so complicated that correlations in elementary texts on fluid flow are not easily applicable to predicting flame behavior. Reactive flows are among the most complex problems in modern engineering.

10.13   ENERGY GENERATION

Energy considerations give the first estimate of the characteristics of combustion. As discussed in Chapters 2 and 5, the adiabatic reaction temperature (frequently called the adiabatic flame temperature) gives an approximation to the final temperature,

$$\Delta T = \frac{-\Delta H_R}{N C_p} \qquad (10.98)$$

We need to be especially careful in using this equation to try to describe combustion accurately, however, because the final temperature is so large that C_p is not at all constant. Further, there are many reactions and products in flames, and one has to use the multiple-reaction version of this equation for accurate calculations. The intermediates, which are at negligible concentrations at low temperature, have significant concentrations at high temperatures. For example, the water dissociation reaction

$$H_2O \rightarrow H \cdot + \cdot OH \qquad (10.99)$$

has an equilibrium constant that predicts considerable H and OH in the high temperature of most flames.

10.13.1   Radiation

Another consideration in flames is radiation. The light that one sees in a flame is mostly fluorescence from the radiation of particular radical species formed in electronically excited states. (The blue color from CH_4 flames is CH· emission.) Gases also radiate blackbody radiation, primarily in the infrared. The glow from burning wood or coal is blackbody emission radiated from the surface.

[Recall that the emission observed from cracks in a burning wood surface is brighter because the emission from a cavity has the equilibrium blackbody distribution, which is independent of the emissvity of the surface. Also recall that most of the heat from a campfire is from radiation if you are beside the fire. The air beside the campfire is close to the ambient temperature because natural convection is drawing it up into the rising flame.]

Most of the emission from combustion processes is in the infrared and is not visible, but a large fraction of the energy carried off from combustion processes is in the form of radiative heat transfer.

10.13.2   Flammability limits

In practice there is only a certain range of compositions where premixed gases will burn. There is obviously a great deal of interest in this composition because it sets limits of operation of many processes. The ignition limit is only loosely defined because it depends sensitively on temperature, flow conditions, and the presence of trace chemicals in the fuel, which can act as flame initiators or inhibitors by enhancing or slowing either r_i or r_t.

A very important consideration in flammability is the presence of surfaces. Cold surfaces act as free-radical quenchers and maintain the gas temperature below the ignition temperature, while hot surfaces can act as ignitors.

Many of the industrial accidents in the chemical industry have originated from reaction systems that were quite stable in small containers but exploded when larger reactors were used.

10.14   COMBUSTION OF LIQUIDS AND SOLIDS

10.14.1   Burning of wood and coal

These solids have been very important fuels since our cave-dwelling ancestors, and both are still important fuels in heating and power generation. Their combustion has characteristics that are common to most solids combustion processes.

10.14.2   Pyrolysis

The first step when coal or wood particles (we call them chunks or powder and logs or chips, respectively) burn is pyrolysis, in which heat causes the water and volatile organics to evaporate. These gases then burn in oxygen in the boundary layer in the first stage of burning.

Coal and wood decompose to give many small organic molecules such as alkanes, aromatics, alcohols, aldehydes and ketones, and organic acids. As an example, methanol is still sometimes called wood alcohol because it was first obtained by heating wood in a kiln and condensing the products, one of which was methanol. In the nineteenth century, coal and wood were the major sources of many chemicals such as acetic acid and acetone, and coal was a major source of chemicals such as benzene. As noted previously, one of the types of products from coal was highly colored chemicals, and these were called coal tars, from which highly colored dyes could be extracted. The search for reliable and cheaper sources of these dyes produced the development of organic chemistry for the systematic search for new colored dyes, and this led to the giant English and German chemical companies such as ICI and the I. G. Farben cartel, as discussed in Chapter 3.

10.14.3 Coke and charcoal

Once the volatile products have been pyrolyzed, one is left with a material that is mostly carbon, and these are called coke (from coal) and charcoal (from wood). Coke was of course used in blast furnaces to make iron (the volatiles would make the process less reproducible in reducing Fe_2O_3) by our ancestors, while charcoal was used in soap (mainly the alkalis in the ash) and as an adsorbent.

Coke and charcoal are highly porous, coke from the holes left behind when the volatile components evaporate, and charcoal remains from the cellular framework of wood fibers. Addition of limited air in pyrolysis produces more volatile products and makes a more porous carbon residue.

These carbon residues then can be burned with air in the second stage of combustion. There are two modes of carbon combustion, with and without flames. If there is no flame over the carbon, then reaction occurs by surface reaction,

$$C_s + O_2 \rightarrow CO_2 \tag{10.100}$$

or to CO

$$C_s + \tfrac{1}{2}O_2 \rightarrow CO \tag{10.101}$$

if the reaction is deficient in O_2. When there is a flame, burning is by both surface and homogeneous reactions. The surface reaction is primarily

$$C_s + CO_2 \rightarrow 2CO \tag{10.102}$$

and the homogeneous reaction is primarily

$$CO + \tfrac{1}{2}O_2 \rightarrow CO_2 \tag{10.103}$$

although this is in fact a free-radical chain reaction with many steps.

This creates a reacting boundary layer over the coke or charcoal particle with reactant and product composition profiles as illustrated in Figure 10–14.

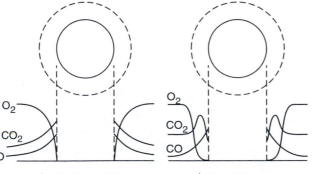

heterogeneous

homogeneous
heterogeneous reaction

Figure 10–14 Concentration profiles around a burning carbon particle, either with or without homogeneous reaction in the boundary layer around the particle.

There is therefore typically a progression of burning of solids involving first a homogeneous reaction, where volatile gases are driven off and burn, followed by a homogeneous–heterogeneous reaction, where both homogeneous and surface reactions occur in a boundary layer, followed by a surface reaction, where glowing coals react only on the surface.

10.14.4 The campfire or charcoal grill

Everyone has seen these processes. When a log is placed in a campfire, it first sputters and pops as water trapped within the solid is released, and then flames are observed around it (flame jets form as the organics escape from cracks in the log) but the surface remains dark. Then after most of these organics have escaped, the wood surface begins to glow as the surface reaction begins with the flame continuing. Finally, after the fire has "died down," one observes no flame at all but just the glowing embers due only to surface reaction. Stirring the coals or blowing on the fire causes the flame to reignite for a while until it returns to the pure surface reaction mode. Finally, the heat release is too small to sustain reaction or the wood turns completely to ashes, and it is time to go home or to bed.

10.14.5 Coal combustion

Our grandparents used coal for home heating, but now its major use is in electric power plants, where fast and efficient combustion is desired. Therefore, coal plants grind large coal chunks into fine powder and then blow it with air into burner jets, where it reacts with O_2 first by pyrolysis, and then the remaining coke burns, leaving behind an ash, which is a few percent of the weight of the original coal.

Most of us have never seen the inside of a coal-fired boiler, but the sparks rising from a campfire are very similar to the second stage of a coal-burning process. When a campfire is stirred, small particles of charcoal are broken off the log, and buoyancy carries them upward. They glow brightly by the heat released by heterogeneous carbon oxidation for several seconds until the carbon is all consumed.

In these processes the reactant O_2 concentration flux is mass transfer limited around the reacting solid sphere, and the temperature of the particle is determined by heat transfer around the sphere, as shown in Figure 10–15.

Figure 10–15 Concentration and temperature profiles around a burning carbon particle in air ($P_{O_2} = 0.2$ atm and $T_o = 25°C$). The concentration of O_2 falls to zero at the surface and the temperature rises to nearly the adiabatic flame temperature.

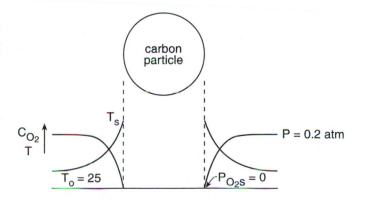

Example 10–2 Everyone has poked a campfire at night and watched the sparks fly up. They glow a bright red, which indicates that they are emitting blackbody radiation from a source with a surface temperature of ~1200°C. You have also noticed (if you were thinking about scientific matters at the time) that their color does not change until they suddenly disappear. The time for this reaction varies, but they seem to glow for a few sec and then suddenly disappear. (a) Estimate the temperature of a campfire spark. (b) How long does it take to consume a 0.1-mm-diameter particle?

Let us construct mass and energy balances on the particle, which we assume to be a sphere of pure carbon of radius R. The reaction is

$$C_s + O_2 \rightarrow CO_2, \qquad r'' = k''C_CC_{O_2}, \qquad \Delta H_R = -94 \text{ kcal/mole} \qquad (10.104)$$

The heat of the reaction is the heat of combustion of graphite. We do not really know the kinetics of the reaction, but it is a surface reaction between carbon on the surface of the particle and O_2 from the air. Since the amount of surface of carbon exposed per unit surface area is constant, we can eliminate it by lumping $k'' = k''C_{Cs}$ to write

$$r'' = k''C_{O_2} \qquad (10.105)$$

A mass balance on O_2 around the particle yields

$$\frac{dN_C}{dt} = \frac{\rho_C}{M_C}\frac{dV}{dt} = \frac{4\pi R^2 \rho_C}{M_C}\frac{dR}{dt} = 4\pi R^2 k''C_{O_2s} = 4\pi R^2 k_m(C_{O_2b} - C_{O_2s}) \qquad (10.106)$$

where we simply equate the loss of moles of solid carbon with the rate of reaction of O_2 at its surface, which is also equal to the rate of mass transfer of O_2 to the surface.

An energy balance on the particle yields

heat generated by reaction = heat lost by conduction + heat lost by radiation

$$4\pi R^2(-\Delta H_R)k''C_{O_2s} = 4\pi R^2 h(T_s - T_o) + 4\pi R^2 \varepsilon\sigma(T_s^4 - T_o^4) \qquad (10.107)$$

where all terms are familiar except the radiation term, with ε the emissivity of the surface, σ the Stefan–Boltzmann radiation constant, T_s the carbon particle temperature, and T_o the ambient air temperature (300 K or whatever it was that night).

We first assume that the O_2 concentration profile is established much more rapidly than R is changing; so that the concentration profiles is in steady state. These equations become

$$k''C_{O_2s} = k_m(C_{O_2b} - C_{O_2s}) \qquad (10.108)$$

$$(-\Delta H_R)k''C_{O_2s} = h(T_s - T_o) + \varepsilon\sigma(T_s^4 - T_o^4) \qquad (10.109)$$

where the area $4\pi R^2$ can be omitted from each term, showing immediately that the temperature is independent of the particle size, as you knew already since there must be a range of particle size in the sparks, and all have the same temperature. We then ignore the radiation term and rearrange these equations to yield

$$C_{O_2b} - C_{O_2s} = \frac{k''}{k_m}C_{O_2s} \tag{10.110}$$

$$T_s - T_o = \frac{(-\Delta H_R)k''}{h}C_{O_2s} \tag{10.111}$$

Before we proceed, note that these equations look identical in form to the adiabatic CSTR equations of Chapter 6,

$$C_{Ao} - C_A = \tau k C_A \tag{10.112}$$

$$T - T_o = \frac{-\Delta H_R}{\rho C_p}k C_A \tag{10.113}$$

Therefore, we immediately suspect that the carbon particle can exhibit multiple steady states, ignited and extinguished.

Returning to the equations describing the particle, we have

$$T_s = T_o + \frac{(-\Delta H_R)k''}{h}C_{O_2s}$$

$$= T_0 + \frac{(-\Delta H_R)k_m}{h}(C_{O_2b} - C_{O_2s})$$

$$= T_0 + \frac{(-\Delta H_R)k_m}{h}C_{O_2b} \tag{10.114}$$

where we have assumed that $C_{O_2s} \ll C_{O_2b}$, meaning that the oxidation of the carbon particle is mass transfer limited. Note that the reaction rate coefficient k'' has disappeared from the equation, as expected if the reaction is very fast.

Now we need k_m and h, the mass and heat transfer coefficients around a sphere. These come from Sherwood and Nusselt numbers, respectively, for flow around a sphere,

$$\text{Sh}_D = \text{Nu}_D = 2.0 + 0.4\text{Re}_D \tag{10.115}$$

Since the fluid (air) is nearly stagnant, these are ~2, but we do not care since they are equal for gases (if Sc=Pr). Therefore, we have

$$\frac{k_m}{h} = \frac{\text{Sh }D_{O_2}/D}{\text{Nu }k_T/D} = \frac{D_{O_2}}{k_T} \tag{10.116}$$

where D_{O_2} is the diffusivity of O_2 in air at atmospheric pressure, $D_{O_2} = 0.1$ cm^2/sec, and k_T is the thermal conductivity of air at atmospheric pressure, $k_T = 0.026$ W/mK.

We can now solve our problem.

$$T_s = T_o + \frac{(-\Delta H_R)k_m}{h}C_{O_2b} \tag{10.117}$$

$$T_s = 300 + \frac{94000 \text{ cal/mole} \times 0.1 \text{ cm}^2 \text{ sec} \times 0.2 \,/\, 22400}{0.026 \text{ W/mK} \times 0.01 \text{ m/cm} \times 1/4.18 \text{ J/cal}}$$

$$= 1650 \text{ K} \tag{10.118}$$

or

$$\boxed{T_s = 1400°C}$$
(10.119)

which is in good agreement with the observed temperature of a spark (bright yellow). In this calculation we used for the concentration of O_2 in air the approximation

$$C_{O_2b} = \frac{0.2}{22400 \text{ cm}^3/\text{mole}}$$
(10.120)

Finally we return to the mass balance to calculate the time needed to burn completely the particle

$$-\frac{4\pi R^2 \rho_C}{M_C}\frac{dR}{dt} = 4\pi R^2 k'' C_{O_2s} = 4\pi R^2 k_m (C_{O_2b} - C_{O_2s})$$
(10.121)

or

$$\frac{dR}{dt} = -\frac{M_C k_m}{\rho_C}C_{O_2b}$$
(10.122)

The mass transfer coefficient is

$$k_m = \frac{Sh_D D_{O_2}}{D} = \frac{2 D_{O_2}}{D} = \frac{D_{O_2}}{R}$$
(10.123)

so the mass-balance equation can be rearranged to give

$$R\,dR = -\frac{M_C D_{O_2} C_{O_2}}{\rho_C}dt$$
(10.124)

This equation can be integrated to yield

$$\int_{R_o}^{R=0} R\,dR = -\frac{M_C D_{O_2} C_{O_2}}{\rho_C}\int_0^\tau dt$$
(10.125)

or

$$\tau = \frac{\rho_C}{2 M_C D_{O_2} C_{O_2}}R_o^2$$
(10.126)

Inserting numbers ($\rho_C = 2.25$ g/cm³, $D_{O_2} = 0.1$ cm² /sec, and $M_C = 12$ amu), we have

$$\tau = \frac{2 \times 2.25 \times 0.05^2}{2 \times 12 \times 0.01 \times 0.2/22400}$$
(10.127)

or

$$\boxed{\tau = 2.6 \text{ sec}}$$
(10.128)

which is a reasonable time for a spark to disappear. [To check out this calculation, take a stopwatch with you the next time you go camping.]

10.15 SOLID AND LIQUID EXPLOSIVES

In combustion of solids such as charcoal, a reaction occurs between carbon and O_2 from the air in diffusion-limited reactions. More than 1000 years ago the Chinese found that when they combined charcoal with sulfur and potassium nitrate (obtained by crystallizing liquid animal wastes in a process too messy to describe), the mixture of solids when ignited had quite different characteristics. This mixture would fizz with a flash of light or, if the solid was confined by wrapping it in a roll of paper, it would become a firecracker. Thus were born explosives, which Westerners were the first to turn into materials of warfare in the fifteenth century.

As mentioned previously, the animal waste sources of fixed nitrogen were too limited for conducting a large-scale war, and chemical engineers developed the Haber process for making ammonia from N_2 and the Ostwald process to oxidize NH_3 into NO and thus to HNO_3, as we have discussed previously. The explosives business then evolved through nitroglycerine, nitrocellulose, and TNT to the modern plastic explosives and rocket fuels. (Recall the history of DuPont in Chapter 3.)

A solid or liquid explosive is simply a chemical or a mixture of solids that contains fuel and oxygen in the condensed state so that no diffusion processes slow the rate. The reaction of TNT can be written approximately as

$$CH_3C_6H_2(NO_2)_3 \rightarrow 6CO + C + \tfrac{5}{2}H_2 + \tfrac{3}{2}N_2, \qquad \Delta H_R = -303 \text{ kcal/mole} \qquad (10.129)$$

There is not enough oxygen in the molecule to convert all C to CO_2 or H to H_2O, but the products quickly are heated to sufficient temperatures that they react instantly with O_2.

If the solid is in an open container, the reaction just fizzes, but if it is confined long enough to build up pressure, it generates a pressure wave, which causes accelerated propagation of the combustion wave, which is heard as a bang. If the container is confined on all but one side (the nozzle), reaction creates a rocket.

The art and engineering of propellant and explosive manufacture consists in finding suitable combinations and treatments of solids that will produce a "stable" compound that will not ignite until desired and will then burn with the appropriate rate to produce the desired blast wave or thrust. Obviously these are dangerous experiments, and they should be attempted after one figures out all the possibilities and what to do about them.

Since an explosive solid contains fuel and oxidant in close proximity, the exothermic reaction can occur very quickly and without diffusion control, just as for separate fuel and oxidant systems. Premixed flames and explosions with gases involve the same basic idea with both fuel and oxidant mixed before reaction begins (and potentially disastrous consequences if they are ignited with inappropriate mixtures and confinement).

Another potentially explosive class of compounds is multiply bonded hydrocarbons. Early in the twentieth century, chemical engineers learned by disastrous accidents that pure acetylene itself can detonate because the reaction

$$C_2H_2 \rightarrow 2C_s + H_2, \qquad \Delta H_R = -54.19 \text{ kcal/mole} \qquad (10.130)$$

is highly exothermic. Many acetylene tanks and pipes exploded before we learned to add stabilizers (chemicals such as acetone, which scavenge free radicals, which would otherwise promote chain-branching reactions). Even then, acetylene is sufficiently dangerous that it is seldom used in large quantities in the chemical industry.

TABLE 10–3
Heats of Formation of Hydrocarbons

C_2H_6	-20.24 kcal/mole
C_2H_6	$+12.50$
C_2H_2	$+54.19$
$C=C-C=C$	$+26.33$
$C=C-C\equiv C$ (VAc)	$+72.8$

Acetylene is widely sold as the fuel for welding torches, and it is stored in large cylinders at high pressures in many welding shops. In fact, this acetylene is mixed with acetone, which has been found to be an effective scavenger of acetylene decomposition, so that these tanks are relatively safe.

Table 10–3 lists the heats of formation of several of these potentially dangerous hydrocarbons. In a homework problem you are asked to calculate the temperatures and pressures of these gases if they suddenly deompose into their elements.

10.16 EXPLOSIONS AND DETONATIONS

In the previous section we described one type of reaction that can cause an explosion, namely, the situation where fuel and oxidant are mixed in solids or liquids that can release large amounts of energy very quickly. These processes can also occur in premixed gases. An example of useful explosions in gases is the repeated explosions that occur in the automobile engine. However, most examples of explosions in gases are in fact disasters in which products burst walls of containers and cause havoc.

An explosion can be defined as a fast, *transient*, exothermic reaction. It needs exothermicity to generate energy and must be fast to generate this energy very quickly in a transient pulse. We can also distinguish between events in which the reaction propagates at subsonic velocity as an explosion and one in which the reaction propagates with sonic or supersonic velocity as a detonation.

In practice, we call an explosion a reaction that makes noise. This can be a "whoosh," a "thump," or a "bang" in order of increasing rate of energy release and time of confinement. The bang is caused by a buildup of pressure that is released very quickly as the confinement system ruptures or the reactor ruptures. The parameter that distinguishes these characteristics is the flame velocity u_f, which is the velocity with which the reaction front propagates down a tube or across a container. The velocity of sound in air at 300 K is approximately 1100 ft/sec or 300 m/sec. [How far away was the lightning bolt if the thunder is heard 5 sec after the flash?] If the velocity of propagation is less than the velocity of sound, the wave is called a *deflagration*, and if it is at or above the velocity of sound, it is called a *detonation*, which can do much more damage because of the shock wave. We call a subsonic reaction wave a deflagration and a supersonic wave a detonation.

If CH_4 and air at the stoichiometric ratio (9.5% CH_4 in air) at atmospheric pressure are ignited at one end of a pipe, the pressure will ultimately rise by a factor of ~ 10, and this pressure wave will travel at 300 m/sec down the pipe. The heat of reaction is -192 kcal/mole of CH_4, and all this heat will be released and the pressure will rise very quickly.

Explosions in solids and liquids can be even more violent than in gases because the energy content per mole is comparable, but the density of solids and liquids is approximately 1000 times the density of gases at atmospheric pressure.

These considerations in fact underestimate the damage an explosion can cause as the flame speed becomes sonic and generates a shock wave. The pressure wave travels so fast that it carries considerable momentum, and this can burst walls even more effectively than a static calculation of wall strength would indicate. Sonic waves also reflect and focus their energy, so that quite unpredictable and extreme localized damage can occur from an explosion.

10.17   REACTOR SAFETY

The chemical industry is one of the safest industries in the United States in terms of accidents and fatalities, with approximately 10 fatalities per year for 100,000 workers. Your chances of getting killed in driving to and from work are much higher than while at work in the most hazardous job you could have in the chemical industry.

Nevertheless, accidents in our industry can be spectacular, with great potential for loss of life and income for many people. Most important, chemical engineers (you) will be responsible for creating safe processes and protecting the lives and health of your colleagues.

A great deal can and should be said about safety, but we have only been able to outline the principles by which chemical reactors may go out of control. We will conclude our discussion of safety by listing some of the notorious accidents that have occurred in the chemical industry (see Table 10–4).

The Texas City Disaster was perhaps the first "modern" chemical disaster, and it is still the U.S. chemical accident with the most casualties (as of this writing). It occurred when a barge of ammonium nitrate (NH_4NO_3 or AN) fertilizer was being unloaded in Galveston Bay near Texas City. A fire started in the crew quarters of the barge. The fire department was spraying water on the fire, and many people on the shore were watching the action. Suddenly the barge exploded, taking with it a sister barge, a sulfur plant on the shore, and two airplanes flying overhead.

No one expected this disaster, and all procedures were being followed according to the "standard safety practices" of 1947. AN was a fertilizer, not an explosive! No one had "done the experiment" until the fire happened to get hot enough with a large enough quantity of AN to cause its detonation.

By current practice, large quantities of AN are never accumulated or exposed to heat or fires. However, fertilizer storage warehouses still explode occasionally, and AN has become a popular, inexpensive, and readily available material for terrorists, for example, in the World Trade Center bombing in New York City in 1992 and the Oklahoma City bombing of 1995.

TABLE 10–4
Some Famous Accidents in the Chemical Industry

Accident	Year	Fatalities	Reaction
The Texas City Disaster	1947	561	NH_4NO_3 explosion
The Texas City Disaster II	1969	0	vinylacetylene explosion
Flixborough, England	1947	0	cyclohexane VCE
Bhopal, India	1986	~2500	methylisocyanate hydrolysis
Phillips polyethylene, Texas	1992	24	isobutane VCE
—	now	—	—

A second explosion occurred in a chemical plant in Texas City, Texas, some years later, and this was called the Texas City Disaster II. This explosion occurred when a distillation column used to purify butadiene from other C_4 hydrocarbons was operating improperly and the feed was shut off, as was the product stream from the reboiler. The column suddenly blew up, sending the column like a rocket $\sim \frac{1}{2}$ mile into the town. After some investigation, it was surmised that the reboiler accumulated a high concentration of vinyl acetylene, which is a strongly endothermic compound. The initial explosion was the reaction

$$CH_2 = CH - C \equiv CH \rightarrow 4C_s + 2H_2, \qquad \Delta H_R = -72.8 \text{ kcal/mole} \qquad (10.131)$$

The hydrogen pressure burst the reboiler and the contents of the column then ignited, turning the distillation column into a primitive but effective rocket that took off. Fortunately, no one was around at either the launch site or landing site. However, the town's residents heard a loud bang, and they had been highly sensitized from the previous one when the AN barges blew up a few years before.

Both Flixborough and the Phillips polyethylene plane explosion were vapor cloud explosions (VCE) of hydrocarbons that escaped through leaks in chemical reactors. In Flixborough, Nylon 6 was being made by oxidation of cyclohexane to produce caprolactam. In the Phillips plant low-density PE was being produced in a process that uses isobutane solvent at very high pressures. These liquids escaped and vaporized without combustion because no ignition sources were nearby. After the hydrocarbon vapors had spread large distances and had become diluted with air (O_2), a cigarette or pilot light somewhere ignited the premixed alkane–air mixture creating a VCE, with disastrous consequences to the plant, to the neighborhood, and in the latter case to people.

Bhopal, the site of the largest chemical accident in the past 50 years, wasn't an explosion but rather an uncontrolled reaction. The plant in Bhopal, India, owned jointly by Union Carbide and the Indian government, was producing methyl isocyanate as an intermediate for a pesticide. Water was admitted into a storage tank containing this chemical, which caused it to hydrolyze and produce CO, HCN, and unreacted MIC, which spread downwind into a crowded slum nearby and caused many deaths and casualties. The desired reaction to synthesize MIC was

$$CH_3HN_2 + OCl_2 \rightarrow CH_3NCO + 2HCl \qquad (10.132)$$

and the very undesired hydrolysis reactions were

$$CH_3NCO + H_2O \rightarrow CO + HCN \qquad (10.133)$$

all of which are highly toxic.

The Bhopal chemical plant was used to produce agricultural pesticides and herbicides. These chemicals are very important in agriculture to maintain insect and weed populations under control so that crop yields can be improved. One wants to produce pesticides and herbicides that are very selective in destroying a particular insect or plant but are not also toxic to other organisms (such as humans). Here the objectives are to produce a chemical that is toxic to specific plants or animals but not to others. Some desirable characteristic of these chemicals is that they (1) attack only the desired pests, (2) do not decompose during storage and shipment, and (3) decompose soon after application so that pesticide residues do not remain on food or contaminate ground water. It is obvious that considerable creative engineering goes into formulating and testing these agricultural products.

Figure 10–16 Reaction steps to prepare SEVIN, a common agricultural pesticide. All of the reactants are quite inert and harmless except phosgene, MIC, and the final product.

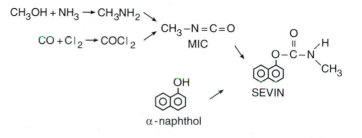

α-naphthol

At Bhopal they were manufacturing SEVIN, also called carbaryl, a broad-spectrum insecticide that was and still is used on many crops throughout the world. The reaction steps to manufacture SEVIN are shown in Figure 10–16.

10.18 SUMMARY

The subject of this chapter has been the peculiar characteristics of chain reactions. This is a very common type of chemical reaction, which chemical engineers need to be able to handle and to be aware of how different a chain reaction can be than an "ordinary" reaction, $A \rightarrow$ products, $r = kC_A$, or its variants. Chain reactions are involved in autooxidation and in combustion, the subjects of this chapter, and also in polymerization, the subject of the next chapter.

A chain reaction process is composed of a sequence of reaction steps whose rates can vary by large factors. Some of these reaction steps slow the overall process, and some accelerate it, and some of these slow steps can have alternate paths.

In combustion processes the reaction is also highly exothermic, and, combined with the high effective activation energy of combustion processes, leads to large temperature dependence. The overall reaction products can also vary (partial oxidation versus total oxidation), and this factor must also be considered when dealing with combustion processes.

Because of these factors, chain reactions are inherently unpredictable. Chemical and thermal autocatalysis make the overall rate $r(C_j, T)$ not a simple function. Chain reactions can also be very fast so that the reaction may be limited by mass transfer processes.

We frequently do not have reliable reaction rate expressions for chain reactions, but we can compensate for this lack by designing and operating the reactor to manage the overall course of the reaction by properly dealing with mixing, mass transfer, promoters and inhibitors, and the presence of surfaces.

10.19 REFERENCES

Crowl, Daniel A., and Louvar, Joseph, *Chemical Process Safety: Fundamentals with Applications*, Prentice-Hall, 1990.
Gardiner, William C., *Gas-Phase Combustion Chemistry*, Springer-Verlag, 2000.
Glassman, Irving, *Combustion*, Elsevier, 1996.
Lewis, B., and von Elbe, G., *Combustion, Flames, and Explosions of Gases*, Academic Press, 1987.
Petroski, Henry, *The Engineer Is Human: The Role of Failure in Successful Design*, St. Martin's Press, 1985.

Petroski, Henry, *Design Paradigms: Case Histories of Error and Judgment in Engineering*, Cambridge University Press, 1994.

Unger, Stephen H., *Controlling Technology: Ethics and the Responsible Engineer*, Wiley, 1994.

Warnatz, J., Maas, U., and Dibble, R. W., *Combustion*, Springer-Verlag, 1996.

PROBLEMS

10.1 Develop a model of the burning of a cigarette. The primary reactions can be regarded as the burning of solid carbon to CO_2 in excess O_2 and to CO if oxygen deficient. In the absence of O_2 the tobacco pyrolyzes to carbon with the vaporization of tars and nicotine, which for unexplained reasons some people like to inhale into their lungs.

(a) What mass- and energy-balance equations are appropriate?

(b) What reactions dominate in different regions of the cigarette? Are these reactions homogeneous or heterogeneous?

(c) Sketch $C_j(z)$ in the cigarette for O_2, CO_2, CO, and tar assuming a steady-state gas velocity and that L changes slowly.

(d) How does the heat generation and absorption vary with position? Sketch $T(z)$.

(e) Illustrate $u(t)$ and $T(z, t)$ during a puff? Sketch $L(t)$.

Estimate transient temperatures, velocities, and concentrations versus time with as much accuracy as possible.

[Caution: Do not do experiments to obtain data for this problem. Cigarette smoking is hazardous to your health.]

10.2 A chain reaction $A \rightarrow B + C$ obeys the rate expression $r = kC_A^3$. What initiation and termination steps will predict these kinetics? Repeat for $r = kC_A^2$.

10.3 In steam cracking the homogeneous pyrolysis of ethane to produce ethylene

$$C_2H_6 \rightarrow C_2H_4 + H_2$$

is carried out in a pipe heated to 850°C at an ethane pressure of 1 atm. The major reactions are thought to be

$$C_2H_6 \rightarrow C_2H_5 + H$$
$$C_2H_5 \rightarrow C_2H_4 + H$$
$$C_2H_6 + H \rightarrow C_2H_5 + H_2$$
$$2C_2H_5 \rightarrow C_4H_{10}$$

(a) What should be the order of this reaction with respect to ethane?

(b) The C–H bond energy is 94 kcal/mole, the first and third reactions have zero activation energy, and the last reaction has an activation energy of 25 kcal/mole. What should be the activation energy of the overall reaction?

(c) It is found that 60% of the ethane is reacted with a residence time of 4 sec. What is the effective rate coefficient and what should be the effective pre-exponential factor? Ignore the mole number change in this reaction.

(d) Suppose it is found that the products contain 99.9 mole % ethylene and 0.1 mole % butane and butylene. What can you say about reaction rate coefficients of reactions 3 and 4?

(e) There are actually many more products in this reaction. Why should one expect formation of butylenes in this reaction? Indicate typical reaction steps.

10.4 In ethane pyrolysis assume that the chain termination step is the adsorption of ethyl radicals on the walls of the tube instead of the homogeneous step listed previously.

(a) What order of reaction should be expected with respect to ethane?

(b) If the tube has a diameter of 1 in. and the flow is laminar ($Sh_D = 8/3$), what is the rate coefficient of this termination step if the adsorption rate on the tube wall is very fast?

(c) With these reaction steps how should the reaction rate change if the tube diameter is 4 in.?

(d) This reaction is strongly endothermic. How should one choose the optimum pipe diameter?

(e) The adsorbed ethyl radical will pyrolyze to coke on the walls of the reactor. If all of it turns into coke and 10^4 pounds of ethylene can be produced per pound of coke formed, what can you say about relative reaction rates in this mechanism?

10.5 Model a campfire spark as a carbon sphere ($\rho_C = 3$ g/cm^3) surrounded by a boundary layer of thickness $2R$ (Sh $= 2$) at 1000°C.

(a) Summarize in several sentences and equations the model of the combustion of a campfire spark.

(b) Why will a small particle be lifted from the fire?

(c) One part of this problem is analogous to a hot air balloon. Formulate a model with which you could calculate the size and density for lift and the velocity of the rising spark particle.

10.6 Ammonium nitrate (AN) is a common fertilizer that is an excellent source of nitrogen.

(a) Sketch the flow sheet to manufacture AN from CH_4, air, and water. (This is a combination of several earlier problems.)

(b) Fertilizer bags list three numbers, N, P, and K percentages. What are the NPK ratings of AN and of KNO_3?

(c) Write a reasonable reaction for the decomposition (explosion) of AN.

(d) How does the energy release of AN compare with TNT?

(e) How many cubic meters of gas would be produced if 1 cubic meter of AN reacted as above and the products were at atmospheric pressure and 20°C? At 2000°C? [Don't try this experiment!]

10.7 A bottle that you thought contained pure diethyl ether actually has autooxidized to 10% diethyl ether hydroperoxide. What is the temperature if the contents of the bottle suddenly react to chemical equilibrium? At what hydroperoxide concentration would the temperature not exceed 100°C? Assume that the heat of decomposition of the hydroperoxide is 30 kcal/mole.

10.8 Dimethyl ether decomposes to methane and formaldehyde. It is assumed that the reaction proceeds according to the elementary steps:

$$CH_3OCH_3 \rightarrow CH_3 + OCH_3$$
$$CH_3 + CH_3OCH_3 \rightarrow CH_4 + CH_2OCH_3$$
$$CH_2OCH_3 \rightarrow CH_3 + HCHO$$
$$CH_3 + CH_2OCH_3 \rightarrow CH_3CH_2OCH_3$$

where the first is the initiation step, the second and third are propagation steps, and the fourth is the termination step. Derive a rate expression for this reaction $r = k[CH_3OCH_3]^n$ assuming pseudo steady state on radical intermediates.

10.9 Phosgene, an intermediate in polyurethane monomer, is made by the homogeneous reaction

$$CO + Cl_2 \rightarrow COCl_2$$

The elementary steps of this reaction are

$$Cl_2 \rightleftarrows 2Cl$$

$$Cl + CO \rightleftarrows COCl$$

$$COCl + Cl_2 \rightleftarrows COCl_2 + Cl$$

where all reactions are reversible.

(a) Identify initiation, propagation, and termination steps.

(b) Show that the rate of this reaction could be

$$r = k[CO][Cl_2]^{3/2}$$

(c) Find an expression for the reverse reaction rate.

10.10 The radioactive isotope ^{235}U undergoes slow radioactive decay, which produces neutrons,

$$^{235}U \rightarrow \text{ decay products } + n, \qquad r_i = k_i C_U$$

When these neutrons bombard ^{235}U, they can produce fission,

$$^{235}U + n \rightarrow \text{ fission products } + \alpha n, \qquad \Delta H_R, \quad r_p = k_p C_U C_n$$

where ΔH_R is millions of calories per mole.

(a) In a nuclear reactor these neutrons produce a controlled fission process by assembling a suitable mass of ^{235}U (using control rods) to control the fraction of the neutrons released by decay that undergo the fission process. Set up a pseudo-steady-state mass balance on C_n and an energy balance to show how chain branching can yield a nuclear reactor.

(b) A nuclear bomb uses these neutrons to produce uncontrolled fission. Show how these equations can predict formation of a bomb if a suitable large amount of ^{235}U (the critical mass) is assembled.

To solve these problems realistically, you would need to examine the absorption processes and neutron capture cross sections in detail. However, for this problem formulate the "termination rate" in terms of neutron escape versus the size of a uranium sphere.

10.11 Consider the combustion of a coal pile of diameter D and height H. Before ignition the air flow is driven by wind, and after ignition by natural convection. The heat of combustion of carbon is -94 kcal/mole. Set up a *qualitative* energy balance (no numbers) to explain the dependence of combustion on weather and on D and H. What size and shape of coal piles do you recommend to minimize spontaneous combustion?

10.12 Consider the chain reaction $A \rightarrow B + C$, which proceeds by the steps

$$A \rightarrow R, \qquad r_i = k_i C_A$$

$$A + R \rightarrow B + C + R, \qquad r_p = k_p C_A C_R$$

$$R \rightarrow X, \qquad r_{t1} = k_{t1} C_{Rb}$$

$$R \rightarrow X, \qquad r_{t2} = k'' C_{Rs}$$

The first termination step occurs homogeneously, while the second occurs by adsorption of R on the walls of the reactor. Formulate an expression for the rate of this reaction expected in a tube of diameter D with laminar flow ($Sh_D = 8/3$) and a diffusion coefficient D_R. What effective rate expressions are obtained in the limits of when

(a) homogeneous termination dominates,

(b) reaction at the wall dominates,

(c) mass transfer to the wall dominates?

10.13 Consider the chain reaction $A \rightarrow B + C$, which proceeds by the steps

$$I \rightarrow R, \qquad r_{iI} = k_{iI} C_I$$
$$A \rightarrow R, \qquad r_i = k_i C_A$$
$$A + R \rightarrow B + C + R, \qquad r_p = k_p C_A C_R$$
$$R \rightarrow X, \qquad r_t = k_t C_R$$

The first initiation step occurs by reaction of an initiator I whose initial concentration is C_{Io}, while the second occurs by decomposition of A whose initial concentration is C_{Ao}.

(a) Formulate an expression for the rate of this reaction in terms of C_A and C_I assuming C_R is small. What are the possible orders of this reaction with respect to C_A?

(b) Repeat part (a) assuming the termination step is $2R \rightarrow X, r_t = k_t C_R^2$.

(c) For the rate in part (a) calculate C_A, C_B, and C_I versus τ in a CSTR assuming $k_{iI} = 0$. What is the conversion when $\tau \rightarrow \infty$ if $k_{iI} = 0$?

10.14 Consider the chain reaction $A \rightarrow B + C$, which proceeds by the steps

$$A \rightarrow R, \qquad r_i = k_i C_A$$
$$A + R \rightarrow B + C + R, \qquad r_p = k_p C_A C_R$$
$$R \rightarrow X, \qquad r_t = k_t C_R$$
$$R + S \rightarrow X, \qquad r_{tS} = k_{tS} C_R C_S$$

The second termination step occurs by reaction of a scavenger S whose initial concentration is C_{So}.

(a) Formulate an expression for the rate of this reaction in terms of C_A and C_s assuming C_R is small. What are the possible orders of this reaction with respect to C_A?

(b) Repeat part (a) assuming the first termination step is $2R \rightarrow X, r_t = k_t C_R^2$.

(c) For the rate in part (a) calculate C_A, C_B, and C_s versus τ in a CSTR assuming $k_t = 0$. What is the conversion when $\tau \rightarrow \infty$ if $k_t = 0$?

10.15 Summarize the reactions used to produce the following chemicals;

(a) Acetic acid from sugar,

(b) Acetic acid from methane.

(c) 1,2-Propanediol from propane using chlorine,

(d) 1,2-Propanediol from propane without chlorine,

(e) Acetone from propane,

(f) Methyl ethyl ketone from butane.

10.16 Consider the chain branching reaction $A \rightarrow B + C$, which proceeds by the steps

$$A \rightarrow R, \qquad r_i = k_i C_A$$
$$A + R \rightarrow B + C + \alpha R, \qquad r_p = k_p C_A C_R$$
$$R \rightarrow X, \qquad r_t = k_t C_R$$

(a) Find an expression for C_R in terms of the ks and C_A assuming steady state.

(b) Find a rate expression $r(C_A)$ in terms of the ks.

(c) Find an expression for C_A versus t in a batch reactor.

(d) For what conditions will a steady state exist in a batch reactor?

(e) Find expressions for C_A versus t in PFTR and CSTR.

(f) For what conditions will no steady state exist in these continuous reactors?

(g) Discuss the significance and consequences of parts (d) and (f).

10.17 If a container filled with each of the molecules in Table 10–3 at room temperature and atmospheric pressure suddenly decomposed into C_s and H_2, calculate the temperatures and pressures in the container.

10.18 A bottle that you thought contained ethanol had oxidized to 10% ethylhydroperoxide. If the contents of this bottle suddenly went to chemical equilibrium, estimate the temperature and pressure in the bottle (for a short time). The heat of formationof ethylhydroperoxide is -41 kcal/mole and the heat of carbon oxidation to CO_2 is -67 kcal/mole.

10.19 A container is filled with a 2:1 mixture of H_2 and O_2, but it may contain trace impurities. Assuming that the initiation step

$$I \rightarrow R, \qquad r_i = k_i C_I$$

is the only activated process in hydrogen oxidation, calculate the relative rates at 25 and 500°C of the reaction with addition of 1% of ethane, methanol, and H_2O_2 compared with pure reactants. Assume that all pre-exponentials are equal and assume that the activation energies of initiation are equal to the bond energies given in the tables in the text and from handbooks.

10.20 Consider the chain reaction $A \rightarrow B + C$, which proceeds by the steps

$$A \rightarrow R, \qquad r_i = k_i C_A$$
$$A + R \rightarrow B + C + R, \qquad r_p = k_p C_A C_R$$
$$R \rightarrow X, \qquad r_t = k_t C_R$$

with $k_i = 0.001$ sec^{-1}, $k_p = 200$ liter/mole sec, $k_t = 0.01$, and $C_{Ao} = 2$ moles/liter.

(a) Find $r(C_A)$ assuming the pseudo-steady-state approximation.

(b) What residence time will be required in a CSTR for 90% conversion for these kinetics?

(c) What residence time will be required in a PFTR for 90% conversion for these kinetics?

10.21 [Computer] Consider the chain reaction $A \rightarrow B + C$, which proceeds by the steps

$$A \rightarrow R, \qquad r_i = k_i C_A$$
$$A + R \rightarrow B + C + R, \qquad r_p = k_p C_A C_R$$
$$R \rightarrow X, \qquad r_t = k_t C_R$$

with $k_i = 0.001$ sec^{-1}, $k_p = 200$ liter/mole sec, $k_t = 0.01$, and $C_{Ao} = 2$ moles/liter.

(a) Plot C_A, C_B, and C_R versus τ in a PFTR for $C_{Ro} = 0, 0.0001, 0.001$, and 0.01. For what residence times will the conversion of A be 90% for these C_{Ro}?

(b) Compare these curves with those obtained using the pseudo-steady-state assumption.

10.22 [Computer] Consider the chain reaction $A \rightarrow B + C$, which proceeds by the steps

$$A \rightarrow R, \qquad r_i = k_i C_A$$
$$A + R \rightarrow B + C + R, \qquad r_p = k_p C_A C_R$$
$$R \rightarrow X, \qquad r_t = k_t C_R$$
$$R + S \rightarrow X, \qquad r_{tS} = k_{tS} C_R C_S$$

with $k_i = 0.001$ sec^{-1}, $k_p = 200$ liters/mole sec, $k_t = 0.01$, $k_{tS} = 100$ liters/mole sec, and $C_{Ao} = 2$ moles/liter.

Plot C_A, C_B, and C_R versus τ in a PFTR for $C_{Ro} = 0$ for different values of $C_{So} = 0, 0.00001, 0.0001, 0.001$, and 0.01. For what value of C_{So} will the residence time for 10% conversion of A be increased by a factor of 10? By a factor of 100?

10.23 A 50-μm drop of diesel fuel is injected into the cylinder of an engine with air at 1500°C. Assume dodecane, $Do_2 = 0.01$cm$^2/S$, $P = 10$ atm, $p = 0.9$ g/cm^3.

(a) Estimate the time it would take for complete combustion of the drop. Use the shrinking sphere models of Chapter 9 and assume that the process is limited by O_2 diffusion to the drop surface.

(b) Suppose the engine designers require complete combustion in 2×10^{-3} sec. What drop size must the fuel injectors produce using the above model?

(c) Formulate a more detailed model of liquid drop combustion which includes a homogeneous reaction and the vapor pressure and boiling point of the fuel.

10.24 There are several possible ways to make 1,2-propanediol. Illustrate the chemical reactions for each of them and indicate possible advantages and disadvantages of each.

(a) methylacetylene + H_2O,

(b) propylene + HOCl,

(c) propylene + HOOH,

(d) propylene + ethylbenzene hydroperoxide,

(e) propylene + O_2.

Chapter 11

POLYMERIZATION REACTIONS AND REACTORS

T he production of monomers and their reaction to form polymers account for more than half of the chemical industry, and the polymers and plastics industries are among the major employers of chemical engineers. Many of the reactions we have considered previously were concerned with methods of preparing the small molecule monomers that are used in preparing polymers. In this chapter we consider the reactors used to polymerize these molecules into polymers. We will see that the reactions we are trying to manage are sufficiently different so that the reactors needed for polymerization processes are qualitatively different than for other chemical processes.

It is not difficult to make many molecules react with each other to form larger molecules with high molecular weights simply by heating small molecules in a closed container. However, most of these systems react by the random breaking of bonds and free-radical chain reactions to form "brown gunk," a solid that is insoluble in all solvents and decomposes or burns before it can be melted.

A major breakthrough in polymers came when scientists and engineers learned how to make *linear polymers*, which consist exclusively of long-chain molecules. These systems will melt and flow under reasonable conditions so that they can be spun into fibers, formed into sheets, or injected into molds. Typical engineering polymers have molecular weights of 10^3 to 10^5 AMU [50 to 1000 monomer units of 28 (ethylene) to 100 AMU linked together], with many polymer molecules possessing very few branching side chains unless they are added intentionally.

The next breakthrough came with the discovery of how to make almost linear polymers with just a few crosslinked connections between molecules. These materials will not flow after crosslinking, and they become *elastomers*. A final type is the heavily crosslinked polymers such as epoxy and urea-formaldehyde, which are processed as monomers and then crosslinked at a suitable time and rate to make paints, adhesives, and very hard and durable plastic materials.

We typically sketch a linear polymer molecule as a long chain that in solution coils into a roughly spherical shape. Polymers are sometimes branched and sometimes crosslinked as shown in Figure 11–1.

432

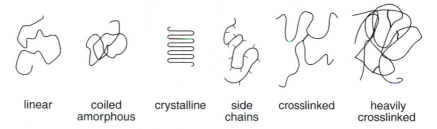

linear   coiled   crystalline   side   crosslinked   heavily
          amorphous   chains   crosslinked

Figure 11–1 Sketches of linear polymer in open and coiled form, linear and with side chains, crystalline and amorphous, and with crosslinking.

The overall goal of producing polymers is to find synthetic chemicals that can be fabricated into materials that duplicate and replace those found naturally. As we have noted previously, we try to find replacements for such natural polymers as leather, wood, silk, cotton, rubber, glue, and paint, which our ancestors discovered have many uses in clothing, housing, and utensils. We want to find ways of making these replacement materials more cheaply with properties we can control and modify for particular applications. These natural polymers are all basically hydrocarbons with appropriate functional groups that are synthesized by biological systems of plants (cellulose) or animals (protein) in materials such as skin, fur, webs, seed pods, wood, and paper.

There are also many inorganic polymers such as rocks and ceramics. We also attempt to duplicate these materials in such products as concrete. High-technology inorganic polymers using organometallic precursors and sol–gel processes are also used to prepare catalysts, coatings, sensors, and linear polymers such as silicones.

We will not attempt to discuss polymers or their properties in any detail in this chapter because this would require several complete courses for comprehensive coverage. Our interest is in polymerization rates and polymerization reactors. These topics are simple extensions of our previous discussion of other reactions and reactors with some important differences. We will only attempt to describe polymers and polymerization in sufficient detail to introduce the notation of polymers and some of the issues arising in polymerization process.

The major issue in designing a polymerization process is the creation of a polymer with specific desired properties such as melting point or melting range, glass transition temperature, viscosity, elasticity, fracture toughness, color, electrical properties, and so on. We will not even discuss the meanings of these terms here. Rather, our goal is to consider the reactor conditions which create these properties cheaply and efficiently.

The stages in producing polymers and plastics can be described as

$$\text{raw materials} \;\rightarrow\; \text{monomers} \;\rightarrow\; \text{polymer} \;\rightarrow\; \text{formed polymer} \qquad (11.1)$$

We have considered the production of monomers throughout this course. In this chapter we consider the second and third stages, namely, the polymerization process, in which monomer is converted into polymer and the forming of this polymer into a final shape in which it will be used.

We will only attempt to outline these processes in order to indicate how some of these goals are accomplished. Only in a specialized course and text can these be considered in any detail. However, other fields such as processing of foods, pharmaceuticals, and materials have many similarities to those of commodity polymers, and from this description of polymer processing the student can perhaps see how these applications are related.

There are basically two major classes of linear polymers: (1) addition and (2) condensation, and most polymers can be divided into these two groups. We will consider these types of polymers in turn because their kinetics and reactors are quite different.

11.1   IDEAL ADDITION POLYMERIZATION

We consider first the reactions in which monomer molecules M add sequentially to an initiator molecule A to form a linear polymer molecule AM_n,

$$A + nM \rightarrow AM_n \tag{11.2}$$

This is the simplest type of polymerization, and it is used to make specialty polymers. Basically one mixes a monomer at initial concentration $[M]_0$ and an initiator at initial concentration $[A]_0$ and allows reaction to occur. The steps are

$$A + M \rightarrow AM, \qquad r_1 = k_1[A][M] \tag{11.3}$$

$$AM + M \rightarrow AM_2, \qquad r_2 = k_2[AM][M] \tag{11.4}$$

$$AM_2 + M \rightarrow AM_3, \qquad r_3 = k_3[AM_2][M] \tag{11.5}$$

$$AM_3 + M \rightarrow AM_4, \qquad r_4 = k_4[AM_3][M] \tag{11.6}$$

$$\vdots$$

$$AM_j + M \rightarrow AM_{j+1}, \qquad r_{j+1} = k_{j+1}[AM_j][M] \tag{11.7}$$

$$\vdots$$

Each step adds one monomer unit to the growing polymer, so that the overall process after n such additions (add all reaction steps together from $j = 0$ to n) is

$$A + nM \rightarrow AM_n \tag{11.8}$$

Thus we have made a single linear polymer molecule consisting of a head of the initiator unit $A-$ and a linear chain of n linear M units,

$$A\text{--}M\text{--}M\text{--}M\text{--}M\text{--}M\text{--}M\text{--} \tag{11.9}$$

The "tail" of this molecule $(M-)$ is the only reactive site on the molecule; so the molecule grows indefinitely as long as there are more free monomer molecules available to react. There will be exactly one polymer molecule formed per initiator molecule, and the average chain length n after all monomers have reacted will be the ratio of the initial concentrations of M and A,

$$\bar{n} = \frac{[M]_0}{[A]_0} \tag{11.10}$$

This type of polymerization process is called "living" polymerization because the polymer molecules grow forever until all monomer is consumed. Ideally, there is no step that terminates growth. One type of this polymer is called "ionic" because the active species to which M adds is an ion such as AM_j^+. An ionic intermediate suppresses termination steps involving reactions between two of these intermediates because both reactive ends have positive charges and will thus repel each other and only allow reaction of the reactive end with the uncharged monomer molecule.

11.1.1 PFTR or batch reactor

In polymerization we play the standard "game" we use for any multiple-reaction system by writing the mass balance for each species using our standard mass-balance equation for a multiple-reaction system,

$$\frac{dC_j}{dt} = \sum_{i=1}^{R} \nu_{ij} r_i \tag{11.11}$$

In a batch reactor or PFTR these equations are

$$\frac{d[A]}{d\tau} = -k_1[A][M] \tag{11.12}$$

$$\frac{d[AM]}{d\tau} = +k_1[A][M] - k_2[AM][M] \tag{11.13}$$

$$\frac{d[AM_2]}{d\tau} = +k_2[AM][M] - k_3[AM_2][M] \tag{11.14}$$

$$\frac{d[AM_3]}{d\tau} = +k_3[AM_2][M] - k_4[AM_3][M] \tag{11.15}$$

$$\frac{d[AM_4]}{d\tau} = +k_4[AM_3][M] - k_5[AM_4][M] \tag{11.16}$$

$$\vdots$$

$$\frac{d[AM_j]}{d\tau} = +k_j[AM_{j-1}][M] - k_{j+1}[AM_j][M] \tag{11.17}$$

$$\vdots$$

[For any complex sets of reactions such as these it is especially important to write down the reactions and rates in our standard form before writing out the mass-balance equations because mistakes are almost inevitable if one attempts to write out these equations by intuition.]

This is an infinite set of equations for the concentrations of $[AM_j]$. Each rate except the first has a term that makes the molecule from the next smaller one $(j - 1)$ and a term by which it is lost to make the next larger one $(j + 1)$.

We also have to write the equation for $[M]$, which is reacted away in each reaction,

$$\frac{d[M]}{d\tau} = -k_1[A][M] - k_2[AM][M] - k_3[AM_2][M] - \cdots \tag{11.18}$$

The equation for $[M]$ is coupled to all the other equations; so one would have to solve the infinite set of equations simultaneously to find $[M](\tau)$ and all the $[AM_j](\tau)$. This appears to be a very difficult problem that can only be solved numerically (and even computers do not like to solve infinite sets of equations).

However, if we assume that all rate coefficients are equal for reaction of all AM_j, $k_1 = k_2 = k_3 = \cdots = k$, then these equations simplify considerably.

The equation for $[M]$ becomes

$$\frac{d[M]}{d\tau} = -k[M]\sum_{j=0}^{\infty}[AM_j] \tag{11.19}$$

which doesn't at first look much better because there are still an infinite number of terms and equations. However, there is a simple trick that makes the problem almost trivial. Let us add all equations for $[AM_j]$. This gives

$$\sum_{j=0}^{\infty}\frac{d[AM_j]}{d\tau} = -k[A][M] + k[A][M] + k[AM][M] - k[AM][M]$$

$$+ k[AM_2][M] - k[AM_2][M] + \cdots \tag{11.20}$$

We see that each term exactly cancels so that this sum becomes

$$\sum_{j=0}^{\infty}\frac{d[AM_j]}{d\tau} = 0 \tag{11.21}$$

and therefore $\sum[AM_j]$ is constant. This immediately shows that the total number of polymer molecules is constant as polymerization proceeds. Furthermore, the sum of all $[AM_j]$ is exactly equal to the amount of A added initially,

$$\boxed{\sum_{j=0}^{\infty}[AM_j] = [A]_\text{o}} \tag{11.22}$$

Our equation for $[M]$ now becomes quite simple,

$$\boxed{\frac{d[M]}{d\tau} = -k[M]\sum_{j=0}^{\infty}[AM_j] = -k[M][A]_\text{o}} \tag{11.23}$$

and, since $[A]_\text{o}$ is known and constant, this equation can be integrated immediately starting with initial monomer concentration $[M]_\text{o}$ to yield

$$\boxed{[M](\tau) = [M]_\text{o}e^{-k[A]_\text{o}\tau}} \tag{11.24}$$

which says that the concentration of M decreases with t or z as a simple exponential. We can now write each of the AM_j equations separately, starting with

$$\frac{d[A]}{d\tau} = -k[A][M] = -k[A][M]_\text{o}e^{-k[A]_\text{o}\tau} \tag{11.25}$$

This equation can be separated to yield

$$\frac{d[A]}{[A]} = -k[M]_\text{o}e^{-k[A]_\text{o}\tau}\,d\tau \tag{11.26}$$

and this can be integrated with a feed concentration of initiator $[A]_\text{o}$ to obtain

$$\ln \frac{[A]_o}{[A]} = -\frac{[M]_o}{[A]_o}(1 - e^{-k[A]_o\tau}) \tag{11.27}$$

or

$$[A] = [A]_o \exp\{-\frac{[M]_o}{[A]_o}[1 - exp(-k[A]_o\tau)]\} \tag{11.28}$$

which is an explicit but not very appealing expression.

We can solve this equation for $[A](\tau)$ and insert it into the equation for $[AM]$

$$\frac{d[AM]}{d\tau} = +k_1[A](\tau)[M] - k_2[AM][M] \tag{11.29}$$

This is a linear first-order differential equation that can be solved for $[AM](\tau)$, and we can proceed to solve each of these successive linear differential equations for $[AM_j]$. Without doing this, we can find the number average polymer length $\bar{n}$ quite simply by noting the definition

$$\bar{n} = \frac{[M]_o - [M]}{[A]_o} = \frac{[M]_o}{[A]_o}(1 - e^{-k[A]_o\tau}) \tag{11.30}$$

These equations can be integrated analytically to find all $[A](\tau)$, although the expressions become quite complicated, and computer solutions are the best way to find species concentrations when the equations become more complicated.

11.1.2 CSTR

For these reactions in a CSTR we write the mass-balance equations using our standard notation

$$C_j - C_{jo} = \tau \sum_i \nu_{ij} r_i \tag{11.31}$$

to obtain the concentrations of all species as functions of τ,

$$[A]_o - [A] = k\tau[A][M] \tag{11.32}$$

$$[AM] = +k\tau[A][M] - k\tau[AM][M] \tag{11.33}$$

$$[AM_2] = +k\tau[AM][M] - k\tau[AM_2][M] \tag{11.34}$$

$$[AM_3] = +k\tau[AM_2][M] - k\tau[AM_3][M] \tag{11.35}$$

$$[AM_4] = +k\tau[AM_3][M] - k\tau[AM_4][M] \tag{11.36}$$

$$\vdots$$

$$[AM_j] = +k\tau[AM_{j-1}][M] - k\tau[AM_j][M] \tag{11.37}$$

$$\vdots$$

How can this set of algebraic equations be solved? Once again, we can add all of these equations to obtain

$$\sum_{j=0}^{\infty}[AM_j] = [A]_o \tag{11.38}$$

as in the batch reactor or PFTR.

The mass balance on the monomer yields

$$[M]_o - [M] = \tau(k[A][M] + k[AM][M] + k[AM_2][M] + \cdots)$$

$$= k\tau \sum_{j=0}^{\infty} [AM_j][M]$$

$$= k\tau[A]_o[M] \tag{11.39}$$

The monomer equation can be immediately solved to yield

$$\boxed{[M](\tau) = \frac{[M]_o}{1 + k\tau[A]_o}} \tag{11.40}$$

which is a very simple expression. The equation for $[A](\tau)$ is

$$[A]_o - [A] = k\tau[A][M] = k\tau[A]\frac{[M]_o}{1 + k\tau[A]_o} \tag{11.41}$$

which gives

$$[A](\tau) = \frac{[A]_o}{1 + [M]_o k\tau/(1 + k\tau[A]_o)} \tag{11.42}$$

We can now proceed to find analytical solutions for all $[AM_j](\tau)$ by solving each algebraic equation sequentially. This gives the simple expression

$$[AM_j](\tau) = \frac{[A]_o(k\tau[M])^j}{(1 + k\tau[M])^{j+1}} \tag{11.43}$$

We will not derive this expression here but rather leave it for a homework problem assignment.

The average polymer length $\bar{n}$ versus τ in the CSTR is also simple to find from the relation

$$\bar{n} = \frac{[M]_o - [M]}{[A]_o} = \frac{[M]_o}{[A]_o}\frac{k\tau[A]_o}{1 + k\tau[A]_o} \tag{11.44}$$

Example 11–1   Monomer at 2 moles/liter and initiator at 10^{-3} moles/liter are reacted in a continuous reactor. What is the polymer chain length for times sufficiently long that the reaction goes to completion?

From the above equation we obtain

$$\bar{n} = \frac{[M]_o}{[A]_o} = 2000 \text{ monomer units} \tag{11.45}$$

If $k = 0.05$ liter/mole min, what times are required to obtain a chain length of 100 monomer units in a PFTR and in a CSTR?

In a PFTR we have

$$\bar{n} = \frac{[M]_o - [M]}{[A]_o} = \frac{[M]_o}{[A]_o}(1 - e^{-k[A]_o\tau})$$ (11.46)

so that

$$\tau = -\frac{1}{k[A]_o} \ln\left(1 - \frac{n[A]_o}{[M]_o}\right)$$

$$= -\frac{1}{0.05 \times 10^{-3}} \ln\left(1 - \frac{100 \times 10^{-3}}{2}\right) = 1025 \text{ min}$$ (11.47)

In a CSTR the above equation

$$\bar{n} = \frac{[M]_o}{[A]_o} \frac{k\tau[A]_o}{1 + k\tau[A]_o}$$ (11.48)

can be solved to yield

$$\tau = \frac{1}{k[A]_o\left\{\frac{[M]_o}{n[A]_o} - 1\right\}} = \frac{1}{0.05 \times 10^{-3}\left\{s\frac{2}{100 \times 10^{-3}} - 1\right\}} = 1052 \text{ min}$$ (11.49)

11.1.3   Chain reactions

This is the ideal addition reaction because the only reactions involve the sequential addition of single monomer units to the polymer, which grows by one monomer unit each time the reaction occurs. The long-chain polymer molecule will grow indefinitely until all monomer is consumed or the reaction is stopped.

This is a chain reaction with many similarities to the chain reactions that occur in free-radical combustion processes in the previous chapter. For the chain reaction $A \rightarrow B + C$ we represented the process as a kinetic chain involving the chain-propagating intermediate R, which was fed and terminated by initiation and termination reactions (Figure 11–2). In the preceding reaction sequence AM_j is involved in a similar chain, but now each time the chain "goes around" the molecule is increased in size by one monomer unit. We can represent this process in Figure 11–3. This reaction system is a *series reaction* in AM_j,

$$A \rightarrow AM \rightarrow AM_2 \rightarrow AM_3 \rightarrow AM_4 \rightarrow \cdots$$ (11.50)

and a *parallel reaction* in M because M adds to all molecules.

Figure 11–2   Sketch of simple ideal chain reaction not involving polymerization (left) and where a long-chain polymer is formed by adding monomer successively (right).

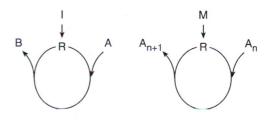

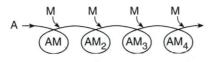

Figure 11–3 Chain reaction for living polymerization reactions. Each time the cycle goes around, a monomer unit is added and the polymer chain length increases.

Thus we can identify several characteristics of polymerization reactions from this simple example of addition polymerization:

1. Polymerization processes involve an infinite set of coupled first-order ordinary equations (PFTR or batch) or coupled algebraic equations (CSTR).

2. While the monomer concentration couples to all equations, the equations for each polymer species are sometimes functions of only species immediately before and after it, so that the equations for each polymer molecule can be solved sequentially.

3. There are frequently relationships among concentrations that simplify the equations.

4. These equations simplify considerably if the rate coefficients can be assumed to be equal and independent of the size of the polymer molecules.

5. On most polymerization processes there are initiation, propagation, and termination rates characterized by k_i, k_p, and k_t.

One can always solve these differential or algebraic equations numerically for any values of the k_js, but analytical solutions assuming all ks equal still provide useful first approximations.

11.1.4 Polymer size distribution

From $[AM_j](\tau)$ one wants to know the amounts of each species in the product for a given τ and for a given reactor flow type. Thus one wants $[AM_j]$ as a function of j (or more important the weight distribution $j[AM_j]$ as a function of j) as shown in Figure 11–4. The weight distribution has an average size and size distribution shown in Figure 11–5. One

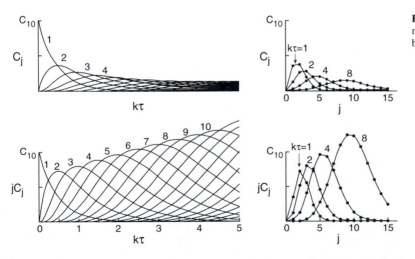

Figure 11–4 Typical distribution of polymer sizes determined on molecule and weight bases.

Figure 11–5 Average polymer size and size distribution.

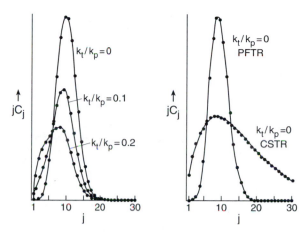

frequently measures these for a given polymerization system and computes the moments of the size distribution as a measure of the properties of a given polymer. It is evident that there are many characteristics of polymers, not simply the average molecular weight but also the distribution of molecular weights that are sensitive functions of the type of reactor used and other process variables.

In any polymerization process one can terminate the reactions by adding a chemical that stops the reaction or by cooling, and this is frequently done in batch processes (the equations and their solutions for the batch reactor are identical to those for the PFTR with τ replaced by t).

In many polymerization processes traces of impurities such a water will stop reaction because just one molecule that attaches to the end of the growing chain will "kill" it. For addition polymerization this will occur whenever a chain-terminating species is added with a concentration higher than the initiator $[A]_o$.

11.2 POLYOLEFINS

The largest-volume polymers are polyolefins, and the kinetics of olefin polymerization are fairly similar to the ideal addition process just considered. All these olefins form condensation products to form a very long-chain alkane such as

$$n\,C_2H_4 \rightarrow (CH_2)_{2n} \qquad (11.51)$$

which is polyethylene. Polyethylene is the highest-volume polymer manufactured, but we can also produce other similar molecules with different properties by simply modifying ethylene by adding a side chain to it. Consider the molecule C=C-X, where $-X$ is a functional group attached to the ethylene. If this side chain does not affect the process, then these reactions may be written as

$$n\,C = C - X \rightarrow \left[\begin{array}{c} C = C \\ | \\ X \end{array} \right]_n \qquad (11.52)$$

Polymerization of these monomer molecules produces monomers and polymers listed in Table 11–1.

TABLE 11–1
Polyolefin Monomers and Polymers

–X	Name of monomer	Polymer
–H	ethylene	polyethylene
–CH_3	propylene	polypropylene
–C_2H_5	butylene	polybutylene
–Cl	vinyl chloride	polyvinyl chloride
–CN	acrylonitrile	polyacrylonitrile
–COOH	acrylic acid	polyacrylic acid
–$COOCH_3$	methyl acrylate	polymethyl acrylate
–C_6H_5	styrene	polystyrene
–$OCOCH_3$	acetate	polyvinyl acetate

TABLE 11–2
Types of Polyethylene

Low density	High density
free radical	catalyst
high pressure	low pressure
solvent	fluidized bed
branched polymers	linear polymers
noncrystalline polymer	crystalline polymer

Almost fitting on this list is methyl methacrylate (MMA), which polymerizes to PMMA with another methyl group attached to methyl acrylate. Another important olefin monomer is perfluoroethylene, which is ethylene, with the four Hs replaced by Fs, and it polymerizes to form Teflon. These monomers and polymers are:

$$H_2C = CH(CH_3)OCOCH_3 \rightarrow PMMA \qquad (11.53)$$

$$F_2C = CF_2 \rightarrow \text{Teflon} \qquad (11.54)$$

There are two types of polyethylene and polypropylene, called low density and high density. High-density polyolefins are made on a catalyst, while low-density polyolefins are made by free-radical polymerization. Characteristics of these polyolefins are summarized in Table 11–2.

11.2.1 Termination

Recall that in ideal addition or "living" polymers, the addition of monomer M to an initiator occurs indefinitely because there are no steps that stop the growth and produce dead polymer. Thus we sketch these as

$$A-/\backslash/\backslash/\backslash/ \cdot + M \rightarrow A-/\backslash/\backslash/\backslash/\backslash/\backslash \cdot$$

$$AM_j \cdot + M \rightarrow AM_{j+1} \cdot$$

and we did not allow any termination steps such as

$$A-/\backslash/\backslash/\backslash/ \cdot + A-/\backslash/\backslash/\backslash/ \cdot \rightarrow A-/\backslash/\backslash/\backslash/\backslash/\backslash/\backslash/\backslash/-A$$

which we can write as

$$AM_j \cdot + AM_{j'} \cdot \rightarrow AM_{j+j'}A \qquad (11.55)$$

by recombination of two free radicals. In ideal addition polymerization the active end is usually an ion rather than a free radical, and ion-ion repulsion retards the recombination of ionic chains so there is no termination of the growing polymer.

In polyolefins, the chain is propagated by an intermediate free-radical species or by an alkyl species adsorbed onto a solid. Both the free radical and the alkyl have the possibility of termination, and this creates the possibility of growth mistakes by chain transfer and chain-termination steps that create dead polymer before all reactants are consumed. The presence of termination steps produces a broader molecular-weight distribution than does ideal addition polymerization.

11.3   FREE-RADICAL POLYMERIZATION

Ethylene is quite stable at very high pressures at low temperatures unless a reaction is initiated by some chemical that acts as a catalyst to initiate the polymerization reaction. This is another example of a chain reaction, but now we mean it in two senses of the word.

1. It involves a chain reaction involving initiation, propagation, and termination steps as in the previous chapter.
2. It involves a set of reactions that generate long-chain polymer molecules.

The formation of polyolefins was first discovered by the occasional and unexpected loss in pressure in high-pressure tanks containing ethylene. The tank did not seem to have a leak; so the scientists cut it open and discovered white solid on the walls. Somehow the tanks occasionally contained certain impurities that initiated free-radical polymerization. This was the first successful commercial polymer, and it is still the largest-volume polymer.

11.3.1   Kinetics

One starts free-radical polymerization by adding an initiator to ethylene or propylene. The initiator is a molecule that readily dissociates into a free radical. A common one is benzoyl peroxide, which dissociates as

$$(C_6H_5COO)_2 \rightarrow 2C_6H_5COO \cdot \qquad (11.56)$$

Let us call this reaction

$$I \rightarrow A \cdot, \qquad r_i \qquad (11.57)$$

where the species $A \cdot$ is a free radical. This molecule is quite reactive and will rapidly react with ethylene (M) to form the new radical $AM \cdot$, which reacts with more ethylene to form $AM_2 \cdot$, etc., in the chain reaction

$$A \cdot + M \rightarrow AM \cdot, \qquad r_{p1} \qquad (11.58)$$

$$AM \cdot + M \rightarrow AM_2 \cdot, \qquad r_{p2} \qquad (11.59)$$

$$AM_2 \cdot + M \to AM_3 \cdot, \qquad r_{p3} \tag{11.60}$$

$$\vdots$$

$$AM_{j-1} \cdot + M \to AM_j \cdot, \qquad r_{pj} \tag{11.61}$$

$$\vdots$$

This process can proceed indefinitely just as in ideal addition polymerization, making an infinite set of free radicals $AM_j \cdot$.

However, there is another possible reaction of free radicals, their reaction with each other in the reaction steps

$$AM_1 \cdot + A \cdot \to P_1, \qquad r_{t10} \tag{11.62}$$

$$AM_1 \cdot + AM_1 \cdot \to P_2, \qquad r_{t11} \tag{11.63}$$

$$AM_1 \cdot + AM_2 \cdot \to P_2, \qquad r_{t12} \tag{11.64}$$

$$\vdots$$

$$AM_1 \cdot + AM_{j-1} \cdot \to P_j, \qquad r_{t1j} \tag{11.65}$$

$$\vdots$$

$$AM_2 \cdot + A \cdot \to P_2, \qquad r_{t21} \tag{11.66}$$

$$\vdots$$

$$AM_3 \cdot + A \cdot \to P_3, \qquad r_{t31} \tag{11.67}$$

$$\vdots$$

$$AM_j \cdot + AM_{j'} \cdot \to P_{j+j'}, \qquad r_{tjj'} \tag{11.68}$$

$$\vdots$$

When two free radicals combine, they form a molecule with all covalent bonds that is a stable and unreactive molecule we call P_n. It consists of n monomer units (M) linked together with an initiator unit (A) at each end, AM_nA. From ethylene, this polymer is the molecule $A(CH_2)_{2n}A$. Note that the product molecule is simply a linear alkane with no double bonds remaining. This molecule is totally unreactive, and it is called the "dead" polymer. If n is large enough, we neglect the end groups and call it simply $(C_2H_4)_n$.

Note that the termination reaction steps forming dead polymer are in fact a double infinite set of steps because j and j' go from 0 to infinity; so we should write all the termination steps simply as

$$AM_j \cdot + AM_{j'} \cdot \to P_{j+j'}, \qquad j = 1, 2, 3, \ldots, \quad j' = 1, 2, 3, \ldots \tag{11.69}$$

We have therefore written this infinite set of reactions as an elementary set of reactions that describe the overall reactions by which monomer M is converted into polymer P_n,

$$nM \to M_n, \quad n = 1, 2, 3, \ldots \tag{11.70}$$

Now we have to try to solve the mass-balance equations for each of these species to obtain $[M](\tau)$, $[AM_j](\tau)$, and $[P](\tau)$, remembering that j and n can go to infinity. We obviously have to solve an infinite number of equations sequentially in the general case. However, we can make the problem quite simple (but obtain a solution that is not entirely accurate) by setting all k_ps and all k_ts equal. Now we can play games with summations and make solution of this reaction set a rather straightforward problem.

11.3.2 Batch reactor or PFTR

We begin by writing the mass balance on all species in batch (t) or PFTR (τ). For monomer $[M]$ we obtain

$$\frac{d[M]}{dt} = -k_p[M] \sum_{j=0}^{\infty} [AM_j \cdot] \tag{11.71}$$

and for the radical intermediates $[AM_j \cdot]$ we obtain

$$\frac{d[AM_j \cdot]}{dt} = k_p[AM_{j-1} \cdot][M] - k_p[AM_j \cdot][M] - \tfrac{1}{2}k_t[AM_j \cdot] \sum_{j'=0}^{\infty} [AM_{j'} \cdot] \tag{11.72}$$

Finally we write the termination reactions as

$$\frac{d[P_{j+j'}]}{dt} = k_t \sum_{k=0}^{j+j'} [AM_k \cdot][AM_{j+j'-k} \cdot] \tag{11.73}$$

These equations can be solved analytically or numerically. In these expressions the rate coefficients were assumed to be k_p and k_t, independent of the chain length. For the situation where k_p and k_t are different for different chain lengths, the k_{pj} and k_{tj} must remain within the summations.

11.3.3 Chain transfer and cyclization

Free-radical polymerization ideally follows the preceding sequence of reaction steps. However, there are also slow but important steps that complicate this simple model. These involve chain transfer steps. We assumed that the only termination involves two radical species reacting with each other to form a stable dead molecule,

$$AM_j \cdot + AM_{j'} \cdot \rightarrow P_{j+j'} \tag{11.74}$$

However, the radical species can also attack a dead polymer to transfer a hydrogen or an alkyl group from one molecule to the other and produce chain transfer. These steps can be written approximately as

$$AM_j \cdot + P_{j'} \rightarrow AM_{j'} \cdot + P_j \tag{11.75}$$

although there are many variants on this sequence and it is best to express them as lines with dots for living chains and lines without dots for dead chains, as in Figure 11–6. Now

Figure 11–6 Sketches of living and dead polymers.

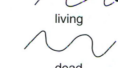

living

dead

we have an almost limitless variety of reactions, and it is difficult to write simple reaction sequences. Further, each class of reaction has different rate coefficients, so there are too many ks that have to be included for simple analytical solutions or even for numerical solutions that are applicable to general situations of olefin reactions. We will leave for a homework problem the corresponding set of mass balances in a CSTR.

In practice, free-radical polymerization of ethylene and other small olefins is usually carried out at very high pressures (because the rate varies approximately as P^2) with a solvent such as isobutane, which forms a supercritical solution at reaction conditions.

11.4 CATALYTIC POLYMERIZATION

Free-radical polyolefin reactions form polymers with many "mistakes" in addition to the ideal long-chain alkanes because of chain-branching and chain-termination steps, as discussed. This produces a fairly heterogeneous set of polymer molecules with a broad molecular-weight distribution, and these molecules do not crystallize when cooled but rather form amorphous polymers, which are called low-density polyethylene.

It was discovered by Ziegler in Germany and Natta in Italy in the 1950s that *metal alkyls* were very efficient catalysts to promote ethylene polymerization at low pressures and low temperatures, where free-radical polymerization is very slow. They further found that the polymer they produced had fewer side chains because there were fewer growth mistakes caused by chain transfer and radical recombination. Therefore, this polymer was more *crystalline* and had a higher density than polymer prepared by free-radical processes. Thus were discovered linear and high-density polymers.

Ziegler and Natta found that Ti alkyls promoted with chloride gave good performance, and scientists from Phillips Petroleum found that Cr alkyls were also effective. Ziegler and Natta received the Nobel Prize for their discovery, but Phillips made more money on the process. The mechanism is thought to involve alkyl ligands R-bonded to the Ti or Cr atom to form a species such as R-Ti and the subsequent bonding and insertion of ethylene to form $R\text{-}C_2H_4\text{-}Ti$ which adds successive ethylenes to create linear polymer of higher molecular weight.

These reactions are typically sketched through an organometallic complex which for Ti is TiX_4 with X a ligand species. We assume that two of the X ligands are tightly bound such as with Cl, and that the alkyl R and the olefin = can be bound on the other two sites, as shown in Figure 11–7.

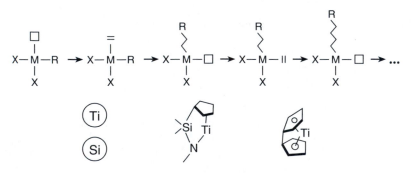

Figure 11–7 Proposed mechanism for catalytic polymerization. An olefin and an alkyl group on an organometallic site react to add the olefin to the growing chain. Also shown are typical polymerization catalysts: Ti/Si solid catalyst and organometallic metallocene catalysts.

This chain will propagate indefinitely until all C_2H_4 is consumed or until the R–Ti bond is hydrogenated to liberate the alkane R–H,

$$R\text{–Ti} + H \rightarrow R\text{–H} + Ti \tag{11.76}$$

which produces "dead" polymer alkane because it has no double bonds for further activity. The average molecular weight and its distribution are controlled by the relative rates of chain propagation and termination, k_p and k_t, and different catalysts, additives, and process conditions give different polymer properties.

11.4.1 Idealized Ziegler–Natta polymerization

We can idealize the preceding reaction steps with the simple sequence

$$A \rightarrow AM \rightarrow AM_2 \rightarrow AM_3 \rightarrow AM_4 \rightarrow \cdots \rightarrow AM_j \rightarrow \cdots$$
$$\downarrow \quad \downarrow \quad \downarrow \quad \downarrow \quad\quad \downarrow$$
$$P_1 \quad P_2 \quad P_3 \quad P_4 \quad\quad Pj$$

which indicates that it is very similar to living polymerization except that there is a termination step that can convert every reactive molecule and thus terminate the chain. We can also sketch this process as

$$
\begin{array}{ccccccccccc}
& M & & M & & M & & M & & M & \cdots & M & \cdots \\
& \downarrow & & \downarrow & & \downarrow & & \downarrow & & \downarrow & & \downarrow \\
A \rightarrow & AM & \rightarrow & AM_2 & \rightarrow & AM_3 & \rightarrow & AM_4 & \rightarrow & AM_5 & \cdots & AM_j & \cdots \\
& \downarrow & & \downarrow & & \downarrow & & \downarrow & & \downarrow & & \downarrow \\
& P_1 & & P_2 & & P_3 & & P_4 & & P_5 & \cdots & P_j & \cdots
\end{array}
$$

which indicates the chain nature of the process.

11.4.2 The Schultz–Flory distribution

If all propagation rate coefficients are equal and all termination rate coefficients are equal, we can obtain a simple expression for the chain length distribution, called the Schultz–Flory distribution.

$$
\begin{array}{ccccccccccc}
A \xrightarrow{k_p} & AM & \xrightarrow{k_p} & AM_2 & \xrightarrow{k_p} & AM_3 & \xrightarrow{k_p} & AM_4 & \rightarrow \cdots & \xrightarrow{k_p} & AM_j \xrightarrow{k_p} \cdots \\
& k_t \downarrow & & k_t \downarrow & & k_t \downarrow & & k_t \downarrow & & & k_t \downarrow \\
& P_1 & & P_2 & & P_3 & & P_4 & & & P_j
\end{array}
$$

The live polymer AM_j can propagate to AM_{j+1} or terminate. We call the fraction that propagates α, which is defined as

$$\alpha = \frac{k_p[M]_o}{k_p[M]_o + k_t[A]_o} \tag{11.77}$$

or if all reactions are unimolecular, then

$$\alpha = \frac{k_p}{k_p + k_t} \tag{11.78}$$

If a process has a high termination rate, then α is small and the average molecular weight will be small. However, if $\alpha \rightarrow 1$, the polymer will grow without much termination (approaching living polymerization), and the molecular weight will be high, It can be shown that the ratio of concentrations of polymer molecules is

$$\frac{C_n}{C_0} = \alpha^{n-1}(1 - \alpha) \qquad (11.79)$$

11.4.3 New Ziegler–Natta catalysts

New and improved polymerization catalysts are continuously being developed to produce polymer with higher rates and with improved properties. Early Ziegler–Natta catalysts were supported on porous alumina or silica catalyst supports. These catalysts had low activity, and large amounts of catalysts had to be used that had to be separated from the molten polymer because otherwise the catalyst would color the polymer. With high-activity catalysts, the catalyst can be left in the polymer, thus saving considerable cost in separation.

One recent development in Ziegler–Natta catalysts was in producing catalyst particles that expanded as the polymerization reaction occurred. In this polymer the catalyst remains dispersed throughout the polymer, retaining its activity. This led to the development of *fluidized bed processes* to make polyethylene and polypropylene in which a sphere of polymer formed around each initial catalyst particle, and the polymer remained solid as the reaction proceeded, rather than requiring a liquid solution. A major class of these catalysts and fluidized bed reactor was developed by Union Carbide and by Shell Oil and called the Unipol process. In this process a very active solid catalyst is introduced into the reactor, and reaction occurs on the catalyst particles, which expand to maintain active sites on the growing polymer sphere.

Another recent change in polymerization has been the development of *homogeneous catalysts* rather than supporting these organometallic catalysts on solid supports, such as the *metallocenes*. These catalysts have exceptionally high activity because there are no mass transfer limitations around the catalyst particle and because the catalyst has a specific formula and thus produces an extremely uniform polymer. These catalysts have ligands consisting of cyclopentadienes, which form specific steric configurations around the metal center to optimize steric and electronic factors in producing the desired polymer.

11.4.4 Elastomers

Butadiene, isoprene, chloroprene, and neoprene are monomers with two double bonds. They form a linear polyolefin polymer that still has double bonds remaining in the chain. These double bonds will crosslink either thermally or by adding a crosslinking agent such as sulfur to form elastomers. Elastomers are made by polymerizing a mixture of monomer with a single bond with a small amount of molecules with two double bonds so that the degree of crosslinking can be controlled.

Automobile tires are the largest-volume rubber material manufactured, and they are formed from natural or synthetic monomers with sulfur compounds added to crosslink the polymer within the tire mold. Tires also contain large amounts (typically 35%) of carbon black particles, which are made by hydrocarbon combustion in excess fuel, as we have considered previously in Chapters 9 and 10.

A new class of elastomers uses *block copolymers*, which contain segments of different monomers. These form crystalline and noncrystalline regions in the polymer, and these polymers have elastomeric properties, as well as high fracture toughness.

11.5 CONDENSATION POLYMERIZATION

In the previous sections polyolefin reactions were shown to occur by addition of a free radical across a double bond or by adding across a double bond in an organometallic complex. This creates an unpaired electron at the end of the chain that reacts with another double bond. The olefin molecules are considered to be inert in reactions with each other, and the process terminates when reactive free radicals are quenched by reaction with each other or by other reactions that produce stable molecules. In these processes the growing polymer can react *only* with the monomer.

In condensation polymerization the monomer molecules can all react with each other. As noted previously, this would ordinarily create "brown gunk" if the process were not controlled to prevent significant crosslinking and form only linear molecules. In condensation polymerization the monomers are bifunctional molecules in which only end groups are reactive, and, as the process proceeds, all molecules continue to still have two reactive end groups.

Organic chemists have long known of bimolecular reactions in which two molecules create a single larger one. The most common of these are formation of esters from acid and alcohol, such as the reaction of ethanol with acetic acid to form ethyl acetate

$$C_2H_5OH + CH_3COOH \rightarrow CH_3COOC_2H_5 + H_2O \qquad (11.80)$$

and the reaction of ethyl amine with acetic acid to form the amide,

$$C_2H_5NH_2 + CH_3COOH \rightarrow C_2H_5NCOCH_3 + H_2O \qquad (11.81)$$

When one places these pairs of reactants in a container and heats, these bimolecular reactions occur and eventually approach thermodynamic equilibrium, which we can write as

$$A + B \rightleftarrows C + H_2O \qquad (11.82)$$

The product ester or amide is unreactive except for hydrolysis back to the reactants.

However, if one starts with molecules that each have two functional groups, then the reaction between them still leaves one functional group from each reactant on the product molecule so that it can continue to react with additional reactants or products. This process can then continue until all molecules with unreacted functional groups have been consumed in forming large polymer molecules.

Nature has long used reactions such as these to produce interesting solids such as cotton (seed pod), hemp (grass), and silk (cocoons for worms while they develop into moths) as fibers that we can strand into rope or weave into cloth. Chemists discovered in the early twentieth century that cellulose could be hydrolyzed with acetic acid to form cellulose acetate and then repolymerized into Rayon, which has properties similar to cotton. They then searched for manmade monomers with which to tailor properties as replacements for rope and silk. In the 1930s chemists at DuPont and at ICI found that polyamides and polyesters had properties that could replace each of these. [Linear polyolefins do not seem to form in nature as do condensation polymers. This is probably because the organometallic

TABLE 11–3
Condensation Polymers

Monomers	Linkage	Common monomers	Common polymers
dialcohol diacid	ester	ethylene glycol terephthalic acid	PET
diamine diacid	amide	hexamethylene diamine adipic acid	Nylon 66
amino acid	amide	caprolactam	Nylon 6
cyclic ether diisocyanate	urethane	tetrahydrofuran toluene diisocyanate	Spandex, Lycra
dialcohol	ether	ethylene glycol	polyethylene glycol
dialcohol $COCl_2$	polycarbonate	bisphenol A phosgene	bisphenol A polycarbonate

catalysts are extremely sensitive to traces of H_2O, CO, and other contaminants. This is an example where we can create materials in the laboratory that are not found in nature.]

In Table 11–3 below are listed the most common of these condensation polymers, the monomers that form them, and some of the common polymers.

When a cyclic monomer such a tetrahydrofuran or caprolactam is used as the monomer, the polymerization can be made to occur primarily by monomer reacting with the polymer rather than all polymers reacting with each other. These kinetics are more like addition polymerization, where only the monomer can react with the polymer. However, we still call this condensation polymerization because it produces this type of polymer.

11.5.1   Nylon

Nylon has a polyamide linkage between an amino group and a carboxylic acid. The amide bond is highly resistant to hydrolysis and is stable up to higher temperatures than other condensation polymers; so Nylon (DuPont's trademark for polyamides) is a strong and high-temperature condensation polymer. Nylon 66 uses two monomers, a bifunctional amine and a bifunctional alcohol, while Nylon 6 uses a single amino acid monomer, but the properties of the polymers are very similar. Kevlar is DuPont's name for an aromatic polyamide that is especially strong. Shown in Figure 11–8 are the steps by which these Nylons are prepared from petroleum feedstocks.

11.5.2   Proteins

Proteins are nature's polyamide condensation polymers. A protein is formed by polymerization of α-amino acids, with the amino group on the carbon atom next to the carboxylic acid. Biologists call the bond formed a peptide rather than an amide. In the food chain these amino acids are continuously hydrolyzed and polymerized back into polymers, which the host can use in its tissues. These polymerization and depolymerization reactions in biological systems are all controlled by enzyme catalysts that produce extreme selectivity to the desired proteins.

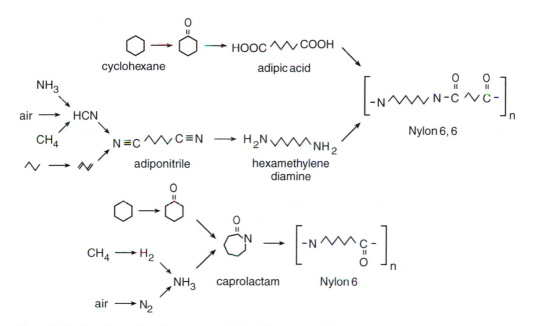

Figure 11–8 Flowsheets of reactions to prepare Nylon 66 and Nylon 6 from petroleum feedstocks.

The most common amino acid industrially is caprolactam, which forms Nylon 6, in which the amino group is six carbons away from the carboxylic acid. Your skin can thus be regarded as Nylon 2. Fortunately your skin is not soluble in water, at least in the absence of strong acids or bases.

11.5.3 Kinetics

Let us consider the general kinetics for each of these reactions. They can all be written as

$$nA + nB \rightarrow (AB)_n \qquad (11.83)$$

if each reactant has the same functional groups or

$$nA \rightarrow A_n \qquad (11.84)$$

if there is only one reactant molecule with different functional groups. Let us consider the latter in detail because it has only one reactant molecule A. The reactions are

$$A_1 + A_1 \rightarrow A_2, \qquad r_{11} = k_{11}[A_1][A_1] \qquad (11.85)$$

$$A_1 + A_2 \rightarrow A_3, \qquad r_{12} \qquad (11.86)$$

$$A_1 + A_3 \rightarrow A_4, \qquad r_{14} \qquad (11.87)$$

$$\vdots$$

$$A_1 + A_j \rightarrow A_{j+1}, \qquad r_{1j} \qquad (11.88)$$

$$\vdots$$

$$A_2 + A_2 \rightarrow A_4, \qquad r_{22} \qquad (11.89)$$

$$A_2 + A_3 \rightarrow A_5, \qquad r_{23} \tag{11.90}$$
$$\vdots$$
$$A_j + A_{j'} \rightarrow A_{j+j'}, \qquad r_{jj'} = k_{jj'}[A_j][A_j] \tag{11.91}$$
$$\vdots$$

Again we have an infinite set of coupled equations. For the PFTR or batch reactor we can write these mass balances for condensation polymerization. The monomer A_1 is lost by reacting with all the polymers,

$$\frac{d[A_1]}{d\tau} = -2k[A_1]\sum_{j=1}^{\infty}[A_j] \tag{11.92}$$

Species A_2 is formed by two A_1 species reacting, and it is lost by reacting with species $[A_2]$ and larger. We have to be careful not to count each reaction twice. We can write these mass balances as

$$\frac{d[A_2]}{d\tau} = +k[A_1]^2 - 2k[A_2]\sum_{j=1}^{\infty}[A_j] \tag{11.93}$$

$$\frac{d[A_3]}{d\tau} = +2k[A_1][A_2] - 2k[A_3]\sum_{j=1}^{\infty}[A_j] \tag{11.94}$$

$$\frac{d[A_4]}{d\tau} = +2k[A_1][A_3] + k[A_2][A_2] - 2k[A_4]\sum_{j=1}^{\infty}[A_j] \tag{11.95}$$

$$\frac{d[A_5]}{d\tau} = +2k[A_1][A_4] + 2k[A_2][A_3] - 2k[A_5]\sum_{j=1}^{\infty}[A_j] \tag{11.96}$$

$$\frac{d[A_6]}{d\tau} = +2k[A_1][A_5] + 2k[A_2][A_4] + 2k[A_3]^2 - 2k[A_6]\sum_{j=1}^{\infty}[A_j] \tag{11.97}$$

$$\vdots$$

$$\frac{d[A_j]}{d\tau} = k\sum_{j'=1}^{j-1}[A_{j'}][A_{j-j'}] - 2k[A_j]\sum_{j=1}^{\infty}[A_j] \tag{11.98}$$

$$\vdots$$

Some of the factors of 2 may appear to contradict our rules of stoichiometric coefficients, but for polymers with two end groups the reaction $A_1 + A_2$ is distinct from the reaction $A_2 + A_1$.

11.5.4 End-group analysis

If we return to the overall mixture, we note that all molecules have a reactive functional group at each end. We denote the total concentration of either reactive functional group as $[C]$. A moment's consideration of the system shows that

$$[C] = \sum_{j=1}^{\infty}[A_j] \qquad (11.99)$$

Let us next try to find the total concentration of these functional groups in the solution in a PFTR or batch reactor. The mass balance gives

$$\frac{d[C]}{d\tau} = -k\sum_{j=1}^{\infty}\left([A_j]\sum_{j'=j}^{\infty}[A_{j'}]\right) = -k[C]^2 \qquad (11.100)$$

Thus the total concentration of functional groups decreases with second-order kinetics. This differential equation can be integrated to give

$$\boxed{[C](\tau) = \frac{[C]_o}{1 + k[C]_o\tau}} \qquad (11.101)$$

where $[C]_o = [A_1]_o = [M]_o$, the initial concentration of monomer.

We can further describe the conversion and average polymer size using end-group analysis. We can write

$$[C] = [M]_o(1 - X) \qquad (11.102)$$

where X is the fractional conversion of the reactive functional groups into polymer. The average polymer size is given by the expression

$$\bar{n} = \frac{[M]_o}{[C]} = \frac{1}{1 - X} \qquad (11.103)$$

For example, if one-half of the end groups have reacted, then the average chain length is 2 (dimers) and $X = \frac{1}{2}$. When $X = 0.9$, $n = 10$, and when $X = 0.99$, $n = 100$.

Example 11–2   Suppose 2 mole % hexylamine is added to caprolactam, which is allowed to polymerize. What is the maximum molecular weight of the polymer if all end groups are equally reactive?

The monofunctional amine has an equal probability of adding to caprolactam as another bifunctional caprolactam molecule, and, once added, that end is dead. Therefore, the reaction stops when all molecules have reacted, and the average chain length is 100 monomer units when the reaction has gone to completion.

We can also use end groups to determine the thermodynamic equilibrium composition. We write the general reaction as

$$-A + -B \rightleftarrows -C- + D \qquad (11.104)$$

where $-C-$ is an unreactive polymer link and D is the other product of polymerization (H_2O for polyester and polyamide). We can write this as a single reaction and write the equilibrium as

$$\frac{[-C-][D]}{[-A][-B]} = K_{eq} \tag{11.105}$$

For a stoichiometric feed ($[-A] = [-B]$), we can further write

$$\overline{n} = \frac{[-A]}{[-C-]}\frac{[-B]}{[D]} = 1 + K_{eq}^{1/2} \tag{11.106}$$

so if the equilibrium constant is 100, then the average polymer size cannot exceed 11.

In any polymerization process one must be concerned with removal of the coproduct (typically H_2O or HCl) so that equilibrium limitations do not limit the polymer size. The removal of the product in condensation polymerization to attain higher polymer lengths is a major consideration in polymerization reactor design. This can be done by withdrawing water vapor or by using two phases so that the water and polymer migrate to different phases.

11.6 FISHER–TROPSCH POLYMERIZATION

We next return to another reaction of a $CO + H_2$ mixture, which we called synthesis gas or syngas. It has this name because it is used to synthesize many chemicals such as methanol. Another synthesis reaction from CO and H_2 is a polymerization process called the *Fisher Tropsch synthesis* of synthetic diesel fuel.

This polymerization process involves the reaction between CO and H_2 to produce hydrocarbon polymers,

$$nCO + 2nH_2 \rightarrow (CH_2)_n + nH_2O \tag{11.107}$$

This process was discovered and developed primarily in Germany during World Wars I and II to provide synthetic liquid fuels to compensate for the Allies blockade of crude oil shipment. [The argument can be made that both world wars were essentially fought over access to petroleum in the Caucusus region of Russia and the Middle East. The U.S. blockade of Japanese access to Far East crude oil was a major factor in Japan declaring war on the United States with the bombing of Pearl Harbor. These arguments are summarized in the book *The Prize* by Daniel Yergin.]

Fisher–Tropsch technology is now practiced on a large scale only in South Africa because of the need for an independent source of liquid fuels because of their political isolation, and they had abundant coal to make syngas. However, these processes are potentially an exceedingly important replacement for liquid fuels to use when crude oil supplies are depleted. Using coal (several hundred years supply in the United States alone) and natural gas (more proven reserves than petroleum), we can be assured of sources of liquid fuels for transportation through this technology long after supplies of crude oil and even natural gas have been used up.

Fisher and Tropsch found that when a mixture of CO and H_2 was heated to $\sim 250°C$ at high pressures over an iron catalyst, polymer would form, and under suitable conditions this had the appropriate molecular weight for gasoline and diesel fuel. Different metal catalysts give different molecular weights (Ni produces CH_4) and different amounts of alkanes (Fe and Re), olefins (Ru), and alcohols (Rh); so catalysts and process conditions can be altered to produce a desired molecular weight and distribution.

The products are essentially all linear molecules, which for olefins and alcohols have the double bond or the OH group on the end carbon (α-olefins and α-alcohols). The

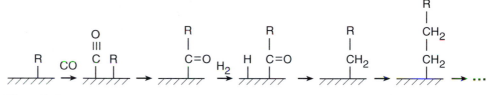

Figure 11–9 Proposed mechanism for Fisher–Tropsch polymerization of CO and H_2 to form alkane polymer.

mechanism of this polymerization process is thought to be similar to (but also very different from) Ziegler Natta polymerization of ethylene and propylene on Ti. It is thought that CO adsorbs and hydrogenates (perhaps to form the CH_2 group) on an adsorption site adjacent to an adsorbed alkyl R, as shown in Figure 11–9. If the CH_2 inserts between the metal and the adsorbed R, one obtains and adsorbed RCH_2-, which can add another CH_2 to form RCH_2CH_2-, and the chain repeats itself indefinitely until the adsorbed alkyl dehydrogenates to olefin, hydrogenates to parrafin, or hydrates to α-alcohol.

While the catalysts and conditions are very different than in ZN (water is a severe poison for ZN but is a product in FT), both produce linear polymers, and the molecular weight distributions are very similar because they are controlled by relative rate coefficients of propagation k_p and termination k_t. If these are nearly independent of molecular weight, one obtains the Schultz Flory size distribution in both processes.

As discussed in earlier chapters, the CO and H_2 (syngas) for FT synthesis were initially made by the gasification of coal,

$$C + H_2O \rightarrow CO + H_2 \tag{11.108}$$

which was plentiful in Germany.

Syngas can also be made using naphtha or other hydrocarbon feedstocks, and now the most promising source of synthesis gas is methane from natural gas,

$$CH_4 + H_2O \rightarrow CO + 3H_2 \tag{11.109}$$

Note that coal produces CO and H_2 in a 1:1 ratio, naphtha in a 1:2 ratio, and methane in a 1:3 ratio. An excess of H_2 is usually desired; so alkanes are the preferred feedstock.

Modern syngas plants operate by direct oxidation of natural gas

$$CH_4 + \tfrac{1}{2}O_2 \rightarrow CO + 2H_2 \tag{11.110}$$

using pure O_2 from a liquid air plant. This process, called autothermal reforming, uses this exothermic reaction in an adiabatic reactor and produces the 1:2 ratio of H_2:CO that is ideal for methanol or FT processes.

This polymerization process may one day be the dominant method by which we will obtain most of our liquid fuels from coal or natural gas. This technology is capable of supplying at least 200 years of liquid hydrocarbons at current consumption rates from known proven reserves of coal.

11.7 POLYMERIZATION REACTORS

Polymerization reactors convert small monomer molecules into very large molecules. Thus we must convert a gas or volatile liquid into a nonvolatile liquid or solid, and do it in large volume with low cost and high reliability. Many of these processes also involve catalysts to promote conversion and selectivity.

Polymerization reactors have several characteristics that are different from most of the reactors we have considered so far.

1. The viscosity of the fluid is a strong function of conversion.
2. Polymerization reactions are strongly exothermic.
3. Most polymerization reactions are strongly influenced by trace impurities, which strongly affect properties.
4. They frequently involve multiple phases.
5. They frequently use catalysts.
6. Condensation polymerization produces a side product such as H_2O, which must be removed from the product and which can cause complications with reversibility.
7. It is frequently desired to produce a blend of several components, and reactors must be designed to accomplish this.
8. Solvents are frequently used to control rates and flow characteristics.

One of the major consideration in polymerization reactors is that the viscosity of the solution changes with degree of polymerization, going from very low viscosity initially to a solid at complete polymerization. This means that tubular reactor must be used very cautiously for polymerization because the viscosity will increase, especially near the wall where the residence time is longest. Many of these fluids also have viscosities that are dependent on the shear rate.

Another characteristic is that most involve multiple phases, the explicit subject of the next chapter. The polymer formed is a liquid or solid, while the monomer is a gas or liquid. In condensation polymerization water is frequently a byproduct; so in many polymerization processes there will be an aqueous phase as well as an organic phase.

Almost all polymerization processes are highly exothermic, and this heat must be removed from the reactor. The temperature must be very tightly controlled in a polymerization process because each step has a highly temperature-dependent rate coefficient; so the properties of the resulting polymer will depend sensitively on temperature and its variations in the reactor.

Solvents can be very important to control the reaction rate and to control the temperature. The reactors listed here are all very important in polymerization processing:

- batch reactors
- reactors in series
- wiped-wall CSTR
- fluidized bed
- screw extruder
- emulsion reactor
- supercritical fluids

11.8 FORMING POLYMERS

Polymerization takes place in reactors such as those described previously. The polymer is then usually cooled to solidify it and then it is formed into pellets or beads of a convenient size for shipping and forming. In this process, the solid is usually melted, and mixed with other chemicals (plasticizers, antioxidants, colorants, solvent) or other polymers to form blends.

The molten polymer is then formed into a desired final shape through one of the following shapes:

- bars
- sheets
- fibers
- composites
- shaped parts

These are accomplished through processes such as the following:

- casting
- reaction injection molding
- blow molding
- rolling and drawing
- tube blowing
- laminating

One can drastically alter the properties of polymers by suitable reactor design during forming of the polymer into its final shape. As examples, crystallinity, anisotropy, and surface characteristics can be tailored by:

- temperature
- cooling rate
- strain rate
- annealing
- solvent removal
- cutting
- surface treatment

11.9 INTEGRATED POLYMER PROCESSING

In this chapter we have examined the reactions and reactors used in polymerizing monomers. In many situations the production of monomers from feedstocks, the polymerization, and the forming of the polymer into products are done in the same chemical plant. In some cases the monomer is too reactive to be shipped to a polymerization plant (isocyanates,

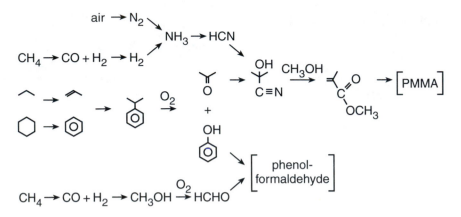

Figure 11–10 Flow sheet of an integrated process to produce PMMA and phenol-formaldehyde polymers simultaneously starting from methane, propane, and cyclohexane through a cumene intermediate.

for example), and in some cases the producer does not want to be dependent on monomer suppliers.

In Figure 11–10 is shown the flow sheet of an integrated process to produce poly-methylmethacrylate (PMMA) and phenol-formaldehyde polymers simultaneously, starting with methane, propane, and cyclohexane.

There is a compelling reason to integrate PMMA and phenol-formaldehyde because the monomers phenol and acetone are both made from cumene oxidation (previous chapter). Therefore, one makes one mole of phenol for every mole of acetone, and a producer would have to sell one of these monomers if he did not have an integrated process to produce both polymers or some other products.

Economics of scale frequently dictate that large and integrated chemical processes are more profitable than small processes where intermediates must be purchased from external suppliers. Supply price fluctuations and availability and variable markets for products frequently give the integrated processor a competitive edge in chemicals and polymers. The negative side of integrated processing is that these plants must be extremely large, and therefore only large companies can afford the investment for an integrated process.

11.10 CRYSTALLIZATION

Crystallization is a polymerization process that has comparable kinetics to the reactions making conventional polymers but whose products are quite different. In crystallization one allows supersaturated monomer to nucleate and grow into crystals. We usually do not regard crystallization as a chemical reaction at all because the process is

$$A_l \rightarrow A_s$$

which is only a phase change.

Crystallization processes are very important in chemical processes whenever there are solid products in a reactor. We saw in Chapter 9 that crystallization and dissolution particle sizes could be handled with the same equations as chemical vapor deposition and reactive etching. We note here that crystallization reactions can be handled with the same equations as polymerization.

Basically, crystallization occurs either by monomer addition to a growing crystal or by coagulation of smaller crystals unto larger crystals. Monomer addition produces more uniform and regular crystals and a narrower crystal size distribution. Coagulation produces irregularly shaped crystals with a wide range in crystal sizes. These processes are obviously addition crystallization and condensation crystallization, respectively. We will not consider these kinetics in any more detail here, but save them for a homework problem.

The types of crystals formed, their perfection and size distribution, and the rates of crystal formation obviously depend on processing conditions, and chemical engineers spend much time and effort in designing crystallization reactors to attain suitable crystal properties. If one wants to form perfect, large crystals, then conditions favoring monomer addition are preferred, while, if one wants polycrystalline aggregates such as in porous catalyst particles, then reactor conditions favoring aggregation should be used. The degree of supersaturation and the stirring and mixing patterns have large influences on crystal properties. The formation of solid polymers and of crystals have many similarities with which chemical engineers must be prepared to deal in making high-quality solid products.

11.11 REFERENCES

Allen, G., and Bevington, J. C., eds., *Comprehensive Polymer Science*, Pergamon, 1989.
Biesenherger, J. A., and Sebastian, D. H., *Principles of Polymerization Engineering*, Wiley, 1983.
Dotson, N. A., Galvan, R., Lawrence, R. L., and Tirrell, M. V., *Polymerization Process Modeling*, VCH, 1996.
Gupta, S. K., and Kumar, A., *Reaction Engineering of Step-Growth Polymerization*, Plenum, 1987.
Odian, G., *Principles of Polymerization*, 3rd ed., Wiley, 1991.
Remmp, P., and Merrill, E. W., *Polymer Synthesis*, Huthig and Wepf, 1991.

PROBLEMS

11.1 (a) Why do the ionic end groups in ideal addition polymerization inhibit termination?

(b) Why does catalytic polymerization have fewer termination processes than free-radical polymerization?

(c) How does one intentionally introduce crosslinking in linear polymers?

11.2 Set up and solve the algebraic equations for the concentrations of all species in ideal addition polymerization in a CSTR.

11.3 Set up the algebraic equations to be solved to find all species concentrations in free-radical polymerization in a CSTR, analogous to those in a PFTR developed in the text.

11.4 Set up and solve the mass-balance equations for condensation polymerization in the CSTR assuming all ks equal.

11.5 Sketch the steps in production of PET from ethane and dimethyl cyclohexanes.

11.6 The polymers and plastics industries are basically concerned with replacing natural flexible and hard one-dimensional and two-dimensional materials.

(a) What were the major soft and rigid containers for liquids before synthetic materials?

(b) What is leather and how is it prepared?

(c) What is glass and how is it prepared?

(d) What is clay and how is it prepared?

(e) What is concrete and how is it prepared?

(f) What are the major natural fibers?

(g) What are the advantages and disadvantages of synthetic polymers as replacements for each of these natural polymers? Note that you probably prefer to wear natural fiber clothing but that cotton tents and silk stockings have totally been replaced by Nylon.

11.7 Suppose we have found a catalyst to polymerize caprolactam, which operates with the monomer in aqueous solution. Starting with 1 molar caprolactam in a batch reactor experiment, we obtain 10% polymerization in 20 min.

(a) What time is required to obtain 90% polymerization in a CSTR starting with 4 molar caprolactam?

(b) How large must the reactor be to obtain 1 kg/h of pure Nylon 6?

(c) The polymerization reaction is exothermic by 25 kcal/mole of monomer reacted. What must be the rate of cooling?

11.8 Polystyrene is made by heating pure liquid styrene with a suitable catalyst.

(a) If 1% of the styrene has polymerized in a 100 liter CSTR with a residence time of 2 min, how much will polymerize in a 100 liter PFTR at this residence time?

(b) The heat of this reaction is −21 kcal/mole. What are the cooling rates of these reactors?

11.9 Write out the complete reactions for polyurethane production starting with toluene, natural gas, air, water, and salt, as sketched in Figure 11-8. Use 1,4-butane diol for the glycol starting from butane. Why is this process preferable for reaction injection molding of automobile parts compared to Nylon or PET?

11.10 Write out the complete reactions for production of PMMA and phenol–formaldehyde, as sketched in Figure 11–11. What are the advantages and disadvantages of an integrated chemical plant compared to separate plants to prepare individual chemicals and polymers?

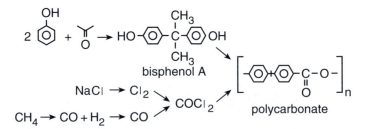

Figure 11–11   Flowsheet of a process to produce polycarbonate polymer.

11.11 [Computer]

(a) For the polymerization reaction $A_1 \rightarrow A_2 \rightarrow A_3 \rightarrow A_4 \rightarrow \ldots$, plot A_j versus $k\tau$ in PFTR for $j < 25$.

(b) Plot (by hand is OK) $C_j(\tau)$ versus $k\tau$ for $k\tau = 2, 5, 10, 15, 20$.

(c) Plot $jC_j(\tau)$ versus $k\tau$. Why is this a more useful description of the polymer distribution?

(d) Repeat these calculation for a CSTR.

(e) Compare the distribution for PFTR and CSTR for given $k\tau$. Why do they have different distributions?

11.12 Polyurethanes are used for most plastic parts in automobiles such as bumpers and nonmetal body panels. They are made from toluene diisocyanate and ethylene glycol. We are designing a chemical plant that makes TDIC and need to evaluate the safety of the units. Phosgene is made by reacting CO and Cl_2 in a 20 atm reactor with a residence time of 10 sec. Toluene is nitrated by reacting with concentrated nitric acid solution at 140°C with a 1 min residence time, and the aqueous and organic phases are separated. The organic phase is distilled to separate dinitrotoluene. The dinitrotoluene liquid is hydrogenated with H_2 at 5 atm at 120°C with a liquid residence time of 20 sec. The product is reacted with phosgene in a vapor-phase reaction at 150°C, and the TDIC is distilled. The TDIC is finally reacted with ethylene glycol in a reaction injection molding process to form the bumper.

(a) This process consists of 5 reactors. Write down the reactions that occur in each.

(b) Sketch the flow sheet of the 5 reactors and 3 separation units with feeds of CO, Cl_2, toluene, H_2, and ethylene glycol.

(c) The process is to produce 10 tons/h. Assume that each reactor produces 100% conversion and that the nitration process produces equal amounts of the three nitrates. What are the sizes of each of these reactors? Approximately how much chemical do they contain?

(d) These 8 units are all connected by valves that will close quickly to isolate all units in the event of an upset of any kind. Describe the potential safety hazards of each unit and the possible things that might go wrong. Include runaway reaction and dangers from venting including chemical toxicity, fire, and explosion. Write ~1/4 page of text on each.

(e) Order the units in terms of decreasing hazard. Which units do you request not to be assigned to?

11.13 In Fisher–Tropsch synthesis of higher hydrocarbons from natural gas, it is desired to produce products in the boiling range for diesel and gasoline rather than lighter alkanes. If the products follow the Schultz–Flory distribution, what weight fractions are in the undesired C_1–C_4 range if $\alpha=0.9$, 0.95, and 0.99?

11.14 In Fisher–Tropsch synthesis from CO and H_2, different catalysts produce different types of product molecules. Referring to the figure of propagation in the FT process, sketch the termination steps that will lead to (a) alkanes, (b) α-olefins, and (c) alcohols.

11.15 In contrast to the Fisher–Tropsch synthesis of higher alkanes from CO and H_2, Ni catalysts produce primarily methane, and Cu/ZnO catalysts produce mainly methanol. Sketch the mechanistic steps that favor these products rather than polymerization?

11.16 (a) "Permanent press" clothing is made from a blend of nylon or polyester with wool or cotton. Explain why blending synthetic and natural polymers can accomplish this property compared to the pure synthetic fabrics.

(b) Wool and cotton shrink when washed. Why do synthetic polymers not shrink?

11.17 Polyethylene terephthalate is unsuitable for carpets because it crystallizes slowly and the fibers therefore break with time and wear. However, if EG is replaced by 1,3-propanediol, the fibers do not crystallize and wear resistance is greatly improved. The 1,2-diol has properties similar to EG. There are several routes to prepare this diol with reasonable selectivity from propane.

(a) Acrolein can be hydrated and reduced, but this produces a mixture of 1,2- and 1,3-diols. Sketch these reactions.

(b) A biological route uses an enzyme to convert glucose to glycerol, and another enzyme "eats" the middle OH to form the nearly pure 1,3-diol. The rate of this process is low. Sketch these reactions.

(c) In a new commercial process 1,3-propanediol is made by reacting ethylene oxide with CO, followed by hydration and reduction. Sketch the flowsheet of the reactions to prepare 1,3-propanediol starting from propane, air, water, and methane.

11.18 1,4-butanediol is simple to prepare, but its melting point is too high to form polyester fibers. However, this diol is widely used in fibers formed from the polyurethane for clothing fibers such as Lycra and Spandex.

(a) Sketch the polymerization reaction.

(b) Sketch the reaction steps to prepare this diol from butane.

(c) Sketch the flowsheet of the reaction steps that could be used to prepare this polymer starting from butane, toluene, nitric acid, air, and water.

11.19 1,4-butanediol can also be prepared by reacting acetylene with formaldehyde, and acetylene can be prepared by partial oxidation of methane. Sketch the reactions by which 1,4-butanediol can be prepared using only methane, air, water, and hydrogen.

11.20 Polycarbonate is a very hard and light plastic used for protection devices such as football helmets. It is made by reacting bisphenol A with phosgene in the reactions shown in Figure 11–11.

(a) Sketch the flowsheet of a polycarbonate process starting with propane, cyclohexane and methane.

(b) Why might it be economical to produce PMMA along with polycarbonate? Sketch this integrated plant.

11.21 [Computer] Consider addition polymerization in a PFTR,

$$A + M \rightarrow AM$$

$$AM + M \rightarrow AM_2$$

$$\vdots$$

$$AM_j + M \rightarrow AM_{j+1}, \qquad r_p = k_p[M][AM_j]$$

$$\vdots$$

with $[M]_o$ constant at 1.0 moles/liter, $[A]_o = 0.01$ moles/liter, and k_p identical for all j.

(a) Plot $[AM_j]$ and $j[AM_j]$ versus $k_p\tau$ for $j < 30$.

(b) From these results plot the molar and weight distributions versus j at $k_p\tau = 1, 2, 5, 10,$ and 20.

11.22 [Computer] Consider addition polymerization in a CSTR,

$$A + M \rightarrow AM$$
$$AM + M \rightarrow AM_2$$
$$\vdots$$
$$AM_j + M \rightarrow AM_{j+1}, \qquad r_p = k_p[M][AM_j]$$
$$\vdots$$

with $[M]_o$ constant at 1.0 moles/liter, $[A]_o = 0.01$ moles/liter, and k_p identical for all j.

(a) Plot $[AM_j]$ and $j[AM_j]$ versus $k_p\tau$ for $j < 30$.

(b) From these results plot the molar and weight distributions versus j at $k_p\tau = 1, 2, 5, 10,$ and 20.

11.23 [Computer] Consider addition polymerization,

$$A + M \rightarrow AM$$
$$AM + M \rightarrow AM_2$$
$$\vdots$$
$$AM_j + M \rightarrow AM_{j+1}, \qquad r_p = k_p[M][AM_j]$$
$$\vdots$$

with $[M]_o$ constant at 1.0 mole/liter, $[A]_o=0.01$ moles/liter, and k_p identical for all j. Add a termination reaction

$$AM_j \rightarrow P_j, \qquad r_t = k_t[AM_j]$$

where P_j is a dead or unreactive polymer. Plot $([AM_j] + [P_j])$ and $j([AM_j] + [P_j])$ versus k_τ for $j < 25$ for $k_t/k_p=0, 0.1, 0.2, 0.5,$ and 1.

From these results plot the molar and weight distributions of $[AM_j] + [P_j]$ for $k_p\tau = 2, 5, 10,$ and 20.

11.24 Consider the reversible condensation polymerization reaction of a hydroxyacid A to form a polyester polymer,

$$A_j + A_{j'} \rightleftarrows A_{j+j'} + H_2O$$

(a) If the equilibrium constant for this reaction is 100, what is the average polymer size at equilibrium if all species remain in a single-phase solution?

(b) In these processes water is usually removed from the solution to obtain a higher degree of polymerization. If the initial monomer concentration is 5 moles/liter, what H_2O concentrations will give average polymer lengths that are 2, 10, and 100 times that calculated in part (a)?

11.25 Consider the condensation polymerization of a bifunctional monomer A in a CSTR with all rate coefficients k equal.

(a) Write out the mass-balance equations for $[A_j]$.

(b) Find an expression for $[A_j](\tau)$ for the jth species in terms of $[A_1]_o$.

(c) Plot $[A_j]$ and $j[A_j]$ versus $k\tau$ for $j < 25$.

(d) In a PFTR $[A_j]$ is given by the expression

$$[A_j](\tau) = [A]_o \left(\frac{1}{1 + k\tau[A_1]_o} \right)^2 \left(\frac{k\tau[A_1]_o}{1 + k\tau[A_1])} \right)^{j-1}$$

Plot this $[A_j](\tau)$ and $j[A_j](\tau)$ versus $k\tau$ for $j < 25$.

(e) Plot molar and weight distributions for CSTR and PFTR for $k\tau[A_1]_o = 0.5$, 1, 2, and 4.

11.26 Consider the reversible condensation polymerization reaction of an amino acid A to form a polyamide polymer A_j and water,

$$A_j + A_{j'} \rightleftarrows A_{j+j'} + H_2O$$

(a) If the equilibrium constant for this reaction is 100, what is the average polymer size at equilibrium if all species remain in a single-phase solution?

(b) To obtain a larger molecular weight of the polymer, the polymerization is sometimes carried out in a two-stage reactor with H_2O from the first stage separated before the second stage. If the first reactor forms only the dimer and the dimer without water is fed to the second reactor, what is the equilibrium polymer chain length obtainable? What chain length can one obtain if the first stage forms the trimer? Why is separation of water simpler with a dimer or trimer than if polymerization is continued to higher chain lengths?

11.27 Suppose we want to produce a polymer with a very narrow molecular weight distribution. Which choices are preferred?

(a) CSTR or PFTR?

(b) Ideal addition or condensation?

(c) Polyester from tetrahydrofuran or 1,4-butanediol?

11.28 Engineers have learned how to prepare genetically engineered DNA molecules (a long-chain polymer) with exactly specified nucleic acid sequences by adding adenine, thymine, cytosine, or guanine (A, T, C, or G) in a DNA synthesizer. These operate by reacting one of these monomer molecules at a time in a sequence of highly automated batch reactors. Sketch the mechanism by which only a single monomer molecule is added in one reactor without mistakes and the "reactor" that accomplishes this process.

11.29 Your body makes specific proteins by reacting single amino acids within cells. What is the process that makes a single protein molecule (a long-chain polymer) without mistakes?

11.30 In the Unipol process of olefin polymerization 0.1 μm particles of Ziegler–Natta catalyst are introduced into a fluidized bed containing ethylene at 250°C. Polymer grows on the surface of this catalyst, and the catalyst expands to remain on the surface of the growing solid particle. When the particle diameter reaches 20 μm, the polymer particles are removed from the reactor by gravity. Calculate the time to form a 20 μm polymer particle assuming that the reaction is mass transfer-limited with a Sherwood number of 2.0. Assume that the pressure of C_2H_4 is 2 atm, $D_{C_2H_4} = 0.1$ cm^2/sec, and the density of polyethylene is 0.99/cm^3.

Chapter 12

BIOLOGICAL REACTION ENGINEERING

12.1 INTRODUCTION

Biotechnology is a rapidly expanding part of chemical engineering, and, by many estimates, biotechnology may be a major part of what chemical engineers do within a few years. Therefore, it is important that all chemical engineering students appreciate some of the principles of biological chemical reaction engineering and how they may differ from more traditional chemical synthesis processes.

An important distinction between "normal" chemical reactions that we have been considering previously and biological systems is that biological systems emphatically do not "want" to make products for humans. They want to survive and replicate, and they will strongly resist any attempts that might lessen their chances of survival and reproduction. Stated differently, biological systems "have a mind of their own" (stored in their DNA) in that even simple viruses and bacteria can only be persuaded to operate under conditions different from those of their ancestors if they have been genetically modified to "like" their new environment. On the other hand, biological systems desperately "want" to survive, and they will therefore try to withstand being placed in a hostile chemical environment that we may choose to force the system to produce a high yield of a desired product.

Living systems consist of plants and animals (and some simpler systems such as bacteria). These form a cycle as sketched in Figure 12–1. Sunlight is used by plants to convert CO_2 and H_2O into organic matter such as sugars. Animals then consume plants and use their energy by oxidation with O_2 back into CO_2 and H_2O, and the cycle is repeated. Energy from the sun is used to form molecules and organisms that we call life. Chemical engineers use this cycle to make chemical products, and one goal in accomplishing this is to formulate quantitative descriptions of this process so that we can manipulate it in desired ways.

Another distinctive feature of bioprocesses is that they are autocatalytic with rates of the form

$$A + S \rightarrow 2A + P, \qquad r = kC_A C_S \qquad (12.1)$$

465

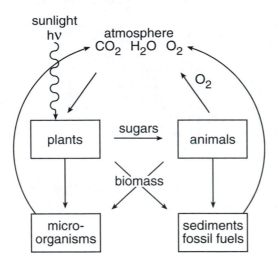

Figure 12–1 "The Big Picture" of biological processes. Plants, animals, and microorganisms interact with molecules on the earth and with sunlight to reproduce themselves, create food for other organisms, and create chemical energy sources.

Here A is the organism, S is the food supply (given the symbol S for substrate by biologists), and P is products. We write this as a "bimolecular chemical reaction" with second-order kinetics, but the process is of course much more complicated than with simple molecules. A sketch of the process described by the above reaction is shown in Figure 12–2. The cell A consumes nutrients S (many compounds) to reproduce itself by cell division and also creates many metabolic products P. We of course do not mean to imply that the stoichiometric coefficients in this reaction are all 1.

We will examine solutions to these equations later. In a simple example, if the food supply S is constant at C_{So}, a species A in a batch environment will grow as

$$C_A = C_{Ao}e^{+kC_{So}t} \tag{12.2}$$

for exponential growth as long as there is no process by which the organism dies. Stated differently, one can say that the yield Y_A of this "reaction," rather than being less than unity as for all nonbiological processes, can in fact approach infinity.

These kinetics have very important and unique consequences for biotechnology. In principle, one needs to prepare just a *single individual* to create a process that can generate an infinite number of organisms. In other words, biological organisms have the ability to replicate, an event that essentially never occurs in nonbiological systems. In fact, a

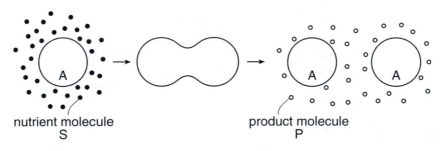

Figure 12–2 Sketch of processes involved in the biological "reaction" $A + S \rightarrow 2A + P$. Species A is a cell (from microbe to human) that divides using nutrient S and produces product P.

reasonable definition is that a biological organism is anything that can replicate itself. The closest nonbiological example is chemical explosions that occur by chain branching processes to increase the concentration of radical species that accelerate the reaction. Chemical explosions are also analogous to biological reactions in being difficult to control.

One consequence of this is that it is possible to select one cell with desired properties out of millions of other similar cells if one can find a method to make that cell reproduce itself while its neighbors do not. One simple method to accomplish this with genetically modified *E. coli* is to add a gene to make a particular antibiotic in that cell. Then one adds a lethal toxin to the system, and only those cells that have the antibiotic survive. A very clever technique and very unique to living systems!

While the biological "soup" of any living organism is an incredibly complex mixture of millions of chemical species at very low concentrations, replication allows the scientist or engineer to find and analyze a particular species by growing it up to a measurable concentration.

The major hazards in conventional chemical production involve escape of a chemical to locations where its toxicity or fire hazard may cause problems. In biotechnology the hazards are much more severe. Escaped chemicals become more dilute, while escaped biological organisms can reproduce and become *more concentrated*. A single escaped toxic bacterium carried on the clothes of an employee could in principle reproduce and poison large populations.

Genetically modified organisms (GMOs) could pose a special threat because they may have been modified explicitly to be resistant to common degradation routes. A cross between a cold virus and an AIDS virus is a disaster too serious to contemplate.

While there is much discussion in the popular press on GMOs, chemical engineers (you) are the people responsible for making sure that problems do not arise or that they can be managed.

Even if you are not actively involved in these subjects, as a chemical engineer you should know enough to assist in decisions regarding the safe use of biotechnology.

12.1.1   Biotechnology products

In this chapter we will focus on some of the major current and perhaps future products made by biotechnology. The major categories are pharmaceuticals (drugs and food supplements) and agricultural chemicals (insecticides and herbicides), but they also include energy sources (biomass for renewable energy, ethanol and hydrogen for fuels) and commodity and specialty chemicals (biodegradable polymers).

Example 12–1   Yeast cells convert sugar into ethanol with ~50% efficiency, the rest forming CO_2 and cell mass. How much sugar is required to form a can of beer assuming 90% sugar conversion?

Assuming 7% alcohol in a 0.3-liter can, this requires $2 \times 300 \times 0.07/0.9 = 47$ grams $= 47/180 = 0.26$ moles of glucose. If the volume of water remains unchanged in the fermentation process, then $C_{glucose,0} = 0.86$ moles/liter. Who says chemical engineering is not practical?

TABLE 12–1
Major Biological Products

Physically processed	Biologically processed	Chemically synthesized
starch*	ethanol*	aspirin
sugar*	soap	Ibuprofin*
protein*	fructose*	Viagra*
rubber*	polyesters*	many drugs
coffee, tea, drugs	citric acid*	most agricultural chemicals
wax	antibiotics	Prozac*
silk	insulin	insulin
wool	human growth factor	
leather	clotting factors	
	lactic acid*	
	acetic acid*	
	lysine	

*Indicates biological chemicals discussed in this book.

Table 12–1 lists some major biological products. These are classified as physically processed (few chemical reactions involved from raw material to product, mostly steps like drying or roasting), biologically processed, where raw materials (usually starch or sugar) are biologically converted into the desired product, and finally chemically processed molecules that are made by nonbiological methods of organic synthesis.

12.1.2   Goals

Biology and biotechnology are of course enormously complicated subjects. We obviously cannot handle them in this chapter in any substantial way. Rather, our intent is much simpler. We want to show how the principles of chemical reaction engineering can be applied to biological processes. Thus we describe bioprocesses as chemical reactions that occur in chemical reactors. This is a much different goal than those of biology texts that describe the mechanisms of these processes in a qualitative and pictorial way. We are interested in how we might manage bioprocesses in a particular situation that we assemble in a configuration to make a desired product. As in the other chapters, we want to examine the features of this type of chemical reaction that are different from the conventional chemical reactor used to produce chemicals.

Another goal of this chapter is to describe biotechnology specifically for the student who has *little knowledge or interest in biotechnology*. Those who plan a career in the field have probably taken courses in these subjects already, and they will be familiar with much of the material presented here. Whether one plans a career in biotechnology or not, the subject will inevitably have an impact on your career and life. First, the biotechnology job market is almost guaranteed to increase at a rate greater than the GNP and be fairly recession-proof. Second, many possible employers in companies with no current biotechnology businesses will consider entering this market because of its growth and profit opportunities, and its employees should be prepared to take advantage of these opportunities. Third, biotechnology will impact all of our lives as consumers and citizens, and literacy in this subject will enable us to make intelligent choices.

Another obstacle in comprehending biotechnology is the maze of language that permeates the subject. As examples, "DNA melting" means that the double-stranded chain opens up rather than that anything actually melts, and all enzymes are named by the molecules they attack followed by "-ase" (proteases destroy proteins and synthases catalyze synthesis reactions). We will try to use very little jargon, attempting to describe bioprocesses in the engineer's language.

Basically, we find that most of biotechnology can be explained by considering reactions to be describable as a sequence of chemical reactions of the form

$$
\begin{array}{ccc}
A & \to B & \to C \\
\downarrow & \downarrow & \\
D & E &
\end{array}
\tag{12.3}
$$

That is, in biotechnology one tries to manipulate a set of series and parallel reaction steps to form a maximum amount of a specific desired product. A major advantage of biological processes is that the k's that determine the selectivities in each of these reaction steps can be controlled over a wide range by changing nutrients or by controlling catalyst (enzyme) concentrations.

More specifically, metabolism to create some familiar products can be written

$$
\begin{array}{ccc}
\text{food} \to \text{glucose} & \to \text{pyruvate} & \to CO_2 \\
\downarrow & \downarrow & \downarrow \\
\text{fructose} & \text{ethanol} & \text{lactate}
\end{array}
\tag{12.4}
$$

We write this reaction in Figure 12–3, which is clearly a network of series and parallel

Figure 12–3   The chain of reactions in glycolysis in all living systems. In biotechnology these rates can be manipulated to produce mainly one product such as those indicated. All are series–parallel reaction networks, and biotechnology involves choosing organisms, conditions, and reactor type to increase or decrease individual k's to optimize production of a desired product.

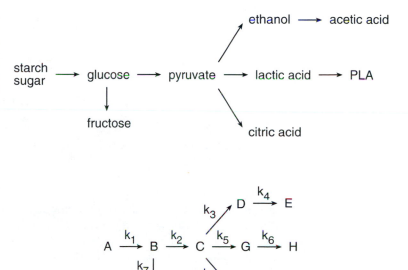

reactions, just as we considered in Chapter 4. We will discuss this reaction chain and products from it in a later section.

The task in biotechnology is to manipulate the k's in these steps to maximize a particular product, such as high fructose corn syrup, alcoholic beverages, or polylactic acid polymer. Manipulation of microbes and their enzymes can make processes that create any of these products.

The chemical reactors used to make these products can also be manipulated to optimize a particular product. The enzymes are all produced within cells of bacteria, yeast, plants, or animals, and each of these organisms has specific preferences in adjusting the k's.

On the other end of reactor complexity, products such as vaccines and blood clotting factors are produced in *live animals* such as mice, cows, and pigs. In a sense the these animal reactors can be regarded by chemical engineers as "simple" because the "chemical reactor" is not difficult to fabricate or expensive to operate.

The methods to accomplish these simple goals are the not-so-simple subject of biotechnology.

12.2 BIOLOGICAL MOLECULES

We first examine the types of molecules that are special to biological systems. While there are many of them, they are of only a few kinds. These molecules are summarized in Figures 12–4 and 12–5.

12.2.1 Proteins and amino acids

Proteins are the basic structures of all living organisms. Proteins are linear polymers made of amino acids that form amide bonds (called peptide bonds by biologists),

$$n\text{H}_2\text{N-}\overset{\text{H}}{\underset{\text{R}}{\text{C}}}\text{-COOH} \rightarrow (\text{-N-}\overset{\text{H H}}{\underset{\text{R}}{\text{C}}}\text{-CO-})_n + n\text{H}_2\text{O}$$

Nylon 6 is made from an amino acid containing six carbon atoms in the reaction

$$\text{H}_2\text{N(CH}_2)_5\text{COOH} \rightarrow (\text{-NH(CH}_2)_5\text{CO-})_n + n\text{H}_2\text{O}$$

All amino acids in proteins have the amino group on the second carbon atom from the carboxylic acid (alpha amino acids), so protein could be called Nylon 2. Your skin (and most of the rest of you) is made of Nylon!

There are only 20 amino acids used in all organisms from bacteria to humans. The formulas of these amino acids are shown in Figure 12–4. These are all α-amino acids, and the only differences are in the R-group attached to the α-carbon. As examples of three amino acids, R=H, CH_3, and isopropyl are called glycine, alanine, and valine, respectively. The α-carbon atom is chiral in all but one of the 20 amino acids, and the only natural isomers are the right-handed or D isomers. Some of these amino acids have only alkyl groups and are hydrophobic, while others have polar or charged side chains and are hydrophilic. Some have acidic or basic side chains that can be charged or covalent, depending on pH.

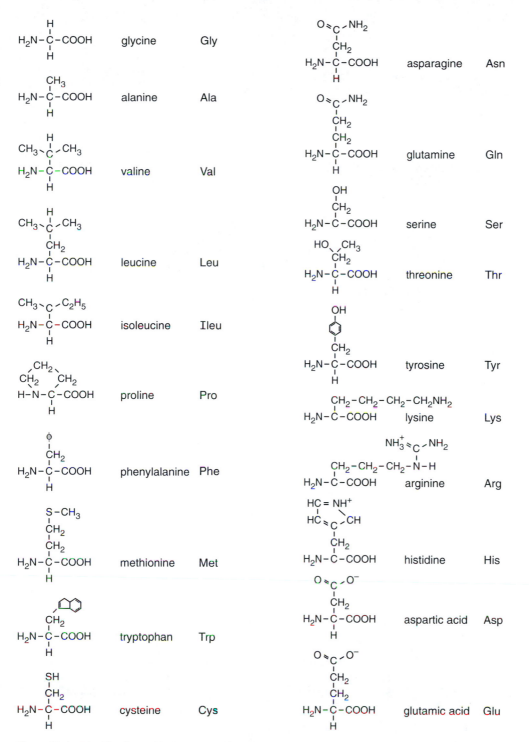

Figure 12–4 The 20 amino acids that polymerize to form proteins that are the major structural and processing components in all living systems.

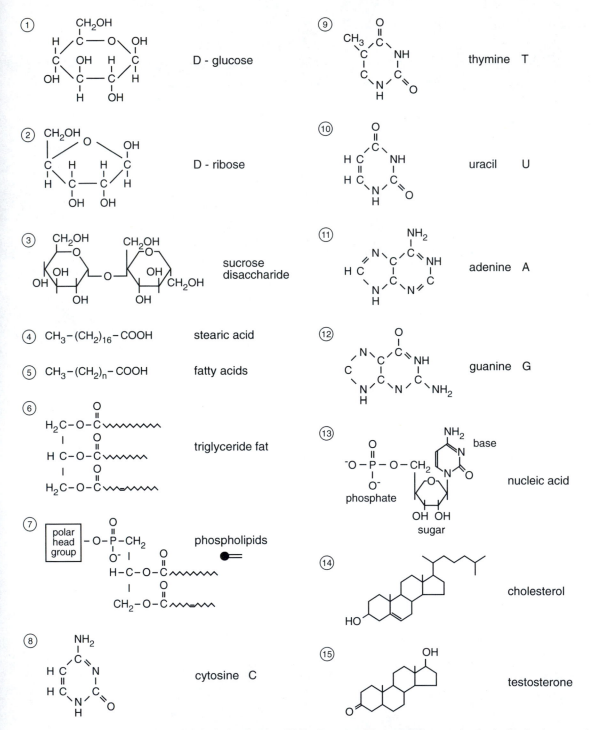

Figure 12–5 The major types of biological molecules. While there are millions of different molecules in the simplest organism, most belong to ~5 classes: proteins, carbohydrates, fatty acids, organic phosphates, and organic bases.

Protein structures are characterized as having hydrophobic and hydrophilic regions, and specific segments of proteins cause formation of α-helices, β-sheets, and folds. These result in very specific conformations that are crucial to their properties and activities.

One important protein is myoglobin, the O_2 carrier in cells of mammals. This protein consists of 78 amino acids in a linear polymer with -NH_2 and -COOH groups on each end (and will be discussed later with Figure 12–10). It consists of many α-helix rod-like regions linked by flexible strands and is folded into a roughly spherical shape. Only a small number of amino acids are involved in binding the Fe-containing heme molecule within the protein. Hemoglobin is a very similar molecule that is in red blood cells (called erythrocytes by biologists). In its active form, it consists of four of these protein molecules in a square configuration. We will discuss myoglobin and hemoglobin later in connection with O_2 transport from lungs to cells in blood.

12.2.2   Enzymes

Of the approximately 120,000 proteins in humans, a majority of these are enzymes, transporters, and signaling proteins whose function is to catalyze and control chemical reactions. These proteins are all fairly large molecules that have small regions where the sites that bind reactants for chemical reactions are located.

One example of an enzyme is glucose isomerase, which is used to convert glucose to fructose

$$\text{glucose} \rightarrow \text{fructose}$$

in production of high fructose corn syrup, as discussed in Chapter 2. (Biologists add the ending -ase to describe the enzyme that runs a reaction of a molecule whose name ends with -ose.)

12.2.3   Carbohydrates

Carbohydrates (formula $(CH_2O)_n$, carbon + water) are the major food source of all organisms (sugar and starch) and are the basic structural building blocks of all plants (cellulose) and insect shells (chitin). The monomers are called sugars, and their polymers are called polysaccharides. Some typical sugars are shown in Figure 12–5. The most common one you know about is sucrose, a disaccharide (dimer) made up of glucose and fructose. There are also sugars containing five carbon atoms, called riboses. Some riboses attach to other functional groups such as in DNA and RNA.

12.2.4   Fats

Fats are basically esters linking polyfunctional alcohols such as glycerol with long chain alkyl carboxylic acids, although most have one or more double bonds (the unsaturated fatty acids). The most common fats are triglycerides. The hydrocarbon tails in these molecules are highly insoluble in water and tend to form globules in cells. Fats are the major chemical storage strategy in animals. Some examples are shown in Figure 12–5.

12.2.5 Phospholipids

Closely related to fats are phospholipids. These molecules have long-chain hydrocarbon carboxylic acid tails and head groups with ester linkages. However, the head groups have phosphate species bonded to other polar groups. A typical phospholipid molecule is shown in Figure 12–5, consisting of two carboxylic acid tails bonded to a glycol, which is in turn bonded to a phosphate with a polar head group. Phospholipids have one major function in biological systems. They form cell walls and membranes connecting compartments in cells through bilayer structures. We will discuss these structures later in descriptions of cells.

12.2.6 Nucleic acids

Another important class of molecules is nucleic acids. There are five of them, as shown in Figure 12–5. These are the building blocks of DNA and RNA, which process genetic information. They consist of purine and pyrimidine bases (cytosine, uracil, thymine, adenine, and guanine), which are cyclic five- and six-member rings containing N atoms that act as weak bases. In DNA and RNA these bases are connected to ribose sugars and phosphates to form the nucleic acids. DNA and RNA differ only by one OH group on the ribose group, but that single OH group gives them very different functions.

12.2.7 Organic phosphates

The early earth environment must have contained significant concentrations of phosphate ion, PO_4^{3-}, and all organisms use phosphates as an essential part of their reaction pathways. Phosphate easily forms compounds with organic molecules to form organic phosphates. In fact, chemical reactions of organic molecules with phosphates are the major pathways of all biological reactions.

Phosphate is also capable of polymerizing with itself to form polyphosphates

$$
\underset{\underset{O}{|}}{\overset{\overset{O}{\|}}{H\text{-}O\text{-}P\text{-}O\text{-}H}} \rightarrow
\underset{\underset{O}{|}\,\underset{O}{|}}{\overset{\overset{O}{\|}\,\overset{O}{\|}}{H\text{-}O\text{-}P\text{-}O\text{-}P\text{-}O\text{-}H}} \rightarrow
\underset{\underset{O}{|}\,\underset{O}{|}\,\underset{O}{|}}{\overset{\overset{O}{\|}\,\overset{O}{\|}\,\overset{O}{\|}}{H\text{-}O\text{-}P\text{-}O\text{-}P\text{-}O\text{-}P\text{-}O\text{-}H}}
$$

which are called monophosphate, diphosphate, and triphosphate. (We omit some charges, although many of these compounds are anions or cations with counterions like Na^+ and Cl^- nearby.)

Phosphates and polyphosphates are also capable of forming bonds with organic molecules, and these intermediates are crucial to many biological reactions. ATP is adenosine triphosphate, which has the structure

$$
\underset{\underset{O}{|}\,\underset{O}{|}\,\underset{O}{|}}{\overset{\overset{O}{\|}\,\overset{O}{\|}\,\overset{O}{\|}}{R\text{-}O\text{-}P\text{-}O\text{-}P\text{-}O\text{-}P\text{-}O\text{-}H}}
$$

where R is adenosine, a molecule consisting of adenine (one of our five bases) and ribose (a sugar containing five carbon atoms). This molecule is one of the most important in many biological chemical reactions, especially those involved in energy transfer.

TABLE 12–2
Biological Polymers

Polymer	Names	Monomer	Linkage
proteins	peptide enzyme	amino acids	amide
carbohydrates	sugar cellulose	monosaccharides	ether
phosphates	organic phosphates	PO_4^{3-}	polyphosphate
cell membranes		phospholipids	hydrophilic and hydrophobic interactions
DNA and RNA	polylactic acid	nucleic acids	polyphosphate
polyesters		hydroxy acids	ester

Phosphate is also capable of forming bonds with more than one organic molecule. This has enormous significance in biological reactions in that phosphate-based polymers form the backbone structures of genetic information in DNA and RNA. These polymers can be written

$$R\text{-}O\text{-}\overset{\displaystyle O}{\underset{\displaystyle O}{P}}\text{-}O\text{-}R\text{-}O\text{-}\overset{\displaystyle O}{\underset{\displaystyle O}{P}}\text{-}O\text{-}R\text{-}O\text{-}\overset{\displaystyle O}{\underset{\displaystyle O}{P}}\text{-}O\text{-}$$

where now R is one of the sugars bound to a nucleic acid. These polymers in DNA and in RNA must be strong enough to make a very long polymer that is stable through many coiling and uncoiling operations, because they carry all genetic information that must not be lost or mistranslated.

Table 12–2 summarizes these types of biological polymers. All are complicated, but their classification is straightforward.

Example 12–2   Several classes of molecules are usually extreme toxins for biological systems: (1) organic phosphates (Malathion, Dieldrin), (2) polychlorinated hydrocarbons (DDT, PCB, Dioxin), and (3) heavy metals (Cr, As, Sb, Se, Pb, Hg). How does each of these classes of molecules interact with biological systems?

Manmade organic phosphates have structures and reactions similar to the organic phosphates that are involved in almost all biological reactions. Therefore they can easily disrupt normal biological functions with deadly consequences. Organic phosphates should never be produced or used without considering their toxicity.

Polychlorinated hydrocarbons are very insoluble in water, but are highly soluble in oil. Therefore they accumulate in fatty tissues and cell membranes. This kills cells by rupturing membranes. Accumulation in fatty tissues is particularly harmful because the chemicals are concentrated as they move up the food chain, from small fish to predator fish to birds to humans.

Heavy metals such as those listed above are not normally found in biological systems, but other metals such as Fe, Mg, Ca, and Zn are essential in many metalloenzymes and other reactions systems. These heavy metals can replace the naturally occurring metals and inactivate their enzymes or change their reactions.

12.3 CELLS

Cells are the "microreactors" in which all biological chemical reactions take place. We sketch a simplified cell in Figure 12–6, showing only the cell wall, nucleus, and several mitochondria. Our major interest as chemical engineers is the metabolism that occurs in a cell, and this requires transport and can generate concentration gradients as indicated in the figure.

These cells can be single-cell organisms such as bacteria like *E. coli* (which do not contain a nucleus or mitochondria) or collections of cells that make a complex organism such as humans. Therefore, before considering how chemical engineers can use cells to create processes and products, we first need to consider the nature of cells and how they function.

12.3.1 Chemical composition

The approximate weight compositions of bacterial and mammalian cells are listed in Table 12–3.

Note that we are not much different in composition from a bacterium, and both of us are 70% water. We will see that the chemical reactions in a bacterium and in ourselves also have many similarities. Note that the dry weight of all cells is about 60% protein. Phospholipids are mainly used in membranes that form the cell wall and internal membranes. DNA, the molecule that carries genetic information, is a very small fraction of the total weight of the cell. Lipids are fats and polysaccharides are sugars that are used primarily for chemical storage.

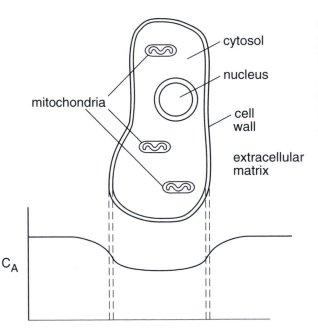

Figure 12–6 A highly simplified sketch of a eukaryotic cell, consisting of a nucleus and mitochondria. Each of these components are divided by bilayer phospholipids membranes. Not shown are other organelles in most cells such as the endoplasmic reticulum, golgi apparatus, cytoskelaton, and (in plants) chloroplasts. Chemical engineers are interested in the transport of reactants and products (called A, B, S, P) to and from cells. This produces concentration gradients because of transport resistances within and without the cell such as that shown for a reactant A migrating into a cell.

TABLE 12–3
Approximate Weight Compositions of Bacterial and Mammalian Cells

Component	E. coli	Mammalian cell
H_2O	70%	70%
inorganic ions (Na^+, K^+, Ca^{2+}, Mg^{2+}, Cl^{2-}, etc.)	1	1
miscellaneous small metabolites	3	3
proteins	15	18
RNA	6	1.1
DNA	1	0.25
phospholipids	2	3
other lipids	—	2
polysaccharides	2	2
total volume	2×10^{-12} cm^3	4×10^{-9} cm^3
diameter	1 μm	20 μm
relative cell volume	1	1000

From Alberts et al. (1994).

TABLE 12–4
Elements in Biological Systems

Atoms in organic molecules	Ions	
	Common	Seldom or never
C	PO_4^{3-}, Cl^-	SO_4^-, NO_3^-
O	Na^+, K^+	Sr^{2+}, Al^{3+}
H	Mg^{2+}, Ca^{2+}	F^-, Br^-
N	Zn^{2+}	B, Be
P	Fe	F^-
S	Mn, Mo, I^-	all $Z > 30$

Only a small fraction of the elements in the Periodic Table are found in significant amounts in living systems. Table 12–4 lists some of the common and less common elements in living systems.

12.3.2 Components of cells

Regions of cells are divided by membranes. These are bilayers of phospholipids as shown in Figure 12–6. These are formed because phospholipids have polar heads and nonpolar tails, just as in soaps and detergents. The latter form a monolayer film around oil droplets in water, because this allows the hydrocarbon end to associate with the oil phase and the polar end to associate with the water phase, thus lowering the surface tension and stabilizing small oil droplets. This is the principle of using soap and detergents in the cleaning of clothing.

Another stable configuration of phospholipid is to form a bilayer with water inside and outside and with the hydrocarbon tails associating in the center of the layer. This structure is called a vesicle. As sketched in Figure 12–6, this bilayer structure can form a roughly spherical compartment dividing regions, and this is the basis of separation of all regions of cells. The material inside of the cell is called the cytosol, and the material outside is called the extracellular matrix.

Prokaryotic organisms such as bacteria have a single compartment within the cell, and they store their DNA in single ring structures called plasmids. However, more complex cells (called eukaryotic organisms) have internal structures. One of these is the cell nucleus which contains the genetic instructions for the cell in DNA molecules. Each nucleus contains one copy of the entire DNA sequence of the organism. For mammals this is packed in the form of chromosomes, 23 for humans.

Most eukaryotic cells have many mitochondria within their cells. These are small structures with two sets of walls, the outer roughly spherical and the inner being convoluted. A major function of mitochondria is to complete the oxidation of sugars into CO_2 and H_2O, an essential step in metabolism in higher organisms. There are thousands of mitochondria in mammalian muscle and liver cells.

Cells need structural components to give them shape and to assist in cell division. Muscle cells obviously need a mechanism to deform as required to motion. This is accomplished in structures called the cytoskeleton. These are rod-like structures called actin and microtubules. Processes that force them to slide past each other make cells expand and contract. This powers the muscles around and in your eyes as you read this page.

Endoplasmic reticulum, Golgi apparatus, and isosomes are some other structures within cells of eukaryotic organisms. They perform complex and essential functions in cells, and some of these functions appear to be poorly understood.

Plant cells also contain chloroplasts. These are similar to mitochondria in terms of overall structure. However, instead of oxidizing sugars to CO_2 and H_2O as occurs in mitochondria, chloroplasts contain chlorophyll and the capability to reduce CO_2 and H_2O to synthesize sugars using energy from sunlight.

12.3.3 Organisms

We will not discuss different types of organisms in detail because they are considered in basic biology courses. They also have very different cellular structures, and each of the components listed above are different for each type of organism. Single-celled organisms are classified as bacteria, fungi, yeast, protozoa, and algae. Multicellular organisms are divided between plants and animals. It is useful to consider their components by function because in biotechnology we frequently want to use these functions or alter them to produce particular products.

The first structure that any organism needs is walls to separate itself from the outside world. This allows the organism to collect and store nutrients and other chemicals that it needs and to expel chemicals that it does not need. Bilayer cell membranes of phospholipids are the walls that separate a cell from its surroundings. Sea animals construct calcium carbonate shells, insects construct chitin polysaccharide shells, and humans construct a skin layer to protect ourselves from the outside.

An organism next needs to secure nutrients and expel waste products. For single-celled organisms this is through diffusion and migration through cell walls, and more sophisticated organisms have developed specialized digestive organs to provide this function, for example by separating solid and liquid waste products.

Many organisms need to move to secure food and escape predators. Even plants need to orient leaves to maximize sunlight input. Some bacteria and protozoa have flagella protruding from their cells that move in more or less random fashion. Even the simplest organisms have chemical sensors to detect whether motion should be toward a food source or away from a predator. These motion capabilities become more complex in more complex organisms. Mammals of course have muscles, and these are extremely specialized into skeletal, heart, digestive, and other functions.

Defense capability is frequently through motion, escaping a predator if necessary. The simplest defense against predators is chemical. By secreting a chemical that is toxic or repulsive to predators, an organism can protect itself. Plants, bacteria, and yeast all produce chemicals that have protective properties, and mammals produce antibodies that protect us from bacterial invasions.

When diffusion is inadequate to provide nutrients and expel wastes at rates necessary to function, an organism develops circulation functions. In mammals, the lungs, blood, lymph system, intestines, and bladder are all designed to move fluids appropriately for needs of the organism.

Finally, no living organism can exist without propagating itself. DNA and RNA are chemicals designed for this purpose. Testosterone, estrogen, and other chemicals provide signals for reproduction. Further sex education is probably not needed for college-level chemical engineers.

12.3.4 Functions of cells

In higher organisms such as humans, cells have specific functions that are also discussed in basic biology courses. Structural cells are skin, collagen, and bones in animals, cellulose and lignin in plants, and chitin for the exoskeletons in insects. Motion is through muscle cells which are voluntary, heart, and peristaltic. Signaling cells are vital to all higher organisms, and these consist of nerve and brain cells.

Three classes of cell functions deserve special consideration in engineering because we can use them for biotechnology: chemical storage, antibodies, and reproduction.

Living organisms need to store material for future use. Plants make primarily starch and sugar for this purpose, and we harvest plants to obtain these chemicals for food and structural materials. Animals cannot store carbohydrates for long times, and we resort to converting carbohydrates into fats for chemical energy storage. The energy density in fats is much greater than in carbohydrates, so this is an efficient method to build reserves within the organism. In higher animals, fat is stored in special cells that contain large globules of fat, a phenomenon you will become increasingly aware of as you grow older. Another chemical storage vehicle in plants is polyester, such as polylactic acid and polyhydroxybutyrate. Plants can utilize these compounds as needed, and we have learned to genetically modify some plants to overproduce these chemicals for biodegradable polymers.

Humans consider many plants and animals as prey. However, bacteria, viruses, and parasites rely on us for food and as ideal locations in which to reproduce themselves. We have developed immune systems to try to counteract each of these. They involve a very complex immune system that produces antibodies that fight off each type of invasion. These are involved in a continual war between predator and prey in which every organism can be on either side of the fence.

You learned all about reproduction in sex education courses, and bacteria and other lower organisms have much simpler and less interesting methods.

12.4 ORIGINS AND CHANGES IN LIVING SYSTEMS

These organisms and processes are amazingly complex, and yet the types of molecules and processes that occur are remarkably similar between the simplest organisms and ourselves. How did these processes begin? Why did these molecules and sequences of reactions become the ones chosen for all life? As engineers we search for explanations for these events with the hope that they may assist us in using and modifying them.

12.4.1 The early earth and its atmosphere

Let us begin by considering (guessing) the origin of the earth and its conditions. The universe is mostly H atoms, formed at the original Big Bang. These H atoms occasionally condensed by gravity into stars, their radiation coming from the nuclear reaction $4H \rightarrow He$ that occurs spontaneously at the high densities of stars including our sun. Heavier elements are formed by successions of reactions of these species such as $3He \rightarrow C$ and $4He \rightarrow O$. (Being chemical reaction engineers, we cannot resist writing every process as a chemical reaction.)

When the earth condensed, it was a collection of elements heavy enough not to escape the earth's gravity. (All H_2 and He that may have been here in the early earth escaped long ago, and that is why the rare gases got their name.) The earth was very hot and still consists basically of a core of molten Fe covered by solid crust now less than 100 miles thick that forms because of radiation cooling of the surface. The early earth crust consisted mostly of heavy cations such as Si, Al, alkalis, alkaline earths, and halogens, with the denser elements such as Fe sinking to the core. These light elements form the minerals that compose the earth's crust.

Oxygen is highly reactive to form rocks, but N and C have few chemical reactions available to them. Nitrogen does not form stable solids, and it formed almost exclusively the dimer N_2 and some "fixed" nitrogen as NH_3 or nitrates. Carbon could form carbides, graphite, or diamond, but its most favorable chemical reaction involves the formation of CO_2, which in turn forms carbonate rocks, such as limestone (which has a biological origin).

The atmosphere of the early earth contained almost no free O_2 because all would quickly react away to form solids and CO_2. The atmosphere of the early earth was mainly CO_2, N_2, NH_3, and H_2O with considerable H_2S (not enough O_2 to oxidize it to SO_2). As described in Genesis 1:2, "And the earth was waste and void," at least for human habitation.

Water is abundant on the earth, and one of the prevalent theories is that it arose primarily from impact of comets that are basically frozen ice balls. Comets do not strike the earth very often, but there were 5 billion years to wait.

12.4.2 Environmental toxins

In these nuclear reactions, the mechanisms lead to very small amounts of some light elements such as Li, Be, and F. While there is considerable Cl in the world, almost all is in the form of chloride ions (the salt in the oceans). As a consequence, living systems contain essentially none of these elements, and, if they are introduced into organism, they are frequently very toxic. In fact, the basis of most herbicides and pesticides is that they contain compounds with these elements. As examples, NaF is rat poison, DDT kills insects and softens the shells of bird eggs, and PCPs, dioxin, and freons are major manmade environmental pollutants. Another class of insecticides involves organic phosphates that mimic the organic phosphates involved in biological reactions.

Nuclear fusion reaction rates generally slow down in the middle of the Periodic Table near Fe, and heavier elements are also rare in the earth. (That is how the rare earth elements got their name.) Biological systems also were not exposed to these elements in their development, and heavy metals are usually very toxic (Hg, Pb, and Pu are notable examples).

12.4.3   Elements in biological systems

We have now formed H, C, and O, the three basic atoms of living systems. As you learned in quantum chemistry, the rich variety of organic molecules is possible because of the close spacing of the $2s$ and $2p$ orbitals of C which lead to tetrahedral bonding and single, double, and triple bonding with only slightly different energies. Tetrahedral carbon is also necessary for chirality, an essential feature of biological molecules.

The earth on which life arose, then, was one of liquid and vapor water covering most rocks and an atmosphere consisting mainly of CO_2 and N_2. The pressure may have been 10 atmospheres or higher. Hydrocarbons and organic compounds (molecules containing C, H, O, S, and N) were abundant in the water. Amino acids are fairly stable, as are alcohols, aldehydes, ketones, and carboxylic acids, as well as olefins. Inorganic ions such as Na^+, K^+, Mg^{2+}, Ca^{2+}, Zn^{2+}, Mg^{2+}, Fe^{2+}, Cl^-, and PO_4^{3-} were also present.

Next we come to the step that no one can explain or duplicate, the transformation of organic soup to living organisms. While tetrahedral carbon leads to chiral molecules, all of these processes lead to racemic mixtures, while in earth biology only the D isomer forms. Once the right-handed isomers started, they could easily reproduce, but their origin is unclear.

12.4.4   Colloids, vesicles, and membranes

Organisms need a "container" or reactor to confine desired molecules and exclude undesired ones. The formation of this type of structure is simple in an aqueous environment, because some of these molecules are surfactants that have a polar head group on one end and a nonpolar tail on the other. These will segregate a hydrophobic liquid from water spontaneously to form colloids (that is how soap and mayonnaise work). Cholesterol is a hydrophobic molecule with a single polar OH group that is very soluble in membranes and has many biological functions.

Structures can also form with H_2O in both sides by forming vesicles, a double layer surfactant with polar groups on the inside and outside as sketched in Figure 12–7. We thus have a plausible mechanism for a cell, an aqueous region (the cytosol) enclosed by the membrane wall consisting of a hydrocarbon tail and a phosphate head on either edge of the membrane. Vesicles can also exist within vesicles, the structures that now occur for the nucleus and other structures within a cell.

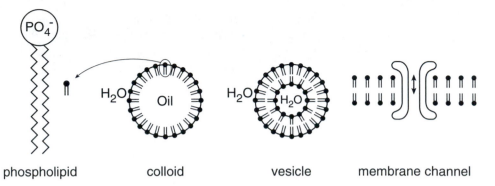

phospholipid   colloid   vesicle   membrane channel

Figure 12–7   Cell membranes are bilayers of phospholipids. These molecules and fats consist of polar or ionic head groups and hydrophobic hydrocarbon tails. These molecules form monolayer (colloids) or bilayer (vesicles) films in oil–water mixtures. Transport of molecules through these membranes is controlled by molecules such as proteins that form in these layers and facilitate migration through them.

12.4.5   Digestion

The organism needs a method to access nutrients and eliminate wastes. This can be accomplished by simple diffusion through the cell wall. The solubility of various compounds in the cell membrane becomes a critical parameter in deciding which molecules can enter and exit the cell. Different chemicals in the cell membrane will obviously alter these reactions and molecules; they should not be too hydrophobic (alkanes), but some hydrophobic character will increase solubility in the interior of the membrane.

The entrance and exit of reactants and products can be made more efficient by the use of specialized channels in cell membranes that pass certain species and exclude others. All organisms have these protein structures in their cells. Thus the cell, in addition to being a chemical reactor, has specialized separation equipment in its piping to improve its efficiency.

One method of incorporating other cells is by surrounding a small cell by the cell wall of a larger cell, a process called endocytosis. Once ingested, the larger cell can use nutrients from the smaller cell as needed.

12.4.6   Defenses and biological toxins

A primary concern of any living organism is to protect itself against its environment, be it from harmful chemicals or predators. This need comes even before its need for food, and certainly before its desire to reproduce. Therefore, biological systems need to create defenses against these threats.

To protect against predators, the organism (if it cannot run or hide) creates toxins that kill or repel. The objective, of course, is to create chemicals that are toxic to predators while not harming the individual. This is probably why yeast produces ethanol as a byproduct of its metabolism of sugar (and as most undergraduates know, ethanol can be quite toxic). Ethanol in high concentrations denatures proteins (causes them to unfold), and yeast cells have higher concentrations of ethanol outside their cells to repel predators while keeping the inside concentration smaller. Yeast cells die at ~16% ethanol, so higher concentrations of ethanol in beverages require distillation.

Plants produce toxins especially in their seeds, which otherwise are very nutritious and tasty to predators because they are high in protein. In some cases these seeds are the spices whose strong tastes we enjoy. All hot peppers produce capsaicin. Viruses such as the cold virus produce toxins that give us runny noses to help them spread to our colleagues.

Biotechnology frequently uses an organism's natural defenses by using these compounds for some desired purpose. The alcohol industry and vaccine production are important examples.

12.4.7   Bacteria, fungi, algae, and yeast

The first and simplest living organisms formed in an aqueous, anaerobic environment. Cyanobacteria appear to be the simplest organisms surviving today. These organisms learned to use sunlight to assist them in metabolism in photosynthesis, another process that is absolutely essential to life. The atmosphere was nearly pure CO_2, which also has a high solubility in water. At present, the CO_2 partial pressure on the earth is very small, 0.03% (but rising again!), but photosynthesis was "invented" in an atmosphere with plenty of CO_2 and little of its product O_2.

The simplest organisms also needed an oxidation–reduction cycle. We now use primarily O_2–H_2O, but this is not possible without atmospheric O_2, and the earliest organisms used the S–H_2S oxidation–reduction cycle instead. Among primitive life forms, sulfur bacteria live in anaerobic environments, such as near deep sea vents that produce H_2S that they use as an energy source.

12.4.8   The atmosphere becomes oxidative

About 2.7 billion years ago a catastrophe occurred for all living organisms. Photosynthesis by algae and primitive plants was converting CO_2 and H_2O into carbohydrates (the food source for themselves, bacteria, and simple animals) and also forming O_2. The oceans were filled with algae, and their photosynthesis was producing enough O_2 to leave significant O_2 remaining in the atmosphere. This caused many organisms to die because they could not deal with the oxidation that occurred. Some bacteria found O_2-free regions where they continue to survive and reproduce in an anaerobic environment, but the rest had to adapt to an O_2-containing environment. However, those that survived could now take advantage of the O_2–H_2O oxidation–reduction cycle that is now the basis of most living systems.

Cells also developed a nucleus, a region surrounded by another membrane in which most DNA resides in the form of chromosomes. Cells without a nucleus such as bacterial are called prokaryotes, and their DNA is in a circular form called plasmids, while those with a nucleus (all plants and animals up to humans) are called eukaryotes (biologists love Greek and Latin words).

12.4.9   Mitochondria and chloroplasts arise

Eukaryotic cells grew in size, and they began to digest bacteria by incorporating them into their cells through their cell walls (endocytosis). These bacteria had the ability to run oxidation processes very efficiently, and it is now accepted that some of them became permanently incorporated as mitochondria, the cell components that now accomplish most oxidation in both plants and animals.

Cyanobacteria can accomplish photosynthesis, and they also became incorporated into eukaryotic cells to form chloroplasts, the location where modern plants accomplish photosynthesis.

The rest is history.

12.5   BIOENERGY AND METABOLIC PATHWAYS

12.5.1   Biological combustion

The major purposes of biological combustion are to generate chemicals, store energy, and create work. In fact, there is one major fuel in all systems, glucose. We repeat this metabolic cycle in Figure 12–8. The reactions can be written as the chemical reaction

$$C_6H_{12}O_6 + 6O_2 \rightarrow 6CO_2 + 6H_2O, \qquad \Delta H = -673 \text{ kcal/mole} \qquad (12.5)$$

Biological combustion does not generate much heat directly. Rather, the primary purpose of these reactions is the generation of chemicals for reproduction and to enable the cell to do work. The major intermediate in chemical synthesis and work is a single molecule, ATP (adenosine triphosphate), that can undergo a reaction to form adenosine diphosphate (ADP) and phosphate ion,

$$ATP + H_2O \rightarrow ADP + PO_4^{3-}, \qquad \Delta G^\circ = -12 \text{ kcal/mole} \qquad (12.6)$$

These molecules are shown in Figure 12–9. Oxidation of one mole of glucose to CO_2 generates about 30 moles of ATP from ADP. Most of this ATP in mammals comes from the citric acid cycle within mitochondria.

The other significant energy transferring molecule is NADH which is transformed to NAD^+. These molecules are also shown in Figure 12–9. This reaction stores electrons for electron transfer reactions.

There are two major stages in glucose oxidation, the first being oxidation to pyruvate:

$$\text{glucose} \rightarrow 2 \text{ pyruvate}$$

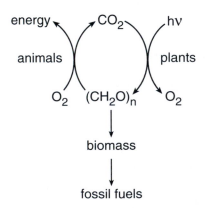

Figure 12–8   The reaction cycle of life. Similar to the reaction cycles of many chemical reaction processes, biological processes can be sketched as a cycle. Carbon cycles between CO_2 and carbohydrate $(CH_2O)_n$ with sunlight input and with energy and biomass as outputs. All chemical energy comes from this cycle.

Figure 12–9 The major energy storing and converting molecules in all living systems. Adenosine triphosphate (ATP) is converted reversibly to adenosine diphosphate (ADP) with release of energy. Another important energy molecule is NADH which is converted to NAD$^+$ with release of energy. This is called an electron storing reaction.

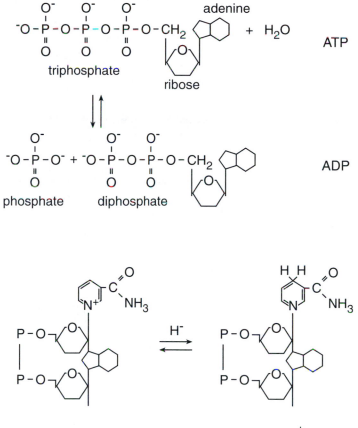

This involves many steps, the first and slowest of which is phosphorylation of glucose to form glucose-6-phosphate. These reactions occur within the cytosol of the cell, and they require the diffusion of glucose through the cell membrane where it is attacked by a series of enzymes with pyruvate the final product.

Pyruvate has at least three major reaction channels: (1) total combustion, (2) conversion to lactate, and (3) conversion to ethanol,

$$\text{pyruvate} \rightarrow CO_2 + H_2O$$
$$\rightarrow \text{lactate}$$
$$\rightarrow \text{ethanol}$$

These processes are important in nature, and they can be used to make chemicals as we will consider later.

Pyruvate cannot be oxidized on to CO_2 in the main region of the cell (the cytosol). Rather, it must diffuse into the interior of mitochondria located within the cell to find the enzymes necessary to finally oxidize it to CO_2. The mitochondrion is actually a double membrane structure in which the intermediate region has a very low pH.

12.5.2   Oxygen transport

The biological combustion reactions described above occur by oxidizing sugars with O_2. The oxygen must be carried from outside our bodies into muscle cells where the mitochondria react in the citric acid cycle. In this section we consider the path by which O_2 goes from air to the muscle cells.

When we breathe, we inhale O_2 into our lungs where it is absorbed into capillaries. The solubility of O_2 is approximately that in water: $\sim$10 ppm from a Henry's Law calculation. This would be far too little to permit us to metabolize at our normal rate. Rather, the O_2 is absorbed onto hemoglobin molecules that are in red blood cells. The solubility of O_2 in hemoglobin creates concentrations many orders of magnitude higher than by solubility in an aqueous solution. The O_2 is then carried in red blood cells as oxyhemoglobin to capillaries near muscle cells where it is released and diffuses through the cell wall to again adsorb on myoglobin molecules within muscle cells. It is stored there until needed for combustion within mitochondria.

The adsorption of O_2 occurs by binding on Fe^{++} ions in the reversible adsorption process

$$Fe + O_2 \rightarrow FeO_2 \tag{12.7}$$

In the free state, Fe^{2+} would be quickly oxidized by O_2 to form Fe^{3+} (basically Fe_2O_3 or rust, which is highly insoluble in water). To keep Fe in the reduced state, it is bound in the center of the heme molecule, a porphyrin whose structure is indicated in Figure 12–10.

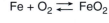

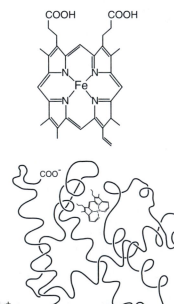

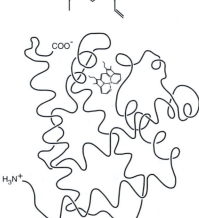

Figure 12–10   The upper panel shows the reversible reaction of O_2 with Fe^{2+} that carries O_2 in the body of mammals by complexing with the prophyrin heme that is bound in the protein hemoglobin (center). The lower panel shows myoglobin, the O_2 carrying molecule in muscle tissue. There are 78 amino acids in this molecule whose conformation is also shown. The heme molecule is bound in a pocket in the protein, and much of the molecule consists of α-helix strands.

This also would not be stable or soluble in blood unless bound to a protein called globin to make hemoglobin and myoglobin.

The myoglobin protein is a 78-amino-acid sequence sketched in Figure 12–10. Only the backbone of the chain is indicated, starting with a free NH_2^+ group on one end and ending with a COO^- group on the other. Much of the molecule is in the form of α-helices, which are sketched as coils. Several regions of the globin protein have functional groups that bind to the Fe in the heme molecule to keep it in place as shown, both with and without O_2 bound to the Fe atom.

The adsorption isotherms of O_2 on hemoglobin and myoglobin are shown in Figure 12–11. These are very cleverly arranged (by nature) to maximize uptake and release as needed. Myoglobin binds a single O_2 in a reaction that can be represented as

$$Mb + O_2 \rightleftarrows MbO_2 \tag{12.8}$$

with an equilibrium

$$K = \frac{[Mb][P_{O_2}]}{[MbO_2]} \tag{12.9}$$

This uptake curve has the same shape as a Langmuir isotherm when plotted as P_{O_2} versus bound O_2 whose concentration is $[MbO_2]$.

$$Y_{O_2} = \frac{[MbO_2]}{[MbO_2]_{max}} = \frac{P_{O_2}}{K + P_{O_2}} \tag{12.10}$$

This uptake equilibrium curve is sketched in Figure 12–11.

Hemoglobin consists of four individual protein molecules bound together so that four O_2 molecules can be bound. Furthermore, the structure of the four-molecule complex changes such that O_2 binding increases as the number of O_2 molecules increases. This is called an alosteric interaction by biologists. The molecule can bind four O_2 molecules in the reactions

Figure 12–11   The oxygen uptake curves of myoglobin and hemoglobin. Hemoglobin is contained in red blood cells, and its function is to pick up O_2 from the lungs and transport it to capillaries where it releases O_2 so that it can be transported into cells for oxidation reactions. In muscle cells, O_2 is stored in myoglobin until needed. The myoglobin uptake has an S-shape because four protein molecules form a cluster in which a single O_2 adsorption facilitates further O_2 uptake. In the lungs the O_2 pressure saturates the hemoglobin, but at the lower capillary pressure O_2 is released.

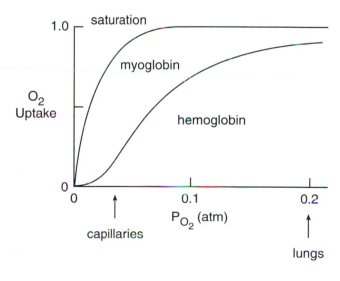

$$Hb + O_2 \rightleftarrows HbO_2, \qquad\qquad K_1 \qquad\qquad (12.11)$$

$$HbO_2 \rightleftarrows Hb(O_2)_2, \qquad\qquad K_2 \qquad\qquad (12.12)$$

$$Mb(O_2)_2 + O_2 \rightleftarrows Hb(O_2)_3, \qquad K_3 \qquad\qquad (12.13)$$

$$Hb(O_2)_3 + O_2 \rightleftarrows Hb(O_2)_4, \qquad K_4 \qquad\qquad (12.14)$$

This gives an uptake curve (adsorption isotherm)

$$Y_{O_2} = \frac{\dfrac{C_{O_2}}{K_1} + \dfrac{2C_{O_2}^2}{K_1 K_2} + \dfrac{3C_{O_2}^3}{K_1 K_2 K_3} + \dfrac{4C_{O_2}^4}{K_1 K_2 K_3 K_4}}{1 + \dfrac{C_{O_2}}{K_1} + \dfrac{C_{O_2}^2}{K_1 K_2} + \dfrac{C_{O_2}^3}{K_1 K_2 K_3} + \dfrac{C_{O_2}^4}{K_1 K_2 K_3 K_4}} \qquad (12.15)$$

This is an equilibrium adsorption isotherm, similar to those for adsorption on surfaces in Chapter 7 and on enzymes as described later in this chapter. We will leave its derivation for a homework problem. This equation predicts an S-shaped O_2 uptake curve for hemoglobin similar to that shown in Figure 12–11.

Oxygen is carried from the lung to the capillaries bound to hemoglobin in red blood cells. Here the O_2 partial pressure is lower because it is being used in metabolism, so O_2 begins to desorb from hemoglobin back into the blood stream. It then diffuses into cells where it is needed.

Muscle cells have a need for large amounts of O_2 on demand, and O_2 is therefore stored in muscles in myoglobin. As noted previously, this protein has a similar structure to hemoglobin, but it is a single molecule and therefore has a different O_2 uptake curve as shown in Figure 12–11. The curve for myoglobin has the normal shape and, when combined with the S-shaped curve for hemoglobin, is ideally designed for O_2 transport into the muscle cell.

12.5.3 Photosynthesis

This is "just" the reverse reaction of glucose combustion

$$6CO_2 + 6H_2O \rightarrow C_6H_{12}O_6 + 6O_2, \qquad \Delta H = +673 \text{ kcal/mole} \qquad (12.16)$$

and this reaction closes the cycle of Figures 12–1 and 12–8. At first thought it might seem simple to run combustion backward. However, the reaction has a very large positive $\Delta G°$ and would be very difficult to accomplish in the laboratory. It is absolutely impossible when we realize that the glucose formed is only the D isomer. The reaction has never been carried out and probably never will be by a nonbiological process. Yet this is the major reaction by which all biological energy is obtained (including coal, petroleum, and natural gas, our only sources of chemical energy). It occurs by a chain of reactions in which sunlight is first used to produce an electron transfer by the molecule chlorophyll. This is a conjugated system that absorbs green light with the help of a Mg porphyrin. Then a sequence of reactions occurs by which CO_2 is bound and reduced to react eventually with H_2O to form carbohydrate. The detailed mechanism of this process has only been recently understood, and some of the multiple pathways are still active subjects of research.

However, modern plants take 0.03% CO_2 (300 parts per million) from the atmosphere and convert it to "fixed" carbon as carbohydrate, which is all of the plant life on the earth.

TABLE 12–5
Language of the Genetic Code

Components	Number	Chemical	Humans
alphabet	4 letters: A, T, G, C	organic bases	
words	all 3 letters, $4^3 = 64$ possible	α-amino acid	20 used plus punctuation
sentences	up to hundreds of words	proteins	~30,000-word vocabulary
chapters	chromosomes		23
book	organism		us

As we noted previously, they probably had an easier task in the primitive earth where the atmosphere consisted primarily of CO_2 with negligible O_2. (As suggested previously, the O_2 in the present atmosphere probably was formed by photosynthesis of CO_2, simultaneously leaving some of the carbon behind as coal and petroleum in the ground.)

It is humbling that we can just begin to understand this process, and we have little hope of ever duplicating it in the laboratory at any scale.

12.5.4   Genetic code

The genetic code was broken in 1953 by Francis Crick and James Watson, who used x-ray diffraction to determine the crystal structure of a biomolecule that had been called DNA (deoxyribonucleic acid). The molecule got its name because it contains a ribose sugar or a deoxyribose, phosphate groups, and organic bases, and it is located primarily in the nucleus of eukaryotic cells. This discovery is frequently regarded as the most important discovery of the twentieth century (along with some discoveries of Albert Einstein) because it opened the key to understanding how biological systems transmit genetic information to their progeny and how changes occur between individuals.

The language of the genetic code can be understood simply from Table 12–5.

12.6   MEASUREMENTS IN BIOLOGICAL SYSTEMS

In any biotechnology process an accurate analysis of reactants and products is essential. Some of these techniques are necessary in development of biological processes, but obviously each requires a different set of tools and degree of accuracy.

A number of processes occur regularly in living systems that never occur otherwise. At the same time, the chemical composition of cells is infinitely more complex than for conventional organic or inorganic solutions. We will not consider many of these techniques in detail because they are discussed in courses in biology and organic chemistry. Some important techniques are

- cell counting
- cell sorting
- cell replication
- chemical characterization
- optical microscopy

- electron microscopy
- fluorescence tagging
- isotope tagging

We leave for a homework problem the description of these processes and their different applications.

12.6.1 Polymerase chain reaction (PCR)

An interesting and very important laboratory growth process is the replication of DNA outside of living organisms through a process called the polymerase chain reaction, sketched in Figure 12–12. This method is used in criminal cases to determine whether a particular sample of DNA matches a specific person. One starts with a small sample of human tissue such as blood or hair found at a crime scene. This is usually a very small quantity, perhaps a single drop of blood or a single hair.

The sample is prepared by rupturing all cells and extracting their DNA, at most a few percent of the original sample, which is far too small to analyze directly. However, it can be amplified by the PCR reaction. One begins by heating the sample to 90°C, a temperature that causes the DNA strands to separate into two single strands. Then one cools to ~50°C and adds a solution containing the four individual nucleic acids and a small piece of DNA called a "primer." The primer has the properties that it codes for starting DNA chain growth of a second strand onto the original single strand. The primer attaches to each strand, and nucleic acid monomers add to the strand, forming a double helix from each single helix. This forms two double strands from the single original strand, as shown in Figure 12–12.

One then heats this sample containing two double strands for each original double strand, cools, adds more nucleic acids and primer, and allows this mixture to react to form four double strands. This sequence is repeated until enough DNA is obtained for analysis. In our "reaction" picture, this is a chemical reaction

$$A \rightarrow 2A \rightarrow 4A \rightarrow 8A \rightarrow 16A \rightarrow \dots \qquad (12.17)$$

with each growth cycle doubling the amount of A.

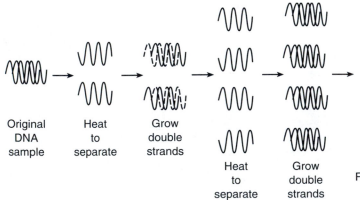

Original DNA sample → Heat to separate → Grow double strands → Heat to separate → Grow double strands → Repeat

Figure 12–12 The polymerase chain reaction (PCR). This is a method to replicate fragments of DNA in the laboratory. Large amounts of identical DNA molecules can be created to be analyzed to identify DNA sequences in biology and to identify traces of cells.

For example, 20 cycles of growth produces $2^{20} = 1 \times 10^6$ times more DNA than the original sample and 40 cycles produces $2^{40} = 1 \times 10^{12}$ times more sample. This produces an extremely large amplification of the number of molecules which in principle could start from just one strand of DNA and produce macroscopic amounts.

This is a simple example of the reaction $A + S \rightarrow 2A$, where now A is a single strand of DNA, $2A$ is a double strand, and S is the (many) nucleic acids that reacted on the single strand to form the double strand. Note that A here is a single molecule rather than a living organism.

Analysis of DNA is usually done by gel chromatography in which one first cleaves the DNA into smaller fragments using specific enzymes. One then separates these fragments of DNA by their mobilities on a porous sheet. This chromatogram is compared with a larger sample taken from the person being tested. Since every person's DNA is different, this test provides an almost perfect test of whether the sample was from the person in question (subject only to lawyers irrational arguing about admissibility as evidence).

12.7 RATES AND KINETICS OF BIOLOGICAL PROCESSES

Biological processes are a complex network of coupled chemical reactions. As engineers, we need to develop simplified reactions and rate expressions to model these processes. We begin with a general rate expression

$$A + n_S S \rightarrow n_A A + n_P P, \qquad r = k C_A C_S \tag{12.18}$$

Here A is the biological species which could be the organism or a number of cells in a multicellular organism, and C_A is the concentration of cells or the number of cells. We could express this in moles/liter or in other units such as cells/cm^3 for cells growing in a uniform volume or cells/cm^2 for cells growing on a planar support.

The quantity S (called substrate in biology) is the nutrient concentration that the cell or organism metabolizes to reproduce, such as sugar in fermentation by yeast. The quantity P in this expression designates metabolic products. For fermentation these products are ethanol and CO_2. In general, these may be waste products, chemical storage products, or toxins used for defense. In biotechnology we frequently want to harvest and sell these products.

We use this expression to model organism growth where $n_A > 1$ as well as metabolism where $n_A = 1$. For the latter we are interested in production of products P. Extracellular enzymes cannot reproduce themselves, so this is a metabolic process with $n_A = 1$, where the enzymes are catalysts for the reaction.

By this expression we do not imply that the process stoichiometry or kinetics are as simple as it suggests. Rather, we will find that a process with this stoichiometry and kinetics can be used to reproduce many qualitative features of biological systems. Table 12–6 lists some examples.

12.7.1 Kinetics of reproduction

We first describe the change in the number or concentration of the biological species A in a reaction process as it feeds on (reacts with) its nutrients S to reproduce itself.

Cells in ideal environments with ample nutrients begin to reproduce, in the mode called exponential growth. We can simplify this as the chemical reaction

TABLE 12–6
Some Examples of Autocatalytic Growth: A + S → 2A + P

Reactant A	Reactor	Nutrient S	Products A and P
yeast	vat	sugar	more yeast and ethanol
yeast	bread loaf	sugar	more yeast and CO_2 bubbles
Lactobacillus	CSTR	starch	more bacteria and lactic acid
glucose isomerase	CSTR	glucose	fructose (HFCS)
bacteria	flask	carbohydrate	more bacteria
bacteria	Petri dish	agar gel	colonies from each individual
algae	pond	CO_2 + sunlight	more algae
plants	field	CO_2 + sunlight	larger and more plants
animals	earth	sugar + protein + O_2	larger and more animals
fetus	mother	sugar + protein + O_2	larger fetus
Pasteurella pestis	human	sugar + protein + O_2	dead humans and more plague
cancer cell	human	sugar + protein + O_2	more cancer cells
DNA	solution	nucleic acids + primer	more identical DNA molecules

Note: In addition to these nutrients, most living organisms require traces of N, P, S, some metals, and vitamins.

$$A + S \rightarrow 2A + P, \qquad r = kC_A C_S \qquad (12.19)$$

because we are interested in reproduction of $A(n_A = 2)$. We can set $n_S = 1$ and we ignore metabolic products P.

It is interesting to note that in previous chapters the same reaction and rate were used to explain autocatalysis in chain reactions such as in free radical processes. We wrote the reaction $A \rightarrow B + C, r = kC_A C_B$ and remarked that product B was necessary to catalyze the decomposition of A, so the elementary reaction would be $A + B \rightarrow 2B + C$. This is exactly the form of the above growth model with P (the product) replaced by A (the growing organism) and A (the reactant) replaced by S (the food supply). In biological systems the food supply is consumed while the organism reproduces or grows.

In excess nutrient, $C_S = C_{So}$, in a batch reactor we obtain a mass balance

$$\frac{dC_A}{dt} = +kC_{So}C_A \qquad (12.20)$$

which can be simply integrated to predict exponential growth,

$$C_A(t) = C_{Ao} \exp(+kC_{So}t) \qquad (12.21)$$

Exponential growth is the "natural" tendency of any living system because it wants to reproduce itself. However, exponential growth is an inherently unstable growth mode because the system will inevitably increase until it exceeds its food supply or fills its container.

Desired examples of exponential growth are fetal growth as the fertilized egg divides repeatedly and replication of DNA molecules in the polymerase chain reaction (PCR). Undesired examples of exponential growth are plagues (populations of viruses, bacteria, or insects) and cancer (uncontrolled growth from a particular mutated cell that has no control mechanism in it).

Once a cell has consumed most of its nutrients or grows to a size where the nutrient supply is not uniform, growth changes with food supply or various control mechanisms within the cell or organism begin to limit growth. As just noted above, this is a very fortunate situation, because uncontrolled growth is usually bad.

The system then switches from exponential to "linear" growth where continued reproduction depends on nutrients or on signals that control continued growth. By linear we do not mean that the number of organisms, number of cells, or volume of the organism necessarily increases linearly with time, but that growth switches from exponential growth, $V \sim e^t$, to some power dependence on time. For example, $V \sim t^1$ would be linear in volume or number density and $V \sim t^3$ represents linear increase of diameter of a spherical organism or mass of cells.

If growth is limited by food supply S, then one must find how the food supply is consumed and solve several species equations simultaneously.

For the above stoichiometry, S and A are related by the relation

$$C_A - C_{Ao} = C_{So} - C_S \tag{12.22}$$

or more generally as

$$\frac{C_A - C_{Ao}}{n_A - 1} = \frac{C_S - C_{So}}{n_S} \tag{12.23}$$

This predicts the growth of A until the food supply S disappears. We will show solutions of these equations in an example.

When a cell or an organism no longer needs to grow or the food supply is depleted, it sends signals that stop cell division. Alternately, in a population of organisms individuals die by old age or predation. Both of these mechanisms yield a steady state population. In our simple kinetic model of a single irreversible reaction, this is explained with $k = 0$ or $C_S = 0$.

We can also develop a simple model that predicts steady state in a population that suffers death. Suppose cell or organism birth is by a first-order process in the same rate expression we wrote above. Now we add a "reverse reaction" representing death as second-order "kinetics":

$$\frac{dC_A}{dt} = r_b - r_d = +k_b C_A - k_d C_A^2 \tag{12.24}$$

Here r_b and r_d are birth and death reaction rates. Comparing this with the simple growth model above, we see that the reaction $A + S \rightarrow 2A$, $r = kC_S C_A$ is the birth rate for constant nutrient supply with $k_b = kC_{So}$.

We can set the rate of change of C_A equal to zero to yield a steady-state concentration

$$C_{As} = \frac{k_b}{k_d} \tag{12.25}$$

This equation can be solved in a batch process to yield $C_A(t)$ for specific C_{Ao}. It predicts that C_A will increase with time if $C_{Ao} < C_{sS}$ and will decrease with time if $C_{Ao} > C_{As}$.

We considered population models in more detail in Chapter 8. These systems can exhibit single or multiple steady states, extinctions, exponential growth (for some time), and oscillatory populations. Especially interesting are prey–predator population balances that can readily exhibit complex dynamics.

12.7.2 Metabolism

Biological systems process nutrients S to produce products P. If the organism A is not reproducing, we set $n_A = 1$ and write the reaction as

$$A + S \rightarrow A + P, \qquad r = kC_A C_S \tag{12.26}$$

or we omit A to obtain

$$S \rightarrow P, \qquad r = kC_A C_S \tag{12.27}$$

If C_A is constant at C_{Ao}, the mass balance on S in a batch reactor becomes

$$\frac{dC_S}{dt} = -kC_{Ao}C_S \tag{12.28}$$

which can be integrated to yield

$$C_S(t) = C_{So} \exp(-kC_{Ao}t) \tag{12.29}$$

and

$$C_P = C_{So} - C_S = C_{So}(1 - \exp(-kC_{Ao}t)) \tag{12.30}$$

This has exactly the same form and kinetics as the unimolecular reaction $A \rightarrow B, r = kC_A$, that was considered in Chapters 2 and 3. It describes simple exponential decay of the substrate S to form product P. The difference from $A \rightarrow B$ is that the rate coefficient is now kC_{Ao}. Since this is a biological reaction catalyzed by an enzyme or organism, there would obviously be no reaction without an organism or enzyme, $C_A = 0$.

As long as C_A is constant, the kinetics appear "normal." A well-known example of these kinetics is the isomerization of glucose to make high fructose corn syrup,

$$\text{glucose} \rightarrow \text{fructose}, \qquad r = k_f C_{\text{glucose}} - k_b C_{\text{fructose}} \tag{12.31}$$

which we considered in Chapter 2 as one of our first examples of simple kinetics. This reaction is reversible, and k_f and k_b are both proportional to the concentration of the enzyme glucose isomerase, which is first grown and then placed on polymer beads in a packed bed reactor so that the enzyme concentration is constant until the catalyst deactivates.

12.7.3 Reactor types

Biological reactors are also "just like" nonbiological reactors in that they are modeled as batch or flow, but they obviously can be much more complicated. Examples of reactors are bacteria on a Petri dish, yeast in a fermenter with batch or continuous cultures, plants such as corn, or animals such as pigs and humans.

Examples of closed uniform systems are (a) cell cultures in Petri dishes or stirred flasks and (b) populations on islands or on the earth (the latter is not uniform). In this approximation, one solves ordinary differential equations with specified initial conditions.

Many systems of interest are only semibatch because nutrients are fed continuously. Closed ecosystems require input of sunlight, so these systems are also semibatch.

Batch reactors can be very attractive for biological systems. Some reasons for choosing a batch reactor over a continuous reactor are as follows.

1. Amounts of material needed are frequently too small to justify continuous operation.
2. Reactions are slow so reaction times are long.
3. Costs of products are high so equipment costs do not dominate process economics.
4. Product testing is easier in batch processes because each batch can be analyzed separately.

The costs of equipment and operation for a batch process compared to a continuous process frequently favor batch processes because of these considerations. For commodities such as large-scale beer production (megabreweries), continuous processes are favored, while microbreweries make better tasting beer in batches. High fructose corn syrup is a large volume commodity where continuous processes have favorable economics because of scale.

Quality assurance is much more reliable with batch processes because each batch can be separately analyzed and discarded if unsatisfactory, while more product must be discarded in a continuous process. Growth of undesired strains of bacteria or yeast in continuous processes are dangerous because they have long times to grow, especially if they grow on the walls of the reactor.

Fermenters are frequently stirred flow reactors (called chemostats) because they permit continuous addition of nutrients and withdrawal of product. For the growth reaction above, the CSTR mass balance is

$$C_A - C_{Ao} = \tau r = \tau k C_A C_S \tag{12.32}$$

(Note the sign difference because $\nu_A = +1$.) If the nutrient concentration C_S is high enough that C_S remains constant at the feed concentration C_{So}, then this equation can be solved to yield

$$C_A(\tau) = \frac{C_{Ao}}{1 - k C_{So} \tau} \tag{12.33}$$

(Note the minus sign.) The equation predicts that $C_A \to \infty$ as $k C_{So}\tau \to \infty$. This is plotted in Figure 12–13.

Tubular reactors are used for biological processes mainly when catalyst can be packed in the reactor because otherwise washout will make the process less stable in a tubular reactor.

For the reaction above in a PFTR

$$\frac{dC_A}{d\tau} = +k C_A C_S \tag{12.34}$$

and if $C_A = C_{Ao}$ at $\tau = 0$ and C_S is constant at C_{So}, the solution becomes

$$C_A(\tau) = C_{Ao} \exp(+k C_{So}\tau) \tag{12.35}$$

just as in the batch reactor.

Figure 12–13   Plot of concentration C_A versus time in a growth process $A + S \to 2A + P$ for different values of nutrient concentration C_{So} in a PFTR. The growth of A is initially exponential but levels out to an approximately linear growth rate and then approaches a constant as all nutrient S is depleted.

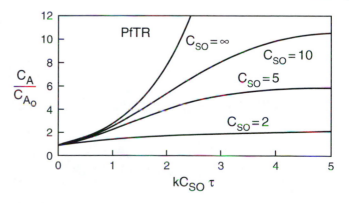

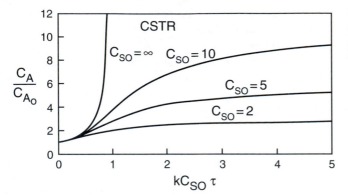

Figure 12–14 Plot of concentration C_A versus time in a growth process $A + S \rightarrow 2A + P$ for different values of nutrient concentration C_{So} in a CSTR. The growth of A is initially very fast but levels out to a constant as S is depleted. The growth rate in a CSTR is much faster than in a PFTR as seen by comparing with Figure 12–11.

This is also plotted in Figure 12–14. It is seen that the production of the organism A is higher in the CSTR than in the PFTR for all τ. Also, while C_A increases rapidly (exponentially) in the PFTR, it does not increase as fast as in the CSTR. Once again, we see that the CSTR gives a higher conversion (or requires a smaller volume) than does a PFTR for an autocatalytic reaction such as growth of living systems. In fact, there is no physical solution for $\tau > (kC_{So})^{-1}$ in a CSTR.

Example 12–3 Biological growth can be modeled through the chemical reaction

$$A + S \rightarrow 2A + P, \qquad r = kC_S C_A \qquad (12.36)$$

where A is the organism and S is its food supply. Find $C_A(t)$ in a batch reactor for $C_{Ao} = 1$ and $C_{So} = 2, 5, 10$, and ∞. Plot results as C_A/C_{Ao} versus $kC_{So}t$.

Since $C_S = C_{So} + C_{Ao} - C_A$ from stoichiometry, the mass balance on A in a batch reactor becomes

$$\frac{dC_A}{dt} = +kC_A C_S = +kC_A(C_{So} + C_{Ao} - C_A) \qquad (12.37)$$

and this equation can be integrated to become

$$t = \int_{C_{Ao}}^{C_A} \frac{dC_A}{kC_A(C_{So} + C_{Ao} - C_A)} \qquad (12.38)$$

(We frequently retain the plus sign to remind ourselves that this is not "normal" kinetics.)

For $C_{So} \gg C_A$, C_S does not change as the reaction proceeds so that

$$\frac{dC_A}{dt} = +kC_A C_S = +kC_{So}C_A \qquad (12.39)$$

which can be integrated to yield

$$C_A(t) = C_{Ao}e^{+kC_{So}t} \qquad (12.40)$$

We plot this as C_A/C_{Ao} versus $kC_{So}t$ as shown in Figure 12–14. This represents exponential growth of A.

If $C_{So} = 2$, the integral becomes

$$t = \int_1^{C_A} \frac{dC_A}{kC_A(3 - C_A)} \tag{12.41}$$

which can be simply integrated to yield the curve shown in Figure 12–14. Also shown in the figure are plots of C_A versus time for $C_{So} = 5$ and 10.

It is seen that all curves initially show exponential growth, and they are initially identical when plotted as $kC_{So}t$. However, they flatten and level off as the nutrient S is depleted. They appear to have a region of nearly linear growth as the nutrient supply decreases because A is increasing while S is decreasing.

Repeat these calculations for a PFTR and a CSTR.

In a PFTR the same equations apply with t replaced by τ, the residence time. In a CSTR with C_S constant as C_{So} we obtain the mass balance

$$C_A = \frac{C_{Ao}}{1 - kC_{So}\tau} \tag{12.42}$$

while if C_S varies, the equation becomes the solution of the quadratic

$$\tau = \frac{C_A - C_{Ao}}{r} = \ldots \tag{12.43}$$

Solutions for these kinetics in a CSTR are plotted in Figures 12–13 and 12–14.

12.7.4   Mass transfer in biological systems

Mass transfer of nutrients to and within the cell or organism can also limit growth. As sketched in Figures 12–15 and 12–16, the variation of concentration C_S of reactant outside and inside a cell can control growth rates. We consider next the metabolism and cell growth limitations caused by mass transfer.

As in Chapter 7, we equate the rate of reaction of a cell A with nutrient (reactant) S with its mass transfer rate

$$r_{mA} = r_R \tag{12.44}$$

$$k_{mS}(C_S - C_{Ss}) = k_R C_{Ao} C_{Ss} \tag{12.45}$$

where C_S is the concentration of S far from the cell and C_{Ss} is the concentration of S at the surface of the cell. Since we are considering the mass transfer and reaction of S near the surface of the cell A, we insert C_{Ao} in the rate.

For external mass transfer control around a sphere in a uniform medium, the Sherwood number is $2.0 + 0.4Re^{1/2}$ from Chapter 7. Under external mass transfer control the rate of a process can be controlled by the stirring rate, and the growth rate can vary with cell size. One can use this expression to predict the maximum metabolic rate of nutrient S if the process is mass transfer-limited.

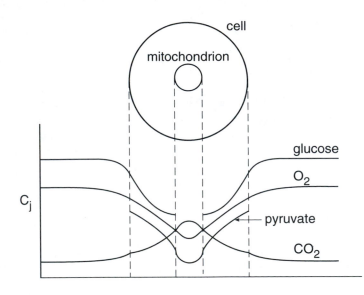

Figure 12–15 Idealized sketches of possible concentration profiles of species involved in cell metabolism. Glucose is converted to pyruvate in the cytosol, while pyruvate is converted to CO_2 only in the mitochondria. For simplicity in sketching profiles, the nucleus is not shown and only a single mitochondrion is shown at the center of the cell.

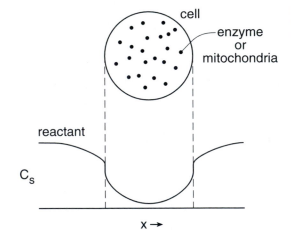

Figure 12–16 Sketch of possible nutrient profile C_S in a cell if limited by mass transfer outside and within the cell. If there are many mitochondria uniformly distributed throughout the cell, the profile is identical to mass transfer limited transport in a catalyst particle, described by an effectiveness factor η which is a function of the Thiele modulus ϕ as described in Chapter 7.

Cell growth can also be controlled by external mass transfer. For a sufficiently small cell or organism, the mass-transfer-limited growth rate is exactly the same as we predicted for growth of solid spheres in Chapter 9. Equating the rate of S incorporation with growth rate we obtain an expression for the rate of growth of a cell

$$R^2 = R_o^2 + \alpha t \tag{12.46}$$

so that $V \sim t^{2/3}$.

Here α is a coefficient that contains diffusion coefficient and stoichiometric relations between S incorporation and cell growth.

There can also be reactant concentration gradients within a cell. Reaction occurs at enzymes that may be distributed throughout the cell. If there are many enzymes in a cell and they are uniformly distributed, the problem is identical to the porous catalyst sphere with catalyst sites uniformly distributed. The concentration profile around a spherical cell is calculated by solving the equation

$$D_S \nabla^2 C_S = \lambda^2 C_S \tag{12.47}$$

in spherical coordinates. This leads to an effectiveness factor $\eta(\phi)$ where ϕ is the Thiele modulus as discussed in Chapter 7.

Oxidation of pyruvate within mitochondria in eukaryotic cells or in bacteria can also be mass transfer-limited as sketched in Figures 12–15 and 12–16. In liver cells there are hundreds of mitochondria within each cell so their concentration can be considered continuous if they are uniformly distributed. The same form of the effectiveness factor $\eta(\phi)$ is also obtained for O_2 (bound to hemoglobin or myoglobin), which must diffuse through the cytosol to enter the mitochondria to react with pyruvate.

Note that O_2 within the cell is bound to myoglobin, so its diffusion coefficient is that of the bound complex rather than that of free O_2. In general, many small nutrient molecules are bound to proteins or other large molecules, and their effective mass transfer rates are therefore lower due to the lower diffusion coefficient of the larger complex.

Glycolysis involves two overall reactions

$$\text{glucose} \rightarrow \text{pyruvate} \rightarrow CO_2 + H_2O$$

The first reaction occurs in the cytosol where glucose diffusion may be limiting. The second reaction is between pyruvate and O_2. Pyruvate is generated in the cytosol while glucose must diffuse from outside the cell. It is clear that there can be many gradients of different reactants in various regions of the cell or organism. Possible concentration profiles for glucose, pyruvate, O_2, and CO_2 within a eukaryotic cell (drawn as having a single mitochondrion at its center) are sketched in Figure 12–15. These sketches are of course very hypothetical, and it is impossible to measure them in most situations. We suspect that it would be rare that transport coefficients of all species be rate-limiting.

Multicellular organisms cannot obtain adequate nutrients externally, and they must develop a vascular system to provide nutrients to cells. Mass transfer from the vascular system to the cell can limit cell metabolism.

12.7.5   Reaction networks

We have discussed above some single reaction systems that can be used to represent biological processes. There are in fact hundreds of thousands of reactions in the simplest living systems. The objective in writing a single reaction was to obtain an overall reaction and rate so that we could use a simple mass balance equation to predict time or concentration dependences. We next consider how more detailed kinetic descriptions should be handled.

A simplified metabolic pathways graph is shown in Figures 12–17 and 12–18. Each dot in Figure 12–18 represents a biological molecule, and each line connecting dots represents a reaction. This diagram roughly represents the reaction chains in higher plants and animals, all of which have the same approximate networks for the reactions shown. Every individual organism contains many additional reaction steps that it has developed for its particular metabolic needs.

As with any reaction network, our task is to write out the individual steps and then solve the resulting mass balance equations with appropriate initial or boundary conditions

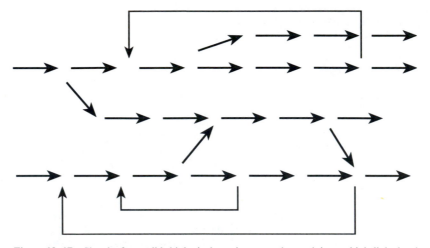

Figure 12–17   Sketch of a possible biological reaction network containing multiple linked pathways and positive and negative feedback loops. All biological processes contain many complex loops as indicated in Figure 12–18.

to find compositions versus time or in steady state. Also, as with any reaction system, we (urgently) seek simplifying assumptions that reduce the complexity of the network.

First we note that all processes are either series or parallel networks of the form

$$A \rightarrow B \rightarrow C \tag{12.48}$$

or

$$A \rightarrow B$$
$$A \rightarrow C \tag{12.49}$$

and we analyze them in terms of the rate coefficients and orders of reactions with respect to all species involved. As in Chapter 4, we usually consider unimolecular reactions because they are easier to solve mathematically. In most reactions of interest, there are many bimolecular reactions, and most are controlled by enzyme catalysts.

The principles of solving these equations are of course the same as in Chapter 4, but now there are many more reactions in the networks, the kinetics are seldom first order, and we usually do not have accurate kinetic expressions for most steps. We still use equilibrium step and pseudo-steady-state assumptions and try to eliminate concentrations of intermediate species wherever possible, but we must be exceedingly cautious to retain the overall structure of the system to the degree that the application demands.

This situation requires all of the skills of the engineer in literature searching, seeking expert advice, and being ingenious to analyze and solve such problems. Accurate solutions have seldom been obtained even for old and industrially important reaction systems such as sugar fermentation to ethanol.

The next characteristic of these systems, besides their complexity, is that they exhibit many and strong feedback loops that essentially turn on and off particular reaction pathways as the system needs require. Understanding and modifying these feedback networks to maximize a particular pathway is the fundamental task of biotechnology. The pathways are

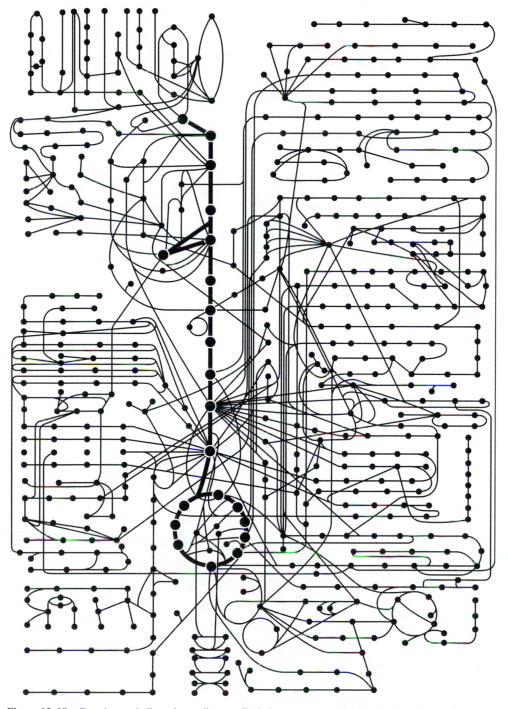

Figure 12–18 Generic metabolic pathway diagram. Each dot represents a biological molecule or molecules, and each line connecting them represents a reaction. This is a highly simplified pathway diagram, which approximates the general reaction pathways in all eukaryotic systems. The heavy line and circle are the glycolysis pathway of glucose to CO_2 which supplies the "engine" for all reactions.

very different for different classes of organisms such as bacteria, yeast, plants, and animals, and we cannot consider them in any detail here.

Figure 12–17 sketches a possible network of reactions. Shown are two reactants and four products, but these interact through crossover and feedback. Feedback loops can be signaling or chemical and positive or negative (promoting or inhibiting). The structures and functions of these interactions is the heart of biological chemical reactions, and their manipulation is the heart of biotechnology.

12.7.6 Glycolysis

We describe one general and simple example where biotechnology is used to generate chemical products, namely the metabolism of glucose to produce several products. Figure 12–3 shows the reactions to produce these products of glucose metabolism. These are shown as series and parallel reaction steps that are tuned by specific organisms to produce high selectivity to the desired product. Sketched in this way, we see that we need to first adjust the k's to maximize a particular desired product.

Listed in Table 12–7 are five products that can be derived from glucose. Production of ethanol and bread are the oldest and most used bioprocesses, and they use natural yeast cells that are designed specifically to optimize these reactions.

Shown in the table are typical selectivities S_B and conversion of glucose X_S that are attained in typical commercial processes. It is seen that with different catalysts (organisms) one can obtain extremely high yields of a single and valuable product.

12.7.7 Enzyme kinetics

Essentially all biological processes are catalyzed and controlled by enzymes. These are catalysts whose mechanisms are essentially identical to the inorganic catalysts we discussed in Chapter 7. There are probably 100,000 enzymes in humans, and each is a complex protein that is designed for a specific reaction. In contrast, in nonbiological reactors we usually try to have just one type of catalyst. In situations where we use two nonbiological catalysts simultaneously in a reactor, we believe that we have created a very "sophisticated" process, but for any biological process, this would be trivially simple.

Biological catalysts are also amazingly specific compared to manmade catalysts. We are usually happy to obtain selectivities of 99% in nonbiological systems, but biological catalysts usually have selectivities much greater than this, essentially making *no mistakes* in running the desired reaction.

TABLE 12–7
Some Useful Metabolic Products in Glycolysis

Product	Uses	Organism	S_B	X_S
fructose	HFCS	glucose isomerase (enzyme)	50%	50%
ethanol	beverages	*Saccharomyces* (yeast)	60%	>90%
	bread	yeast		
acetic acid	food	*Acetobacter* (bacterium) $+O_2$	90%	90%
lactic acid	biodegradable polymer	*Lactobacillus* (bacterium)	>90%	>90%
citric acid	beverages	*Aspergillus niger* (fungus)		

Figure 12–19 Sketch of some enzyme reaction process. The upper left figure shows a simple process in which reactant S binds to the enzyme E where it is catalytically converted into product P. Binding constants K_S and K_P control the populations in the enzyme, which control the rates. An inhibitor I can bind competitively with S and block reaction. At the upper right is a sketch of a situation for a bimolecular reaction $S_1 + S_2 \rightarrow$ products where each reactant binds on distinct sites. The lower left panel shows a situation where a cofactor C binds to another site on the enzyme and either activates or deactivates it. The lower right panel shows a situation where a reaction $2S \rightarrow P$ requires two sites.

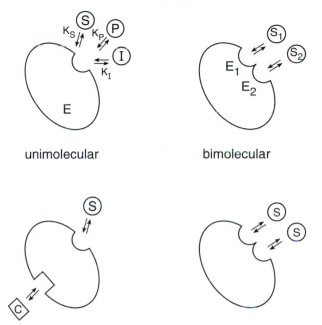

unimolecular   bimolecular

cofactor   bimolecular

The mechanisms of enzyme reactions are basically the same as for manmade catalytic reactions. The reactant (substrate S) binds (adsorbs) reversibly on the enzyme (E), and then it reacts to form the product P and leaves a free enzyme site for further reaction. Several possible configurations of enzymes and reactants are shown in Figure 12–19. The round object is the enzyme protein, and one or more pockets are sketched which are the binding sites for reactants, cofactors, and products.

The process for a reaction with a single reactant

$$S \rightarrow P \tag{12.50}$$

(substrate S reacts to form product P) can be represented through the two-step reaction sequence

$$S + E \rightleftarrows ES, \qquad r_S = k_{fS}C_SC_E - k_{bS}C_{ES} \tag{12.51}$$

$$ES \rightarrow P + E, \qquad r_R = k_R C_{ES} \tag{12.52}$$

Now we assume the first step is in thermodynamic equilibrium

$$r_S = k_{fS}C_SC_E - k_{bS}C_{ES} = 0 \tag{12.53}$$

to obtain

$$C_{ES} = \frac{k_{fS}}{k_{bS}}C_SC_E = \frac{1}{K_S}C_SC_E \tag{12.54}$$

Since the total concentration C_{E0} of empty (C_E) and filled (C_{ES}) enzyme sites is constant,

$$C_{Eo} = C_E + C_{ES} = C_E + \frac{C_E C_S}{K_S} = C_E \left(1 + \frac{C_S}{K_S}\right) \tag{12.55}$$

we can solve this equation to find an explicit expression for the concentration of empty sites,

$$C_E = \frac{C_{Eo}}{1 + \dfrac{C_S}{K_S}} \tag{12.56}$$

The rate of the overall reaction $S \to P$ in terms of C_S and total enzyme concentration C_{Eo} alone is therefore

$$r_R = k_R C_{ES} = \frac{k_R C_E C_S}{K_S} = \frac{k_R C_{Eo}}{K_S} \frac{C_S}{1 + \dfrac{C_S}{K_S}} \tag{12.57}$$

This is just like the catalytic reaction rate expression we derived in Chapter 7

$$r_R = k_R \theta_A = \frac{k_R K_A P_A}{1 + K_A P_A} \tag{12.58}$$

except for several confusing differences in notation. We used the symbol A, and biologists use S for reactant. We used reactant partial pressure P_A and they use C_S for concentration of reactant. We use coverage of A, θ_A, and they use C_{ES} for bound enzyme concentration. Finally, we use K_A as the equilibrium adsorption–desorption constant, while biologists use its inverse, which they call K_S.

This difference in notation is a historical coincidence. Catalyst people refer to these models as Langmuir–Hinshelwood kinetics, while biologists refer to them as Michaelis–Menten kinetics, named after the first scientists in each community to write down these expressions. Chemical engineers must be able to translate between these notations without hesitation.

Consider next the enzyme reaction

$$S1 + S2 \to P \tag{12.59}$$

Assume two sites on the enzyme $E1$ and $E2$, sketched in the upper right panel of Figure 12–19. The elementary steps are

$$S1 + E1 \rightleftarrows ES1, \qquad r_{S1} = k_{fS1} C_{S1} C_{E1} - k_{bS1} C_{ES1} \tag{12.60}$$

$$S2 + E2 \rightleftarrows ES2, \qquad r_{S2} = k_{fS2} C_{IS2} C_E - k_{bS2} C_{ES2} \tag{12.61}$$

$$ES1 + ES2 \to P + 2E, \qquad r_R = k_R C_{ES1} C_{ES2} \tag{12.62}$$

Now we assume both binding steps are in thermodynamic equilibrium, $r_{s1} = r_{s2} = 0$, to yield the relations

$$C_{ES1} = \frac{k_{fS1}}{k_{bS1}} \qquad C_{S1} C_{E1} = \frac{C_{S1} C_{E1}}{K_{S1}} \tag{12.63}$$

$$C_{ES2} = \frac{k_{fS2}}{k_{bS2}} \qquad C_{S2} C_{E2} = \frac{C_{S2} C_{E2}}{K_{S2}} \tag{12.64}$$

Since the total concentrations C_{E1o} and C_{E2o} are constant

$$C_{E1o} = C_{E1} + C_{ES1} \tag{12.65}$$

and

$$C_{E2o} = C_{E2} + C_{ES2} \tag{12.66}$$

we can eliminate the concentration of empty sites (C_{E1} and C_{E2}) to obtain an expression for the concentration of filled reactive sites,

$$C_{E1} = \frac{C_{E1o}}{1 + \dfrac{C_{S1}}{K_{S1}}} \tag{12.67}$$

and

$$C_{E2} = \frac{C_{E2o}}{1 + \dfrac{C_{S2}}{K_{S2}}} \tag{12.68}$$

The rate of the overall bimolecular reaction $S_1 + S_2 \rightarrow P$ is therefore

$$r = k_R C_{ES1} C_{ES2} = \frac{k_R C_{E2o} C_{E1o}}{K_{S1} K_{S2}} \frac{C_{S1} C_{S2}}{\left(1 + \dfrac{C_{S1}}{K_{S1}}\right)\left(1 + \dfrac{C_{S2}}{K_{S2}}\right)} \tag{12.69}$$

Another class of enzyme reactions is where a reaction

$$S \rightarrow P \tag{12.70}$$

occurs but where the enzyme is inactive unless a cofactor molecule is present, as sketched in the lower left panel in Figure 12–19. The reaction sequence might look like

$$C + E \rightleftarrows EC, \qquad r_{S1} = k_{fC} C_C C_{E1} - k_{bC} C_{EC}, \qquad \frac{C_E C_C}{C_{EC}} = K_C \tag{12.71}$$

$$S + EC \rightleftarrows ECS, \qquad r_{S2} = k_{fS} C_S C_E - k_{bS} C_{ECS}, \qquad \frac{C_{EC} C_S}{C_{ECS}} = K_1 \tag{12.72}$$

$$ECS \rightarrow P + EC, \qquad r_R = k_R C_{ECS} \tag{12.73}$$

where E is the site that binds the cofactor C and EC is the site that binds the substrate S.

Next we make a balance on the total enzyme concentration C_{Eo} in terms of empty and full sites to obtain

$$C_{Eo} = C_E + C_{EC} + C_{ECS} \tag{12.74}$$

$$C_{Eo} = C_E + \frac{C_E C_C}{K_C} + \frac{C_{ECS} C_S}{K_1} \tag{12.75}$$

$$C_{Eo} = C_E + \frac{C_E C_C}{K_C} + \frac{C_E C_C C_S}{K_1 K_C} \tag{12.76}$$

or, solving for the concentration of empty sites, we obtain

$$C_E = \frac{C_{Eo}C_{CS}}{1 + \dfrac{C_C}{K_C} + \dfrac{C_C C_S}{K_1 K_C}} \tag{12.77}$$

The rate expression becomes

$$r = k_R C_{ECS} = k_R \frac{C_E C_C C_S}{K_1 K_C} = \frac{kC_{Eo}}{K_1 K_C} \frac{C_C C_S}{1 + \dfrac{C_C}{K_C} \dfrac{C_C C_S}{K_1 K_C}} \tag{12.78}$$

which predicts a range of possible reaction orders in C_S and C_C;

$$r \sim C_S^1 C_C^1$$

$$\sim C_S^1$$

$$\sim C_S^0 C_C^0$$

Thus we see that the reaction is strongly affected by a chemical whose only function is to bind with the enzyme to activate it for the desired reaction. A similar opposite situation is where the cofactor binds to the enzyme to *deactivate* an otherwise active enzyme. The rates of this reaction would be

$$E + C \rightleftarrows EC, \qquad \frac{C_E C_C}{C_{EC}} = K_C \tag{12.79}$$

$$E + S \rightleftarrows ES, \qquad \frac{C_{EC} C_S}{C_{ECS}} = K_1 \tag{12.80}$$

$$ES \rightarrow P + E, \qquad r = kC_{ES} \tag{12.81}$$

Next we consider a bimolecular reaction $2S \rightarrow P$ where the enzyme has two sites for the same reaction as sketched in the lower right panel in Figure 12–19. The elementary reaction steps are

$$E + S \rightleftarrows ES, \qquad \frac{C_E C_S}{C_{ES}} = K_1 \tag{12.82}$$

$$ES + S \rightleftarrows ES_2, \qquad \frac{C_{ES} C_S}{C_{ES2}} = K_2 \tag{12.83}$$

$$ES_2 \rightarrow P + 2E, \qquad r = kC_{ES2} \tag{12.84}$$

Solving for the concentration of vacant sites in terms of total site, we obtain

$$C_{Eo} = C_E + C_{ES} + C_{ES}^2 \tag{12.85}$$

$$C_{Eo} = C_E + \frac{C_E C_S}{K_1} + \frac{C_{ES} C_S}{K_2} \tag{12.86}$$

$$C_{Eo} = C_E + \frac{C_E C_S}{K_1} + \frac{C_E C_S^2}{K_1 K_2} \tag{12.87}$$

$$C_E = \frac{C_{Eo}C_S^2}{1 + \dfrac{C_S}{K_1} + \dfrac{C_S^2}{K_1 K_2}} \tag{12.88}$$

so that the reaction rate becomes

$$r = kC_{ES2} = k\frac{C_E C_S^2}{K_1 K_2} = \frac{kC_{Eo}}{K_1 K_2}\frac{C_S^2}{1 + \dfrac{C_S}{K_1} + \dfrac{C_S^2}{K_1 K_2}} \tag{12.89}$$

This can give $r \sim C_S^2$, $\sim C_S^1$, or $\sim C_S^0$, depending on the relative sizes of K_1 and K_2.

Enzyme control of biological processes enables their manipulation in ways that are impossible with "normal' reactions. We discussed the roles of inhibitors and promoters of catalytic reactions in Chapter 7. For a unimolecular enzyme reaction such as that above, addition of an inhibitor species I that binds competitively to the same site as S and P leads to the reactions

$$S + E \rightleftarrows ES, \qquad r_S = k_{fS}C_S C_E - k_{bS}C_{ES} \tag{12.90}$$

$$P + E \rightleftarrows EP, \qquad r_P = k_{fP}C_P C_E - k_{bP}C_{EP} \tag{12.91}$$

$$I + E \rightleftarrows ES, \qquad r_I = k_{fI}C_I C_E - k_{bI}C_{EI} \tag{12.92}$$

$$ES \rightarrow P + E, \qquad r_R = k_R C_{ES} \tag{12.93}$$

These reaction steps are also sketched in the upper left panel of Figure 12–19. In the above sequence we also assume that the product P binds competitively with S and I on this site. Now we assume all three binding steps are in thermodynamic equilibrium, $r_S = r_P = r_I = 0$, to yield the relations

$$C_{ES} = \frac{k_{fS}C_S C_E}{k_{bS}} = \frac{C_S C_E}{K_S} \tag{12.94}$$

$$C_{EP} = \frac{k_{fP}C_P C_E}{k_{bP}} = \frac{C_P C_E}{K_P} \tag{12.95}$$

and

$$C_{EI} = \frac{k_{fI}C_I C_E}{k_{bI}} = \frac{C_I C_E}{K_I} \tag{12.96}$$

Since the total concentration C_{Eo} of empty (C_E) and filled ($C_{ES} + C_{EP} + C_{IE}$) enzyme sites is constant

$$C_{Eo} = C_E + C_{ES} + C_{EP} + C_{EI} \tag{12.97}$$

we can solve explicitly for the concentration of empty sites (C_E) to obtain

$$C_E = \frac{C_{Eo}}{K_S}\left(1 + \frac{C_S}{K_S} + \frac{C_P}{K_P} + \frac{C_I}{K_I}\right) \tag{12.98}$$

The rate of the overall reaction $S \rightarrow P$ with product and inhibitor competitively bound with S is therefore

$$r = k_R C_{ES} = \frac{k_R C_{Eo}}{K_S} \frac{C_S}{1 + \dfrac{C_S}{K_S} + \dfrac{C_P}{K_P} + \dfrac{C_I}{K_I}} \qquad (12.99)$$

This shows that an "inert" species I can inhibit the reaction by binding competitively with the reactant S.

Example 12–4   In the above reaction suppose $K_S = 10^{-6}$ moles/liter $K_P = 10^{-8}$ moles/liter and $K_I = 10^{-12}$ moles/liter. The rate coefficient in dilute S with no P or I is 10^{-2} min^{-1}. What is the rate and its concentration dependence for the following situations?

(a) $C_S = 10^{-8}, C_P = 10^{-10}, C_I = 10^{-14}$
(b) $C_S = 10^{-4}, C_P = 10^{-10}, C_I = 10^{-14}$
(c) $C_S = 10^{-8}, C_P = 10^{-5}, C_I = 10^{-14}$
(d) $C_S = 10^{-8}, C_P = 10^{-10}, C_I = 10^{-10}$

Our general expression including substrate, product, and inhibitor is

$$r = \frac{k_R C_{Eo}}{K_S} \frac{C_S}{1 + \dfrac{C_S}{K_S} + \dfrac{C_P}{K_P} + \dfrac{C_I}{K_I}}$$

If the first-order rate coefficient is 0.01 min^{-1}, then

$$\frac{k_R C_{Eo}}{K_S} = 0.01 \text{ min}^{-1}$$

Since we are given all K's, the rate with all coefficients given is

$$r(\text{min}^{-1}) = 0.01 \frac{C_S}{1 + \dfrac{C_S}{10^{-6}} + \dfrac{C_P}{10^{-8}} + \dfrac{C_I}{10^{-12}}}$$

We can now insert the concentrations in the four cases.
For case (a) the denominator is

$$\frac{1}{1 + \dfrac{10^{-8}}{10^{-6}} + \dfrac{10^{-10}}{10^{-8}} + \dfrac{10^{-14}}{10^{-12}}} = \frac{1}{1 + 0.01 + 0.01 + 0.01}$$

The first term in the denominator dominates so near these concentrations the rate expression is

$$r(\text{moles liter}^{-1} \text{ min}^{-1}) = 0.01 C_S$$

to give first-order kinetics in S and a rate independent of C_P or C_I.
For case (b) the denominator is

$$\frac{1}{1 + \dfrac{10^{-4}}{10^{-6}} + \dfrac{10^{-10}}{10^{-8}} + \dfrac{10^{-14}}{10^{-12}}} = \frac{1}{1 + 100 + 0.01 + 0.01}$$

The term proportional to C_S is largest, so C_S cancels and the rate is zeroth-order $r(\text{moles liter}^{-1} \text{ min}^{-1}) = 0.01/100 = 10^{-4}$.

For case (c) the denominator is

$$\frac{1}{1 + \dfrac{10^{-8}}{10^{-6}} + \dfrac{10^{-5}}{10^{-8}} + \dfrac{10^{-14}}{10^{-12}}} = \frac{1}{1 + 0.01 + 1000 + 0.01} = 10^{-3}$$

The third term dominates to give a rate that is first order in C_S and -1 order in C_P

$$r(\text{moles liter}^{-1}\,\text{min}^{-1}) = 10^{-10}\frac{C_S}{C_P}$$

For case (d) the denominator is

$$\frac{1}{1 + \dfrac{10^{-8}}{10^{-6}} + \dfrac{10^{-10}}{10^{-8}} + \dfrac{10^{-10}}{10^{-12}}} = \frac{1}{1 + 0.01 + 0.01 + 100}$$

The last term dominates to give a rate proportional to C_S and to C_I^{-1}. Thus the rate becomes

$$r(\text{moles liter}^{-1}\,\text{min}^{-1}) = 10^{-14}\frac{C_S}{C_I}$$

We see that we can obtain very different kinetics and values of r depending on concentrations of S, P, and I. Furthermore, the magnitudes of rates are strongly affected by trace amounts of products or inhibitors.

The four cases in this example thus give very different rates and rate expressions:

(a) $C_S = 10^{-8}, C_P = 10^{-10}, C_I = 10^{-14}, r = 0.01C_S$
(b) $C_S = 10^{-4}, C_P = 10^{-10}, C_I = 10^{-14}, r = 0.0001$
(c) $C_S = 10^{-8}, C_P = 10^{-5}, C_I = 10^{-14}, r = 10^{-10}C_S/C_P$
(d) $C_S = 10^{-8}, C_P = 10^{-10}, C_I = 10^{-10}, r = 10^{-10}C_S/C_I$

12.7.8   Chemical feedback

Biological reactions are almost always under tight control. They seldom operate simply as "normal" unimolecular and bimolecular processes with rates determined by rate coefficients and orders. Rather, they have strong feedback that modulates rates by turning them on and off as required by needs of the system.

We have discussed the autocatalysis mode above in connection with growth of biological systems. The reaction

$$A + S \rightarrow 2A, \qquad r = kC_A C_S \tag{12.100}$$

is identical in form to the autocatalytic reaction

$$A \rightarrow B, \qquad r = kC_A C_B \tag{12.101}$$

because products catalyze the rate of the forward reaction. The analogy can be seen simply by calling the above reaction $A+B \rightarrow 2B$ so that A in the autocatalytic reaction is identified with nutrient S, and B is identified with organism A in biological systems.

Considering the enzyme reaction above with S, P, and I competitively bound on the same site, we obtain the rate expression

$$r = \frac{k_R C_{E_0}}{K_S} \frac{C_S}{1 + \dfrac{C_S}{K_S} + \dfrac{C_P}{K_P} + \dfrac{C_I}{K_I}} \qquad (12.102)$$

As shown in the example above, this can lead to approximate rate expressions

$$r = \frac{k_R C_{E_0} K_P}{K_S} \frac{C_S}{C_P} \qquad (12.103)$$

$$r = \frac{k_R C_{E_0} K_I}{K_S} \frac{C_S}{C_I} \qquad (12.104)$$

Both of these situations lead to rates of reactions that are strongly inhibited by product species or by inhibitors that do not enter into the stoichiometry of the reaction.

12.8 BIOCHEMICAL ENGINEERING

Biochemical engineering is generally concerned with creation of products such as pharmaceuticals by biological reactions or by chemical synthesis that are then used by biological systems.

Biomedical engineering is more concerned with the reaction processes in humans and production of devices that improve human health.

12.8.1 Fermentation

Many chemicals, beverages, and foods are prepared by reacting various feedstocks (frequently sugars) with single-celled organisms, and all of these processes are called fermentation. We have discussed these processes previously.

12.8.2 Antibody production

Most vaccines are made by injecting an animal or human with a toxin and allowing the animal's immune system to make antibodies which fight the toxin. With animals these antibodies are withdrawn from the animal, concentrated, and injected into humans.

One of the first of these was the smallpox vaccine, which was discovered by noting that people who milked cows in England (milkmaids) were immune to smallpox, which was a severely disfiguring and often fatal disease. This was because cows have a related but much milder disease called cowpox. Continued contact with these cows gave the milkmaids cowpox, and the antibodies they produced effectively immunized them against smallpox. By injecting the cowpox toxin into people, they were effectively immunized against smallpox. Within the past 20 years, immunization programs have been so successful that smallpox has been eliminated worldwide.

12.8.3 Isoprene and its products

One of the strategic issues in World War II was the struggle for the control of Asian rubber plantations because they were necessary to provide tires for trucks used to transport soldiers.

Natural rubber is a sap produced by a certain species of tree that grew naturally in Malaysia. This tree produces a sticky substance to heal wounds in the bark of the tree to prevent insect attack, and its culture involves cutting gashes in the tree and harvesting the gum that forms. The monomer that forms this gum is isoprene, which has the formula

$$H_2C=C\text{-}CH=CH_2$$
$$\underset{CH_3}{|}$$

The molecule is methyl 1,3-butadiene (butadiene with a methyl group attached). However, this molecule is rather difficult to produce in high purity in the laboratory, and the desired polymer is the cis isomer, which is nearly impossible to produce economically by chemical synthesis. High-performance airplane tires and high-grade latex gloves are still made from natural rubber even though its cost is much higher than synthetic rubber (usually a styrene–butadiene mix). Polyisoprene forms by the polymerization reaction

$$n\text{-}H_2C=\underset{CH_3}{\underset{|}{C}}\text{-}CH=CH_2 \rightarrow (\text{-}CH\text{-}\underset{CH_3}{\underset{|}{C}}=CH\text{-}CH_2\text{-})_n$$

with all of the double bonds in the cis configuration. These molecules and their reactions to form natural rubber are shown in Figure 12–20.

A fascinating aspect of isoprene is that it is an important component in all biological systems, from bacteria to humans. We do not have enough isoprene in us to heal cuts by bleeding isoprene. Instead we have a mechanism (a chain of many even more complex reactions) to clot blood instead. However, we use polymers made from isoprene for several crucial chemical synthesis reactions, in particular the formation of steroids and optical molecules such as β-carotene and retinal. Some of the reactions of isoprene that form color and scent molecules are shown in Figure 12–21.

In all cases, the reactive initiator in isoprene reactions is isopentenyl phosphate (Figure 12–20), which is in turn formed from acetyl coenzyme A, which is formed at the start of the citric acid cycle. This is another example where phosphates bound to organic molecules cause biological reactions. In all cases, in addition to phosphate, enzymes are involved in all steps.

Cholesterol is the prototype of all steroids and hormones. It contains 27 carbon atoms in a polycyclic structure. Almost magically, it forms by polymerization of isoprene to form squalene which then rearranges to form cholesterol.

$$CH_3COO^- \rightarrow C_5 \rightarrow C_{10} \rightarrow C_{15} \rightarrow C_{20} \rightarrow C_{25} \rightarrow C_{30}$$
$$\qquad\qquad\qquad\qquad\qquad\quad \downarrow \qquad\qquad \downarrow$$
$$\qquad\qquad\qquad\qquad\qquad\text{squalene} \quad \text{cholesterol}$$
$$\qquad\qquad\qquad\qquad\qquad\quad \downarrow \qquad\qquad \downarrow$$
$$\qquad\qquad\qquad\qquad\quad \beta\text{-carotene} \quad \text{testosterone}$$

This sequence is shown very schematically in the Figure 12–20. These processes are all controlled by enzymes, and they can "tailor" their reactions to form slightly different side chains on cholesterol. These form hormones such as testosterone, estrogen, and all of the other hormones that keep us active.

A third class of isoprene-derived compounds is involved in photoinduced chemical reactions, and several of these are also shown in Figure 12–21. They are not isoprene

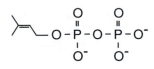

isopentenyl pyrophosphate

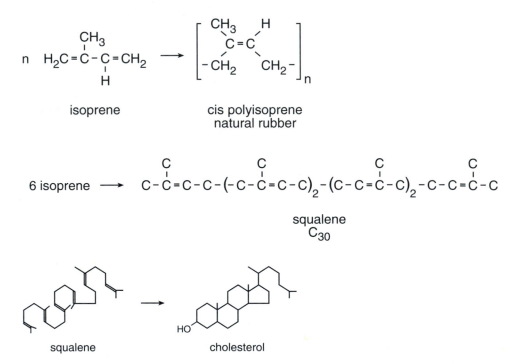

isoprene

cis polyisoprene
natural rubber

6 isoprene

squalene
C_{30}

squalene   cholesterol

Figure 12–20   Polymerization of isoprene forms natural rubber in plants. The actual monomer is isopentenyl diphosphate The polymer is the cis isomer, which is nearly impossible to produce in the laboratory. Six isoprene molecules polymerize to form a 30-carbon polymer called squalene which forms cholesterol. Shown below in the figure is a polymer that contain 6 isoprenes that is called squalene. This molecule rearranges (with the action of many enzymes) to form cholesterol. This is related in structure to hormones such as testosterone and estrogen.

but instead are linear alkanes with conjugated double bonds that absorb light at visible wavelengths. These are required for photosynthesis and for vision. Chlorophyll has a polyisoprene-derived "tail" that acts as an antenna to capture sunlight (maximum sensitivity at green wavelengths) to induce an electronic excitation in the chlorophyll molecule. Vision similarly requires a phoinduced chemical reaction for all colors in the visible for humans, and in the human eye this is from the molecule retinal, which is caused to transfer from the cis to the trans isomer by the interaction with light. (There is a lot more going on in vision, but this is the initial step.) Humans cannot synthesize retinal, and we must consume its precursor, called vitamin A.

One of the photosensitive molecules involved is β-carotene, another polymer formed from isoprene. Another photosensitive polymer is melanin, which is formed in the skin

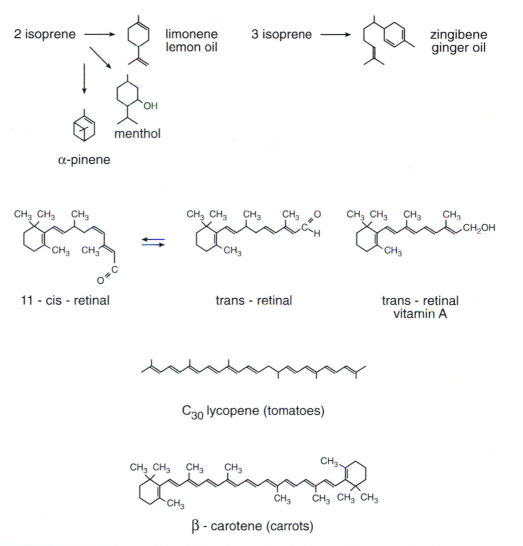

Figure 12–21 Some isoprene-derived biological molecules, called isoprenoids, that are major color and scent molecules. The upper molecules are cis and trans retinal, the primary isomerization involved in vision. At the upper right is shown the structure of vitamin A, a molecule that converts to retinal. At the center is shown the structures of lycopene, a chromophore molecule found in ripe tomatoes that makes them red, and an isomer called β-carotene that is the major chromophore that makes carrots orange. The lower part of the figure shows some dimers and trimers of isoprene that are major olfactory molecules made by plants.

upon exposure to sunlight and protects the lower layer of skin from radiation damage. This chemical absorbs light and gives us suntans.

12.8.4 Biomedical engineering

Chemical engineers must be extremely concerned with the hazards of the chemicals they produce. Flames and explosions are an obvious danger, but chemical and biological poisons are potentially much worse and more insidious because there may be long delays before one knows of the problem, either that an exposure occurred or that it was dangerous.

Since experiments cannot be done intentionally on humans, biological hazards must be determined on animals. Here one exposes the animal to a toxin and determines any harmful effects. There are many problems with such tests. First, there may be long delays before problems are evident, with cancer and genetic defects being the obvious examples. Second, statistical fluctuations must be accounted for in any animal experiments, and the sample size must be large enough to demonstrate statistical significance. Errors in measurement go as $N^{-1/2}$, so the number N of individuals tested must be large, especially if the incidence is small.

Toxins can be classified as acute (immediate response and rapid recovery), chronic (long-lasting harm), delayed response (no evidence of harm for some time such as for most cancers), and genetic response (harm detected only in fetuses or progeny), which are called teratogenic toxins.

Another problem in animal tests is that significant harm can only be studied at much higher levels of exposure and for shorter times than would be encountered in the human environment. For example, a substance that causes cancer in 1 in 1000 humans would require too many animals, and so tests are run at concentrations where 1 in 10 animals show damage. How does one extrapolate injury versus concentration and time? The obvious model is linear: a 100-fold decrease in exposure for the situation above. However, the response may not be linear and may exhibit a threshold or any exposure may show significant damage. Expressed as a chemical reaction

$$A \text{ (chemical)} \rightarrow B \text{ (harm)} \qquad (12.105)$$

one asks if the rate is first order in total amount and in concentration.

12.8.5 Human cell reproduction and death

The unique characteristic of living systems is that they can reproduce themselves. In the individual, many of the cells in our bodies must reproduce themselves regularly. Table 12–8 indicates some of the cell lifetimes in humans. These of course vary greatly with the age of the individual. Some cells such as our heart and brain were formed before we were born and will never be replaced. Others, such as the skin lining our mouth, are replaced in a day. How long does it take for your tongue to heal when you bite it? How long does it take for a skin cut or burn to heal?

Cell reproduction can take place by cell division or by transformation of stem cells into differentiated cells. Stem cells are formed in bone marrow and are transformed into red blood cells, white cells, and other cells throughout the body. Your skin also has a few "immortal" stem cells at its base that supply the continuously renewed skin cells.

Cells also need a mechanism of cell death because uncontrolled cell reproduction results in cancer. Apoptosis is the name for controlled cell death, and the process for it is programmed into cells to occur in an orderly fashion that uses the material as nutrients for new cells. Necrosis is the term used for unprogrammed cell death such as in a burn. Here the body creates emergency repair mechanisms that involve clotting, inflammation, and regeneration.

TABLE 12–8
Lifetimes of Some Human Cells

Type of cell	Lifetime
heart	life of individual
retinal rod cells	life of individual
nerve	life of individual
hearing	life of individual
mouth mucus	<1 day
intestine wall	3–5 days
skin	2–4 weeks
photodisks in retina	1 month
olfactory nerves	1 month
red blood cells	120 days
bacteria in intestine	20 min
embryo	1 h
new human	~20 years

We repeat that these are "just" chemical reactions that we write as $A \rightarrow 2A$ for reproduction and $A \rightarrow X$ for death. There are billions of individual reactions incorporated in these simple overall reactions, and it is the subject of biology to describe them. It is the task of the engineer to understand their organization and function in order to manage and improve them.

12.9   CHEMICALLY SYNTHESIZED BIOLOGICAL MOLECULES

We have so far considered chemical synthesis that occurs inside living organisms or from natural enzymes extracted from organisms. In biological processes we allow the organism to generate the enzymes that catalyze the reactions we desire, just feeding the reactor with desired raw materials and harvesting and purifying products. In enzyme processes we grow enzymes in another reactor and introduce them into a reactor where they catalyze a desired reaction such as isomerization of glucose into fructose.

In biological processes we must take whatever products the cell provides, although through genetic engineering we can frequently modify the cell to produce much more of a desired product than it would do naturally in the "wild" state.

However, there is a large class of processes in which we use organic synthesis reactions (reactions among "dead" molecules) rather than biological or enzyme systems to produce the desired reactions. Several problems arise when we try to make molecules similar to those in nature in the laboratory without using cellular engineering.

First, most biological molecules are *chiral*, possessing one or many centers where only one isomer is biologically active. It is very difficult to duplicate chiral synthesis in the laboratory, and other isomers may be inactive biologically or have undesired activity. Since glucose has 6 chiral centers, only 2^{-6} of the isomers of chemically synthesized glucose are biologically active. All others pass through the body without digestion. Generally, one requires a chiral reactant or enzyme to create a chiral product, and the presence of many chiral centers or imperfect yields of chiral products can quickly produce mixtures of many products with low overall activity.

Second, biologically active molecules are frequently very large, and the number of steps in a chemical synthesis can be very large. This is the problem discussed previously of a number of reaction steps in series. If each step has a yield Y_i, the total yield Y from an N-step synthesis is $Y = \Pi Y_i$. Therefore if each step has 80% yield, the total yield from a 10-step synthesis is $0.8^{10} = 0.1$, a rather poor yield.

A third problem is that intermediate products must frequently be separated from reactants and undesired byproducts before the next reaction stage. Separation stages can also be complex and expensive.

Thus, the yield from a multistep synthesis with several chiral centers quickly becomes unattractive compared to natural biosynthetic processes.

However, a large fraction of pharmaceutical products are now made by organic synthesis, and this trend will probably continue. Antibodies and very large proteins will always be made biologically, but organic synthesis has many attractive features.

Biological products are usually a single molecule with few byproducts. However, if one wants a modified molecule (slightly different activity or selectivity or particular solubility in a given solvent), it is difficult to modify a biological product except by generic engineering, and this can be a complicated trial-and error process. With organic synthesis, chemical modifications (an ethyl side chain instead of a methyl side chain, for example) is frequently a trivial modification.

We consider below three examples of chemically synthesized pharmaceutical molecules.

12.9.1   Ibuprofen

The most common over-the-counter drug after aspirin is ibuprofen, a simple pain relieving molecule. As sketched in Figure 12–22, this molecule is made in approximately two steps, starting with isobutyl benzene. The first step is acetylation with acetyl chloride and the second (rather more complicated than indicated) involves converting a ketone to a carboxylic acid.

12.9.2   Prozac

The most common tranquilizer S-fluoxetine which is sold under the trade name Prozac. This is a typical drug made by synthetic organic chemical methods. One chemical preparation method is indicated in Figure 12–23.

This can be made by a four step preparation which starts with chloroethyl phenyl ketone. By a sequence of steps the ketone is reduced to the alcohol using BH_3, the chloride is reacted to the iodide which is then reacted with methyl amine to make a secondary amine, and finally p-chlorotrifluorotoluene is added by replacing the Cl with OH.

12.9.3   Viagra

The molecule with the common name Viagra is sildenafil citrate. It is made by a multistep process, a simplified version of which is sketched in Figure 12–24. This is a rather complex sequence of synthesis reactions with several different kinds of organic reactions.

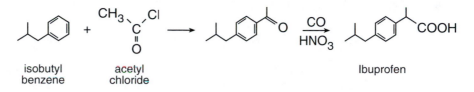

isobutyl
benzene

acetyl
chloride

Ibuprofen

Figure 12–22 A chemical synthesis reaction route to ibuprofen, a simple pain reliever. These steps are highly simplified, but they indicate the process by which this chemical is manufactured in industry.

Figure 12–23 A chemical synthesis reaction route to *S*-fluoxetine, the tranquilizer called Prozac. These steps are highly simplified, but they indicate the process by which this chemical is manufactured in industry.

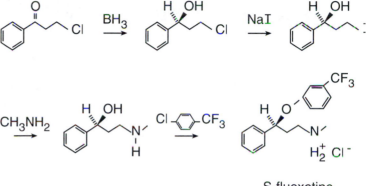

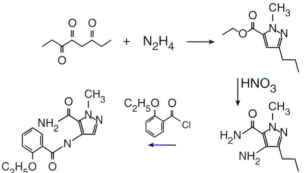

S-fluoxetine
Prozac

Figure 12–24 A chemical synthesis reaction route to the pharmaceutical sildenafil, whose trade name is Viagra. These steps are highly simplified, but they indicate the process by which this chemical is manufactured in industry.

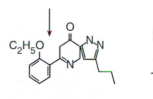

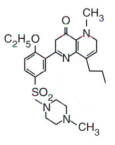

sildenafil
Viagra

12.10 ECONOMICS OF BIOPROCESSES

Bioprocesses have very different economics than do petroleum, commodity chemicals, or even most specialty chemicals, and the chemical engineer should be aware of these differences in designing a successful process.

12.10.1 Small quantities

Quantities of bioproducts are usually small, and this suggests that batch processes may be preferable. Costs of batch processes are generally higher, but when startup and shutdown are frequent and demand can fluctuate, batch processes can be more economical.

12.10.2 Certification

Biological processes must frequently be tested for purity to be certified for human consumption, and batch processes are much superior because each "lot" of product can be tested separately and discarded if necessary. In a continuous process there is always a problem that the product is mixed, and one does not know how far back one must discard product that does not meet specifications.

On the other hand, some bioprocesses are extremely large. Minnesota adds 10% of corn-derived ethanol to its gasoline in the winter, so this requires processing equipment much more than 10% of the size of refineries because the corn-to-ethanol processing plants are much less efficient than petroleum refineries. Similarly, the volume of soft drinks consumed (especially among students) suggests that large amounts of fructose and citric acid are needed.

Beer production is a continuous process for the major national brands, but it occurs in batches in microbreweries. Consumers frequently like the different tastes caused by the variations in small batches of beer (or not).

12.10.3 Product and process specifications

Product quality requirements are frequently quite different in bioprocesses. While conventional chemicals frequently must have extremely high purities (polymer-grade ethylene can contain less than 10 ppm of CO), chemicals intended as food and especially drugs must meet extremely stringent purity requirements. Drugs must also undergo extensive and time-consuming clinical trials where they are tested on small groups of people under carefully supervised and documented conditions.

Products prepared biologically can never be guaranteed to be identical over long times because biological feedstocks (such as corn) cannot be absolutely controlled. Therefore, pharmaceuticals must meet federal standards on the chemical process as well as the product. Once a process is approved, it cannot be changed without reapproval. Thus, even if engineers know that an improvement is possible, it is frequently not implemented because of the cost and time required for FDA approval.

Approval issues with biologically prepared pharmaceuticals provide another reason for sometimes electing to use an organic synthesis route instead of biological synthesis because the former makes a product with precisely defined byproducts, and the approval of process improvements can be much shorter and cheaper.

12.10.4   Intellectual property

Profit margins in pharmaceuticals are extremely high compared to almost all chemical products except computer processor chips. (Computer memory chips are a low-profit product, and the prices on old processor chips falls rapidly to commodity levels.) However, the costs of drug discovery and clinical trials necessary to create a new product frequently require that high profit margins (sales price versus cost of manufacturing) must be very high for profitability of a product.

In the drug discovery industry, patent protection is crucial. In many other industries, speed to market, lower costs of production, and market share can keep competitors out long enough to make profits. These are seldom adequate in pharmaceuticals because clinical trials take so much time, and production costs are a small fraction of overall costs.

12.11   BIOLOGICAL REACTORS

We next inquire what types of chemical reactors are suitable for biological processes. In a previous section we considered mass balances in batch reactor, PFTR, and CSTR for biological reactions where growth occurs. In this section we consider generally some of the overall issues in bioreactors (see Table 12–9).

As noted above, bioprocesses are frequently slow, so the reactors are large. Fermentation requires days to weeks, and production of antibodies in pigs may require the lifetime of the pig.

The energy balance is seldom important in biological processes since they are extremely slow, so heat management is straightforward, and they take place either at room temperature or at 38°C for animal cells.

TABLE 12–9
Types of Biological Chemical Reactors

Type	Examples	Reactors	Harvesting
homogeneous enzyme	starch hydrolysis	continuous	purification
immobilized enzyme	glucose to fructose	continuous	settling and purification
fermenter			
bacteria	sewage treatment	series of CSTRs	settling
yeast	beer and wine	batch	clarification
	fuel ethanol	CSTR	distillation
	breadmaking	batch	bake and eat
fungi	citric acid	CSTR	extraction
plant cells			burst cells and extract
mammalian cells			burst cells and extract
aerobic fermenter	vinegar		distill
animals	human growth factor	pig	extract from blood
	clotting factors	cow	extract from milk
plants	biodegradable polymers	crops	extract from plants

12.11.1   Immobilized enzyme reactors

This is a common type of reactor that looks exactly like a heterogeneous catalytic reactor because it is a catalytic reactor as described in Chapter 7 that uses a biologically derived catalyst. The process can be run in batches for small production rates or can be run continuously (packed bed or slurry) for larger production rates.

This is an ideal reactor if the enzyme is stable enough to survive outside of the cell where it was created. One must then provide the proper nutrients and environment, such as a suitable pH for the cell to survive and operate.

Enzymes outside of cells cannot reproduce, and therefore they cannot regenerate themselves. Enzyme catalysts are therefore subject to the same constraints that inorganic catalysts are: loss of activity. However, enzymes can have selectivities much greater than any inorganic catalysts because they are truly a single molecule with a single type of active site. An interesting question is whether the enzyme is living. We say that it dies when it deactivates, which usually occurs by poisoning of the active site or denaturation of the protein. The protein can only be made by a living organism, but it is just a large organic molecule.

A key economic advantage of immobilized enzyme reactors is that they require minimal separation after the reactor because the catalyst remains within the reactor. Major efforts are required to find suitable supports, which may be porous inorganic or resin structures (spheres are a common shape). Methods must be found to anchor the protein to the substrate chemically without modifying its catalytic activity, and this is a key research and development step in these reactors.

Since this is always a liquid aqueous reactor, mass transfer to the support and within the porous support can be important in enzyme reactor design. Sherwood numbers and effectiveness factors must be calculated in designing an enzyme reactor, just as discussed in Chapter 7.

12.11.2   Fermentation reactors

The most common type of biological reactor is the fermenter. A fermenter involves cells distributed in the reactor and can be operated in batches or as a continuous CSTR. If cells can be immobilized, the reactor can be operated in plug flow, but keeping immobilized cells alive for long times can be difficult. It is frequently desired to seed the reactor with a few cells (sometimes by product recycle) and allow natural cell growth to maintain the cell concentration.

Cells generate enzymes that they use in digestion. Sometimes these enzymes pass naturally through the cell walls where they can catalyze reactions in the homogeneous phase, or they may require the reactant to diffuse through the cell wall.

If reaction occurs within the cell, a crucial step is the transport of the product back to the homogeneous phase. If this does not occur, the cells must be broken apart (liced) to extract the product. If products diffuse out of the living cells, the process can be run continuously without killing cells to extract product.

12.11.3   Plants and animals

We finally come to the ultimate biological reactor, the individual organism. Animals have been used to prepare vaccines and antibodies for many years. One simply injects a disease-causing bacterium or virus into the animal and harvests the antibodies it produces. These are concentrated and given to people to protect them against the disease.

In a modified version of this technique, the animal is *genetically modified* to express a particular protein or other chemical. Examples are proteins such as clotting factors and growth hormones that are inserted so that they are produced in the animal such as in cow's milk.

Another application is in genetically modifying higher plants to express a particular product. For example, if corn or soy beans can be made to produce a particular chemical, its seeds need only be planted and harvested to extract the product. While biodegradable polymers are now made in *E. coli*, they could in principle be made in genetically modified plants that can be planted and harvested.

These techniques suggest an important and potentially revolutionary change in chemical processing. They do not require a chemical plant, because farm animals and plants make the product in feedlots or fields. Feeding and harvesting are agricultural processes that are not subject to the rules of the chemical industry. Byproducts in chemical synthesis are fertilizer or animal feed. These processes eliminate the need for all of the conventional facilities in chemical processing, and they are subject to none of the restrictions involved in handling hazardous chemicals.

The ultimate biological reactor is of course ourselves. By genetically modifying humans, we can create disease resistances without the need for preparing drugs separately. We can also repair damage to ourselves by using our own cell mechanics to create or eliminate specific structures. God created humans, but He gave us brains and engineers capable of devising repair methods. Many dangers but incredible prospects may become common in a short time from applications of biological reaction engineering.

12.12 SUMMARY

In this brief chapter we have tried to present an introduction to biological chemical reactions and biological chemical processes. We have necessarily simplified the subject extensively, and we have without question been quite wrong in some situations. We have discussed some of the processes that only occur in nonbiological systems such as molecules reproducing themselves and making chiral products in high purity.

The knowledge of biologists regarding these subjects is expanding rapidly. A few years ago the human genome was largely unknown, while it is now almost complete. The ways that genes code proteins are now very incomplete for higher organisms, but in a few years these issues will probably be well understood.

While chemical engineers will lag in time and complexity from the knowledge of biologists, we also need to keep pace with these discoveries to adapt these ideas into reactor and reaction design. All of these issues are becoming increasingly sophisticated, such that within a few years the present knowledge of biological processes will appear extremely naïve.

This is the nature of biotechnology: extremely complicated, changing rapidly, and having incredible potential to change our lives and our vocations.

12.13 REFERENCES

Alberts, A., Bray, D., Lewis, J., Raff, M., Roberts, K., and Watson, J. D., *The Molecular Biology of the Cell*, Garland, 1994.

Bailey, T. J., and Ollis, D., *Biochemical Engineering*, 2nd ed., McGraw-Hill, 1987.

Diamond, Jared , *Guns, Germs, and Steel: The Fates of Human Societies*, W. W. Norton, 1999.

Kirk-Othmer Encyclopedia of Chemical Technology, 27 vols., Wiley, 1991–1998.

Shuler, M. L., and Kargi, F., *Bioprocess Engineering*, Prentice-Hall, 1992.

Stryer, L., *Biochemistry*, 4th ed., W. H. Freeman, 1988.

Tanaka, T., Tosa, T., and Kobayashi, T., *Industrial Applications of Immobilized Enzymes*, Marcel Dekker, 1993.

Voet, D., and Voet, J., *Biochemistry*, 2nd ed., Wiley, 1995.

PROBLEMS

12.1 (a) If the cells in your body have an average diameter of 20 μm, how many of them do you have?

(b) How many brain cells do you have?

(c) A DNA molecule has a monomer length of 3.4 nm. If stretched into a single chain, how long is all of your DNA, which consists of 3×10^9 base pairs?

(d) One copy of your DNA is in each of your cells. How long are all of these strands?

12.2 Humans generate heat at a rate of 100 watts in the resting state, such as while reading this book, which thermodynamically requires negligible work.

(a) How much glucose do you need to consume per day to generate this energy?

(b) How many liters of O_2 do you need to consume per day?

(c) The air you exhale contains $\sim$16% O_2. What is the volumetric flow rate of your lungs in liters/minute?

(d) What is the CO_2 content of your exhaled breath?

(e) How many calories per day do you need to consume? Compare this with the recommended calorie intake of 2000 calories/day. These intake numbers in the popular literature are actually in kilocalories.

(f) How many pounds of CO_2 do you personally produce per day? per year?

12.3 During vigorous exercise humans generate approximately 300 watts. The excess energy in running is consumed in muscles which comprise $\sim$20% of body mass.

(a) Lung capacity can only increase by a factor of 2 from the resting state. What is the maximum steady state energy consumption rate?

(b) In anaerobic exercise at the maximum possible level, how much glucose and O_2 must be stored in muscle cells for 20 seconds of activity?

12.4 What types of toxins are produced by the following diseases, what is the strategy for optimum growth of the disease organism, and what public health practices minimize their spread? (Estimate answers only by discussion with colleagues.)

(a) common cold virus

(b) stomach flu

(c) giardia

(d) *E. coli* in beef and chicken

(e) smallpox

(f) ebola

(g) hoof and mouth disease

(h) anthrax

(i) venereal disease (omit this answer)

12.5 A bacterium 1 μm in diameter digests glucose from the surrounding nutrient to produce the body mass needed to divide. Calculate the time required for cell division if cell growth occurs by the reaction

$$glucose \rightarrow mass + byproducts$$

if the growth rate is limited by diffusion of glucose ($D_A = 1 \times 10^{-6}$ cm^2/sec) around the cell which is in a stagnant environment with a bulk glucose concentration of 5×10^{-7} grams/cm^3. Assume that the density and molecular weight of the cell are those of water and that 5 grams of cell mass are formed for every gram of glucose consumed.

12.6 A certain drug at 3 moles/liter in aqueous solution (55 moles/liter) hydrolyzes to an inert product at a rate of 2% per day. We need to concentrate and dry this drug (by vacuum distillation and freeze drying) in order to develop a product with a room temperature shelf life (90% of original effectiveness) of 2 years. The solid has a molecular weight of 500 and a density of 1.0 g/cm^3. If the kinetics of hydrolysis are the same in solution as in the solid, what mass fraction of water may remain in the product?

12.7 A 10-μm-diameter cell metabolizes glucose to produce the cell mass it needs for cell division.

(a) If a cell in 1 mole/liter glucose solution converts 5% of the mass of glucose into cell mass, estimate the time for cell division assuming mass transfer limitation. Assume that the process is mass transfer limited in a stagnant solution with $D_{glucose} = 10^{-6}$ cm^2/sec and all densities of 1 g/cm^3. The cell mass roughly doubles in mass while retaining a spherical shape, and then the cell quickly divides. Assume constant diameter in this calculation.

(b) Show how you would solve the problem with variable diameter, still assuming a spherical shape throughout the growth process.

12.8 Yeast cells 1 μm in diameter convert $\sim$50% of glucose into ethanol, the rest forming CO_2 and cell mass. After an initial growth stage, cells stop dividing when the cell population reaches $\sim$1% by volume.

(a) Starting with 1 mole/liter glucose containing 1% yeast cells (when cells stop reproducing), estimate the ethanol concentration when all sugar is converted.

(b) Yeast cells die when the ethanol concentration reaches 10% by volume. What is the glucose conversion?

(c) How long would be required for 90% conversion of sugar in 1% yeast cells if the process is mass transfer limited with $D_{glucose} = 10^{-6}$ cm^2/sec?

12.9 A layer of 20 μm cells is growing on a plate. The cells are in a simple cubic array with equal height and width. The cells metabolize glucose oxidatively with a reaction limited by the dissolved O_2 which is at 2×10^{-5} moles/liter. If 10% of the process produces cell mass, what is the cell division time of the top layer of cells?

12.10 From Figure 12–4 identify the amino acids that are covalent, polar, charged in basic solution, or charged in acidic solution and that have no chiral centers.

12.11 Biotechnology has been around for a long time. What are the raw materials, functions, and reactions in the following processes?

soap	milk product storage	hair curling
rope	meat storage	TNT
wool	guncotton	scurvy prevention
leather	gunpowder	gout prevention
cotton	cellulose acetate	penicillin
rubber	rayon	aspirin

12.12 Describe the following methods for drug discovery:

evolution
directed evolution
genetic modification
combinatorial selection

12.13 Describe the following techniques used for characterizing biological systems:

cell counting
cell separation
cell selection
cell chemical characterization

12.14 Sketch concentration profiles expected for a cell metabolic process that occurs in mitochondria if the process is limited by the following steps:

1. reaction
2. cell wall transport
3. mitochondrion wall migration
4. external mass transport
5. cytosol transport

12.15 A human begins as a 20-μm fertilized single cell and becomes an 8-pound baby 9 months later.

(a) If growth were exponential at constant rate throughout gestation, what would be the doubling time? How much would you weigh now if you continued to grow exponentially?

(b) Human fetuses grow approximately exponentially for the first month and then linearly for the next 8 months. Describe the kinetics and weight versus time.

(c) Sketch the typical weight versus time of a human from −9 months to death at 80 years.

12.16 The liver of mammals is a major organ of glucose metabolism, and each of its cells contains hundreds of mitochondria where oxidation occurs uniformly dispersed throughout the cell, as sketched in Figure 12–16.

(a) Sketch possible concentration profiles of oxygen in the cell, spatially averaging around individual mitochondria.

(b) Assuming a sphere, what is the effectiveness factor for oxidation within a cell? Find an expression to estimate the Thiele modulus.

12.17 The uptake of O_2 by hemoglobin in the human lung occurs by the steps

$$Hb + O_2 \rightleftarrows HbO_2, \qquad\qquad K_1$$
$$HbO_2 \rightleftarrows Hb(O_2)_2, \qquad\qquad K_2$$
$$Mb(O_2)_2 + O_2 \rightleftarrows Hb(O_2)_3, \qquad K_3$$
$$Hb(O_2)_3 + O_2 \rightleftarrows Hb(O_2)_4, \qquad K_4$$

because the hemoglobin protein has four sites that can bind O_2, and the initial binding of an O_2 increases the strength of successive O_2 binding. This is equivalent to saying that $K_1 > K_2 > K_3 > K_4$.

(a) Show that the total uptake of O_2 from this sequence is

$$Y_{O_2} = \frac{\dfrac{C_{O_2}}{K_1} + \dfrac{2C_{O_2}^2}{K_1 K_2} + \dfrac{3C_{O_2}^3}{K_1 K_2 K_3} + \dfrac{4C_{O_2}^4}{K_1 K_2 K_3 K_4}}{1 + \dfrac{C_{O_2}}{K_1} + \dfrac{C_{O_2}^2}{K_1 K_2} + \dfrac{C_{O_2}^3}{K_1 K_2 K_3} + \dfrac{C_{O_2}^4}{K_1 K_2 K_3 K_4}}$$

(b) If $K_1 > K_2 > K_3 > K_4$, show that this expression can be approximated as

$$Y_{O_2} = \frac{4C_{O_2}^4}{K_1 K_2 K_3 K_4 \left(1 + \dfrac{C_{O_2}^4}{K_1 K_2 K_3 K_4}\right)}$$

(c) Plot the above expression as Y_{O_2} versus C_{O_2} to see that it has an S-shape as is observed for hemoglobin uptake in the lung.

Chapter **13**

ENVIRONMENTAL REACTION ENGINEERING

I n this chapter we consider the reaction engineering related to chemical processes that affect the environment. Many of the issues are the same as those considered in previous chapters, but some are quite different.

Let us first consider what we mean by "environment." We are all concerned with making the environment around us as pleasant and safe as possible. This usually means having air and water that are free of noxious chemicals, but this is difficult to define because air and water can be quite unpleasant without manmade changes. Swamps smell bad because of the products of anaerobic fermentation of plants, and the Smoky Mountains are smoky because of organic molecules (called terpenes which are carcinogenic) emitted by the trees in Appalachia. Arsenic and selenium in drinking water are frequently natural products of geology.

Definitions are therefore somewhat ambiguous, but we generally mean that we want our environment to be as clean as if human activity and waste products were not present. This may be as local as trace chemicals in the room and as widespread as mercury in the Great Lakes.

In any case, everyone is concerned with maintaining a clean environment, and chemical manufacturing is regarded as a major cause of chemical pollution.

13.1 ONLY CHEMICAL ENGINEERS CAN SOLVE ENVIRONMENTAL PROBLEMS

While everyone is concerned with the environment, only chemical engineers are qualified to understand and solve environmental problems.

First, most people are not capable of understanding the fundamental issues. The difference between concentrations of parts per thousand and parts per trillion is not obvious to many people (such as Liberal Arts majors). Laws frequently require "no" pollutants, but this merely specifies that the level be lower than the minimum detectable limit using whatever techniques are currently available. Chemical engineers must use the principles in this course to understand the issues and find solutions.

526

Second, the chemical process industry creates much of the waste products we call pollutants. Since chemical engineers design and operate these processes, we are responsible for controlling them. This involves designing processes that produce less byproducts as well as cleaning up pollutants that are produced.

A third issue is that the problems with the environment usually involve the interactions of chemicals with humans. This involves biological reactions, the subject of the previous chapter. We must understand the nature of toxicity before we can do anything to control it. This is why all chemical engineers must be at least literate in biological reaction processes.

Since we are the only people who understand environmental issues, we are responsible for educating the public about them. Anyone can carry a placard denouncing some polluter, but most of these people do not understand much about them. However, our response must be not to simply denounce these people as ignorant and uninformed. Rather, we must

1. take responsibility to educate the public about environmental issues,

2. make sure that the chemical processes in our control are safe, and

3. make sure that our employers adopt responsible policies regarding the environment.

These are formidable tasks that are the responsibility of every chemical engineer (you).

13.2   GREEN CHEMISTRY

The most obvious impact that we can make on the environment is to produce less undesired products. This is of course the core of this entire course, producing high conversion and selectivity to a desired product. We consider next some of the concepts that can reduce the impact of chemical manufacturing on the environment.

13.2.1   More efficient processes

Undesired products are reduced by increasing selectivity, and unreacted reactants are reduced by increasing conversion and using reycle. The maximization of conversion, selectivity, and yield has been the major subject of this course, and this is obviously the major way to reduce undesired emissions from chemical processing. Reactor type, feed composition, reactor temperature, and use of catalysts are primary variables that should be chosen to maximize yield of the desired product in any chemical process.

Recycling of reactants and undesired products is an obvious way to reduce emissions from the process if they can be reacted to improve overall conversion and selectivity. The next alternative to recycle is to send byproducts to an incinerator where they can be used to generate heat needed in the process. A third alternative is to find some use for byproducts.

13.2.2   New processes

When a process has been optimally tuned by adjusting these variables as far as possible, the next alternative is to find a new process. New processes provide revolutionary options that can make major improvements in process economics that could drastically improve yield while reducing byproducts and, sometimes most important, increase the profitability of a process. The goal in searching for new products is "shut down economics," devising a process that is so economically superior that it forces the competition to shut down their competing processes and either go out of that business or pay you royalties for the process you have discovered and developed.

Examples of revolutionary new processes in the petroleum industry are catalytic cracking replacing thermal cracking, fluidized bed reactors rather than fixed bed reactors, catalytic reforming rather than catalytic cracking, and hydrotreating rather than just discarding the heavy components of crude oil. All of these changes have increased the efficiency of petroleum processing such that a modern refinery now converts as much as 90% of the energy and carbon of crude oil into fuels and chemicals.

Many new processes have been introduced in the chemical process industry. Ethylene glycol is now produced almost exclusively by direct oxidation over a silver catalyst,

$$C_2H_4 + \tfrac{1}{2}O_2 \rightarrow C_2H_4O \tag{13.1}$$

However, 30 years ago most ethylene oxide was produced by a two-step process, which involved reacting ethylene with hypochlorous acid, HOCl, in the reaction

$$C_2H_4 + HOCl \rightarrow HOC_2H_4Cl \tag{13.2}$$

which is called ethylene chlorohydrin. This can be hydrolyzed to yield ethylene oxide

$$HOC_2H_4Cl + H_2O \rightarrow C_2H_4O + HCl \tag{13.3}$$

Because the HCl must be reacted with NaOH before disposal, the old process produced several pounds of salt for every pound of ethylene oxide, and it also generated considerable organic chlorides that had to be disposed of. The new process has such better efficiency and lower cost that the chlorohydrin process has been completely replaced by direct oxidation with considerable reduction in pollution from the process.

Another example of a revolutionary chemical process is the ammoxidation of propylene to produce acrylonitrile, a major monomer in acrylic fibers and in acrylic paints. Before 1970, this was produced mostly by reacting ethylene with HCN, which was made from methane,

$$CH_4 + NH_3 + O_2 \rightarrow HCN + H_2O \tag{13.4}$$

followed by the reaction of ethylene with HCN

$$C_2H_4 + HCN \rightarrow C_2H_5CN + H_2O \tag{13.5}$$

and then dehydrogenation

$$C_2H_5CN \rightarrow C_2H_3CN + H_2 \tag{13.6}$$

a three-step process. Sohio Petroleum (later bought out by Amoco which was then bought out by BP) discovered a process that did this in one step by reacting NH$_3$ with propylene in the presence of O$_2$ (called ammoxidation)

$$C_3H_6 + NH_3 + O_2 \rightarrow C_2H_3CN + H_2O \tag{13.7}$$

This was accomplished in a fluidized bed that provided improved heat transfer in the reactor, with O$_2$ making the overall process exothermic. The major breakthrough in the process was the discovery of a new catalyst that involved a mixture of bismuth and antimony oxides.

Essentially all acrylonitrile is now made by the Sohio process, and all older processes have been shut down because of the lower costs of the direct process.

Major challenges await solutions in the chemical industry. For example, all olefins (needed in both of the processes described above) are exclusively produced by steam cracking. This required a flame in a tube furnace to attain the desired temperature, and this process produces more than 30% of the pollution from the chemical industry because of NO_x and unburned hydrocarbons and CO from the furnace. Processes that would eliminate furnaces in chemical processing would make major impact on both efficiency and pollution.

13.2.3 New feedstocks

There is continuous effort to find new feedstocks to increase the efficiency and economics of processes. Reducing steps in a process is always a goal because the overall selectivity is the product of individual step selectivities.

Major efforts are directed toward using alkanes directly rather than first converting them into olefins because the losses in olefin formation are eliminated. This is no easy task because alkanes are unreactive compared to olefins so that their activation usually required extreme conditions that are harmful to subsequent steps in producing a final product.

An example of a new feedstock would be the direct reaction of propane with ammonia to make acrylonitrile,

$$C_3H_8 + NH_3 \rightarrow C_2H_3CN + 4H_2 \tag{13.8}$$

which would eliminate the dehydrogenation of propane to propylene, which is the first step in the current acrylonitrile process. This cannot be done in a single reactor because it is strongly endothermic and has an unfavorable equilibrium. However, there is no reason in principle why the oxidative reaction

$$C_3H_8 + NH_3 + \tfrac{1}{2}O_2 \rightarrow C_2H_3CN + H_2O + 3H_2 \tag{13.9}$$

might not work if one could find a suitable catalyst and reactor conditions. This process could have "shutdown economics" compared to the current propylene process to make acrylonitrile.

13.2.4 New solvents

If a reactant is fed in a solvent, the solvent obviously must exit the reactor and be separated and handled. Reducing the amount of solvent by increasing the concentration of reactant or changing the solvent can have large impacts on the economics and environmental impact of a process.

Essentially no solvent (including air and water) can be emitted from a process, so eliminating or reducing the amount of solvent used in a process can be very important. Recycling of solvents and cleaning up air and water waste streams are essential and can be very expensive.

13.2.5 Biodegradable products

Products must be stable for the conditions and time of their desired usage, but their long-term stability in landfill is a major environmental problem. Plastics such as polyolefins, polyesters, and polyvinyl chloride are major components of landfills that stay in the landfill for decades with no degradation. It is an obvious paradox to desire stability in use but instability after use. Biodegradable products frequently satisfy these requirements if their uses operate in different environments than upon disposal. Products such as polylactic acid and paper have characteristics that make them rapidly biodegradable when exposed to water and the bacteria that occur naturally in soil.

Chlorine containing chemical products are almost always a potential problem because they are not attacked by biological organisms, and therefore chlorine atoms remain in chemical products forever and form HCl if incinerated. Infamous examples are DDT, Freons, PCBs, and tetrachlorodioxin.

13.3   RENEWABLE CHEMICAL RESOURCES

Essentially all chemicals and chemical energy now comes from fossil fuels. Coal, crude oil, and natural gas are used for over 90% of worldwide needs. The major reasons for using fossil fuels are that (a) they are now the most economical resources and (b) an infrastructure exists for using them. Any new energy source must compete with these compelling advantages.

There is considerable discussion in the general public and among technologists that the world should switch from use of fossil fuels to renewable resources. This will be a long-term process because of the issues above. However, several factors suggest that renewable resources have a significant future:

1. Fossil fuels inevitably cause pollution such as SO_x, NO_x, and solid wastes.
2. Fossil fuels require mining or drilling that harm the environment both in extraction and from spills and accidents.
3. Fossil fuels inevitably release CO_2, the dominant greenhouse gas that appears to be responsible for global warming.

The alternatives to fossil fuels are wind, water, direct solar, nuclear, and biomass. Wind and water generate power by conversion of mechanical energy into electricity, direct solar converts photons from the sun into electricity, and nuclear converts nuclear reactions into heat that drives turbines that make electricity.

None of these except biomass appears to use chemical energy, but all are potentially applicable to chemical resources because electricity can be converted into hydrogen through electrolysis of water,

$$H_2O \rightarrow H_2 + \tfrac{1}{2}O_2 \tag{13.10}$$

and the resultant chemical energy in H_2 and O_2 can be used for other reactions that make chemicals. One notable use of this technology is to produce, ship, and store H_2 for the widely discussed Hydrogen Economy.

13.3.1   Revolutionary chemical processes

What is required to discover and develop a new chemical process? Obviously, all companies and academic researchers have been searching for new processes for the past 100 years of the chemical industry, so most easy alternatives have been suggested and tried. A new process might have the following characteristics:

1. A new catalyst that could in principle cause any thermodynamically possible process to run in a single reactor with high selectivity to the desired product.

2. Exotic reactor conditions that no one has explored before. Extremely high pressures and temperatures and contact times could change reaction conditions to give a new product. New solvents and supercritical fluids could produce very different chemistry than has been observed in the past.

3. New feedstocks that have never been explored are a promising route to new and more economical processes.

4. Combined processes where intermediates and byproducts are intentionally formed to react in successive reactions. An example of this is the production of phenol and acetone coproducts from isopropylbenzene:

$$C_6H_5C_3H_7 + O_2 \rightarrow C_6H_5OH + CH_3COCH_3 \qquad (13.11)$$

We discussed reactions such as these in Chapter 9 as free radical chain reactions. The major issue is that one makes phenol from a very unlikely reactant by producing a coproduct acetone, both of which have considerable value.

The discovery of new chemical processes is one of the most exciting and creative areas in chemical engineering. It offers the prospect of making a process with favorable economics (from which the inventor might profit) as well as conserving resources and protecting the environment (from which we will all profit).

13.3.2 Chemicals from biomass

Biomass has a particularly important role in renewable resources because it directly forms carbon energy through the reaction of CO_2 with water and sunlight:

$$nCO_2 + nH_2O \rightarrow (CH_2O)_n + nO_2 \qquad (13.12)$$

This is of course photosynthesis, the most important chemical reaction to us because it permits life, creating plants and O_2 that are essential for animal existence. Plant formation from CO_2 is the only practicable large-scale method to react CO_2, although other methods of "CO_2 sequestration" such as burying it underground or in the oceans or reacting it somehow to form a solid product have been proposed. Many technologists regard CO_2 sequestration as highly improbable because of the large amount of CO_2 that must be reacted to counteract the current emissions of CO_2 from burning fossil fuels. Furthermore, these techniques will cause their own problems with environmental damage or accidents.

The fundamental question with biomass conversion is, "How does use of biomass reduce CO_2 since oxidation reactions of biomass generate CO_2 just as do fossil fuels?" The answer is in the definition of "renewable." The biological cycle of living systems is basically as shown in Figure 13–1. The photosynthesis reaction is coupled with combustion of carbohydrate either biologically or in flame combustion.

$$(CH_2O)_n + nO_2 \rightarrow nCO_2 + nH_2O \qquad (13.13)$$

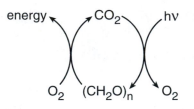

Figure 13–1 The photosynthesis–combustion cycle by which CO_2 is consumed and O_2 is generated in producing biomass with energy from sunlight. Then the biomass reacts with O_2 to produce CO_2. This cyclic process produces no net CO_2 as long as it is completed on a short time scale.

to form the basic biological cycle of all living systems. In fossil fuels, biomass was converted into either coal $(CH)_n$, petroleum $(CH_2)_n$, or natural gas CH_4 that are formed and stored underground. The basic issue is that these chemicals, while very efficient storage processes, all occurred hundreds of millions of years ago, so there is no way to renew the reactions that would place the CO_2 we produce from fossil fuels back into a storable carbon source.

The major issue in biomass is thus the time required for CO_2 conversion back into a carbon source. For plants, this time is one year for annual crops and a few years for trees, and therefore any biomass we convert into chemicals can easily be converted back into stored carbon just as quickly as our plants can be regrown. Some of these times are indicated in Table 13–1.

Thus, while there are many silly proposals for eliminating fossil fuels that violate mass conservation (carbon remains carbon in any chemical form) or the First Law of Thermodynamics (essentially all reactions that react CO_2 are strongly endothermic), the basic idea that renewal resources do not contribute to CO_2 at all is basically correct. Furthermore, while biomass contains traces of S and N that form SO_x and NO_x when processed or burned, these elements are reformed into the biomass that replaces that used initially.

Thus one can make the fundamental case that biomass derived chemicals when considered on a global scale will fundamentally eliminate the CO_2 and pollutants associated with fossil fuels. The basic issues in our switch from fossil fuels to biomass feedstocks are economic in the costs of production, global warming, and pollution. It seems almost inevitable that this conversion will accelerate such that all students will see major shifts in resources during their engineering careers.

The future of chemical reaction engineering will certainly involve biomass. The basic question is whether the conversion from fossil fuels to renewable fuels will occur in 5 or 50 years. The best educated guess is that within 50 years our reliance on fossil fuels will be much less than today. Both political and technical issues remain to be solved for this conversion from fossil to renewable fuels to become a reality, and chemical engineers will play a major role in dealing with these issues.

TABLE 13–1
Recycle Times for Chemical Feedstocks

Chemical	Recycle time
algae	1 month
crops	1 year
trees	5–50 years
oil and coal	200 million years

13.3.3   Polylactic acid

One environmentally friendly chemical process is the production of polymers from biological sources, and the production of polylactic acid from corn starch is being actively pursued commercially in pilot and full-scale operations. This is a polyester that has properties similar to those of polyethylene terephthalate (PET), the most common polyester used in many products from clothing to milk bottles. PET is made from petroleum, and it is very slowly degraded in landfills.

Lactic acid is readily produced from sugar by *Lactobacillus*, the same bacterium that makes yogurt from milk. Its structure, shown in Figure 13–2, is that of a three-carbon carboxylic acid with a hydroxyl group on the carbon next to the COOH group. Note that the middle carbon is chiral (designated by a *), and fermentation of sugar produces only the D isomer. The chiral center promotes continued bacterial action on either the monomer or polymer, so the polymer slowly degrades back into CO_2 in if exposed to soil and water in a landfill.

The processes can be written as

starch → sugar → lacticacid → polylacticacid → polymerproducts → CO_2

Production starts with enzymatic conversion of corn starch to sugar, followed by fermentation with *Lactobacillus* to produce lactic acid. The polymerization of lactic acid forms the ester bond and produces a mole of H_2O for each ester bond formed. This cannot be run directly in a single reactor because equilibrium with water limits the process to a polymer containing approximately eight lactic acid monomer units.

However a two-stage process has been developed to produce high polymers. Lactic acid readily forms the cyclic dimmer, called the lactide, as shown at the bottom of Figure 13–2. This is a reversible reaction, but by distillation in the reactor, the dimmer and water can be separated and the lactic acid monomer can be recycled. The lactide can then be

Figure 13–2   Polylactic acid, a renewable and biodegradable polyester made from starch or sugar which is derived from corn. Production of this polymer does not use fossil fuels and it does not contribute to CO_2 emissions in the atmosphere because CO_2 from its biodegradation is recycled back into the next crop of corn.

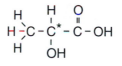

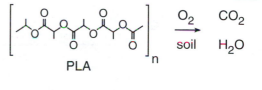

PLA

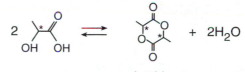

lactide

polymerized to PLA in another reactor by simply opening the lactide rings by addition polymerization (Chapter 11) and adding to the growing polymer.

The lactide dimmer, being chiral, has two isomers, and these have different boiling points, so that the D and L isomers can be separated by distillation of the dimer. This permits preparation of a polymer containing only the D isomer or the mixed D–L polymer, and these have very different crystallization and other properties, so that different polymer properties can be designed into polylactic acid polymers for different applications.

13.4 REGULATIONS

Chemical emissions are tightly regulated by federal, state, and local laws, and these regulations must be met in all chemical processing. Regulations typically limit overall emissions (in pounds or parts per million) of particular classes such as total organic compounds, salts, metals, and chlorine. This frequently allows the chemical process industry considerable flexibility in designing processes because only the total emissions are counted.

Permits are required in starting any chemical process, and these limits must be met at all times, including limits that tighten with time. This is a complex situation for any chemical manufacturing operation because engineering limitations and unforeseen definitions of toxic emissions can cause emissions that exceed current standards at any time. Regulations require reporting emissions that exceed regulations, and fines and terribly bad publicity frequently accompany these emissions.

13.4.1 Automotive emissions

The automobile industry represents an interesting example of how emissions control has required continual response from the manufacturing industry. Before 1970, emissions were essentially unregulated, but since then laws have continuously tightened allowable emissions of hydrocarbons, CO, and NO_x. Table 13–2 summarizes these limits versus time for passenger vehicles in the United States. California frequently has tighter emissions standards and sets regulations for the rest of the country.

The automobile industry has responded by changing engines and fuels and by adding emission control devices. Notable among these are exhaust gas recycle (called EGR, which involves recycle of part of the exhaust from the engine to allow higher conversions), use

TABLE 13–2
Emission Standards for Automobiles

Year	Hydrocarbons	CO	NO_x
1960	10.6 g/mile	84 g/mile	4.1 g/mile
1970	4.1	34.0	5.0
1975	1.5	15	3.1
1980	0.41	7.0	2.0
1983	0.41	3.4	1.0
1994	0.25	3.4	0.4
2001	0.125	3.4	0.2
2004	0.09	1.7	0.07

of sensors to tightly control the fuel/air ratio (control of feed composition), and addition of catalytic converters (another chemical reactor added to the automobile).

Many of the issues in the automotive catalytic converter were considered in Chapter 7. As described there, this involves a catalytic wall reactor with noble metals on extruded ceramic honeycomb catalysts in the three way catalytic converter.

13.5 ACCIDENTS

A significant aspect of chemical reaction engineering is the consideration of possible accidents that could occur in chemical processes in order to consider all possible modes by which the system can malfunction.

We generally design a chemical reactor system to give maximum yield of a desired product with the simplest and least expensive equipment and the lowest operating costs. However, we also need to consider whether that design is also sufficiently robust that it will be stable for any unforeseen situations that may arise. We need to consider operating modes outside of the regime of optimal design and determine that all possible situations in which the system may be operated will not lead to uncontrollable consequences.

13.5.1   Simulating accidents

A first step in accident prevention is to consider the reactor behavior for all possible feed conditions. What are ranges of pressure, temperature, feed, and other variables that the system may possibly experience, and what will be their consequences? The engineer needs to continuously be aware of these upsets.

Handling upsets requires consideration of steady-state and transient behavior of the reactor. We have only discussed steady-state behavior except for (a) brief considerations in Chapter 3 and (b) thermal transients in CSTRs in Chapter 6 because the mathematics required to model transients is beyond what we expect in this course.

One should build a model of a chemical process early in process design and use this model to simulate all possible upsets. After each possible upset has been considered individually, combinations of upsets need to be considered. What happens if temperature, pressure, and feed concentration suddenly rise at the same time?

13.5.2   Flames, fires, and explosions

In Chapter 9 we considered some of the issues involved in flames, fires, and explosions, and these are some of the major issues that must be considered in accident prevention. Where are the most likely trouble spots, and what are the consequences of an incident?

13.5.3   Biological hazards

After fires and explosions, the major accident that can occur in chemical processing is the discharge of a chemical that is hazardous to humans. Air and water can become contaminated with chemicals that cause harm. We discussed some of these issues in Chapter 12 on biological processes. We note again that this is a major reason that all chemical engineers should be at least literate in biological reaction engineering. These hazards can be acute (causing immediate and critical harm to people as in the Bhopal accident in India) or chronic (ground water contamination which can require years and millions of dollars for cleanup).

13.5.4 Probabilities, redundancy, and worst-case scenarios

There is an extensive literature on accidents and loss prevention in the chemical industry. This involves extensive use of statistics and probabilities in assessing damage and probable outcomes which are beyond what can be covered in this course.

A critical issue is how one designs a chemical plant with sufficient redundancy to be able to respond to perturbations that may occur. A primary issue in reactor design is minimization of costs, and any redundancy will obviously add cost and complexity to a chemical process. Design with transients and accidents in mind is obviously an extremely complex subject that requires careful analysis and consultation before deciding on a final design of a process.

It is also important to consider the possibilities that could arise if the worst possible sequence of events occurred. Hurricanes can be accounted for and responded to, but tornadoes are more difficult because they occur so quickly. Sabotage and intentional damage caused by insiders are very difficult to factor in because they require a response to conscious damage by people who are potentially well qualified to understand the maximum damage that could occur. How should one respond to a terrorist attack?

Figure 13–3 shows one of the most spectacular chemical plant accidents ever. It occurred at an ammonium nitrate factory in Ludwigshafen, Germany, in 1921. The blast

Figure 13–3 A BASF ammonium nitrate chemical plant following a massive explosion in 1921. You would not have wanted to be in charge of this plant or be anywhere near it when it blew up.

dug a large crater (which filled with water) and destroyed all buildings in the area. Everything within approximately half a mile of the blast was destroyed.

13.6 WASTE TREATMENT

Another subject that goes beyond the scope of this course is waste treatment. All chemical processes must have waste treatment reactors. Water streams can contain organic and inorganic contaminants. Organics can frequently be oxidized by aeration ponds or adsorption beds. Metals must generally be adsorbed or captured on ion exchange columns. Precipitation and settling or filtration is used to capture particulate matter in water streams.

Air streams frequently are high volumes containing low levels of volatile organics and particulates. Catalytic reaction and filtration are most commonly used to clean up air streams. For higher concentrations of organics, incineration is commonly used, frequently by adding a fuel to the stream and burning it to simultaneously incinerate the undesired organic compounds. Incinerators can also cause pollution in flames such as NO_x, CO, and unburned hydrocarbons.

All of the principles in this course are applied to waste treatment systems, but now the concentrations and constraints are very different. Concentrations go from percent to parts per billion, measurements are complicated, and emissions are frequently transient rather than being steady state as in most production reactors. These issues have the same principles, but they require special considerations in chemical reaction engineering.

13.7 MODELING THE ENVIRONMENT

In any chemical plant one needs to know how unexpected releases of chemicals might cause their dispersal into the surroundings. One wants to estimate these effects before they occur to be able to take appropriate action quickly. One interesting application of the ideas in this chapter is in constructing models of the concentrations of chemicals in various situations.

13.7.1 Water pollution

We are interested in $C_j(t)$ following a toxic source emission $C_{jo}(t)$ to find out (1) if maximum concentrations exceed safe levels and (2) how much time will elapse following a spill before concentrations return to safe levels. The systems of major interest are water and air pollution. These are both *flow systems with chemical reactions*, so the previous equations apply.

Consider a chemical plant located on the bank of a river. Rivers and lakes are common locations for industrial plants because of the ready availability of cooling water (and, it must be admitted, an easy place to dump waste without detection). A map of the region might look as in Figure 13–4.

Chemical spills are described as $C_j(t)$, much like the injection of tracers into a reactor. Simple spills are *steady leaks* that give steady-state concentrations $C(t)$ or *pulse leaks* that give a dispersed pulse $C(z, t)$ that propagates downstream. We can describe these as a series of chemical reactors (which are the rivers and lakes) through which the water with chemicals flows, with exactly the $p(t)$ calculations we developed in Chapter 8.

There are three water systems of concern: rivers, lakes, and ground water.

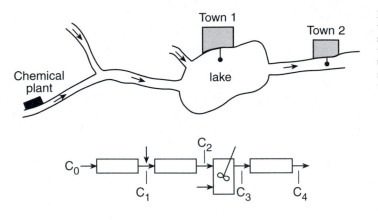

Figure 13–4   Map of chemical plant located on a river which feeds into a lake. One town obtains its water supply from the lake and a second town from a river, which drains the lake. We must be concerned with the concentration profiles (z, t) following a chemical spill from the plant and particularly $C(t)$ near the town water intakes.

A river to a chemical engineer is of course just a tubular reactor, which at low flow rate is laminar and at high flow rate is turbulent. The flow rate is generally steady over short time periods, but flow rates vary seasonally from high in the spring to low in the fall (and they all freeze in Minnesota).

A lake mixes on some time scale, so it might be modeled as a stirred reactor for times longer than the mixing time, and it is a tubular reactor for shorter times. Most lakes turn over seasonally because of the density minimum of water at 4°C, so they are mixed on at least a one-year time scale. Small lakes are mixed effectively by wind on shorter time scales. Deep lakes exhibit an upper layer that turns over and mixes and a lower layer that does not. The inlets to a lake are all of the rivers that flow into it and also water coming from rainfall, and the outlets are all of the rivers that flow out of it and evaporation.

Ground water is basically lakes and rivers underground flowing through sand and gravel. Aquifers are major drinking water supplies, and the adsorption and reaction processes in their flow through sand is important for purification and controls the mineral content of drinking water. This is of course just flow through a packed bed, but the size and residence time can vary from minutes for a septic tank field to centuries for major aquifers.

Almost any chemical not arising from natural sources is considered to be a pollutant in water. Agricultural wastes from fertilizers and pesticides are the major pollution sources nationally. Fertilizers add nutrients that promote algae growth whose decay can cause the lake or river to become anaerobic through the overall reactions

$$\text{nutrients} \rightarrow \text{algae} \tag{13.14}$$

and

$$\text{algae} + O_2 \rightarrow \text{decay products} \tag{13.15}$$

Thus the processes are the *autocatalytic growth* of single-celled plants where the nitrogen and phosphorous nutrients are effectively catalysts for algae growth. When these algae blooms die, their *decay* occurs by air oxidation, which is catalyzed by bacteria. When there are large concentrations of algae, their decay causes depletion of the O_2 dissolved in the water, which kills animal life such as fish.

A major problem with water pollution from these sources is the smell from the decay of these organisms. Aerobic decay produces SO_2 and N_2 from the S and N in living systems.

The N in algae comes primarily from proteins that are derived from nitrates in fertilizer which in turn were made by nitrogen fixation to NH_3 and its oxidation to HNO_3 (the Haber and Ostwald processes). Anaerobic decay (by a different set of bacteria) converts S and N in these organisms to H_2S (the rotten egg smell) and to ammonia and amines (the dead fish smell). Thus, it is the *transients* from rapid input of nutrients into water that create a major problem in water pollution from fertilizers. The rapid bloom of algae and its decay depletes O_2 and causes unpleasant odors.

The other source of water pollution is *toxic chemicals*, of which the chemical and petroleum industries are a major source. The oil spill from the Exxon Valdez in Alaska and the methyl isocyanate release from Bhopal, India, are the most widely known examples. Here the overall problem is that chemicals are released which are not normally found in nature, and consequently plants and animals have no defenses against them.

The most notorious of these water pollutants are polychlorinated biphenyl (PCBs), dioxin (Agent Orange), and heavy metals such as Hg. Organic chlorine compounds and heavy metals do not exist significantly in nature, but they are excellent chemical intermediates and catalysts respectively in the chemical industry. PCBs were widely used as heat exchanger fluids because they are so *inert* and will not decompose or polymerize when circulated through heat exchange equipment. The inertness of PCBs is exactly their problem, however, because they are not rapidly degraded by bacteria as are most organic compounds, and thus they accumulate in sediments in lakes and rivers.

Agricultural chemicals such as herbicides, insecticides, and crop and animal growth regulators are a major chemical industry. Here a major feature of a suitable product is that it is rapidly degraded to harmless chemicals after it accomplishes its function. For example, the weed killer Roundup quickly kills all plants it contacts, but it is rapidly oxidized by air so that within hours it is harmless.

The removal of most chemical pollutants is primarily *biological*, either rather rapidly by aerobic reactions or more slowly by anaerobic reactions. Except for rapid chemical cleanup by collection, we must rely on bacterial activity to remove most organic pollutants. The most beneficial removal of oil from the Exxon Valdez spill occurred by bacterial decay, and the most effective assistance in this was by spreading N and P nutrients so the bacteria could grow faster and by spraying bacteria cultures over the water. The worst response to the oil spill was the steam cleaning of rocks on the shores because this sterilized the area of the naturally occurring bacteria and slowed the biological remediation process.

The major principles for avoiding water pollution in chemical processing in order of preference are (1) contain all chemicals, (2) avoid or minimize the use of chemicals whose spill is especially troublesome, (3) rapidly contain and remove spills, and (4) maximize the conditions for bacterial digestion of chemicals by supplying nutrients, bacterial cultures, or O_2.

13.7.2   Air pollution

Urban air pollution is a problem for cities which have mountains or other features that recirculate the air and thus trap it. The air flow in a city might be approximated a CSTR. The sources are NO_x and hydrocarbons from automobiles, power plants (the major producers of SO_2), and industries, and these mix rapidly (within a few hours) so the composition is fairly uniform.

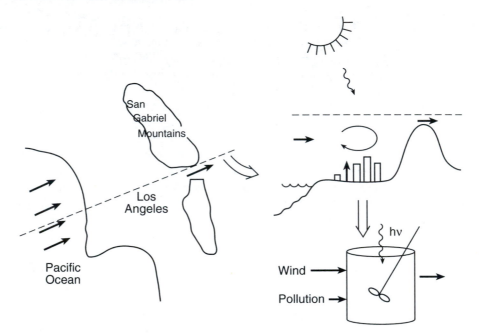

Figure 13–5   Sketch of Los Angeles and simplified CSTR model of air in the LA basin.

Sketched in Figure 13–5 are cross-sectional and top views of the Los Angeles basin. The wind blows primarily from the Pacific Ocean and is trapped by the San Gabriel mountains with narrow valleys draining the basin. The system is highly transient because automobile emissions are concentrated about morning and evening rush hours. It is also strongly affected by *photochemical reactions*, which are also time-dependent.

Smog is strongly enhanced by photochemical reactions, so this reaches a maximum in midday. The reactants of photochemical smog are primarily NO_2, olefins, and aldehydes. NO_x forms by high-temperature combustion, which effectively promotes the reaction

$$N_2 + O_2 \rightarrow 2NO \tag{13.16}$$

in the engine cylinder. [Recall the discussion of the automotive catalytic converter in Chapter 7. We also considered these reactions in Chapter 10 as combustion processes.]

This NO rapidly oxidizes in air to form NO_2:

$$NO + \tfrac{1}{2}O_2 \rightarrow NO_2 \tag{13.17}$$

which absorbs visible light and forms the brown haze that we associate with photochemical smog. Olefins, aldehydes, and ketones are other polluting products of incomplete combustion of alkanes in the engine.

NO_2 is dissociated by sunlight to form O atoms that react with O_2 to form ozone,

$$NO_2 \overset{h\nu}{\rightarrow} NO + O \tag{13.18}$$

$$O + O_2 \rightarrow O_3 \tag{13.19}$$

Olefins and aldehydes react chemically and photochemically to form peroxy compounds in reactions such as

$$C_2H_4 + O_3 \rightarrow CH_3CHO + O_2 \tag{13.20}$$

Ozone, aldehydes, and peroxy acids are strong lachrymators.

One such reaction in smog formation is the formation of the acetyl radical such as by sunlight photolysis of acetaldehyde

$$CH_3CHO + h\nu \rightarrow CH_3CO \cdot + H \cdot \tag{13.21}$$

followed by the multistep reaction of acetyl with O_2 to form peroxy acedic acid:

$$CH_3CO \cdot + O_2 \rightarrow CH_3COOOH$$

The next air pollution problem is *smog*, which is a combination of smoke and fog, which is caused when pollutants combine with water droplets. NO_2 and SO_2 are further oxidized and hydrated to acids,

$$3NO_2 + H_2O \rightarrow 2HNO_3 + NO \tag{13.22}$$

and

$$SO_2 + \tfrac{1}{2}O_2 \rightarrow SO_3 \tag{13.23}$$

$$SO_3 + H_2O \rightarrow H_2SO_4 \tag{13.24}$$

Nitric and sulfuric acids readily dissolve in water drops, lowering the vapor pressure of the water and stabilizing the smog aerosols. Ammonia from animal feedlots also dissolves in these drops, creating ammonium sulfate and ammonium nitrate.

Thus, smog is a combination of NO_2 and aerosols containing salts and acids which reduce visibility and cause acid rain. Smog also consists of O_3 and partially oxidized hydrocarbons which irritate our respiratory passages and damage plants.

These processes are both natural and manmade. In fact, the Los Angeles basin was called by the early Native American inhabitants the "land of the smokes," and salt spray from oceans is a major source of Cl in the atmosphere. In many situations, people have only exaggerated the natural chemicals and reactions which were present before we and our technology arrived. The Smoky Mountains are an example of natural smog caused by chemicals such as isoprene (the natural rubber monomer) and terpenes, which are emitted by trees.

13.7.3 The troposphere and the stratosphere

On a global scale the air layers within a few kilometers of the earth's surface are rapidly mixed by wind action. This region is called the troposphere. Natural and manmade sources of chemicals such as CH_4 and other hydrocarbons, CO, SO_x, NO_x, ozone, and chlorine are emitted into the troposphere. Most of these are removed or reacted away to form harmless products by dissolving in rain, adsorption on solids, and chemical reactions.

Above the troposphere is the stratosphere, which extends many kilometers above the earth. Chemicals slowly migrate into the stratosphere where there are few reactions to remove them. The stratosphere is also irradiated by high intensities of ultraviolet light, which

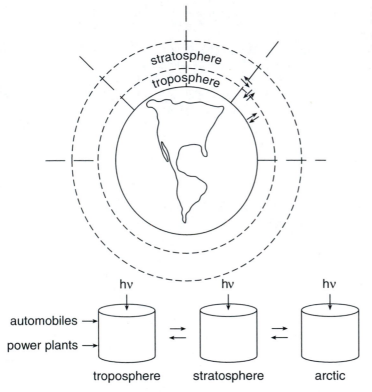

induces photochemical reactions such as production of O_3 from O_2. (In the troposphere O_3 is bad, but in the stratosphere it is good because it absorbs ultraviolet radiation from the sun and prevents it from reaching the earth's surface.)

An extremely simplified model of the global environment is shown in Figure 13–6. This is a highly lumped model of the world, but it is sometimes necessary to simplify in this way to have workable models in which to insert complex chemistry. It is usually a good approximation to assume that no mixing occurs between hemispheres, and the troposphere, stratosphere, and arctic each mix rapidly but exchange mass between regions slowly so that each region is approximately a CSTR for simple environmental modeling. Versions of these models are used in weather forecasting in which microscopic models (where the weather comes from the west in approximately plug flow but coriolis forces create vortices) and the jet stream in the lower stratosphere are major driving forces for weather in the troposphere.

13.7.4 Stratospheric reactions

Ozone is formed by dissociation of O_2 by very short wavelength solar ultraviolet light,

$$O_2 \overset{h\nu}{\to} 2O \qquad (13.25)$$

$$O + O_2 \to O_3 \qquad (13.26)$$

Chlorocarbons are very effective in destroying ozone. This occurs by a chain reaction

$$RCl \xrightarrow{h\nu} R + Cl \qquad (13.27)$$

followed by the chain reaction

$$Cl + O_3 \rightarrow ClO + O_2 \qquad (13.28)$$

$$ClO + O_3 \rightarrow Cl + 2O_2 \qquad (13.29)$$

These two reactions add up to the overall ozone destruction reaction

$$2O_3 \rightarrow 3O_2$$

This is a *chain reaction* as shown in Figure 13–7 and which was considered in detail in Chapter 10.

Chlorine is thus a very effective *catalyst* in promoting ozone decomposition without reacting away the chlorine. The primary problem is that chlorocarbons are very inert and will escape the troposphere into the stratosphere where the short-wavelength UV radiation creates Cl. The above reaction between Cl and ClO can destroy many O_3 molecules for one Cl atom. Molecules such as HCl and Cl_2 are not a problem in ozone depletion because they are rapidly dissolved in rain in the troposphere and never escape into the stratosphere, while inert chlorocarbons escape without dissolving in water. Freons have been important chemicals for refrigerants, for aerosol sprays, and for foaming agents in polymer and fire fighting foams, and their value in these applications is that they are chemically inert. There is considerable research and development in finding replacement molecules that are inert enough for these uses and yet reactive enough to not escape into the stratosphere.

We considered *chain reactions* such as that above in Chapter 10. Each Cl atom introduced into the chain can cycle for 10^4 to 10^5 times, so that 1 chlorine atom can destroy up to 10^5 ozone molecules in the stratosphere.

Thus, we see that environmental modeling involves solving *transient mass balance equations* with appropriate flow patterns and kinetics to predict the concentrations of various species versus time for specific emission patterns. The reaction chemistry and flow patterns of these systems are sufficiently complex that we must use approximate methods and use several models to try to bound the possible range of observed responses. For example, the chemical reactions consist of many homogeneous and catalytic reactions, photoassisted reactions, and adsorption and desorption on surfaces of liquids and solids. Is global warming real? (Minnesotans hope so.) How much of smog and ozone depletion are manmade? (There is considerable debate on this issue.)

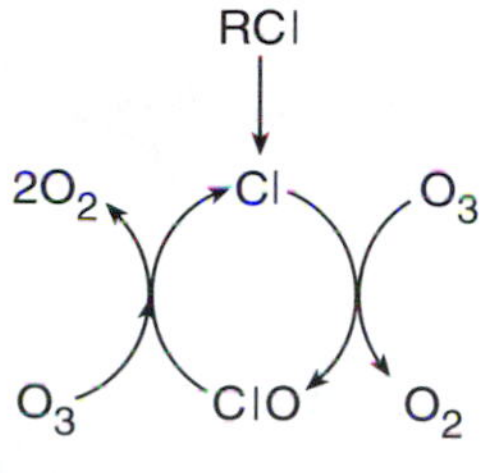

Figure 13–7   The chain reaction by which small concentrations of organic chlorine compounds produce ozone destruction in the stratosphere.

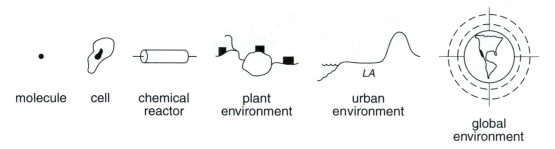

molecule　　cell　　chemical　　　plant　　　　urban　　　　
　　　　　　　　　　reactor　　environment　　environment

global
environment

Figure 13–8　Comparison of systems which can be described by the models of chemical reaction engineering. Reactions include those of molecules and living systems. Possible chemical reactors include conventional reactors as well as the environment.

The notions of different combinations of ideal reactors and residence time distributions are essential in analyzing these problems and in suggesting appropriate solutions. We summarize the many applications of chemical reaction engineering in Figure 13–8, which indicates the types of "molecules," "reactors," and "reactors" which we can handle.

13.8　ECOLOGICAL MODELING

We can also use chemical kinetics to attempt to model populations of living systems such as *single-celled organisms* or plant, animal, and human *populations*. We describe the density of individuals (species A) as C_A, which might be in individuals/volume in a three-dimensional nutrient solution or a pond, and might be individuals/area on a glass slide or on the surface of the earth. We could use moles/liter, but living organisms seldom reach molar densities (fortunately), so we just count the individuals.

Thus far, we have considered *the reaction engineering of molecules* in containers that conserve mass. The same equations describe the populations of living systems in their environment, a "container" operating in batch or flow mode with inputs, outputs, disturbances, catalysts, internal recirculation patterns, and even diffusion (swimming, walking, flying) (Figure 13–9).

We assume that an organism reproduces by a first-order process

$$A \rightarrow 2A, \qquad r_b = kC_A \tag{13.30}$$

so in a batch process the rate of growth would be

$$\frac{dC_A}{dt} = +k_b C_A \tag{13.31}$$

which integrates to yield

$$C_A(t) = C_{Ao} e^{+k_b t} \tag{13.32}$$

Note the $+$ sign in the differential equation ($\nu_A = +1$) and in the exponent. We *never* observe these kinetics in "ordinary" reaction systems. Population growth is a highly *autocatalytic* process in that organisms "react" with each other to reproduce more organisms with a first-order birth rate coefficient k_b.

Figure 13–9   Comparison of "reactions" of molecules, single-celled organisms, and more complex organisms. We can regard reproduction as a reaction such as those listed in the figure for chemicals and for living organisms.

MOLECULES

$$A \longrightarrow B + C$$

CELLS

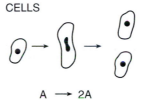

$$A \longrightarrow 2A$$

ANIMALS

$$F + M \longrightarrow F + M + nC$$

13.8.1   Continuous reactors

The steady-state solutions for these reactions in continuous reactors have interesting characteristics. In a PFTR the solution is simply the solution for the batch reactor,

$$C_A(t) = C_{Ao}e^{+k_b t} \tag{13.33}$$

However, for the CSTR (called a chemostat by biologists) the mass balance equation is

$$C_{Ao} - C_A = -\tau k_b C_A \tag{13.34}$$

which looks like the equation we have seen many times for a first-order reaction in a CSTR *except for the minus sign*. Solving for C_A, we obtain

$$C_A(\tau) = \frac{C_{Ao}}{1 - k_b \tau} \tag{13.35}$$

Clearly, this problem is different than we have encountered previously because $C_A \rightarrow \infty$ when $k_b \tau = 1$. These $C_A(\tau)$ are plotted in Figure 13–10.

The solution to these kinetics is clearly unstable because it predicts unlimited exponential growth. Therefore, we need to assume that there is some sort of a "death reaction." One form of this might be

$$A + A \rightarrow A, \qquad r_d = k_d C_A^2 \tag{13.36}$$

A simple explanation of these kinetics might be that the species is cannibalistic and that when two A organism encounter each other, one is eaten or at least killed. A more benign interpretation of this rate expression is that food competition limits populations.

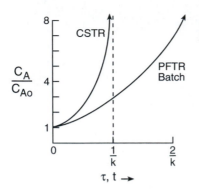

Figure 13–10   Plot of $C_A(\tau)$ for the reaction $A \rightarrow 2C$, $r = k_b C_A$ in a PFTR and in a CSTR. Reaction is always faster in a CSTR, and the concentration is predicted to go to infinity when $k_b \tau = 1$.

If we assume both reactions, we obtain in a batch reactor

$$\frac{dC_A}{dt} = +k_b C_A - k_d C_A^2 \tag{13.37}$$

We can integrate this for given C_{Ao} to obtain $C_A(t)$. However, we can simply look at the long time solution, $t \rightarrow \infty$. If we set $dC_A/dt = 0$, we obtain

$$k_b C_A = k_d C_A^2 \tag{13.38}$$

or, solving for C_A, we obtain

$$C_{AS} = \frac{k_b}{k_d} \tag{13.39}$$

which is the steady-state solution for the population of a species that obeys these kinetics even in a batch reactor. This looks like the equilibrium long time solution to reversible reactions, but *there is no equilibrium in living systems*.

There is clearly something strange about this "reaction system," which we never encounter when we deal with molecules. This species can reproduce itself, and therefore it *does not obey mass conservation*. While we assumed a batch reactor, the species needs nutrients to grow and reproduce, and living systems have no interesting solutions (or life) unless they are *open* to their surroundings. However, remember that crude oil comes ultimately from sunlight, so biological reactions are involved as both the source of our chemicals and energy as well as in their ultimate disposal back to CO_2 and H_2O.

13.8.2   Kinetics of living organisms

While a kinetic analysis may provide a first approximation to population densities, we clearly need to modify our kinetic descriptions. First, the age distribution of individuals must be considered—at least as juveniles, adults, and aged—because the reproduction and death kinetics of living systems depend on the individual's age. For humans and for valves this *age distribution* might look as shown in Figure 13–11.

From installation or birth, the valve or human has a probability of surviving which is the integral of the lifetime probability which we can call $p(t)$,

$$\int_0^t p(t)\, dt \qquad\qquad (13.40)$$

as defined and used in Chapter 8.

The average of this function is the mean residence time, the replacement time of a valve, or the life expectation of a human. In this picture the world is our "reactor" that we enter and exit at appropriate times.

In order to describe reproduction and death rates of living organisms, we need to consider the "rate constants" to be functions of the age of the individual, as shown in Figure 13–12.

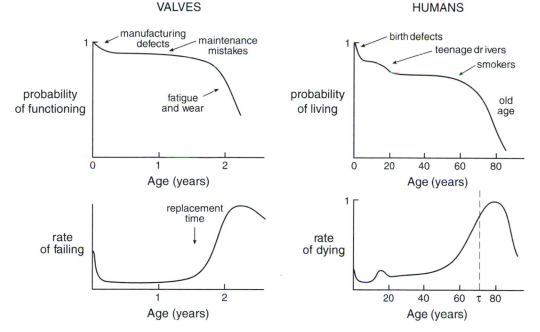

Figure 13–11 Age distribution of valves which might be installed in a chemical plant and in a living organism such as humans.

Figure 13–12 Effective "rate constants" of reproduction k_b and death k_d of living organisms.

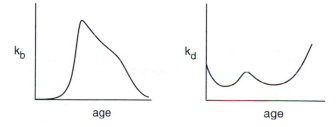

Example 13–1   Consider the rabbit population on an isolated island.

(a) The rabbits eat grass, and they reproduce at a rate proportional to the supply of grass and die of natural causes. The grass grows at a rate proportional to the solar energy density (concentration C_S). Formulate reasonable rate expressions for the rabbit population C_R and grass C_G, assuming these kinetics. What are the steady-state populations?

The equations for rabbits and grass are

$$\frac{dC_R}{dt} = k_{bR} C_G C_R - k_{dR} C_R \tag{13.41}$$

$$\frac{dC_G}{dt} = k_{bG} C_S - k_{dG} C_G C_R \tag{13.42}$$

(b) Suppose that foxes are introduced onto the island at concentration C_{Fo}, and they reproduce at a rate proportional to their food supply and die of natural causes. Add C_F to this ecosystem, and examine the solution.

When foxes are added, the equations become

$$\frac{dC_G}{dt} = k_{bG} C_S - k_{dG} C_G C_R \tag{13.43}$$

$$\frac{dC_R}{dt} = k_{bR} C_G C_R - k_{dR} C_R C_F \tag{13.44}$$

$$\frac{dC_F}{dt} = k_{bF} C_F C_R - k_{dF} C_F \tag{13.45}$$

It is preferable to be at the top of the food chain.

13.8.3 Periodic reproduction

Many species reproduce only at certain times of the year, and their populations are then described by *difference equations* (designated by $C_{A,t}$ and $C_{A,t+1}$) rather than differential equations (designated by dC_A/dt). These systems are in fact more unstable than differential equations, and long-term oscillations and chaotic population densities are easily predicted by the difference equations,

$$C_{A,t+1} = k_b C_{A,t} - k_d C_{A,t}^2 \tag{13.46}$$

In this formula, k_b is the birthrate and k_d is an overcrowding factor, which causes a decrease in population size anytime the species gets too populous for its resources. The formula can be made simpler by choosing the units of population in the most convenient way. With the right choice of units, the model can be simplified to read

$$C_{A,t+1} = k_b \left(CA, t - C_{A,t}^2 \right) \tag{13.47}$$

One of the interesting features of difference equations such as these is that they are much more unstable than the differential equations we have used throughout this text. If the reproduction rate exceeds a certain value, the system is predicted to exhibit population *oscillations* with high and low populations in alternate years. At higher birthrates, the system exhibits *period doubling*, and at even higher birthrates, the system exhibits *chaos* in which the population in any year cannot be predicted from the population in a previous year. Similar phenomena are observed in real ecological systems, and even these simple models can capture this behavior.

A homework problem considers these equations and their solutions.

13.9   SUMMARY

In this chapter we have tried to summarize some of the issues involved in the chemical reactions that occur in the environment. This is an extremely important subject for all chemical engineers, no matter what their opinions on environmental considerations, because these aspects must be considered in all chemical processes. Your job and the future of chemical engineering depend on intelligent consideration of environmental issues.

The basic principles of chemical reaction engineering are the same in chemical synthesis and in environmental situations, and in this chapter we have mostly noted some of these differences in emphasis.

Finally, we chemical engineers have the opportunity to accomplish more for the environment than almost any other discipline because only we understand the issues in sufficient detail to make rational, quantitative, and creative decisions on how changes should be made to protect and improve our environment.

PROBLEMS

13.1 Answer the following simple but important questions.

(a) The population of the world is increasing at 2% per year. At this rate, what will be the world's population when your children are your age? Your grandchildren?

(b) There are 30 million species of plants and animals in the world, and it is estimated that they are going extinct at a rate of 2% per year. At this rate, how many species will there be when your children are your age? Your grandchildren?

(c) If the economy continues to grow at 3% annually, what will be your salary when you retire, assuming you have the same job? Salaries are now growing at 5% per year in biotechnology and 3% per year in the chemical industry. Estimate your salary at retirement in these positions.

(d) Why do the world's petroleum reserves tend to decrease with approximately zeroth order kinetics? Why will it revert ultimately revert to first order as the supply decreases?

13.2 Write a half page commentary on each of the following statements. Any serious response from strong approval to outrage will be acceptable.

(a) The United States has 5% of the world's population but consumes 30% of its energy reserves and emits 30% of its CO_2.

(b) Approximately 95% of our energy is now derived from nonrenewable resources.

(c) The major energy source of this country in 20 years will still be petroleum.

(d) The solution to pollution is dilution.

(e) The world produces enough food to feed itself for the foreseeable future. The problem is transportation and the inability of much of the world to buy food.

(f) "The secret to pollution is dilution."

13.3 Estimate the amount of CO_2 you will produce every time you fill your gasoline tank. Assume 15 gallons of octane with a density of 8 lb/gallon. How much CO_2 does your car liberate into the atmosphere each year?

13.4 A chemical plant discharges a reactive chemical at concentration C_{Ao} into a river that is flowing in plug flow with constant velocity u. The chemical decomposes with first-order kinetics with a rate constant k.

(a) Find an expression for the steady-state concentration of the chemical in a town a distance L downstream for the following a *steady leak*.

(b) What equation must you solve for the concentration versus time for a steady leak that starts at $t = 0$?

(c) Sketch solutions versus time for the concentration versus time for a *pulse leak* assuming (1) plug flow, (2) laminar flow, and (3) plug flow with dispersion.

13.5 A sewage treatment plant consists of several tanks in series with inlet and outlet pipes at opposite sides. The liquid and sludge are stirred by a weir that rotates slowly to keep solids suspended and to incorporate oxygen.

(a) Sketch $p(t)$ if rotation is too slow for complete CSTR mixing. Indicate the time interval between passage of the weir blades and the residence time.

(b) Suppose there are blades that divide the tank into 10 nearly isolated compartments and that the rotation time is much shorter than the residence time in each compartment. Sketch $p(t)$ for this reactor.

13.6 Your colleague at the large company you are working for rushes into your office during lunch and announces that he has discovered a process to make synthetic fructose. He claims that his process makes it for only 10% of the cost of producing it from sugar beets or sugar cane! He admits, however, that his process makes a racemic mixture. There are six chiral centers of natural glucose, and only one has nutritional value or taste.

(a) He wants you to go in with him in commercializing this product since he invented it on his own time, and therefore you and he can get a patent. Should you quit your day job or just finish your sandwich and return to work?

(b) Your colleague uses 1 teaspoon of sugar in his coffee. How much of his product would he have to use?

(c) He comes in later to say that he has a process which makes 80% of the desired isomer of each chiral center. Now how many teaspoons does he need for each teaspoon of sugar?

(d) The next week he says that he just tasted a product he was making (a highly unapproved practice) and found that it tasted just like sugar but had no nutritional value. Ignoring his poor track record, might this product be interesting?

13.7 You have just been hired as the chemical engineer for a small soup company. They have just found that their cans of Cream of Spinach soup have become contaminated with a bacteria whose Latin name is *Retchiosis barfitosum*, which fortunately is not lethal but causes serious discomfort. The cans had been sterilized by heating for 1 h in an oven at 200°F, but this procedure apparently was not adequate. We need to sterilize the prepared soup cans to kill the bacteria, which will multiply on the shelf. Bacteria concentrations are determined by coating a known amount of the material on a Petri dish and culturing it so that each bacterium forms a colony that can be counted on the dish.

(a) What kinetics do you suppose should describe the thermal destruction of a single cell organism?

(b) Your predecessor, a careful but not very imaginative engineer, made the following culture counts:

0.1 cm^3 of original soup sample	253 colonies
0.01 cm^3 heated for 1 h at 230°F	86 colonies
0.002 cm^3 heated for 1 h at 250°F	38 colonies

Find the original number of bacteria in the sample.

(c) Find a reasonable rate expression for sterilization.

(d) We need a bacteria count of 1 bacterium per million cans of soup. However, the soup tastes like Cream of Grass Clippings if heated over 280°F. How should the soup cans be sterilized?

(e) Discuss the statistical problems of sterilization and suggest some inexpensive remedies in assuring quality control.

[Note: All food processing companies live in mortal fear of such incidents and take great care to minimize such problems.]

13.8 You have just been hired as the chemical engineer for a small company which has been recently been making excellent profits on a very old process that was begun at the start of World War I to make 10 gallons/day of product but has been retrofit and debottlenecked to product 250 gallons/day at present. You are asked to determine the safety of the very old equipment and make recommendations. You discover from old records that they had a very serious explosion in 1917 in which five people were killed. Evidently the 50 gallons of butane solvent you still use leaked completely from the reactor and dispersed in air, causing a VCE. Examination of the records indicates that there was no wind that day and the county coroner's office indicated that all casualties lived less than 1/8 mile from the site. Examination of the census records indicates that the population density of the city is the same as it was in 1917.

(a) How many deaths would result today if the VCE forms a hemisphere, there is no wind to spread the cloud downstream, and the butane concentration at the time of the explosion is the same now as it was then?

(b) Your boss decides to take the risk and make no changes. However, you note that he lives in a valley in a wealthy suburb located 5 miles to the east of the plant. What should be your next calculation and presentation?

13.9 Cell culture growth by cell division can be modeled by the "reaction"

$$A + B \rightarrow 2A, \qquad r = kC_A C_B$$

where A is the cell density in moles/liter and B is the nutrient concentration.

(a) In a PFTR find $C_A(\tau)$ for $C_A = C_{Ao}$ assuming that C_{Bo} is large enough that the nutrient is not depleted.

(b) Repeat for a CSTR. What happens when $\tau \rightarrow 1/kC_{Bo}$? Why is the solution very different in a CSTR?

13.10 A more realistic model of cell growth assumes a rate expression

$$r = k_b C_A - k_d C_A^2$$

(a) What is a reasonable "mechanism" for the reverse reaction?

(b) Formulate expressions for C_A in batch reactor, PFTR, and CSTR.

(c) What happens to these solutions as $\tau \to \infty$? This appears to be very strange because the concentration (cell population) approaches a constant (in fact the same) value in any reactor. What is the significance of this long-time behavior? Why is this different from normal reactions that conserve mass?

13.11 On an isolated island the foxes and rabbits exist in a prey–predator relationship. Foxes reproduce at a rate proportional to the food supply, and foxes die a natural death with first-order kinetics. Rabbits reproduce with first-order kinetics, and they die by being eaten by foxes.

(a) Formulate expressions for each species density (in animals/area because this is a surface reaction problem if we ignore leaping and burrowing) for these kinetics.

(b) What are the steady-state populations in terms of the four rate coefficients?

(c) Rabbits and foxes occasionally arrive on the island by swimming from the mainland. Formulate expressions for the population densities on the island for this situation. Why does migration stabilize this ecosystem?

(d) Suppose the rabbits all die suddenly from an epidemic. Describe the fox population versus time following this extinction.

(e) Suppose an island initially has no animal life until a pregnant rabbit swims ashore. Describe the rabbit population versus time, ignoring effects of inbreeding.

13.12 A simpler model of fox and rabbit populations might be

$$r_F = k_{bF} C_R - k_{dF} C_F$$

$$r_R = k_{bR} C_R - k_{dR} C_F$$

(a) Formulate a "mechanism" that might give these "kinetics."

(b) How would you solve these equations in a batch reactor for given initial conditions? Why do you expect these equations to predict unstable population densities?

(c) Solve these equations in a CSTR and discuss the stability of the solutions.

(d) Why can equations such as these sometimes predict instabilities where our conventional kinetics are usually stable?

13.13 In a pond the frogs reproduce by natural methods and are eaten by bass, who die of natural causes.

(a) Formulate expressions for C_F and C_B.

(b) How should these equations be modified to account for fishermen who do not practice catch-and-release?

13.14 Chlorine and ClO cause ozone destruction in the stratosphere by catalyzing the reaction

$$2O_3 \to 3O_2$$

as discussed in this chapter. NO in the stratosphere can do similar damage. Sketch the reactions and show how they lead to a catalytic cycle.

13.15 A 10-μm-diameter water drop is blown off the Pacific Ocean into Los Angeles with a wind velocity of 5 mph from the west. The drop flows through the basin where the SO_3 is 0.01 ppm and the NH_3 is 0.02 ppm.

(a) Calculate the concentration of $(NH_4)_2SO_4$ in the water drop by the time the drop reaches the desert which is 10 miles from the shore, assuming that the absorption of both species is mass transfer-limited with $D_A = 0.1$ cm^2/sec. Assume that the drop diameter does not change as the solute dissolves. [Answer: $\sim$4 moles/liter.]

(b) When the particle reaches the desert, the pollutants disappear and the drop evaporates in the dry air to form solid $(NH_4)_2SO_4$ which then settles in Death Valley. What is the diameter of the solid particle? [Answer: $\sim$8 μm.]

(c) Set up a model to calculate the time required for complete drying of the particle. You will need to use the ideas in Chapter 9 and know the vapor pressure of ionic solutions and the solubility of the salt. If you are ambitious or your instructor is tough, estimate this time.

13.16 A Simple Model of Chaos. The following model from population biology provides probably the simplest and best known model of how chaos develops. The model is for a single species with generations that do not interact with each other; a hypothetical insect population, for instance, that hatches in the spring lives through the summer, and then it dies after it lays its eggs in the fall. One can count the number of insects alive at one particular time each year and get a yearly record: N_1 insects the first year, N_2 the second, and so on. The number of insects alive one summer N_t will determine the number alive the following summer, N_{t+1}. The exact relationship between populations in succeeding years is hard to determine, but the simplest mathematical formula that seems somewhat close to reality is

$$N_{t+1} = A * N_t - B * N_t^2$$

In this formula, A corresponds to the birthrate and B is an overcrowding factor, which causes a decrease in population size anytime the species gets too populous for its resources. The formula can be made simpler by choosing the units of population in the most convenient way. With the right choice of units, A is equal to B, and the model simplifies to read

$$N_{t+1} = A * \left(N_t - N_t^2\right)$$

For this choice of units, one must have $0 < N_t < 1$ or N_{t+1} will become negative, which implies that the population becomes extinct. [Taken from *Science* **24**, January 20, 1989.]

The following Basic program solves this equation and plots $N(t)$ on a Mac or PC:

```
CLS
T=0:N0=0:T0=0
N=.01
A=3.58
10 N=A*(N-N*N)
LINE (5*T0,250*(1-N0))-(5*T,250*(1-N))
T0=T:N0=N
T=T+1
GOTO 10
```

(a) Solve this equation for different values of the growth constant A. You will find very different characteristics near $A = 3$, 3.4, 3.57, and 4. Describe the behavior.

(b) How does $N(t)$ depend on N_0 after long times for different values of A? Show several screen prints to illustrate selected behaviors.

13.17 A certain animal on an island reproduces at a rate coefficient of 4/year and dies by second-order kinetics with a rate coefficient of 0.2 miles²/animal year.

(a) What is the steady-state density?

(b) If the initial population is 2 animals/mile², what equation describes $C_A(t)$ assuming batch conditions?

(c) Single-celled organisms are to be grown in a steady-state chemostat (a CSTR). If the process is $A \rightarrow 2A$, $r = kC_A$ with $C_{Ao} = 0$, find the condition for steady state with finite concentration.

(d) Cans of vegetables are to be sterilized by heating in an autoclave. If there are 10^6 bacteria initially and 10^3 after heating for 10 min, how long must the can be heated under these conditions to have a probability of 10^{-6} of finding a single bacterium in a can?

13.18 We need to calculate the concentration of a pollutant A following a spill into a river which flows by our chemical plant at 10^5 gal/h at 5 mph. There is a waterfall just upstream, so the water is saturated with air at $C_{O_2} = 1.5 \times 10^{-3}$ moles/liter. The river bottom is rocky so that the flow is well mixed radially. This chemical is oxidized by dissolved O_2 with a rate

$$r = kC_A C_{O_2}$$

with $k = 2000$ liters/mole h and r in moles/liter h.

(a) Consider a steady leak of 10 gal/h at $C_A = 0.1$ moles/liter. What is C_{Ao} in the river after mixing?

(b) If trout are killed if $C_A > 10^{-6}$ moles/liter, estimate the downstream distance at which the trout be killed. [You don't need to solve a second-order equation.]

(c) Suppose that the leak is 100 gal/h of 10 molar A. What is C_{Ao} in the river after mixing?

(d) If carp are not harmed by A but they die if $C_{O_2} < 10^{-4}$ moles/liter, estimate the downstream distance at which the carp be killed for the spill in part (c), assuming no replenishment of the O_2. [Again, you do not need to solve a second-order equation.]

13.19 Our chemical plant discharges a chemical (A, of course) at a steady-state concentration of 100 ppm into a creek from which we take drinking water 500 feet downstream. A decomposes to harmless products with a rate coefficient of 2.0 h^{-1}.

(a) Estimate the steady-state concentration of A at the drinking water intake if the velocity of the creek is 100 ft/h.

(b) How might you estimate the concentration more accurately?

(c) The creek drains into a small well-mixed pond just below the water intake. The creek flows at 1000 gals/h and the pond has a volume of 10,000 gals. Estimate the steady-state concentration of A in the pond.

Chapter 14

MULTIPHASE REACTORS

14.1 TYPES OF MULTIPHASE REACTORS

W
e finally arrive at the type of reactor usually encountered in industrial practice: the multiphase reactor. Obviously, a reactor has multiple phases whenever its contents do not form a single-phase solution. For two phases these may be gas–liquid, liquid–liquid, gas–solid, and liquid–solid. Listed in Table 14–1 are the names of some of the common important multiple-phase reactors.

We have encountered many of these reactor types in previous chapters. In Chapter 2 all the reactors in the petroleum refinery were seen to be multiphase, and we will close this chapter by returning to the reactors of the petroleum refinery to see if we can now understand how they operate in a bit more detail.

Catalytic reactors are multiphase if solid catalysts are used as discussed in Chapter 7. Reactions that form or decompose solids were discussed in Chapter 9, and many polymerization reactors are multiphase, as discussed in Chapter 11. Biological reactions usually take place in multiphase reactors.

In this chapter we summarize the characteristics of these reactors and discuss the features that differ from single-phase systems. There are several obvious distinguishing features of multiphase reactors.

1. Essential mass transfer steps between phases always accompany reaction steps, and these frequently control the overall rate of the chemical reactions.
2. Mass- and energy-balance equations must be written and solved for each species in each phase.
3. Gravity is frequently important in controlling flow patterns, as lighter phases tend to rise and denser phases tend to fall in the reactor.
4. Mixing within phases and mixing between phases frequently has a dominant effect.
5. Solubilities and phase distributions of species between phases require careful application of principles of thermodynamic phase equilibrium.

TABLE 14–1
Types of Multiphase Reactors

Gas–Liquid and Liquid–Liquid Reactors
- spray tower
- bubble column
- sparger
- oil burner
- emulsion reactor
- membrane reactor
- catalytic distillation column

Gas–Solid and Liquid–Solid
- slurry
- packed bed
- moving bed
- fluidized bed
- riser
- circulating FB
- CVD
- spin coater
- fiber spinner
- crystallizer
- conveyer
- kiln
- ore refiner
- coal or wood burner
- injection molding
- reaction injection molding
- membrane reactor
- chromatographic reactor

Three Phase
- multiphase solid
- trickle bed
- gas–liquid slurry with solid flow
- membrane reactor

We might first ask why one should use multiphase reactors at all because they obviously introduce additional complexities compared to single-phase reactors. First, we have no option whenever we need to react gases, liquids, and solids with each other. For example, whenever we want to react a liquid organic molecule with H_2 or O_2, we have to find a way to introduce these gases into the liquid. Most solids processes require reacting with gases or liquids, and products may also be gases or liquids.

The second reason for using multiphase reactors is that we may want to use multiple phases (a) to attain conversions higher than attainable from equilibrium constraints in a single-phase process and (b) to improve mixing within in the reactor by allowing gravity to cause flows of phases relative to each other.

14.2   MASS TRANSFER REACTORS

Multiphase reactors are basically mass transfer reactors. Students have been introduced to these reactors in courses on separation processes, but in those courses they were called "separation units" rather than "chemical reactors." Thus essentially all the reactors listed have been described and their mass balances have been derived in courses on mass transfer and separations. We now "simply" add chemical reactions to these separation units to describe multiphase reactors.

Multiphase reactors can also "look" very similar to separation units, but there are important differences or additions. First, reactions are frequently catalyzed by solids; so the solid walls or structures in the reactor are often catalysts whose activity and stability are important as well as the mass transfer enhancements they provide.

14.2.1   Heat generation and absorption

A second important complication in chemical reactors compared to conventional mass transfer equipment is that the heat generation or absorption caused by chemical reaction can be much larger than the heat liberated or absorbed in phase changes. The heat of vaporization of most liquids is only a few kilocalories/mole, while heats of reaction are frequently much larger than this. Recall also that chemical reaction processes frequently involve multiple reactions with different ΔH_R, and we must design for the possibility that the heat release increases dramatically if the selectivity unexpectedly switches to a more exothermic reaction. This means that the heat transfer design of a multiphase reactor must be considered much more carefully with chemical reactions occurring, and designs must be much more conservative to assure stable operation. Heat release always stabilizes a mass transfer process because all mass transfer operations are endothermic and require process heat addition. Chemical reactors may be either exothermic or endothermic, and heat effects can frequently destabilize a multiphase chemical process.

This chapter will be a brief summary of how multiphase reactors should be considered and how the material from separations should be incorporated into reactor design. If one were actually designing a multiphase reactor, the notions from separations are in fact probably more useful than the notions of chemical reactions in developing a workable chemical reactor. However, we would argue (being somewhat prejudiced) that separation processes only become really interesting when chemical reactions are involved.

14.3   MASS BALANCE EQUATIONS

In a single-phase reactor we have to solve a mass-balance equation for each reaction. In a multiphase reactor we have to solve one mass balance equation for each reaction in each phase so for R reactions in P phases we would have to solve $R \times P$ simultaneous mass-balance equations.

This seems like a formidable problem, but we note immediately that we can describe these situations simply as two or three of our old single-phase reactors coupled together (Figure 14–1). Here the superscripts α, β, and γ are the designation of the phase, and $A^{\alpha\beta}$ is the interfacial area between phases. We use superscripts to designate phases in this chapter because we have run out of space for additional subscripts. Superscripts α and β (or g, ℓ, and s for gas, liquid, and solid) in this chapter do not signify exponents.

Chemical species can transfer between phases, and this represents the coupling between the mass-balance equations. This geometry looks like a membrane reactor in which

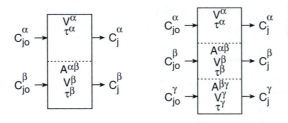

Figure 14–1 Sketch of two- and three-phase reactors. Variables must be specified in each phase α, β, and γ.

a permeable area $A^{\alpha\beta}$ (dashed lines) separates the phases, but all multiphase reactors can be described by this notation.

As before, we consider only two ideal continuous reactors: the PFTR and the CSTR, because any other reactor involves detailed consideration of the fluid mechanics. For a phase α the mass balance if the fluid is unmixed is

$$u^\alpha \frac{dC_j^\alpha}{dz} = \sum v_{ij} r_i + \text{[gain or loss to other phases]} \tag{14.1}$$

If the fluid is mixed, the transient CSTR equation becomes

$$V^\alpha \frac{dC_j^\alpha}{dt} = v^\alpha (C_j^\alpha - C_{jo}^\alpha) + V^\alpha \sum v_{ij} r_i + \text{[gain or loss to other phases]} \tag{14.2}$$

where the superscript α indicates the phase. In steady state the CSTR equation becomes

$$0 = v^\alpha (C_j^\alpha - C_{jo}^\alpha) + V^\alpha \sum v_{ij} r_i + \text{[gain or loss to other phases]} \tag{14.3}$$

We can similarly define the residence time τ^α in phase α as

$$\tau^\alpha = \frac{V^\alpha}{v^\alpha} \tag{14.4}$$

and in phase β

$$\tau^\beta = \frac{V^\beta}{v^\beta} \tag{14.5}$$

For the PFTR this also becomes

$$\tau^\alpha = \frac{V^\alpha}{v^\alpha} = \frac{L^\alpha}{u^\alpha} \tag{14.6}$$

with constant pipe diameter and fluid density. In a multiphase reactor there are distinct v, V, u, and L that must be specified in each phase.

These equations are only valid if there is no density change in the reactor, because otherwise $v_o \neq v$ and C_j is not an appropriate variable. With mass transfer to and from a phase, one expects the volumetric flow rates in and out of the phase to not be identical. For this situation one must write the equations in terms of the molar flow rates F_j in and out of the reactor. However, we promised that we would use only concentration as the variable, and we will therefore use only the preceding equations in this chapter.

14.3.1 Flow configurations

There are therefore two alternate types of equations in each phase, and for two phases there are three combinations: mixed–mixed, mixed–unmixed, and unmixed–unmixed. If both phases are unmixed, cocurrent and countercurrent give distinctly different behaviors. These combinations are shown in Figure 14–2.

The first situation involves two algebraic equations, the second involves an algebraic equation (the mixed phase) and a first-order ordinary differential equation (the unmixed phase), and the third situation involves two coupled differential equations. Countercurrent flow is in fact more complicated than cocurrent flow because it involves a two-point boundary-value problem, which we will not consider here.

For any more complex flow pattern we must solve the fluid mechanics to describe the fluid flow in each phase, along with the mass balances. The cases where we can still attempt to find descriptions are the nonideal reactor models considered previously in Chapter 8, where laminar flow, a series of CSTRs, a recycle TR, and dispersion in a TR allow us to modify the ideal mass-balance equations.

14.3.2 Reaction only in one phase

In most multiphase reactors the reactions occur only in one phase, with the other phase serving to

1. supply a reactant,
2. remove a product, or
3. provide a catalyst.

The catalytic reactor is an example where reaction occurs only at the boundary with a solid phase, but, as long as the solid remains in the reactor and does not change, we did not need to write separate mass balances for the solid phase because its residence time τ_s is infinite. In a moving bed catalytic reactor or in a slurry or fluidized bed catalytic reactor where the catalyst phase is continuously added and removed from the reactor, we actually have a multiphase reactor where the reaction occurs on the catalyst surface.

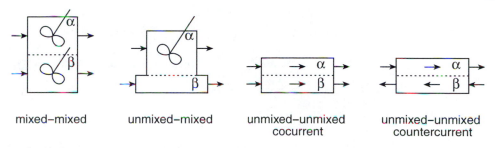

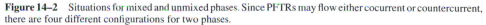

Figure 14–2 Situations for mixed and unmixed phases. Since PFTRs may flow either cocurrent or countercurrent, there are four different configurations for two phases.

14.4 INTERFACIAL SURFACE AREA

Thus we see that these multiphase reactors can be quite complex because the volumes of each phase can change as reaction proceeds and, more important, the interfacial area between phases $A^{\alpha\beta}$ can change, depending on conversion, geometry, and flow conditions. Thus it is essential to describe the mass transfer rate and the interfacial area in these reactors in order to describe their performance.

It is evident that in many situations the reaction rate will be directly proportional to the surface area between phases whenever mass transfer limits reaction rates. In some situations we provide a fixed area by using solid particles of a given size or by membrane reactors in which a fixed wall separates phases from each other. Here we distinguish planar walls and parallel sheets of solid membranes, tubes and tube bundles, and spherical solid or liquid membranes. These are three-, two-, and one-dimensional phase boundaries, respectively.

In some multiphase reactors, stirring with an impeller or the flow pattern caused by gravity will control the interfacial area. By suitably designing and positioning propellers and reactant injection orifices or by using static mixers, it is possible to provide very efficient breakup of liquids into drops and bubbles. A factor of two decrease in drop or bubble size means a factor of four increase in interfacial area.

In some cases one does not want drops or particles to be too small, however, because emulsions, foams, and suspensions can be difficult to separate into distinct phases. Misting and entrainment can be problems with gas–liquid systems, but separation of liquid–liquid phases can be much more difficult because of the possibility of stable emulsions.

Surface tension can be very important in determining drop and bubble sizes and shapes. This ultimately controls the size of drops and the breakup of films and drops. The presence of surface active agents that alter the interfacial tension between phases can have enormous influences in multiphase reactors, as does the surface tension of solids and the wetting between solids and liquids.

14.5 MASS TRANSFER BETWEEN PHASES

In the preceding sketches the reactors are coupled through the interface between them. In a single-phase system the only ways that a species could change were by flow in entrance and exit pipes and by reaction. Now we also have the possibility of a species entering or leaving the reactor by transferring between phases within the reactor.

14.5.1 Mass transfer to or from a mixed phase

For the CSTR the mass transfer term for phase α in contact with phase β can be written as

$$V^\alpha \frac{dC_j^\alpha}{dt} = v^\alpha (C_j^\alpha - C_{jo}^\alpha) + V^\alpha \sum v_{ij} r_i + A^{\alpha\beta} k_m (C_j^\beta - C_j^\alpha) \tag{14.7}$$

that is, we add the term

$$\text{mass transfer rate of species } j = \text{area} \times \text{flux} = A^{\alpha\beta} \times J_j = A^{\alpha\beta} k_m (C_j^\beta - C_j^\alpha)$$

to the standard mass-balance equation. The dimension of each term in the preceding equation is moles/time, and

$$J_j = k_m(C_j^\beta - C_j^\alpha) \tag{14.8}$$

is simply the flux (in moles/time area) of species j from phase β to phase α. The flux J_j of species j will be out of phase α into phase β if $C_j^\beta < C_j^\alpha$. If these are liquid phases, the equilibrium concentrations are not equal, and we will discuss this modification in the next section.

We can consider mass transfer to or from a phase as equivalent to a surface reaction at the boundary between phases,

$$v_j r'' = J_j = k_m(C_j^\alpha - C_j^\beta) \tag{14.9}$$

Thus we can regard mass transfer as the "reversible surface reaction"

$$A_j^\alpha \rightleftarrows A_j^\beta \tag{14.10}$$

with a "surface rate coefficient" k_m. The "stoichiometric coefficient" of species j is -1 for the phase from which it is being removed and $+1$ for the phase into which it is moving. Thus we can write mass transfer as equivalent to a "pseudohomogeneous rate"

$$r = \left(\frac{A}{V}\right)r'' = \frac{A^{\alpha\beta}}{V^\alpha}r'' = \frac{A^{\alpha\beta}}{V^\alpha}k_m(C_j^\beta - C_j^\alpha) \tag{14.11}$$

which we have to insert into the previous expressions with (A/V) adapted from our discussion of catalytic reactions in Chapter 7.

Therefore the steady-state mass balance for a reactant A in a mixed phase α becomes

$$\boxed{C_{Ao}^\alpha - C_A^\alpha = \tau^\alpha r + \tau^\alpha \frac{A^{\alpha\beta}}{V^\alpha}k_m(C_A^\beta - C_A^\alpha)} \tag{14.12}$$

which looks exactly like the CSTR equation with a mass transfer "reaction" term added.

This shows how catalytic reactions compare with other interfacial reactions. In a fixed bed reactor the catalyst (in phase β) has an infinite residence time, which can be ignored in the expressions we derived in previous chapters. For a moving bed reactor in which catalyst moves through the reactor, we have a true multiphase reactor because the residence time of the catalyst phase is not infinite.

14.5.2 Mass transfer to or from an unmixed phase

For the PFTR the mass transfer area is simply the total interfacial area between the phase in question and the other fluid, while for the PFTR, where C_j is a function of position, [area/volume] is the area per unit volume of that phase at position z. The steady-state mass balance is therefore

$$u^\alpha \frac{dC_j^\alpha}{dz} = \sum v_{ij}r_i + [\text{area/volume}]\, k_m(C_j^\beta - C_j^\alpha) \tag{14.13}$$

where area per volume is simply the perimeter of the "tube" through which the fluid flows in this phase divided by its cross-sectional area.

For reactant A in phase α this equation becomes

$$u^\alpha \frac{dC_A^\alpha}{dz} = -r - \frac{[\text{area}]}{[\text{volume}]} k_\text{m} (C_A^\beta - C_A^\alpha) \qquad (14.14)$$

The area per unit volume in this expression may not be simply the total area between phases $A^{\alpha\beta}$ and volume V^α of the phase, but rather the differential area per unit volume, which may be more complicated than this ratio.

14.6 MULTIPHASE REACTOR EQUATIONS

The preceding equations must be solved to determine the performance of any multiphase reactor. We repeat them as

$$(C_{Ao}^\alpha - C_A^\alpha) = \tau^\alpha r + \tau^\alpha \frac{A^{\alpha\beta}}{V^\alpha} k_\text{m} (C_A^\beta - C_A^\alpha) \qquad (14.15)$$

$$\uparrow \qquad \uparrow \ \uparrow$$

if the phase is mixed and

$$u^\alpha \frac{dC_A^\alpha}{dz^\alpha} = -r - \frac{[\text{area}]}{[\text{volume}]} k_\text{m} (C_A^\beta - C_A^\alpha) \qquad (14.16)$$

$$\uparrow \qquad \uparrow \qquad \uparrow$$

if the phase is unmixed. We must solve two of these equations (three if we have a three-phase reactor) simultaneously to find C_A^α and C_A^β for given inlet conditions and phase volume. In fact, all multiphase reactors involve solving essentially identical equations like these. The essential components of these equations (arrows) are the following.

14.6.1 Interfacial area

The quantity $A^{\alpha\beta}$ couples the two equations. For a membrane reactor this is simply the area of the membrane, but for other multiphase reactors the interfacial area may vary with conditions.

14.6.2 Mass transfer coefficients

The equations are also coupled by k_m the mass transfer coefficient. This is the standard mass transfer problem we encountered with catalytic reactions. As with catalytic reactors, processes are very often limited by mass transfer so that the kinetics become unimportant in predicting performance.

14.6.3 Reaction rates

Next we need the reaction rates, and chemical reaction is the only aspect that differentiates a multiphase reactor from a separation unit. As with all chemical reactors, kinetics are the most difficult quantities to obtain in describing a reactor accurately. Frequently the rates are zero in all but one phase, usually the reaction is catalytic and occurs on a catalyst surface in one of the phases. In fact, there are usually several chemical reactions, and therefore there is a mass-balance equation for each reaction, a formidable mathematical problem even before we begin to worry about the energy-balance equation and temperature variations.

This is the basic idea of this chapter. We need to solve several mass-balance equations simultaneously to predict the performance of a multiphase reactor. This is a complex numerical problem for any realistic reactor geometry and reaction system, but our goal here is not to show solutions of these equations, but rather to indicate how one would go about setting the problem up to begin to construct a solution. We will also see that there are simple qualitative arguments that permit us to estimate how varying operating parameters affect performance without going through complete solutions of the equations.

14.7 EQUILIBRIUM BETWEEN PHASES

A species frequently maintains phase equilibrium while it is reacting in one phase. An example is hydrocracking of heavy hydrocarbons in petroleum refining, where H_2 from the vapor dissolves into the liquid hydrocarbon phase, where it reacts with large hydrocarbons to crack them into smaller hydrocarbons that have sufficient vapor pressure to evaporate back into the vapor phase. As long as equilibrium of the species between phases is maintained, it is easy to calculate the concentrations in the liquid phase in which reaction occurs.

There is an equilibrium distribution of a solute species A between phases α and β if the phases are in contact for a long time. We consider this a "reaction"

$$A^\alpha \rightleftarrows A^\beta \tag{14.17}$$

between phases α and β with an equilibrium constant

$$K_{eq}^{\alpha\beta} = \frac{C_{Aeq}^\beta}{C_{Aeq}^\alpha} \tag{14.18}$$

In mass transfer courses this equilibrium constant is given different names such as *partition coefficient*, but we will refer to it as an equilibrium constant and use the symbol $K_{eq}^{\alpha\beta}$. It is important to note that in gas–liquid and liquid–liquid systems the concentrations of solute A are not usually equal in both phases at equilibrium, but rather they are related by the relation

$$C_{Aeq}^\beta = K_{eq}^{\alpha\beta} C_{Aeq}^\alpha \tag{14.19}$$

and the mass transfer between phases should be given by an expression of the form

$$r'' = k_m (C_A^\beta - K_{eq}^{\alpha\beta} C_A^\alpha) \tag{14.20}$$

For gases this is usually expressed as the Henry's Law constant H_j,

$$P_j = H_j x_j \tag{14.21}$$

but because of the complicated notation this introduces, we will not use it here (Figure 14–3).

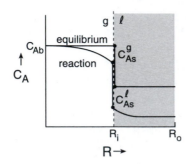

Figure 14–3 Concentration profiles across a gas–liquid interface in equilibrium and with mass flow from left to right. There is a discontinuity in C_A between the gas–liquid and liquid–liquid phases because of differences in equilibrium solubilities.

14.8 MEMBRANE REACTORS

We define a membrane reactor as one in which two phases are separated by a wall through which only one species can permeate. It is also common to have catalyst within the membrane or coated on the surface of the membrane so we can achieve reaction and separation simultaneously. Thus the membrane reactor is a multiphase reactor with a specified wall area $A^{\alpha\beta}$ between phases.

14.8.1 Partial oxidation reactor

One example of membrane reactors is oxidation, in which oxygen from one phase diffuses from one side of an oxygen-permeable membrane to react with a fuel on the other side of the membrane. This avoids a high concentration of O_2 on the fuel side, which would be flammable. A catalyst on the fuel side of the membrane oxidizes the fuel to partial oxidation products. One important process using a membrane reactor is the reaction to oxidize methane to form syngas,

$$CH_4 + \tfrac{1}{2}O_2 \rightarrow CO + H_2 \qquad (14.22)$$

which occurs on noble metal catalyst surfaces. A membrane for this reactor, which is permeable to oxygen, is ZrO_2 stabilized by yttria to form an oxide film through which oxygen is transported as O^{2-} ions. However, the membrane is impermeable to all other species except oxygen. A membrane reactor for methane oxidation might look as shown in Figure 14–4, in which the chamber contains CH_4 and the inner surface of the ZrO_2 is coated with platinum. Oxygen in the annulus outside the ZrO_2 tube contains O_2 (usually air). In this reactor the CO and H_2 form on the inside space, while N_2 from the air is left on the outside. Thus the membrane reactor acts as a separator to keep N_2 out of the product syngas, thus avoiding the need for pure O_2, which would be required in a single-phase reactor.

The membrane also keeps the O_2 concentration low in the CH_4 compartment, which favors syngas over combustion to CO_2 and H_2O and prevents a flame that would form at higher O_2 concentrations.

14.8.2 Hydrogenation and dehydrogenation reactor

Another example of a gas-phase membrane reactor is a palladium tube through which only hydrogen can permeate. This can be used to run the reaction

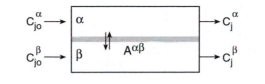

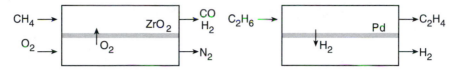

Figure 14–4   Membrane reactor in which a catalyst promotes reaction in the membrane and maintains reactants and products separate. Examples shown are CH_4 oxidation to syngas and C_2H_6 reduction to C_2H_4. The membrane eliminates N_2 from the syngas and produces C_2H_4 beyond equilibrium by removing H_2 in these applications.

$$\text{alkane} \rightleftarrows \text{olefin} + H_2 \tag{14.23}$$

This reaction is reversible; so it is frequently desirable to have high H_2 pressures to convert olefins to alkanes and low H_2 pressures to convert alkanes to olefins. Hydrogen passes readily through a hot Pd metal membrane while other gases do not, and therefore the H_2 can be added or removed from the reaction chamber. This configuration also permits the use of a H_2 stream that contains impurities in hydrogenation reactions with the olefin on the inside and the H_2 on the outside, a configuration similar to that described for oxidation reactions. The Pd is also a good catalyst for this reaction. Configurations to either produce alkane from olefin or olefin from alkane are shown in Figure 14–4.

Neither of these processes is yet practiced commercially, mainly because of materials durability problems. However, they indicate the types of processes in which use of a membrane could improve reactor performance compared to a single-phase reactor.

14.8.3   Rate-limiting steps

Reaction in a membrane reactor requires the migration of reactant to the membrane, its diffusion through the membrane, and its migration within the second phase, as well as chemical reaction on the membrane surface. As discussed previously at the end of Chapter 9, the rate might be given by an expression

$$r'' = \frac{1}{\dfrac{1}{k_m^\alpha} + \dfrac{1}{k_m^\beta} + \dfrac{\ell}{D_A} + \dfrac{1}{k''}} C_{Ab} \tag{14.24}$$

Different steps can be rate limiting, and these give the concentration profiles shown in Figure 14–5.

The effective reaction rate expression will contain the rate coefficient of the slowest step in this series process.

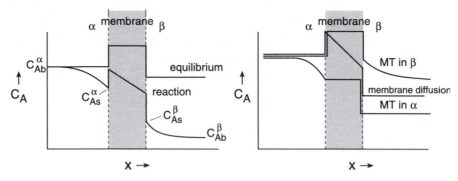

Figure 14–5   Concentration profiles of a reactant that migrates into one phase through a membrane and into a second phase for reaction.

14.8.4   Membrane reactor configurations

It is important to attain as high an area as possible for a membrane reactor. Configurations with multilayer planar membranes, coiled membranes, or as multiple tubes also can be used for similar processes with potentially very high surface areas, as sketched in Figure 14–6.

Fuel cells and batteries are examples of membrane reactors in which a conducting membrane separates the anode and cathode compartments, which supply fuel and oxidant, respectively. With fuel cells we have the added complexity that we need an ion-conducting electrode, which is also a catalyst at each electrode so that we can extract electrical power from the energy of the reaction. A battery is similar to a fuel cell except that now the fuel and oxidant are stored and supplied within the cell rather than being supplied externally. A fuel cell is usually operated as a continuous-flow reactor, while a battery is a rechargeable batch reactor.

Crucial factors in designing membrane reactors are (a) high area for high mass transfer, (b) membranes that have a high permeability to only one species, (c) membranes that are catalytic, and (d) membranes that have no holes that leak reactants from one compartment to the other. With any process the reactor will not function properly if all these criteria are not met, and for oxidation reactions any leaks between compartments can be disastrous.

It should be evident, therefore, that the design of membrane reactors is largely concerned with finding membranes that have sufficient area, catalytic activity, mechanical strength, and freedom from cracks, holes, and clogging.

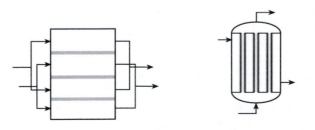

Figure 14–6   Sketches of membrane re-actors to attain high areas for transfer of a species.

14.9 FALLING FILM REACTOR

In the membrane reactor a wall of area $A^{\alpha\beta}$ separates the phases, and this area is generally fixed by the geometry of the reactor using planar or cylindrical membranes. However, most multiphase reactors do not have fixed boundaries separating phases, but rather allow the boundary between phases to be the interfacial area between insoluble phases. This is commonly a variable-area boundary whose area will depend on flow conditions of the phases, as shown in Figure 14–7.

Before we deal with these situations, it is instructive to consider a fixed-area version of a membraneless reactor, the falling film reactor. We cannot think of many applications of this reactor type because one usually benefits considerably by using configurations where the surface area is as large as possible, but the falling film reactor leads naturally to the description of many variable-area multiphase reactors.

We consider a vertical cylindrical tube of length L and diameter D_o (radius $2R_o$) with liquid admitted at the top such that it forms a *falling film* that coats the walls of the tube. We also add a gas into the top of the tube (cocurrent) or into the bottom (countercurrent). This is a standard unit in extraction processes called a *wetted-wall column*.

In analysis of the wetted-wall column one finds the velocity profile in the film $u(R)$ and then places a solute A in one of the phases and calculates its extraction to the other phase by first integrating across the liquid film to find $C_A(R)$ and the average concentration, and then integrating along the column to find $C_A(z)$ and then $C_A(L)$. When used as a chemical reactor, we assume that species A is a reactant that reacts with another solute in the liquid or with a catalyst in the liquid or on the walls of the column.

Since we do not want to consider the fluid mechanics of this reactor in detail, we will assume that the liquid falls with a constant average velocity u^ℓ and forms a liquid film of thickness $R_o - R_i$, with R_i the radius in the tube at the surface of the liquid film. If these assumptions hold, then u^ℓ and $R_o - R_i$ are independent of the position z in the tube of length L (actually the height, since the tube must be vertical). Similarly, if the liquid film thickness is constant, then the cross section occupied by the gas is constant, and the velocity of the gas u^g is also independent of z if the density of the gas is constant. These distances are sketched in Figure 14–8.

The interface between the gas and liquid is at R_i, and its area is

$$A^{g\ell} = 2\pi R_i L$$

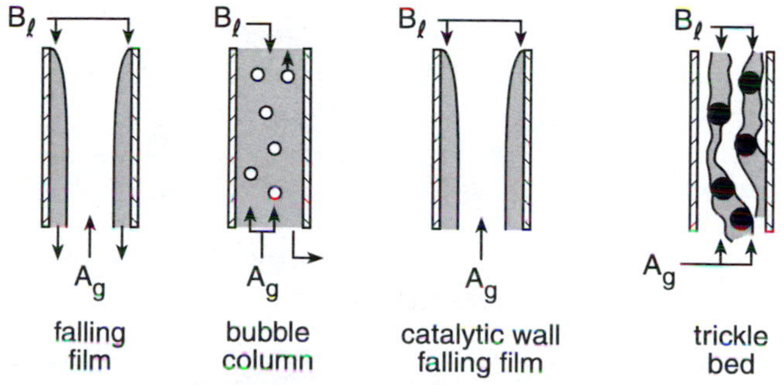

Figure 14–7 Sketches of four types of gas–liquid multiphase reactors.

falling film

bubble column

catalytic wall falling film

trickle bed

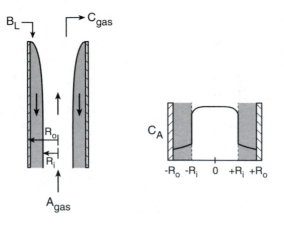

Figure 14–8 Sketch of a falling film reactor with positions R_o and R_i indicated. At the right are sketched possible concentration profiles of a reactant A, which is supplied in the gas but reacts in the liquid.

Since the phases are gas and liquid in this section, we use g and ℓ rather than α and β to designate phases. The cross section of the tube is πR_o^2, the gas will occupy a cross-sectional area πR_i^2, and the liquid will occupy a cross-sectional area $\pi(R_o^2 - R_i^2)$. The volumes occupied by the phases are

$$V^\ell = \pi(R_o^2 - R_i^2)L \tag{14.25}$$

$$V^g = \pi R_i^2 L \tag{14.26}$$

so that the residence times are

$$\tau^\ell = \frac{L}{u^\ell} = \frac{V^\ell}{v^\ell} = \frac{\pi(R_o^2 - R_i^2)L}{v^\ell} \tag{14.27}$$

$$\tau^g = \frac{L}{u^g} = \frac{V^g}{v^g} = \frac{\pi R_i^2 L}{v^g} \tag{14.28}$$

Thus we have defined all the parameters for the reactor mass balances in gas and liquid phases in terms of the inputs to the tube. In all this analysis we make the assumption that all parameters are independent of height z in the column.

Next we add solute A to the gas phase and solute B to the liquid phase and allow the reaction

$$A_g + B_\ell \rightarrow C_g + D_\ell \tag{14.29}$$

to occur to form products C and D. We assume that the reactions are irreversible, and we usually assume that the products can easily escape from this phase so the specific products are not important. We further assume that reaction occurs only in the liquid phase with a rate

$$r^\ell = k_\ell C_A^\ell C_B^\ell \tag{14.30}$$

so that $r^g = 0$.

The reactant A must be transferred from the gas phase into the liquid phase in order to react with B. The rate of mass transfer per unit of area is

$$r_{mA} = k_{mA}(C_A^g - K^{g\ell} C_A^\ell) \tag{14.31}$$

where $K^{g\ell}$ is the equilibrium distribution of A between gas and liquid phases, as discussed previously. The area of this interface is $2\pi R_i L$.

We can now write the mass-balance equations for the phases as

$$u^g \frac{dC_A^g}{dz} = 0 - \frac{2\pi R_i}{\pi R_i^2} k_{mA}(C_A^g - K^{g\ell} C_A^\ell) \tag{14.32}$$

$$u^\ell \frac{dC_A^\ell}{dz} = -k^\ell C_A^\ell C_B^\ell + \frac{2\pi R_i}{\pi(R_o^2 - R_i^2)} k_{mA}(C_A^g - K^{g\ell} C_A^\ell) \tag{14.33}$$

We included the term $r^g = 0$ to indicate that there is no reaction in the gas phase. The mass transfer rates obviously have opposite signs, and we have to multiply the mass transfer flux by [area/volume], where the volume is that occupied by that phase. Note that the mass transfer term after dividing out becomes proportional to R_i^1. Since the reactor volume is proportional to R_o^2 while the surface area for mass transfer is proportional to R, the falling film column obviously becomes less efficient for larger reactor sizes. This is a fundamental problem with the falling film reactor in that small tubes give high mass transfer rates but low total production of product.

The mass transfer coefficient is usually much lower in the liquid phase, and therefore C_A^ℓ is a function of R, the distance from the wall to the interface. One would have to solve for the steady-state profile $C_A(R)$, and find its average $C_A(z)$ to insert into the PFTR mass-balance equations simultaneously to find $C_A(L)$ in each phase.

Example 14–1   An aqueous solution containing 10 ppm by weight of an organic contaminant of molecular weight 120 is to be removed by air oxidation in a 1-cm-diameter falling film reactor at 25°C. The liquid flows at an average velocity of 10 cm/sec and forms a film 1 mm thick on the wall, while the air at 1 atm flows at an average velocity of 2 cm/sec. The reaction in the liquid phase has the stoichiometry $A + 2O_2 \rightarrow$ products with a rate $r^\ell = k^\ell C_A^\ell C_{O_2}^\ell$.

(a) What are the compositions when the reaction has gone to completion?

We assume that both phases are in nearly plug flow, and we need first to calculate the flow rates of each reactant.

$$v^\ell = A_c^\ell u^\ell = \pi(0.5^2 - 0.4^2) \times 10 = 2.83 \text{ cm}^3/\text{sec} \tag{14.34}$$

$$v^g = A_c^g u^g = \pi \times 0.4^2 \times 2 = 1.005 \text{ cm}^3/\text{sec} \tag{14.35}$$

The feed concentrations are

$$C_{Ao}^\ell = \frac{10^{-5}}{120} = 8.33 \times 10^{-8} \text{ moles/cm}^3 \tag{14.36}$$

assuming the density of water is ~ 1 g/cm³ and

$$C_{O_2o}^g = \frac{P_{O_2}}{RT} = \frac{0.2}{82 \times 300} = 8.58 \times 10^{-6} \text{ moles/cm}^3 \tag{14.37}$$

Therefore, the molar flow rates are

$$F_{Ao}^{\ell} = v^{\ell}C_{Ao}^{\ell} = 2.83 \times 8.33 \times 10^{-8} = 2.36 \times 10^{-7} \text{ moles/sec} \qquad (14.38)$$

$$F_{O_2o}^{g} = v^{g}C_{O_2o}^{g} = 1.005 \times 8.58 \times 10^{-6} = 8.62 \times 10^{-6} \text{ moles/sec} \qquad (14.39)$$

Thus we see that there is an excess of O_2; so all A will be consumed at complete conversion. This gives

$$F_{O_2o} - F_{O_2} = 2F_{Ao} \qquad (14.40)$$

(there are 2 moles of O_2 consumed per mole of A reacted), so that if the reaction goes to completion, we have

$$C_A = 0 \qquad (14.41)$$

$$C_{O_2^g} = 8.14 \times 10^{-6} \text{ moles/cm}^3 \qquad (14.42)$$

which we will assume to be a negligible change in O_2.

(b) Write the equation(s) that must be solved to find the reactor length required for a given organic removal.

At these flow rates the O_2 in the air is not depleted, and we do not need to worry about the mass balance of the air in the gas phase. We only need to solve the equation for the organic in the liquid phase, which is

$$u^{\ell}\frac{dC_A^{\ell}}{dz} = -r^{\ell} \qquad (14.43)$$

(c) How many tubes in parallel are required to process 100 liters/min of water? What will be the diameter of the tube bundle if the tubes are packed in a square array and the tube wall thickness is neglected?

One tube processes 2.83 cm³/sec; so to process 100 liters/min we will need

$$\frac{100 \times 10^3}{2.83 \times 60} = 590 \text{ tubes} \qquad (14.44)$$

If these are packed in a square configuration with a negligible tube wall thickness, each tube occupies an area 1 cm² in the tube bundle. The cross-sectional area of the tube bundle is 590 cm²; so the diameter of the column is

$$D = \left(\frac{590 \times 4}{\pi}\right)^{1/2} = 27.4 \text{ cm} \sim 1 \text{ ft} \qquad (14.45)$$

(d) Estimate the reactor length needed to reduce the organic concentration to 1 ppm if the reaction rate in the liquid is infinite and the reaction is limited by the mass transfer of O_2 in the gas. Assume $D_{O_2} = 0.1$ cm² sec.

This doesn't seem to be a good approximation because mass transfer in gases is almost always much faster than in liquids. If it were true, the Sherwood number would be $\frac{8}{3}$ for laminar flow of the air flowing in the 0.8-cm-diameter tube. Thus we have

$$k^g_{mO_2} = \frac{Sh_D D_{O_2}}{D_i} \tag{14.46}$$

We have to solve for A in the liquid phase,

$$u^\ell \frac{dC^\ell_A}{dz} = r^\ell = \frac{A}{V} k_{mO_2} C_{O_2} \tag{14.47}$$

The interfacial area per volume of reactor is

$$\frac{A}{V} = \frac{\pi D_i L}{\frac{\pi}{4}(D^2_o - D^2_i)L} = \frac{4D_i}{D^2_o - D^2_i} \tag{14.48}$$

and the equation becomes

$$\frac{dC^\ell_A}{dz} = \frac{4Sh_D D_{O_2} C_{O_2}}{[D^2_o - D^2_i]u^\ell} \tag{14.49}$$

The right-hand side is constant because C_{O_2} does not change significantly, so this equation can be simply integrated to yield

$$L = \frac{(C_{Ao} - C_A)u^\ell[D^2_o - D^2_i]}{4Sh D_{O_2} C_{O_2}} = 0.031 \text{ cm} \tag{14.50}$$

This answer is clearly nonsense because we would need a reactor length of less than 1 millimeter!

(e) Estimate the reactor length needed to reduce the organic concentration to 1 ppm if O_2 mass transfer is sufficiently fast that the liquid solution remains saturated with O_2 and the rate coefficient k^ℓ is 1×10^2 liter/mole sec. The Henry's Law constant for O_2 in water is 1×10^5 atm.

Now mass transfer of O_2 is assumed to be fast so that the liquid remains saturated with O_2, and the process is assumed to be limited by the reaction in the liquid phase. Therefore, we have to solve the equation

$$u^\ell \frac{dC^\ell_A}{dz} = -r^\ell = k^\ell C^\ell_A C^\ell_{O_2} \tag{14.51}$$

We need the O_2 concentration in the liquid phase. We assume Henry's Law

$$P_{O_2} = H_{O_2} x_{O_2} \tag{14.52}$$

so that

$$x_{O_2} = \frac{P_{O_2}}{H_{O_2}} = \frac{0.2}{10^5} = 2 \times 10^{-6} = \frac{C_{O_2} \times 32}{10^3} \tag{14.53}$$

which gives $C_{O^\ell_2} = 5.56 \times 10^{-7}$ moles/liter.
Recall that C_{O_2} is nearly constant so this equation can be rearranged to yield

$$L = \frac{u^\ell}{k^\ell C^\ell_{O_2}} \int_{C_{Ao}}^{C_A} \frac{dC_A}{C_A} = \frac{10}{10^5 \times 5.56 \times 10^{-7}} \ln 10 = 415 \text{ cm} = 5.15\text{m} = 16 \text{ ft}$$

$$(14.54)$$

a reasonable column height.

(f) Estimate the reactor length needed to reduce the organic concentration to 1 ppm if the reaction actually occurs on a catalyst on the wall of the reactor tube. Assume that the reaction rate on the wall is infinite and the process is limited by the diffusion of O_2 through the liquid, with $D_{O2} = 10^{-5}$ cm²/sec, and that the concentration profile is linear.

Now we assume that reaction occurs on the exterior surface of the tube. The equation to be solved is

$$u^\ell \frac{dC^\ell_A}{dz} = -\frac{A}{V} r'' = -\frac{A}{V} \frac{k_m (C_{O_2} - C_{O_2 s})}{2} = -\frac{A}{V} \frac{k_m C_{O_2}}{2}$$

$$= \frac{A D_{O_2} C_{O_2}}{2V(R_o - R_i)} = -3.47 \times 10^{-10} \text{ moles/liter sec}$$

$$(14.55)$$

The area for mass transfer per unit volume of reactor in the liquid is given by

$$\frac{A}{V} = \frac{4D_O}{D_O^2 - D_i^2} = 11.1 \text{ cm}^{-1}$$

$$(14.56)$$

We now solve the preceding differential equation, noting that all terms on the right-hand side are constants to obtain the reactor length needed,

$$L = \frac{u(C_{Ao} - C_A)}{3.47 \times 10^{-10}} = 2160 \text{ cm} = 22 \text{ meters} = 70 \text{ ft}$$

$$(14.57)$$

This is a very tall column.

(g) How would the answers to (c), (d), and (e) change if the pressure of air were increased from 1 to 2 atm?

Doubling C_{O_2} will double the rate of the reaction and the rate of O_2 mass transfer to the wall of the reactor. Therefore, the reactor length needed will be half of those calculated in each of the previous situations.

14.10 BUBBLE COLUMN REACTORS

We just noted that the falling film reactor has the problem that the surface area increases as the perimeter between gas and liquid while the volume increases as the cross section

$$\frac{\text{area}}{\text{volume}} = \frac{A^{\alpha\beta}}{\text{gas volume}} = \frac{2\pi R_i}{\pi R_i^2} = \frac{2}{R_i} = \frac{1}{D}$$

$$(14.58)$$

In order to design an efficient reactor using a falling film reactor, we would need to have many small tubes in parallel so that the interfacial area can be large. This is difficult to

Figure 14–9 Bubble column and spray tower reactors. Large drop or bubble areas increase reactant mass transfer.

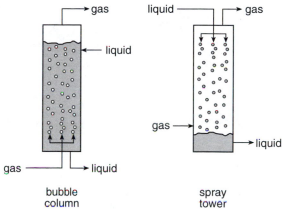

accomplish with flow down tubes, but it is easy to accomplish with rising bubbles or falling drops. The interfacial area is not now the area of the cylinder between gas and liquid but rather the area of the bubbles or drops, which can be large in a large-diameter tube.

Consider a liquid flowing downward in a tube with gas bubbles introduced at the bottom. The bubbles rise because they are less dense than the liquid. At the top the gas is separated from the liquid in the head space, while pure liquid is withdrawn at the bottom, as shown in Figure 14–9.

An important example of bubble column reactors is in partial oxidation of organics, such as p-xylene to make terephthalic acid, a monomer for polyester (PET),

$$p\text{–xylene} + O_2 \rightarrow \text{terephthalic acid} \tag{14.59}$$

where p-xylene and terephthalic acid are liquids and O_2 is a gas. In partial oxidation such as these the reaction occurs homogeneously (either with free-radical intermediates or with a homogeneous catalyst) in the liquid phase, and the desired product remains in the liquid phase, where it is separated from the other organic. It is usually important to have no O_2 in the product because it can be explosive; so the reactor must be designed so that all O_2 in the bubbles has reacted and none reaches the head space, where it could react explosively with the organic products.

Another example of bubble columns is hydrogenation of olefins and other organics, such as reduction of dinitrotoluene

$$\text{dinitrotoluene} + H_2 \rightarrow \text{diaminotoluene} \tag{14.60}$$

which is an intermediate in production of polyurethanes and in the hydrogenation of fats and oils,

$$C_n H_{2n} + H_2 \rightarrow C_n H_{2n+2} \tag{14.61}$$

In this case the reaction occurs only in the liquid phase (the oils have negligible vapor pressure), but reaction requires a solid catalyst such as finely divided Ni, which is suspended in the liquid. Thus we see that this is in fact a three-phase reactor containing H_2 gas, organic liquid reactants and products, and solid catalyst.

The concentration profile for a reactant A which must migrate from a drop or bubble into the continuous phase to react might be as shown in Figure 14–10. There is a

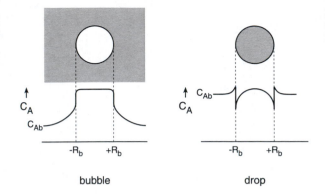

Figure 14–10 Sketches of reactant concentration C_A around a spherical bubble or drop that reacts after migrating from the gas phase into the liquid phase in bubble column and spray tower reactors.

concentration drop around the spherical drop or bubble because it is migrating outward, but, as with a planar gas-liquid interface in the falling film reactor, there should be a discontinuity in C_A at the interface due to the solubility of species A and a consequent equilibrium distribution between phases.

Bubble columns rely on nozzles, mixing plates, and impellers within the reactor to control the bubble size, which determines the interfacial area between gas and liquid phases. Clearly, the interfacial area can be varied over a wide range by suitable design of the mixer and flow pattern.

Consider a bubble phase occupying a fraction ε^g in a reactor with volume

$$V^g = \varepsilon^g V \tag{14.62}$$

Suppose all bubbles have radius R_b, so that each bubble has a volume $\frac{4}{3}\pi R_b^3$. If there are N^g bubbles per unit volume of reactor, then the volume occupied by bubbles is

$$V^g = N^g \frac{4\pi R_b^3}{3} \tag{14.63}$$

If the bubbles supply reactant into the liquid phase, then the bubbles are decreasing in size because of reaction; so we need to find $R_b(t)$ or $R_b(z)$ of the bubbles as they rise in the reactor. Thus we have the problem of a sphere that varies in diameter as the reaction proceeds. We considered this in Chapter 9 where we were concerned primarily with reaction of solid spheres instead of liquid spheres. The bubbles usually have a distribution of sizes because larger bubbles usually rise faster than small ones, and they can coalesce and be redispersed by mixers. However, to keep the problem simple, we will assume that all particles have the same size. For the reactant A supplied from the bubble, we have to solve the equation

$$4\pi R_b^2 \rho \frac{dR_b}{dt} = -k_m 4\pi R_b^2 (C_A - C_{As}) \tag{14.64}$$

just as we did for a reacting solid concentration profiles of reactant A and product B reacting around a spherical particle are shown in Figure 14–11.

As with reactions of solids, we first assume a steady-state concentration profile of reactant to calculate $C_A(x)$ and then calculate the variation of R_b versus time as the reaction proceeds.

Figure 14–11   Sketches of concentration profiles of a reactant A around a drop where the reaction is in the surrounding continuous phase. Arrows indicate directions of reactant mass flow.

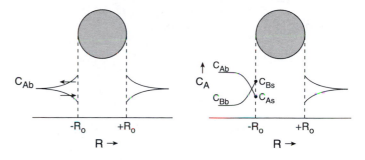

Example 14–2   An aqueous solution contains 10 ppm by weight of an organic contaminant of molecular weight 120, which must be removed by air oxidation in a 10-cm-diameter bubble column reactor at 25°C. The liquid flows downward in the tube at an average velocity of 1 cm/sec. The air at 1 atm is admitted at 0.1 liter/sec and is injected as bubbles 1 mm diameter, which rise at 2 cm/sec. Assume no coalescence or breakup and that both gas and liquid are in plug flow. The reaction in the liquid phase has the stoichiometry $A + 2O_2 \rightarrow$ products with a rate $k^\ell C_A^\ell C_{O_2}^\ell$.

(a) What are the compositions when the reaction has gone to completion?

The volumetric flow rates are

$$v^\ell = u^\ell A_c^\ell = 1 \times \pi \times 5^2 = 78.5 \text{ cm}^3/\text{sec} \tag{14.65}$$

The feed concentrations are

$$C_{Ao}^\ell = \frac{10^{-5}}{120} = 8.33 \times 10^{-8} \text{ moles/cm}^3 \tag{14.66}$$

$$C_{O_2o}^g = \frac{P_{O_2}}{RT} = \frac{0.2}{82 \times 300} = 8.58 \times 10^{-6} \text{ moles/cm}^3 \tag{14.67}$$

The molar feed rates are therefore

$$F_{Ao}^\ell = v^\ell C_{Ao}^\ell = 6.5 \times 10^{-6} \text{ moles/sec} \tag{14.68}$$

$$F_{O_2o}^g = v^g C_{O_2o}^g = 100 \times 8 \times 10^{-6} = 8 \times 10^{-4} \text{ moles/sec} \tag{14.69}$$

This gives

$$F_A = 0$$
$$F_{O_2} = F_{O_2o} - 2F_{Ao} = 7.8 \times 10^{-4} \text{ moles/sec} \tag{14.70}$$

This means that C_{O_2} decreases by only ~3% in the reactor if the reaction goes to completion, so we can assume $C_{O_2}^g$ is constant throughout the reactor.

(b) Write the equation(s) that must be solved to find the length for a given organic removal.

Since the O_2 concentration remains constant, we do not need to solve a mass-balance equation in the gas phase, and cocurrent and countercurrent flows should give the same answer if velocities remain constant. The equation to be solved is therefore

$$u^\ell \frac{dC_A^\ell}{dz} = -r^\ell \tag{14.71}$$

(c) Assuming that mass transfer of O_2 in the liquid around the bubbles limits the rate, calculate the reactor length required to reduce the contaminant to 1 ppm.

Since mass transfer in the liquid around the bubbles limits the rate, the mass-balance equation on A in the liquid phase becomes

$$u^\ell \frac{dC_A^\ell}{dz} = -r^\ell = -\frac{A}{V} r'' = -\frac{A}{2V} k_{mO_2}^\ell C_{O_2}^\ell \tag{14.72}$$

for flow over a sphere we assume $Sh_D = 2.0$, so that

$$k_{mO_2^\ell} = \frac{Sh_D D_{O_2}^\ell}{D} = 2 \times 10^{-4} \text{ cm}^2/\text{sec} \tag{14.73}$$

The right-hand side of the differential equation is constant; so it can be integrated to yield

$$L = \frac{(C_{Ao}^\ell - C_A^\ell)u^\ell}{(A/2V)k_{mO_2}^\ell C_{O_2\ell}} \tag{14.74}$$

We next need A/V, the mass transfer area per unit volume of the reactor. The residence time of a bubble is L/u^g. The volume of each bubble is $\frac{\pi}{6}D_b^3$, and the surface area of a bubble is πD_b^2. The volumetric flow rate of gas is

$$v^g = \frac{N_b(\pi/6)D_b^3}{L/u^g} \tag{14.75}$$

Therefore, the surface area of bubbles per unit volume of liquid in the reactor is

$$\frac{A}{V} = \frac{\pi D_b^2 N_b}{(\pi/4)D^2 L} = \frac{24v^g}{\pi D_b u^g D^2} = 38.2 \text{ cm}^{-1} \tag{14.76}$$

Substituting, we obtain

$$L = \frac{8.33 \times 10^{-8} \times 0.9 \times 1}{\frac{1}{2} \times 38.2 \times 2 \times 10^{-4} \times 1.11 \times 10^{-7}} = 177 \text{ cm} = 1.77 \text{ m} \tag{14.77}$$

(d) Estimate the reactor length required to reduce the organic concentration to 1 ppm if the concentration of O_2 remains at equilibrium in the solution.

If the O_2 remains in equilibrium, the equation to be solved is

$$u^\ell \frac{dC_A^\ell}{dz} = -r^\ell = -k^\ell C_A^\ell C_{O_2}^\ell \tag{14.78}$$

Since $C_{O_2}^\ell$ is constant, the differential equation can be integrated to yield

$$L = \frac{u^\ell}{k_m^\ell C_{O_2}^\ell} \ln \frac{C_{Ao}}{C_A} = \frac{\ln 10}{10^5 \times 1.11 \times 10^{-7}} = 207 \text{ cm} = 2.07 \text{ m} \qquad (14.79)$$

14.10.1   Mixing in drop or bubble phases

If an isolated drop or bubble rises or falls in the reactor, then the flow pattern in this phase is clearly unmixed, and this phase should be described as a PFTR. However, drops and bubbles may not have simple trajectories because of stirring in the reactor, and also drops and bubbles can coalesce and breakup as they move through the reactor.

These possible flow patterns of a drop or bubble phase are shown in Figure 14–12. At the left is shown an isolated bubble which rises (turn the drawing over for a falling drop), a clearly unmixed situation. If the bubble is in a continuous phase which is being stirred, then the bubble will swirl around the reactor, and its residence time will not be fixed but will be a distributed function. In the limit of very rapid stirring, the residence time of drops or bubbles will have a residence time distribution

$$p(t) = \frac{1}{\tau} e^{-t/\tau} \qquad (14.80)$$

which is the residence time distribution in a CSTR. We derived this expression in Chapter 8 where we considered the distribution of residence times of *molecules* in the reactor, but now we consider the residence time of isolated bubbles or drops. If bubbles flow generally upward, but continuously break up and coalesce, the residence time distribution of the species in this pase is narrow or roughly that of a PFTR,

$$p(t) = \delta(t) \qquad (14.81)$$

However, with stirring and coalescence and breakup, both effects tend to mix the contents of the bubbles or drops, and this situation should be handled using the CSTR mass balance equation. As you might expect, for a real drop or bubble reactor the residence time distribution might not be given accurately by either of these limits, and it might be necessary to measure the RDT to correctly describe the flow pattern in the discontinuous phase.

Figure 14–12 Sketches of possible flow patterns of bubbles rising through a liquid phase in a bubble column. Stirring of the continuous phase will cause the residence time distribution to be broadened, and coalescence and breakup of drops will cause mixing between bubbles. Both of these effects cause the residence time distribution in the bubble phase to approach that of a CSTR. For falling drops in a spray tower, the situation is similar but now the drops fall instead of rising in the reactor.

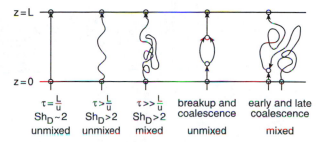

14.11 FALLING FILM CATALYTIC WALL REACTOR

Next we complicate this situation further by allowing reaction, not throughout the liquid, but on a film of porous catalyst that coats the wall. This film has thickness ℓ and a surface area per volume $\rho_c S_g$ as in Chapter 7. We assume that $\ell \ll R_o$, so that the definitions given of volume and τ apply. This reactor is sketched in Figure 14–13.

Now the reaction occurs by A migrating from the gas in the center of the tube, through the falling liquid film, and into the porous catalyst, where it reacts with B from the liquid phase to form products that must then migrate out of the catalyst and into the flowing gas or liquid streams to escape the reactor. The reactant concentration profile across the reactor $C_A(R)$ now is modified because A reacts in the porous catalyst rather than within the liquid film discussed in the previous section, as shown in Figure 14–13.

This problem can be recognized as a variation of the tube wall reactor of Chapter 7, where the reactants flowing down the tube had to migrate to and into the porous catalyst on the wall to react. The only difference here is that the reactant must first migrate through the film, which coats the wall before it can enter the catalyst and react.

If the second reactant B is in high concentration (liquids have much higher densities than gases even at high pressures) and the reaction is not limited by its mass transfer, then this becomes a pseudo-first-order reaction $r'' = k''C_A$. If all mass transfer steps are fast compared to the reaction, then this problem would simplify to be identical to the tube wall catalytic reactor, which gave

$$C_A(z) = C_{Ao} \exp\left(-\frac{S_g \rho_c^\ell k'' \eta}{d} \frac{z}{u}\right) \tag{14.82}$$

We can continue to write other expressions for different limiting cases, for example, with reaction on a nonporous catalyst film, which coats the wall.

In the preceding expression we include an effectiveness factor η to account for pore diffusion limitations of A. In fact, if the catalyst film thickness on the wall of the reactor is small enough that we can assume it planar, then the effectiveness factor becomes

$$\eta = \frac{\tanh \phi}{\phi}$$

where the Thiele modulus $\phi = (\rho_c S_g k''/d_p)^{1/2}\ell$, which we also derived in Chapter 7.

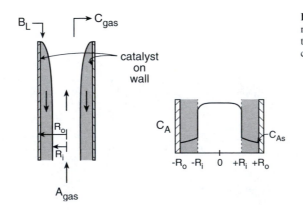

Figure 14–13 A falling film catalytic wall reactor in which reactant in the gas must diffuse through a liquid film to react on a catalyst which coats the wall of the reactor.

It is evident, however, that this problem can be much more complicated than either the wetted wall column or the catalytic wall reactor, because it combines the complexities of both. In fact, there are numerous additional complexities with this reactor beyond those simplified cases.

14.11.1 Mass transfer and pore diffusion

Recall than both A and B must diffuse into the catalyst for a bimolecular reaction to occur. This can create fairly complex concentration profiles of the two reactants within the catalyst, and the overall effectiveness factor is more complex than with the assumptions above.

Example 14–3   An aqueous solution contains 10 ppm by weight of an organic contaminant of molecular weight 120, which must be removed by air oxidation in a *sparger reactor* at 25°C. The liquid is admitted at 1 liter/sec. The air at 1 atm is admitted at 0.5 liter/sec. An impeller disperses the air into bubbles of uniform 1 mm diameter and mixes gas and liquid very rapidly. The reaction in the liquid phase has the stoichiometry $A + 2O_2 \rightarrow$ products with a rate $r^\ell = k^\ell C_A^\ell C_{O_2}^\ell$.

(a) What are the compositions when the reaction has gone to completion?

The volumetric flow rates are

$$v^\ell = 10^3 \text{ cm}^3/\text{sec} \tag{14.83}$$

$$v^g = 0.5 \times 10^3 \text{ cm}^3/\text{sec} \tag{14.84}$$

The feed concentrations are

$$C_{Ao}^\ell = \frac{10^{-5}}{120} = 8.33 \times 10^{-8} \text{ moles/cm}^3 \tag{14.85}$$

$$C_{O_2o}^g = \frac{P_{O_2}}{RT} = \frac{0.2}{82 \times 300} = 8.58 \times 10^{-6} \text{ moles/cm}^3 \tag{14.86}$$

The molar feed rates are therefore

$$F_{Ao}^\ell = v^\ell C_{Ao}^\ell = 8.33 \times 10^{-5} \text{ moles/sec} \tag{14.87}$$

$$F_{O_2o}^g = v^g C_{O_2o}^g = 0.00429 \text{ moles/sec} \tag{14.88}$$

As before, the O_2 changes little when the reaction has gone to completion, so we may assume that C_{O_2} is constant.

(b) Write the equation(s) that must be solved to find the reactor volume required for a given organic removal.

Since the O_2 concentration remains constant, we do not need to solve the mass-balance equation in the gas phase. Both phases are mixed, so we must solve the CSTR equation for C_A in the liquid phase,

$$C_{Ao\ell} - C_{A\ell} = \tau^\ell r^\ell \tag{14.89}$$

(c) Assuming mass transfer of O_2 in the liquid around the bubbles limits the rate, calculate the reactor volume required to reduce the contaminant to 1 ppm.

We assume that $Sh_D = 2$ to obtain

$$k^\ell_{mO2} = \frac{Sh_D D_{O_2}}{D} = \frac{2 \times 10^{-5}}{0.1} = 2 \times 10^{-4} \text{ cm}^2/\text{sec} \tag{14.90}$$

The O_2 concentration in the liquid at the bubble surface is

$$C^\ell_{O_2} = \frac{P_{O_2}}{H_{O_2} M_{H_2O}} = \frac{0.2}{10^{-5} \times 18} = 1.1 \times 10^{-7} \text{ moles/cm}^3 \tag{14.91}$$

We must solve the equation

$$C_{Ao\ell} - C_A = \tau^\ell \frac{A}{2V} k^\ell_{mO_2} C^\ell_{O_2} \tag{14.92}$$

(d) Estimate the reactor volume required to reduce the organic concentration to 1 ppm if the concentration of O_2 remains at equilibrium in the solution.

Now reaction limits the rate; so the mass-balance equation becomes

$$C^\ell_{Ao} - C^\ell_A = \tau^\ell k^\ell C^\ell_{O_2} C^\ell_A = \frac{V^\ell}{v^\ell} k^\ell C^\ell_{O_2} C^\ell_A \tag{14.93}$$

with $C^\ell_{O_2}$ assumed constant. Solving this equation for V^ℓ, we obtain

$$V^\ell = \frac{v^\ell (C^\ell_{Ao} - C_A)}{k_{mO_2} C^\ell_{O_2} C^\ell_A} = \frac{9 \times 10^{-3}}{10^5 \times 1.11 \times 10^{-7}} = 8.1 \times 10^5 \text{ cm}^3 = 810 \text{ liters} \tag{14.94}$$

The residence times of gas and liquid in a completely stirred reactor are equal,

$$\tau^\ell = \frac{V^\ell}{v^\ell} = \tau^g = \frac{V^g}{v^g} \tag{14.95}$$

Therefore, the total reactor volume is

$$V = V^\ell + V^g = V^\ell \left(1 + \frac{v^\ell}{v^g}\right) = 810 \left(1 + \frac{1}{0.5}\right) = 1215 \text{ liters} \tag{14.96}$$

14.12 TRICKLE BED REACTOR

As with the falling film reactor, the rate of mass transfer to the catalyst goes as R, while the size of the reactor goes as R^2, so this reactor becomes very inefficient except for very small-diameter tubes. However, we can overcome this problem, not by using many small tubes in parallel, but by allowing the gas and liquid to flow (trickle) over porous catalyst pellets in a trickle bed reactor rather than down a vertical wall, as in the catalytic wall reactor.

Figure 14–14   Sketch of trickle bed reactor in which liquid from the top of the reactor and gas rising from the bottom of the reactor contact stationary catalyst particles to produce a catalytic reaction. At the right is a sketched concentration profile for the cross section shown.

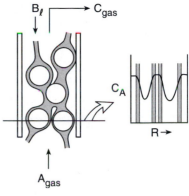

In this multiphase reactor a tube or tank (a very large tube) is filled with catalyst pellets packed into a bed and a liquid flows down over the catalyst while a gas flows up or down in countercurrent or cocurrent flow. A cross section of this reactor is shown in Figure 14–14.

The petroleum hydrotreating reactor is the most important example of this reactor. As discussed in Chapter 2, the hydrogenation and cracking of the heavy fraction of crude oil to produce lighter and volatile products occurs in an overall reaction

$$\text{heavy oil (liquid)} + H_2 \rightarrow \text{lighter oil (gas)} \tag{14.97}$$

which takes place over a sulfided cobalt–molybdenum catalyst at $\sim$500°C with a H_2 pressure of $\sim$50 atm. We wrote this as a single prototype reaction

$$C_{24}H_{50} + 2H_2 \rightarrow 3C_8H_{18} \tag{14.98}$$

where we cracked the large molecule into three smaller alkanes with the molecular weight of gasoline. In this reaction the products are more volatile than the reactants; so they migrate to the vapor phase and are carried out of the reactor with the excess H_2.

14.13   MULTIPHASE REACTORS WITH CATALYSTS

In many of the examples we have considered and will consider later in this chapter, a catalyst is present to promote the reaction. If this is a homogeneous catalyst in one of the reactant phases such as an acid, base, or a dissolved organometallic complex, there are no mass transfer resistances to reaction, and the process behaves as a true homogeneous reaction.

However, with slurry or trickle bed reactors a solid catalyst is added, and this constitutes an additional phase, which can introduce additional mass transfer resistances, as sketched in Figure 14–15.

Consider the overall reaction

$$A_g + B_\ell \rightarrow \text{products} \tag{14.99}$$

where A is fed into the reactor as a gas that must first dissolve in the liquid and then migrate through the liquid to the catalyst surface, where it reacts with B. The steps in this process can be written as

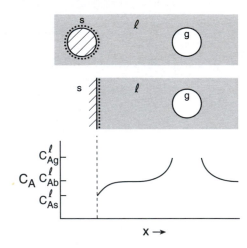

Figure 14-15   Sketch of concentration profiles between a spherical bubble and a solid spherical catalyst particle in a continuous liquid phase (upper) in a gas–liquid slurry reactor or between a bubble and a planar solid wall (lower) in a catalytic wall bubble reactor. It is assumed that a reactant A must migrate from the bubble, through the drop, and to the solid catalyst surface to react. Concentration variations may occur because of mass transfer limitations around both bubble and solid phases.

$$A_g \rightleftarrows A_{soln}, \qquad r'' = J_A = k_m(C_A^g - K^{g\ell}C_{A,soln}), \qquad \text{dissolution of } A$$

$$A_{soln} \rightleftarrows A_s, \qquad r'' = k_{aA}C_{A,soln}, \qquad \text{adsorption of } A \text{ from solution}$$

$$B_{soln} \rightleftarrows B_s, \qquad r'' = k_{aB}C_{B,soln}, \qquad \text{adsorption of } B \text{ from solution}$$

$$A_s + B_s \rightarrow \text{ products}, \qquad r'' = k''C_{As}C_{Bs}, \qquad \text{reaction on catalyst}$$

The mass transfer of A from the bubble to the solution and from the solution to the catalyst surface is described by an overall coefficient k_m. Any of these steps can be rate limiting, and the overall reaction rate will be a function of each of these coefficients.

14.14   OTHER MULTIPHASE REACTORS

14.14.1   Spray tower and oil burner

If we simply turn the drawing of the bubble column upside down, we have a *spray tower reactor*. Now we have dense liquid drops or solid particles in a less dense gas; so we spray the liquid from the top and force the gas to rise. The same equations hold, but now the mass transfer resistance is usually within the liquid drop.

We have the same geometry for reacting solid particles such as the burning of pulverized coal that was discussed in Chapter 10, but now the mass transfer resistance is in the gas phase as the reaction is limited by the diffusion of O_2 to the surface.

In combustion of liquid oils such as heavy diesel fuels, the fuel is sprayed through nozzles into air. After ignition, a flame forms around the evaporating drops, driven by the vaporization of the drop in the hot boundary layer. The latent heat of vaporization strongly affects the temperature of the drop so that the heat release by the reaction is compensated by the heat absorbed in vaporizing the fuel and the heat lost to the cooler gas in the air. The composition profiles we might expect are shown in Figure 14–11, but the temperature will also vary around the particle.

If reaction occurs only in the boundary layer around the drop of radius R, then a steady-state energy balance around the drop yields

$$4\pi R^2[(\Delta H_R - \Delta H_{vap})r'' - k_m(T_{drop} - T_{air})] = 0 \qquad (14.100)$$

The rate of combustion is controlled either by the evaporation rate of the fuel into the reacting boundary layer or the mass transfer of the O_2 from the surrounding air. Note the signs of ΔH_R and the heat of vaporization ΔH_{vap}. Since vaporization is always endothermic, it will tend to cool the drop and thus slow reaction.

14.14.2 Sparger reactors

The bubble column and spray tower depend on nozzles to disperse the drop or bubble phase and thus provide the high area and small particle size necessary for a high rate. Drop and bubble coalescence are therefore problems except in dilute systems because coalescence reduces the surface area. An option is to use an impeller, which continuously redisperses the drop or bubble phase. For gases this is called a *sparger reactor*, which might look as shown in Figure 14–16.

By placing the impeller within a *draft tube* within the reactor, the fluids are forced to pass through the impeller, where the bubbles are redispersed by impacting on the impeller surfaces. The draft tube is placed in the center of the reactor; so the fluids recirculate repeatedly (a recycle reactor) to allow bubbles to be repeatedly redispersed in the draft tube. The overall reactor becomes well mixed and is therefore described by the CSTR equations. The rapid flow of this reactor enhances the mass transfer rate and thus increases the overall reaction rate if it is limited by mass transfer of a reactant from the liquid phase into the bubbles.

14.14.3 Emulsion reactors

We can generalize this to any two-phase reaction where both phases are mixed. One of the most common examples is the emulsion reactor, where immiscible liquids (usually one aqueous and one nonpolar) are fed into a tank that is stirred sufficiently fast that both phases break into drops or one phase is continuous and the other forms drops. We assumed that

Figure 14–16 Sketch of sparger reactor in which an impeller continuously redisperses the bubble phase to maintain small bubble sizes and increase mass transfer. The impeller also recirculates the fluids in the reactor by placing it in a draft tube, which forces the fluids past the impeller at high velocities and further enhances mass transfer.

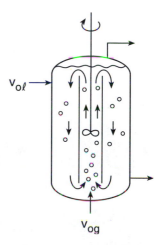

reaction occurs only in one phase but that reactant is fed from the other. Saponification of an ester fed from an organic phase into an aqueous phase where acid catalyzed the decomposition is an example of this type of process. Many polymerization reactors use emulsions between liquid phases.

The area between phases $A^{\alpha\beta}$ is the surface area of the drops. It will clearly be a strong function of the stirring characteristics (we assume that stirring is always fast enough to mix both phases). The presence of surfactants, drop size distributions, stirrer design, and circulation patterns. Interfacial area is frequently an unknown in emulsion reactors, but the above formulation should be applicable. Another complication in emulsion reactors is the fact that mass transfer coefficients depend strongly on drop size and stirring rate. The relevant parameter in an emulsion reactor is $A^{\alpha\beta}k_m$ with neither factor known very well.

In all these reactors gravity plays an important role. To obtain good contact between phases, we need to overcome the separations that will be driven by gravity whenever the phases have different densities. This requires that the flow conditions and drop, bubble, or particle sizes are properly chosen. It is also important that the phases be separated after the reactor, and mists, emulsions, and dust in separation units can cause major problems in design of these multiphase reactors. Reactor orientation plays an obvious role in any multiphase reaction processes.

14.14.4   Slurry reactor and fluidized bed reactor

The above reactors can be reduced to those of the slurry reactor of Chapter 7, in which a solid powder is the catalyst on which the reaction occurs. Now the fluid phase is unreactive (the α phase in the previous derivation), and reaction occurs on the catalyst (the β phase). In some cases the solid powder is prevented from leaving the reactor, which involves setting $\tau^\beta = \infty$ so that we do not need to consider C_{Ao}^β explicitly.

The solid could also be withdrawn and returned to the reactor using a cyclone, and in this case (still assuming complete mixing) the complete equations for the solid phase will have to be solved because τ is finite.

14.14.5   Crystallization reactors

An example of a gas–solid reactor in which both phases are mixed is crystallization, where a supersaturated liquid crystallizes onto growing solid nuclei. The crystal area and the mass transfer characteristics now depend on crystal size (discussed in Chapter 9), and the objective is usually to control conditions in such a way as to obtain solid particles of a particular size, size distribution, crystallinity, solvent incorporation, etc.

One example of this type of reactor is in the synthesis of catalyst powders and pellets by growing porous solid oxides from supersaturated solution. Here the growth conditions control the porosity and pore diameter and tortuosity, factors that we have seen are crucial in designing optimal catalysts for packed bed, fluidized bed, or slurry reactors.

14.14.6   Separation

In all these reactors gravity plays an important role. To obtain good contacting between phases, we need to overcome the separations, which will be driven by gravity whenever the phases have different densities. This requires that the flow conditions and drop, bubble, or particle sizes be properly chosen. The addition of surfactants can drastically affect these reaction rates by lowering the surface tension and thus stabilizing small bubbles and drops. Surfactants adsorbed at interfaces can also add a mass transfer resistance, thus lowering the mass transfer rates between phases.

It is also important that the phases be separated after the reactor, and mists, foams, emulsions, and dust in separation units can cause major problems in design of these multiphase reactors. Separation of phases after the reactor is easier for gas–liquid and gas–liquid systems than for liquid–liquid and liquid–solid systems because of the larger density differences for the former. Reactor orientation plays an obvious role in any multiphase reaction processes.

14.15   ANALYSIS OF MULTIPHASE REACTORS

In the previous sections we described gas–liquid and gas–liquid–catalyst multiphase reactors and described how these problems should be handled with various approximations of the rate-limiting step. In this section we consider the solution of some of the simplest of these cases.

14.15.1   Both phases mixed

Let us consider a simple example of multiphase reactors where the reaction $A \rightarrow B$ occurs in a CSTR, but now A enters the reactor in phase α but does not react until it enters phase β, where the homogeneous rate is kC_A^β. As an example A could be an ester in an oil phase α, which will hydrolyze into an alcohol and an acid when it comes in contact with a water phase β.

We write the mass balances on A in phase β as

$$C_A^\beta - C_{Ao}^\beta = \tau^\beta k C_A^\beta - \tau^\beta \frac{A^{\alpha\beta} k_m}{V^\beta} (C_A^\beta - C_A^\alpha) \tag{14.101}$$

For phase α the expression is

$$C_A^\alpha - C_{Ao}^\alpha = +\tau^\alpha \frac{A^{\alpha\beta} k_m}{V^\alpha} (C_A^\beta - C_A^\alpha) \tag{14.102}$$

Note that these are the "conventional" CSTR equations in α and β phases with the mass transfer "reactions" between phases added.

These two algebraic equations are coupled by the mass transfer term. One can solve for C_A^α and insert it into the β phase equation and solve for C_A^β. This gives

$$C_{A\beta} = \frac{C_{Ao}^\beta + \frac{A^{\alpha\beta} k_m}{V^\beta} C_A^\alpha}{1 + k\tau^\beta + \frac{A^{\alpha\beta} k_m}{V^\beta}} \tag{14.103}$$

We can now solve for C_A^α to yield

$$C_A^\alpha = \cfrac{C_{Ao}^\alpha + \cfrac{A^{\alpha\beta} k_m}{V^\alpha} \cfrac{C_{Ao}^\beta}{1 + k\tau^\beta + \cfrac{A^{\alpha\beta} k_m}{V^\beta}}}{1 + \cfrac{A^{\alpha\beta} k_m}{V^\alpha} - \cfrac{\cfrac{(A^{\alpha\beta} k_m)^2}{V^\alpha V^\beta}}{1 + k\tau^\beta + \cfrac{A^{\alpha\beta} k_m}{V^\beta}}} \qquad (14.104)$$

This expression does not look very simple or instructive. However, let us check the expression if mass transfer is very fast compared to reaction, $(A^{\alpha\beta} k_m / V^\alpha) \gg k$. In this case the expression becomes

$$C_A^\beta = \frac{C_A^\alpha}{1 + k\tau^\beta}$$

which is exactly the expression for the first order irreversible reaction in the CSTR except the reactant is being fed from the α phase.

These equations give expressions for the exit concentrations of reactant in a two-phase reactor for a first-order irreversible reaction in terms of the two residence times τ^α and τ^β, the two feed concentrations C_{Ao}^α and C_{ao}^β, and the reaction rate coefficient, the mass transfer coefficient k_m and the area between the two phases $A^{\alpha\beta}$.

We have in fact had to make many approximations to obtain even these complicated expressions, and it is evident that numerical solutions to most multiphase chemical reactor problems can only be obtained after considerable computations and approximations.

14.16 REACTOR–SEPARATION INTEGRATION

We typically think of the chemical reactor as the unit in which reactions are run and that the reactor is preceded and followed by separation units such as distillation, extraction, centrifugation, etc., to remove reactants from products and to prepare products with required purities. In fact, a crucial feature of many chemical reactors is that they involve the integration of chemical reaction and separation in a single chemical reactor.

Integrated reaction–separation units can achieve several desirable goals.

1. Remove product immediately while retaining reactants in the reactor.

2. Achieve product yields beyond those predicted by thermodynamic equilibrium.

3. Provide for efficient heat removal and generation by heat transport in individual phases.

4. Provide for heat transfer by latent heats of interphase transfer such as by boiling or condensing a component.

Figure 14–17 sketches some of the reaction–separation reactors that we consider in this chapter. We sketch most of these for the reversible reaction

$$A \rightleftarrows B + C, \qquad K_{eq} = \frac{C_B C_C}{C_A} \qquad (14.105)$$

with K_{eq} small so that a simple reactor with a feed of pure A cannot give a conversion greater than $C_B = K_{eq}^{1/2} C_A^{1/2}$.

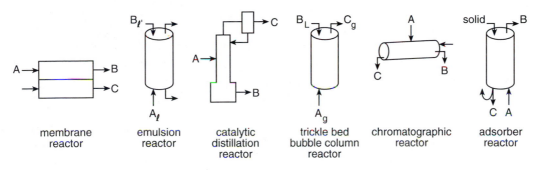

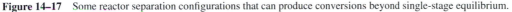

Figure 14–17   Some reactor separation configurations that can produce conversions beyond single-stage equilibrium.

14.17   CATALYTIC DISTILLATION

Another multiphase reactor that achieves reaction with separation is catalytic distillation. In this reactor a catalyst is placed on the trays of a distillation column or packed into a distillation column, as shown in Figure 14–18.

Consider the reaction

$$A \rightleftarrows B + C \tag{14.106}$$

which has an equilibrium constant on the order of unity, so that in a single-stage reactor one cannot attain a conversion beyond that predicted by equilibrium. However, if the reactant is fed into a distillation column containing a catalyst that promotes the reaction, then the equilibrium conversion is assumed to occur in each stage.

Now we assume that A, B, and C have different boiling points such that B is more volatile than C. Therefore, B will rise to the top of the column toward the condenser, while C will move downward toward the reboiler. Thus we have a *separation* of the two products away from each other. This prevents their reaction together to form A, and we thus can have complete conversion of A to B and C even though equilibrium predicts a low conversion.

The catalytic distillation reactor is in effect a multistage reactor, where each tray achieves equilibrium at its temperature and composition, with the temperature being lower

Figure 14–18   Catalytic distillation reactor in which catalyst in the distillation column combines chemical reaction with vapor–liquid equilibrium in the column to achieve conversions higher than obtainable with a reactor alone.

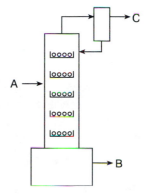

at the top, where there is more B and higher at the bottom, where there is more C. With a high reflux ratio in the condenser to return any A or C to the column, one can attain essentially complete reaction and separation of the products from each other.

Catalytic distillation is now used for the production of most MTBE produced as a gasoline additive. The reaction

$$CH_3OH + i\text{-}C_4H_8 \rightarrow (CH_3)_3COCH_3 \qquad (14.107)$$

is ideal for catalytic distillation because the boiling points are 80, 40, and 70°C, respectively. Thus the products are separated, and pure MTBE can be withdrawn from the reactor.

In catalytic distillation the temperature also varies with position in the column, and this will change the reaction rates and selectivities as well as the equilibrium compositions. Temperature variations between stages and vapor pressures of reactants and products can be exploited in designing for multiple-reaction processes to achieve a high selectivity to a desired product with essentially no unwanted products.

14.18 CHROMATOGRAPHIC REACTORS

14.18.1 Chromatography

Chromatography is the primary analytical method in chemical analysis of organic molecules. This technique is used to analyze reaction products in most of the processes we have been describing. The analysis of reaction products and of the efficiency of separation units usually is done by analytical chemists (who earn lower salaries), but chemical engineers need to be aware of the methods of analysis used and their reliability.

In a chromatographic column a tube is filled with a powder of a solid that acts as an adsorbent for the molecules to be identified. As shown in Figure 14–19, a pulse of the gas mixture is injected into the tube in a flowing stream of an inert carrier gas (which does not adsorb on the solid) such as He or H_2, and the products leaving the column are detected using a thermal conductivity or flame ionization detector and recorded as a signal versus time in the chromatogram. Each species is identified by its residence time on a particular column, and the amount of that species is proportional to the area under the peak.

First we note that a chromatographic column is a PFTR with carrier gas (an inert) flowing at velocity u for a residence time $\tau = L/u$, which is the residence time for the inert carrier. However, the gases to be analyzed have longer residence times than the carrier because they are adsorbed on the solid powder in the column. The average residence time for species j is

$$\tau_j = \tau + t_{\text{ads},j}$$

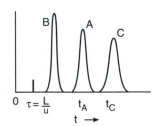

Figure 14–19 Pulse injection of gases or liquids into a chromatographic column with carrier gas or liquid for species A, B, and C. The chromatogram is recorded as the signal versus retention time in the column.

and its residence time in the chromatographic column is

$$u_j = \frac{L}{\tau + t_{adsj}} \tag{14.108}$$

where $t_{ads,j}$ is the average time that the species spends adsorbed on the solid and τ is the time spent flowing. For liquid chromatography the carrier and the molecules to be analyzed are liquids flowing at high pressures, and adsorption occurs from the liquid to the solid.

The next thing we note about chromatography is that it is equivalent to tracer injection into a PFTR. Whereas in Chapter 8 we used tracer injection to determine the residence time distribution in a reactor, here we have nearly plug flow (with the pulse spread somewhat by dispersion), but adsorption from the fluid phase onto the solid reduces the flow velocity and increases the residence time to be much longer than τ.

An important feature of chromatography is that it needs very small amounts of mixtures for analysis. This makes it poor for separating large amounts of chemicals, although *preparative chromatography* can be used to separate mixtures by withdrawing and separating products at the appropriate residence times. This process does not scale up for large quantities very well, but it is occasionally used to separate expensive biological molecules, and there are now several industrial processes that use this technology.

14.18.2 Reaction chromatography

Combining chromatography with a chemical reactor can be used to achieve reaction and separation within the same reactor, and this can be used to generate products beyond the normal thermodynamic equilibrium limitation.

Consider the reaction

$$A \rightleftarrows B + C \tag{14.109}$$

which has a small equilibrium constant so that with a feed of pure A, the conversion is limited by equilibrium, which does not give a high conversion. However, if the retention times of A, B, and C are different, they will move with different velocities in the reactor. However, with steady flow the reaction will still come to equilibrium because, even though the residence times are different, the exit concentrations will be at equilibrium.

Figure 14–20 shows the concentration profiles of these three species in the reaction above following pulse injection of pure A into the reactor, which is packed with a

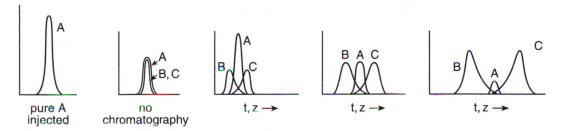

pure A   no   t, z →   t, z →   t, z →
injected   chromatography

Figure 14–20 Chromatogram of species A, B, and C injection of pure A with the reaction $A \rightarrow B + C$. Separation of B from C reduces the rate of the back reaction and produces complete conversion even though equilibrium is not favorable.

chromatographic solid. Without chromatography, the mixture comes to equilibrium, but with chromatographic separation species B moves faster than species A, while species C moves slower. Therefore, species B and C are spatially separated and cannot react back to form A.

As with the catalytic distillation reactor, the chromatographic reactor functions as a multistage reactor. The chromatographic reactor is essentially a *batch* reactor, and we need to adapt this configuration into a continuous process to develop a large-scale and economic process.

14.18.3   Moving-bed chromatographic reactor

We can achieve conversion beyond equilibrium in a reaction such as this and run continuously by using a *moving bed* of chromatographic and catalytic solid. We assume that the fluid flows with a velocity u_{fluid}, while the solid moves upstream at a velocity u_{solid}. If we adjust these velocities with respect to the velocities of the reactant and products, then a product that is strongly adsorbed can be made to move *backwards* in the tube.

We can design a reactor to separate the products and achieve complete conversion by admitting pure A into the *center* of the tube with the solid moving *countercurrent* to the carrier fluid. We adjust the flows such that A remains nearly stationary, product B flows backward, and product C flows forward. Thus we feed pure A into the reactor, withdraw pure B at one end, and withdraw pure C at the other end. We have thus (a) beat both thermodynamic equilibrium and (b) separated the two products from each other.

14.18.4   Rotating bed and simulated moving chromatographic reactors

A moving-bed reactor is not simple to implement because it is complicated to configure the reactor flows and achieve high production rates. Another configuration that provides separation is the rotating annulus. Various configurations such as these are being explored to find economical ways to construct reaction chromatographic reactors, such as by using multiple reactors with fixed catalyst but switching the flows of reactants in simulated countercurrent chromatographic reactors (Figure 14–21).

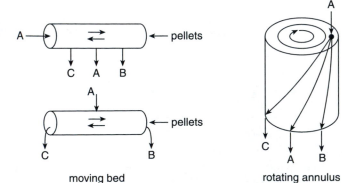

moving bed   rotating annulus

Figure 14–21   Countercurrent moving bed and rotating annulus reactors. Chromatographic separation of species in a continuous chemical reactor can be accomplished with a moving bed tubular reactor or a rotating annulus reactor that separates A, B, and C by carrying a product species counter to the flow direction because it is adsorbed on the moving catalyst.

14.18.5 Complications of reaction–separation integration

The reactors described here provide interesting alternatives to conventional reactors with interstage separation, and they may have distinct economic advantages in equipment and in production of a desired product. However, each of these systems has distinct problems, and few have as yet received wide acceptance. This is because each of these systems requires a compromise between optimal reactor design and optimal separation design. Thus one must adjust the temperature in a chromatographic reactor to attain both good separation and good kinetics. These seldom coincide, and the degradation in performance can overcome the advantages of integration.

In membrane reactors plugging is an ever-present problem because any membrane is also a good filter. In bubble, drop, emulsion, and trickle bed reactors surface-active agents can cause formidable problems with foaming. Traces of soap in liquid feeds are difficult to avoid, and their result is similar to too much detergent in a washing machine.

14.19 IRON ORE REFINING

An example of an incredibly complex multiphase chemical reactor is iron ore refining in a blast furnace. As sketched in Figure 14–22, it involves gas, liquid, and solid phases in countercurrent flows with complex temperature profiles and heat generation and removal processes.

The reactor is fed by admitting solid iron oxide as taconite pellets, which are $\sim \frac{1}{4}$-in.-diameter spheres and solid coke particles, along with air or pure O_2. These undergo many reactions, with the dominant reactions being

$$C^s + O_2^g \rightarrow CO^g \tag{14.110}$$

$$Fe_2O_3^s + CO^g \rightarrow 2Fe^\ell + CO_2^g \tag{14.111}$$

Carbon oxidation produces CO, which reduces the solid iron oxide to liquid iron. This reaction is strongly exothermic, and this supplies the heat necessary to heat the reactor contents above the melting point of iron metal. Iron ore also contains silicate slag, and this melts along with the Fe to form a second liquid phase. The oxide reduction reaction step is especially interesting because it involves a gas reacting with a solid to form a liquid and another gas.

This reactor contains at least two solid phases, two liquid phases, and a gas phase. The flows are largely driven by gravity caused by the density differences of the solid and

Figure 14–22   Phases and flows in iron ore refining in a blast furnace.

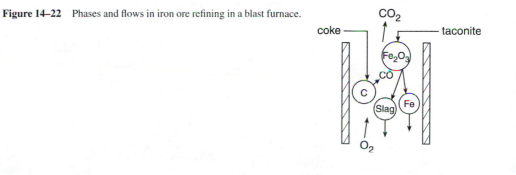

liquid phases. Taconite and coke are admitted at the top of the reactor and O_2 at the bottom. Liquid Fe and slag are withdrawn at the bottom of the reactor. The liquid iron is either cast into ingots in molds or directly passed from the reactor through rolling mills to process it into sheets.

This process has evolved considerably from the start of the Iron Age, when our amateur engineer ancestors developed batch processes, through the continuous processes that fueled the Industrial Revolution, down to the twentieth century, where specialty steels are the desired products. As might be expected, these processes evolved primarily by trial-and-error methods, because no detailed analysis of reactor flows and reactions are productive or even possible.

14.20 THE PETROLEUM REFINERY

Finally, we return to the four major reactors in petroleum refining, which we introduced in Chapter 2. All of these reactors are sketched in Figure 14–23.

14.20.1 Fluidized catalytic cracking

This is the workhorse reactor in a petroleum refinery, processing up to half of the total petroleum from the primary distillation of the crude oil. It consists of a fluidized bed riser reactor coupled with a regenerator to burn off the coke continuously from the catalyst and send fresh catalyst back to the reactor. The feed in the riser flows with approximately plug flow along with the catalyst, while in the regenerator both gas and catalyst are nearly mixed.

In FCC there are therefore four separate residence times, and the riser and regenerator are each described by two coupled mass-balance equations. The mass flow rates of catalyst between the reactors are the same since catalyst is recycled.

Energy coupling between the reactor and regenerator are crucial in designing the FCC reactor, because the heat liberated from burning off the coke from the catalyst supplies the heat to maintain the temperature in the reactor where reactions are endothermic. Therefore,

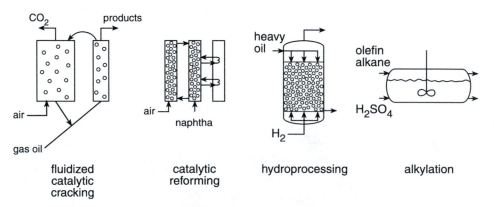

Figure 14–23 Sketches of the four major chemical reactors in petroleum refining. These were considered first in Chapter 2, but now they are seen to be multiphase catalytic reactors using catalysts and requiring careful design for heat transfer and mass transfer.

the energy balance equations and the description of flow of fluid and solid phases must be considered carefully in this reactor.

14.20.2   Catalytic reforming

This reactor creates high-octane gasoline and the aromatics needed for petrochemical feedstocks. In modern reactor configurations it is a moving bed where the catalyst moves slowly down the reactor so that it can be removed from the bottom of the reactor and regenerated before being returned to the top of the reactor.

Since these reactions are also strongly endothermic, the reactor must be heated, and this is usually accomplished by piping the fluid into a furnace in interstage heating, where it is heated back to the desired temperature and returned back to the reactor. In a catalytic reforming process, the furnace and the catalyst recycle equipment are frequently larger than the actual reactor.

14.20.3   Alkylation

This is a *three-phase reactor* with aqueous and organic liquid phases and a gas phase. The C_3 and C_4 alkanes are fed into the reactor as gas and liquid, the catalyst is sulfuric or hydrofluoric acid in water, and the products are liquid organics.

Thus this reactor requires mass transfer between the gas and liquid phases and between the organic and aqueous liquid phases. Stirring to mix phases, make small drops and bubbles, and increase interfacial mass transfer is crucial in designing an alkylation reactor.

These reactors operate below room temperature to attain a high equilibrium yield, and refrigeration equipment is a major component of an alkylation process.

14.20.4   Hydroprocessing

These processes are becoming more and more important in petroleum refining as heavier feedstocks with large amounts of sulfur, nitrogen, and heavy metals are increasing and as regulations require decreased amounts of these in the products. Hydroprocessing is being used to prepare feeds for FCC and for catalytic reforming by removing sulfur and heavy metals, which would poison the catalysts.

This is usually accomplished in a trickle bed reactor because the reactants have molecular weights too high to boil, and H_2 is the other reactant. Catalysts are carefully designed to provide proper catalytic activity and stability as well as appropriate catalyst geometries for contacting gas and liquid. Pressures in this fixed bed reactor are high, and residence times are fairly long; so large reactors with thick walls are required.

The disposal of sulfur removed from the crude oil is a major part of hydroprocessing. This is removed from the feed as H_2S, which is a highly toxic gas. Usually some of this H_2S is oxidized to SO_2, and this is reacted with the rest of the H_2S to form solid elemental sulfur, which can be safely shipped. Some refineries process the sulfur to make H_2SO_4, which is the usual final product formed from sulfur in crude oil. In some refineries this H_2SO_4 is reacted with limestone ($CaCO_3$) to make gypsum ($CaSO_4$), which can be safely disposed as landfill. Disposal of spent catalyst containing heavy metals is a major concern in hydroprocessing. In general, solid waste disposal is a problem because the amount of solid continues to accumulate on the plant site and eventually will fill up all space available.

14.21 SUMMARY

While we promised in Chapter 2 that we would finally explain the details of all these complicated reactors that are essential in the petroleum refinery, we will in fact not even try to do this because their details are actually much more complicated than can be considered in a simple or short fashion. (Many books and monographs have been written on each of these reactors.) Rather, we shall conclude by remarking that throughout this book we have considered all the principles by which one would describe the reactors in a petroleum refinery.

The reactors you will be called on to deal with in your job as a chemical engineer will be just as complicated as those in the refinery, and we hope that this book has given you the necessary principles with which to begin to analyze and design the processes for which you will be given responsibility. [At least, we hope that you know what you need to know not to embarrass yourself or harm you or your employer.]

PROBLEMS

14.1 For the following continuous multiphase chemical reactors construct a table indicating the phases, whether they are mixed or unmixed, the major reactions, and in what phase(s) the reaction is occurring. A sketch may be helpful.

   oil drop burner
   coal burner
   cement kiln
   iron ore refiner
   wood chip burner
   concrete pavement laying
   spin coater
   fiber spinner
   catalytic hydrogenation of vegetable oil
   oxidation of organic sewage
   homogenization and pasteurization of milk
   fluidized catalytic cracking of gas oil
   trickle bed hydrotreating of residual oil

14.2 Listed below are examples of multiphase reactors used in industry. For each reactor, answer the following.

(a) Draw a sketch and identify all the phases.

(b) Which is the dispersed phase and which is the continuous phase?

(c) How would you model each phase, mixed or unmixed?

   slurry reactor
   spray tower
   emulsion reactor
   bubble column
   packed bed

gas–solid fluidized bed
fluidized catalytic cracker
trickle bed

14.3 In a well-mixed emulsion reactor A in a continuous aqueous phase reacts with B in an organic drop phase in the reaction $A + B \rightarrow$ products. The reaction is very fast and is limited by the diffusion of A into the organic phase. The feeds are 1 liter/min of 9 molar A and 1 liter/min of 2 molar B. The stirring is such that the drops are monodisperse at a diameter of 200μm.

(a) Find the surface area per liter.

(b) Find the pseudohomogeneous rate $r(C_A, C_B)$ if the Sherwood number is 2.0 and $D_A = 10^{-5} \text{cm}^2/\text{sec}$.

(c) Find the residence time and reactor volume for 90% conversion.

14.4 A 2×10^4-cm^3 slurry reactor has 1000 *porous* catalyst particles/cm^3 on which the reaction $A \rightarrow B$ occurs with a surface rate r''(moles/cm^2 sec) $= 4 \times 10^{-8}C_A$, with C_A in moles/cm^3. Each particle is 0.02 cm diameter and has a surface area of 500 cm^2. (The catalyst occupies negligible volume in the reactor.) The volumetric flow rate is 400 cm^3/min. Note that all lengths are given in cm.

(a) What is the reactor residence time?

(b) What is the pseudohomogeneous rate if the surface reaction is rate limiting?

(c) Calculate the conversion in this reactor if the surface reaction is rate limiting. [Answer: 50%.]

(d) Calculate the conversion in this reactor if the Thiele modulus for pore diffusion is 10.

14.5 (a) Spherical crystals 1 mm in diameter are dissolved in a liquid solvent that is stirred so the mass transfer coefficient is constant. If 1% dissolves in 1 min, when will the crystals completely dissolve?

(b) If a 1-mm sheet has a 0.1-mm-thick oxide layer on it after 1 week, when will the rust penetrate through the sheet if oxygen diffusion through the growing oxide controls the process?

14.6 Styrene is made by bubbling ethylene vapor into pure liquid benzene with a suitable catalyst. We find that 1.0 liters/sec of ethylene (measured at STP) completely reacts when bubbled into a 100-gallon tank of benzene.

(a) What must be the benzene flow rate if the product is to be 80% styrene and the density is independent of conversion?

(b) The heat of benzene alkylation is -20 kcal/mole. What is the reactor cooling rate?

14.7 Kitchen food processors have several blades to mix, chop, grind, and puree solids and to add air to liquids.

(a) Call your mother and ask her to describe all of the blades in her food processor and describe what each is designed to accomplish. Your mother will be pleasantly surprised by your call and will think (mistakenly) that you are finally learning something practical.

(b) What mixer designs are optimum for maximum heat and mass transfer to the reactor walls and for maximum incorporation of air into the liquid?

(c) Why does a kitchen mixer work best when the bowl is off center of the beaters and the bowl is rotated?

14.8 A student who graduated several years ago called the office to inquire about "fluidized beds." He is working for a water purification company that is developing a process to soften well water. They bubble air through a column 6 ft high and 1 ft in diameter, and water flows up through the column, which contains $\sim$1 ft^3 of MnO_2 catalyst in the form of 1/16-in. particles. The water they treat contains 3 ppm of Fe^{2+}. Hard water contains mostly Fe, which comes from wells as Fe^{2+}. This ion is very soluble, but Fe^{3+} is much less soluble and precipitates to form the orange–brown scum in your bathtub.

(a) What type of reactor are they dealing with?

(b) What are the reactions?

(c) Sketch the process including separation systems for catalyst and precipitate.

(d) What reaction steps are involved in this process?

(e) They measure 20 ppm of dissolved O_2 in the water, both at the exit and in the middle of the reactor. The equilibrium solubility of O_2 in water at 25°C is 30 ppm. What process is definitely not limiting the rate? What reaction steps might be rate limiting?

(f) The present process uses 10 psig air. How should it be improved if the pressure is increased to 60 psig?

(g) Why might smaller-diameter MnO_2 particles give better removal, and what problems might arise?

(h) What ratio of air to water flow rates would you assume might be satisfactory? How might you reduce the air rate required?

(i) What is the ratio of precipitate to water flow rates, assuming complete removal as carbonate?

14.9 The reaction $A_g \rightarrow B_g$ occurs on the surface of a flat plate with a rate $r'' = k''C_{As}$. The concentration of A above the boundary layer of thickness δ is C_{Ab}, and the mass transfer coefficient is D_A/δ.

(a) What is the relation between C_{Ab} and C_{As} in terms of these quantities?

(b) Find $r''(C_{Ab})$ in terms of these quantities.

(c) Suppose the plate in this problem is carbon burning in air with a very fast rate through the reaction $C_s + O_2 \rightarrow CO_2$. The partial pressure of O_2 is 0.2 atm so that $D_A = 0.1$ cm^2/sec, and $\delta = 0.1$ cm. The density of solid carbon (MW=12) is $\rho_s = 2.25$ g/cm^3. What time will be required to burn off 1 mm of carbon from this plate?

14.10 A liquid A in a solvent is being hydrogenated by bubbling H_2 at atmospheric pressure through the solution to form another liquid B in the reaction $A + H_2 \rightarrow B$. The column is 10 cm in diameter and 2 meters high, and the bubbles rise with a constant velocity of 10 cm/sec. The flow rate of the liquid is 15.7 cm^3/sec. The gas bubbles occupy negligible volume in the reactor.

(a) Calculate the residence time of gas and liquid in the reactor.

(b) Write the mass-balance equation that would be used to solve for the *liquid*-phase concentration C_A assuming that the rising bubbles stir the liquid and that the reaction occurs homogeneously in the liquid with a rate r (moles/liter sec) $= kC_AC_{H_2}$, where C_{H_2} is the concentration of H_2 in the liquid solution.

(c) It is found that C_{H_2} is 1×10^{-3} moles/cm^3 in the liquid solution and $k = 4$ cm^3/moles sec (second-order rate constant). Find the conversion. [Answer: 80%.]

14.11 In the previous problem the hydrogen bubbles are initially 0.1 cm in diameter, and they contain pure H_2 gas at a concentration of 0.05 moles/cm^3 (which does not change as the bubbles rise). Hydrogen is transferred from the bubbles to the liquid with a rate limited by mass transfer ($k_m = D_{H_2}/R$) in the liquid around the bubble with $D_{H_2} = 1 \times 10^{-5}$ cm^2/sec. There is no reaction in the bubbles.

    (a) Set up the differential equation that should be solved for the radius of the bubbles R versus height z in the reactor for an initial radius $R_o = 0.05$ cm. The equilibrium solubility of H_2 in the liquid solution (at the bubble surface) is 3×10^{-6} moles/liter.

    (b) Find the hydrogen bubble radius at the top of the reactor.

14.12 The reaction $A + B \rightarrow$ products occurs in a well-mixed emulsion reactor with B entering in the continuous aqueous phase and A in the organic phase. Reaction occurs by A transferring from the organic phase into the aqueous phase, where reaction occurs. The organic phase forms drops 0.1 cm in diameter, and the flow rate of the organic phase is 1 liter/min at $C_{Ao} = 2$ moles/liter. The flow rate of the aqueous phase is 9 liters/min at $C_{Bo} = 1$ mole/liter.

    (a) Find an expression for the pseudohomogeneous reaction rate in moles/liter min versus C_A and C_B if the reaction is limited by mass transfer of A from the organic phase with the reaction instantaneous in the aqueous phase. Assume a Sherwood number of 2 with $D_A = 10^{-5}$ cm^2/sec.

    (b) What volume of reactor is required for 90% conversion of A?

14.13 The reaction $A + B \rightarrow$ products occurs in an unmixed flow reactor, with B entering in the continuous aqueous phase and A in the organic phase. Reaction occurs by A transferring from the organic phase into the aqueous phase, where reaction occurs. The organic phase forms drops 0.1 cm in diameter, and the flow rate of the organic phase is 1 liter/min at $C_{Ao} = 2$ moles/liter. The flow rate of the aqueous phase is 9 liters/min at $C_{Bo} = 1$ mole/liter.

    (a) Find an expression for the pseudohomogeneous reaction rate in moles/liter min versus C_A and C_B if the reaction is limited by mass transfer of A from the organic phase with the reaction instantaneous in the aqueous phase. Assume a Sherwood number of 2 with $D_A = 10^{-5}$ cm^2/sec.

    (b) What volume of reactor is required for 90% conversion of A?

14.14 Suppose you are hired by NASA to examine the feasibility of developing chemical processing applications for the space station. What reactor and separation units would have to be abandoned, and what processes might be improved in a zero-gravity environment? [Your first list will probably be much longer than your second, even before you consider transportation costs.]

14.15 Ethyl benzene is to be converted into the hydroperoxide (used to make styrene and propylene oxide) by bubbling air through a 1 molar solution in cyclohexane at 25°C in the reaction

$$C_6H_5C_2H_3 + O_2 \rightarrow C_6H_5C_2H_2OOH, \qquad r^\ell = k^\ell C_A^\ell C_{O_2}^\ell$$

in a bubble column reactor operating at 25°C.

    (a) What volumetric flow rate of air will be necessary to attain complete conversion if the liquid flow rate is 100 liters/sec?

    (b) Using an air flow rate 10 times that calculated in (a) so that $C_{O_2}^\ell$ can be assumed constant, find the reactor volume necessary to attain 90% conversion if O_2 remains in equilibrium in the liquid phase at a pressure of 1 atm, assuming that the liquid flows in plug flow.

The Henry's Law constant of O_2 in cyclohexane is 1×10^5 atm, and the rate coefficient $k^\ell = 1 \times 10^2$ liters/mole sec.

(c) Using an air flow rate 10 times that calculated in (a) so that $C_{O_2}^\ell$ can be assumed constant, find the reactor volume necessary to attain 90% conversion if O_2 remains in equilibrium in the liquid phase at a pressure of 1 atm, assuming that the reactor contents are stirred rapidly. The Henry's Law constant of O_2 in cyclohexane is 1×10^5 atm, and the rate coefficient $k^\ell = 1 \times 10^2$ liters/mole sec.

14.16 We have a stream of liquid cyclohexane that contains 0.1% cyclohexene, which must be removed by hydrogenation in the reaction

$$C_6H_{10} + H_2 \rightarrow C_6H_{12}, \qquad r = k^\ell C_{H_2}^\ell C_A^\ell$$

in a bubble column.

(a) What H_2 flow rate will be required to reduce the H_2 concentration by 10% for complete olefin conversion for a liquid flow rate of 50 liters/sec?

(b) What liquid volume in the reactor will be required for 99% conversion if H_2 remains in equilibrium in the liquid solution if the Henry's Law constant of H_2 in cyclohexane is 3×10^5 atm and $k^\ell = 2 \times 10^2$ liter/mole sec for a H_2 pressure of 1 atm?

(c) For what H_2 pressure will the liquid volume in the reactor be 200 liters?

14.17 An aqueous solution contains 1.0 ppm by weight of an organic olefin contaminant of molecular weight 80, which must be removed by hydrogenation by a 10% H_2-in-N_2 stream in a 1-cm-diameter falling film reactor at 25°C. The liquid flows at an average velocity of 10 cm/sec and forms a film 1 mm thick on the wall, while the H_2 at 1 atm flows at an average velocity of 2 cm/sec. The reaction in the liquid phase has the stoichiometry $A + H_2 \rightarrow$ products with a rate $r^\ell = k^\ell C_A^\ell C_{H_2}^\ell$.

(a) What are the compositions when the reaction has gone to completion?

(b) Write the equation(s) that must be solved to find the reactor length required for a given organic removal.

(c) How many tubes in parallel are required to process 50 liters/min of water? What will be the diameter of the tube bundle if the tubes are packed in a square array and the tube wall thickness is neglected?

(d) Estimate the reactor length needed to reduce the organic concentration to 0.1 ppm if the reaction rate in the liquid is infinite and the reaction is limited by the mass transfer of H_2 in the gas. Assume $D_{H_2} = 0.1$ cm^2 sec.

(e) Estimate the reactor length needed to reduce the organic concentration to 0.1 ppm if H_2 mass transfer is sufficiently fast that the liquid solution remains saturated with H_2 and the rate coefficient k^ℓ is 1×10^2 liters/mole sec. The Henry's Law constant for H_2 in water is 3×10^5 atm.

(f) Estimate the reactor length needed to reduce the organic concentration to 0.1 ppm if the reaction actually occurs on a catalyst on the wall of the reactor tube. Assume that the reaction rate on the wall is infinite and the process is limited by the diffusion of H_2 through the liquid, with $D_{H_2} = 10^{-5}$ cm^2 /sec, and that the concentration profile is linear.

14.18 An aqueous solution contains 1.0 ppm by weight of an organic olefin contaminant of molecular weight 80, which must be removed by hydrogenation by a 10% H_2-in-N_2 stream in a 10-cm-diameter bubble column reactor at 25°C. The liquid flows downward in the tube at an average velocity of 1 cm/sec. The gas at 1 atm is admitted at 0.1 liter/sec and is injected as bubbles

0.5 mm in diameter that rise at 2 cm/sec. Assume no coalescence or breakup and that both gas and liquid are in plug flow. The reaction in the liquid phase has the stoichiometry $A + H_2 \rightarrow$ products with a rate $k^\ell C_A^\ell C_{H_2}^\ell$.

(a) What are the compositions when the reaction has gone to completion?

(b) Write the equation(s) that must be solved to find the length for a given organic removal.

(c) Assuming that mass transfer of H_2 in the liquid around the bubbles limits the rate, calculate the reactor length required to reduce the contaminant to 0.1 ppm.

(d) Estimate the reactor length required to reduce the organic concentration to 0.1 ppm if the concentration of H_2 remains at equilibrium in the solution.

14.19 An aqueous solution contains 1.0 ppm by weight of an organic olefin contaminant of molecular weight 80, which must be removed by hydrogenation by a 10% H_2-in-N_2 stream in a sparger reactor at 25°C. The liquid is admitted at 1 liter/sec. The air at 1 atm is admitted at 0.5 liter/sec. An impeller disperses the gas into bubbles of uniform 0.5 mm diameter and mixes gas and liquid very rapidly. The reaction in the liquid phase has the stoichiometry $A + H_2 \rightarrow$ products with a rate $r^\ell = k^\ell C_A^\ell C_{H_2}^\ell$.

(a) What are the compositions when the reaction has gone to completion?

(b) Write the equation(s) that must be solved to find the reactor volume required for a given organic removal.

(c) Assuming mass transfer of H_2 in the liquid around the bubbles limits the rate, calculate the reactor volume required to reduce the contaminant to 0.1 ppm.

(d) Estimate the reactor volume required to reduce the organic concentration to 0.1 ppm if the concentration of H_2 remains at equilibrium in the solution.

14.20 The reaction

$$A \rightarrow B, \qquad r = kC_A$$

occurs in the organic phase of a two-phase well-mixed reactor. Reactant A is introduced into the reactor only in the aqueous phase, B is insoluble in the aqueous phase, and A is always in equilibrium between the two phases.

$$C_{Ao}^{aq} = 2 \text{ moles/liter}$$

$$C_{Ao}^{org} = 0$$

$$C_{Bo} = 0$$

$$v^{aq} = 5 \text{ liters/min}$$

$$v^{org} = 1 \text{ liter/min}$$

$$k = 0.01/\text{sec}$$

$$K^{org/aq} = \frac{C_A^{org}}{C_A^{aq}} = 4$$

(a) Write the reaction rate expression for A in terms of C_A^{or}.

(b) The *overall* conversion (moles A reacted/moles A fed) is 0.9. Write down an expression for the appropriate total mass balance equation in terms of the concentrations of species A.

(c) Find C_A^{org}. [Answer: $\sim$0.5 mole/liter.]

(d) Find the total volume required for this reactor. [Answer: $V^{org} \sim$ 30 liters.]

14.21 Consider a slurry reactor in which the reaction

$$A^g + B^l \rightarrow C^l, r'' = k'' C_{As}^l$$

occurs on the surface of porous solid catalyst particles that are suspended in the liquid phase in which A is dissolved. A enters the reactor in the gas phase as bubbles with concentration C_{Ao}^g, and the bubbles do not coalesce or break up in the reactor. A migrates through the liquid to react on the porous catalyst.

(a) Sketch the reactor, indicating inlet and outlet concentrations of all species, flows, and volumes in each phase.

(b) Assuming that the gas and liquid phases are completely mixed, write the mass balances for A in the g and l phases. Assume that the gas–liquid interfacial area in the reactor is A^{gl}, the catalyst surface area is A^{ls}, and that phase volumes are V^g and V^l. Assume that the density does not change significantly so that $v_o = v$ in both phases. Assume that mass trnsfer in the liquid around the bubble is limiting with a coefficient k_m. The concentrations of A across the interface are related by the expression $C_A^g = K C_{As}^l$

$$\frac{C_{Ao}^g - C_A^g}{\tau^g} = \text{(one term)}$$

$$\frac{C_{Ao}^l - C_A^l}{\tau^l} = \text{(two terms)}$$

14.22 For the gas–liquid slurry reactor of the previous problem label the C_A versus distance curves shown around bubble and catalyst particles for the following approximations.

(a) There is no reaction.

(b) All mass transfer coefficients are much larger than the reaction rate coefficient.

(c) Pore diffusion within the catalyst pellet strongly limits the rate.

(d) External mass transfer around the catalyst pellet strongly limits the reaction.

(e) External mass transfer around the bubble strongly limits the reaction.

(f) All mass transfer resistances are comparable.

14.23 In the slurry reactor of the previous problem indicate which of the above situation(s) might give the following observations. Most have multiple situations, and some may have none. (This problem will be graded as number correct minus number incorrect.)

(a) The temperature is very high.   _____

(b) The temperature is very low.   _____

(c) Stirring has no effect on the conversion as long as the bubble size does not change.   _____

(d) The catalyst particle diameter has no effect on the conversion for constant area.   _____

(e) The bubble size has no effect on the conversion for constant bubble area.   _____

(f) Larger pore catalyst has no effect on the conversion for constant catalyst area.   _____

(g) Increasing the gas pressure has no effect on the conversion as long as the bubble area remains constant.   _____

(h) Increasing the gas pressure has no effect on the conversion as long as the molar flow rate of A is constant.   _____

(i) Doubling the number of catalyst particles doubles the rate.   _____

(j) Doubling the number of catalyst particles does not change the rate.   _____

14.24 An oil drop of radius R in an emulsion reactor is undergoing the reaction

$$A_o + B_w \rightarrow \text{polymer}$$

where A is only in the oil at concentration C_{Ao}. B is initially in the water phase at concentration C_{Bw}, but B migrates to the oil drop where it reacts instantly at the surface to form polymer that remains in the oil phase. Find the number of moles of B that has migrated to the oil drop as a function of time.

(a) Assume that A is in excess so that C_{Ao} does not change and that R remains constant.

(b) Show how you would solve this problem allowing C_A and R to change as conversion changes.

(c) Write the mass balances for oil emulsion and water phases in terms of v_o, v_w, C_{Aoo}, and C_{Bwo}, assuming that the oil drops do not mix and the water phase is well mixed.

14.25 We need to scavenge a solute A that is dissolved in water at concentration $C_{Ao} = 0.1$ moles/liter. We do this by placing it in a reactor containing oil drops. A is highly soluble in the oil such that its concentration near the oil surface is zero. If the Sherwood number around the drop is 2.0, $D_A = 2 \times 10^{-5}$ cm^2/sec, and $R = 0.05$ cm, what is the concentration of A in the oil drop after 100 sec if R remains unchanged?

Appendix A

INTEGRATING DIFFERENTIAL EQUATIONS

In this text all numerical problems involve integration of simultaneous ordinary differential equations or solution of simultaneous algebraic equations. You should have no trouble finding ways to solve algebraic equations with a calculator, a spreadsheet, a personal computer, etc.

For differential equations the numerical procedure is to replace the differential equation by a difference equation:

$$\frac{dx}{dt} = f(x)$$

is replaced by

$$\frac{\Delta x}{\Delta t} = f(x)$$

This is solved by writing the difference equation as

$$x_{n+1} = x_n + f(x_n)\Delta t$$

and, starting with initial condition x_0, successively applying this equation for increasing n until the solution reaches a constant value. The step size Δt must be chosen to give sufficient accuracy, while not requiring too long a time for the computation. With packaged equation solvers, the step size and higher-order corrections in the preceding expansion are used for greater speed and accuracy in calculations.

Possible methods you may want to use are

1. a spreadsheet such as Excel,
2. a programmable calculator,
3. a computer language such as FORTRAN or BASIC,
4. a packaged equation solver such as Mathematica, Polymath, Maple, or MathCad,
5. a program given to you by your instructor.

We suggest that you find your own way to do these problems. Once you have done it once, the more complicated problems require simple changes to the lines in the program involving the functions.

A key feature of any numerical calculation is the incorporation of graphics for display of data. We strongly encourage real-time graphics rather than postcalculation graphics, which involve setting up a table for plotting. This is time consuming, it frequently does not allow simultaneous plotting of many curves on one graph or curves that are multivalued, and it does not give the visual display of a computer experiment as you watch the time change on the computer screen.

Real-time graphics are preferred to compiled graphics programs because you can see the solution evolve (a computer experiment) rather than just a prepared graph. Much inferior is a program in which you generate a data table and then insert the data into a separate graphics program, because this is very cumbersome and frequently won't allow you to display multiple curves on one graph or display curves that are multivalued.

Submitting computer homework problems

In all computer homework problems you must submit a copy of the program you used with your name and the date appearing on the graph so the graders can be sure you didn't just photocopy someone else's graph.

Mathematica   The following "Mathematica" commands will calculate and plot the concentration of A, B, and C in a PFTR where the reaction $A \rightarrow B \rightarrow C$ takes place isothermally. We assumed here that $r_1 = k_1 C_A, r_2 = k_2 C_B, k_1 = 1, k_2 = 2, C_A(o) = 1, C_B(o) = C_C(o) = 0$ and we solve from $\tau = 0$ to 10.

At the prompt of UNIX environment (or any environment that provides "Mathematica") type:

```
mathematica&
```

Then, once you are in the program, you type:

```
Clear [t, ca, cb, cc, k1, k2];
k1=1;
k2=2;

NDSolve[
{ca' [t]==-k1 ca[t],
cb' [t]==k1 ca[t]-k2 cb[t],
cc' [t]==k2 cb[t],
ca [0]==1,
cb [0]==0,
cc[0]==0}, {ca, cb, cc}, {t, 0, 10}];
s1 = Plot[ Evaluate[ca[t], {t, 0, 10}];
s2 = Plot[ Evaluate[cb[t], {t, 0, 10}];
s3 = Plot[ Evaluate[cc[t], {t, 0, 10}];
Show [s1, s2, s3]
```

Polymath   To solve $A \rightarrow B \rightarrow C$ using Polymath first select NEW from the FILE menu. Then enter the following equations in the EQUATIONS window:

```
d(a)/d(t)=k1*a
```

```
d(b)/d(t)=k1*a-k2*b
d(c)/d(t)=k2*b
k1=1
k2=2
```

Next, enter the following values in the INITIAL AND FINAL VALUES window:

```
t₀=0
a₀=1
b₀=0
c₀=1
t_f=5
```

$t_o=0$
$a_o=1$
$b_o=0$
$c_o=1$
$t_f=5$

Next choose SOLVE from the PROBLEM window. Finally, use the NEW command from the GRAPH menu to create graphical output for the problem or create tabular output for more accurate numerical solutions.

The following simple BASIC program will integrate simultaneous differential equations to find $C_j(t)$. Just type it into any PC or Mac with BASIC, and it will draw a grid and find concentrations in $A \rightarrow B \rightarrow C$. Honest! Use "Print Screen" to make a hard copy of your graphs.

```
'****INTEGRATION PROGRAM FOR MACINTOSH****
'ENGINEERING OF CHEMICAL REACTIONS
'GRAPH GENERATION
  'SET SIZE AND ORIGIN
  IX=100:SX=450:IY=20:SY=230   'ORIGIN AND STEP SIZE IN X
    AND Y
  TMIN=0:TMAX=1
  'DRAW GRAPH AND TICK MARKS
  LINE (IX+SX,SY+IY)-(IX,SY+IY):LINE -(IX,IY)
  FOR I=0 TO 9
    LINE (IX,SY*I/10+IY)-(IX+10,SY*I/10+IY)
  NEXT I
  FOR T=TMIN TO TMAX STEP .1
    X=(T-TMIN)/(TMAX-TMIN)
    LINE (IX+SX*X,SY+IY)-(IX+SX*X,SY+IY-10)
  NEXT T

  'RATE COEFFICIENTS AND INITIAL CONCENTRATIONS
  K1=5:K2=5:XO=1:YO=0:ZO=0:T=0
  'INTEGRATION STEP SIZE
  DT=.005

  'DOCUMENTATION AND COMMENTS
  LOCATE 2,30:PRINT ''CHE 5301, YOUR NAME&DATE''
  LOCATE 3,30:PRINT ''A->B->C, TMAX='';TMAX
  LOCATE 17,35:PRINT ''TIME''
  LOCATE 9,2:PRINT ''Concentration''

 'INTEGRATE DX/DT=-KX
 10 X=XO-K1*XO*DT
 Y=YO+K1*XO*DT-K2*YO*DT
 Z=1-X-Y
 T=T+DT
 TON=(T0-TMIN)/(TMAX-TMIN)
 TN=(T-TMIN)/(TMAX-TMIN)
```

```
            'DRAW LINES ON GRAPH
LINE (IX+SX*TON,IY+SY*(1-XO))-(IX+SX*TN,IY+SY*(1-X))
LINE (IX+SX*TON,IY+SY*(1-YO))-(IX+SX*TN,IY+SY*(1-Y))
LINE (IX+SX*TON,IY+SY*(1-ZO))-(IX+SX*TN,IY+SY*(1-Z))
XO=X:YO=Y:ZO=Z:T0=T
20 IF T>TMAX THEN GOTO 20 'PAUSE
GOTO 10
```

Appendix B

NOTATION

α	superscript indicates phase α in a multiphase reactor
A	area
A_c	surface area between a reactor and coolant
A_t	cross-sectional area of a tube
ad	subscript indicates adiabatic reactor
a_j	activity of species j
b	subscript indicates bulk value of a quantity
β	superscript indicates phase β in a multiphase reactor
β_j	order of an irreversible reaction with respect to species j
β_{fj}	order of forward reaction with respect to species j
β_{bj}	order of reverse reaction with respect to species j
β_{ij}	order of ith reaction with respect to species j
c	subscript signifies core or catalyst
C_A	concentration of key reactant species with stoichiometric coefficient $\nu_A = -1$
C_j	concentration of species j, usually in moles/liter
C_p	heat capacity per unit weight
D	diameter of a reactor
D_A	diffusion coefficient of species A in a bulk fluid
$\mathbf{D}_A$	diffusion coefficient of species A in a pore
d_p	diameter of a pore
E	activation energy of an irreversible reaction
E_f	activation energy of forward reaction

E_b	activation energy of reverse reaction
F	total molar flow rate
$f(x)$	function of variable x
F_j	molar flow rate of species j
F	total molar flow rate
G	free energy
$\mathbf{G}(T)$	dimensionless rate of heat generation in a reactor
g	subscript signifies gas
GHSV	gas hourly space velocity
β_{fj}	order of reverse reaction with respect to species j
H	enthalpy
H_R	heat of reaction
H_{Ri}	heat of the ith reaction
h	heat transfer coefficient
I	electrical current
k''	rate coefficient of surface reaction per unit area
k	reaction-rate coefficient for the forward reaction in a single-reaction system
k'	rate coefficient of surface reaction per weight catalyst
k_b	reaction-rate coefficient for the back or reverse reaction in a single-reaction system
k_f	reaction-rate coefficient of forward reaction
k_{eff}	effective rate coefficient
K_{eq}	equilibrium constant
K_i	equilibrium constant of ith reaction
k_i	reaction-rate coefficient for the ith reaction
k_o	pre-exponential factor of single reaction
k_{ob}	pre-exponential of reverse reaction
k_{oi}	pre-exponential of ith reaction
K_T	equilibrium constant at temperature T
k_T	thermal conductivity
L	length of tubular reactor
λ	eigenvalue
ℓ	length of a pore or half-diameter or a porous catalyst slab
ℓ	subscript signifies liquid
LHSV	liquid hourly space velocity
M_j	molecular weight
μ_j	chemical potential of species j
N	total moles in a system

n	order of a reaction
N_j	number of moles of species j
ν_j	stoichiometric coefficient of species j in the reaction $\sum \nu_j A_j$
Nu	Nusselt number
$_0$	subscript 0 always signifies reactor feed parameters
π	product
$p(x)$	function of variable x
P	product in biological reaction
P_j	partial pressure of species j, $P_j/RT = C_j$ for ideal gases
Q	heat generation in a process
Q	heat generation per mole
R	gas constant
R	position in a sphere or cylinder or number of reactions in a multiple-reaction system
$\mathbf{R}(T)$	dimensionless rate of heat removal
ρ	density
r	reaction rate of a single reaction in moles/volume time
r''	reaction rate of a surface reaction in moles/area time
r'	reaction rate of a surface reaction in moles/weight catalyst time
r_b	rate of back reaction
Re	Reynolds number
r_f	rate of forward reaction
r_i	reaction rate of the ith reaction in a multiple-reaction system
R_k	electrical resistance of kth resistor in series or parallel
R_o	radius of a sphere or cylinder
S	entropy or number of species in a reaction system
S	substrate in biological reaction
$\sum$	summation
s	subscript signifies solid or surface
Sh	Sherwood number
S_j	selectivity of formation of species j
s_j	differential selectivity of species j
ST	space time in a continuous reactor
SV	space velocity (1/ST) in a continuous reactor
τ	residence time in a constant-density reactor
t	time
τ_k	residence time in the kth reactor in series
u	average velocity in plug-flow tubular reactor

V	reactor volume, usually in liters
v	volumetric flow rate, usually in liters/time
V_k	voltage of kth resistor
W_s	shaft work
X_A	fractional conversion of key species A
X_j	fractional conversion of reactant species j
X_i	fractional conversion of the ith reaction
x	position perpendicular to the direction of flow z in a reactor or the position down a pore
x	variable of integration
x_j	mole fraction of jth species in a liquid mixture
y	variable of integration
Y_j	yield of species j
y_j	mole fraction of jth species in a gas mixture
z	position along a tubular reactor

Appendix C

CONVERSION FACTORS

Length

$$1 \text{ cm} = 10^{-2} \text{ meter}$$
$$= 0.3937 \text{ inch}$$
$$= 0.03281 \text{ foot}$$
$$= 6.214 \times 10^{-6} \text{ mile}$$

Weight

$$1 \text{ g} = 10^{-3} \text{ kg}$$
$$= 0.002205 \text{ lb}$$
$$= 0.03527 \text{ oz}$$
$$= 1.102 \times 10^{-6} \text{ ton}$$
$$= 10^{-6} \text{ metric ton}$$

Volume

$$1 \text{ liter} = 10^3 \text{ cm}^3$$
$$= 10^{-3} \text{ m}^3$$
$$= 1 \text{ dcm}^3$$
$$= 0.008386 \text{ barrel (U.S. liquid)}$$
$$= 0.03532 \text{ ft}^3$$
$$= 61.02 \text{ in.}^2$$
$$= 0.2642 \text{ gal}$$

Energy

$$1 \text{ cal} = 10^{-3} \text{ kcal}$$
$$= 4.186 \text{ joule}$$
$$= 4.186 \times 10^7 \text{ erg}$$
$$= 1.163 \times 10^{-6} \text{ kWh}$$
$$= 3.968 \times 10^{-3} \text{ BTU}$$
$$= 1.559 \text{ hp h}$$

Power

1 watt $=$ 1 joule/sec

$\quad\quad\quad = 0.7376$ ft lb/sec

$\quad\quad\quad = 3.413$ BTU/h

$\quad\quad\quad = 0.001341$ hp

Pressure

1 atm $=$ 1.013 bar

$\quad\quad\quad = 1.013 \times 10^5$ pascal

$\quad\quad\quad = 760$ torr

$\quad\quad\quad = 14.7$ psi

$\quad\quad\quad = 2116.3$ pounds per square foot

Some important quantities

$R = 1.9872$ cal/mole K

$\quad = 8.31$ J/mole K

$\quad = 0.082059$ liter atm/mole K

$\quad = 82.059$ cm^3 atm/mole K

$\quad = 0.7302$ ft^3 atm/lb mole F

$\quad = 1.9872$ BTU/lb mole F

$k_B = R/N_o = 1.38 \times 10^{-23}$ J/molecule K

1 electron $= 1.602 \times 10^{-19}$ coulomb

$N_o = 6.023 \times 10^{23}$ molecules/mole

INDEX